A. E. Watson, M. O. Sampson

Electrician And Mechanic

Volume 24 January-June 1912

A. E. Watson, M. O. Sampson

Electrician And Mechanic
Volume 24 January-June 1912

ISBN/EAN: 9783741182570

Manufactured in Europe, USA, Canada, Australia, Japa

Cover: Foto ©Andreas Hilbeck / pixelio.de

Manufactured and distributed by brebook publishing software
(www.brebook.com)

A. E. Watson, M. O. Sampson

Electrician And Mechanic

Electrician and Mechanic

Incorporating

Bubier's Popular Electrician, Established 1890
Amateur Work, Established 1901
Building Craft, Established 1908
The Collins Wireless Bulletin, Established 1908

EDITED BY

F. R. FRAPRIE, M.Sc. Chem.
A. E. WATSON, Ph.D.E.E.
M. O. SAMPSON

VOLUME XXIV
January, 1912 --- June, 1912

SAMPSON PUBLISHING COMPANY
221 Columbus Ave., Boston, Mass.

INDEX

QUESTIONS AND ANSWERS

INDEX

ELECTRICIAN & MECHANIC

VOLUME XXIV   JANUARY, 1912   NUMBER 1

ENGINEERING LABORATORY PRACTICE—Part II
Steam Engine Testing

P. LE ROY FLANSBURG

One of the primary objects in making a steam engine test is to determine the power developed by the engine. While there are various ways of measuring the power, this article will describe but one, namely, that in which a steam engine indicator is used. The reason for selecting this particular method of testing is

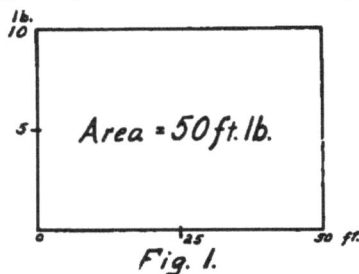

Fig. 1.

that the indicator diagram obtained shows not only the power developed, but also many other things that it is important for the engineer to know, among which may be mentioned the arrangement of the valves for admission, cut-off, release and compression of the steam in the cylinder. There are, however, several objections to the use of the indicator for engine testing, as the indicator, even when in good order, is liable to an error of from 2 to 3 percent and can seldom be run at more than 350 or 400 revolutions per minute.

Since work is the product of Force by Distance, it is possible to represent work by an area. For instance, the lifting of a weight of 10 lbs. through a distance of 50 ft. might be represented by such a diagram as is shown in Fig. 1.

In this diagram the weight lifted is represented by the ordinate AB, and the distance through which the weight moves by the abscissa AC. The area $ABDC$ representing the number of foot-pounds of work done.

In a similar manner the work done by the steam in an engine cylinder can be represented by an area in which one set of co-ordinates is proportional to the piston travel, while the other set of co-ordinates bears a constant ratio to the pressure of the steam within the cylinder during a single cycle of the engine. This steam pressure, which is acting upon the piston, of course varying at different points of the stroke.

The easiest way of determining this work area is by the use of an instrument known as the steam engine indicator. This instrument consists substantially of a carefully adjusted piston of known area (usually ½ in.) moving without sensible friction in a cylinder. The bottom of the piston is subjected to the pressure of the steam in the engine cylinder, while to resist the upward movement of the indicator piston, a spring of known

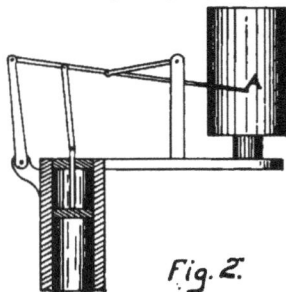

Fig. 2.

resilience is employed. This spring is
very carefully calibrated, so that if a
pressure of 10 lbs. lifts the indicator piston
½ in., a pressure of 20 lbs. will raise it
1 in. Fastened to the top of the piston
is a rod which is allowed to project upward through the cylinder and to which
is attached a lever arm. Any movement
of the indicator piston either upward or
downward is transferred by means of a
parallel motion to a brass point. This
point is marked "A" in Fig. 2.

The length of the lever arm is so proportioned that the motion of the piston
will be magnified a certain amount. The
brass point traces a line upon a special
sheet of chemically treated paper.

The paper upon which the record is
made is wound upon a drum, which is so
connected to the cross-head of the engine
that for every displacement of the cross-
head the drum revolves a certain amount.
As the piston moves up or down inside of
the indicator, the drum will be rotated
and the brass point will trace a line on the
indicator card. Since the rotation of the
drum is proportional to the motion of the
piston of the engine, every point on the
line drawn will correspond to the pressure
inside the engine cylinder at that displacement of its piston.

When testing an engine, the indicator
is attached to the engine cylinder by
tapping a hole in the cylinder at the clearance space. The engine is then started
and the drum of the indicator is connected
by a cord to the cross-head of the engine.

As steam is admitted to the cylinder,
the pressure rises rapidly and the brass
point of the indicator draws a line similar
to A B in Fig. 3. If the steam is admitted
quickly enough, the line A B will be practically vertical.

During the time that the valve is open,
the curve traced will be one resembling
BC. At the moment of cut-off, the point
C is obtained, and from this point to
point D, the steam is expanding in the
cylinder. This curve CD will approximate an equilateral hyperbola, and is the
expansion curve for all true gases. The
point of release (namely, the point where
the exhaust valve opens) is shown by
point D on the diagram. From this
point, there is a loss of pressure until we
reach the lower limit of cylinder pressure.
The line EF represents the pressure
against which the piston must act during

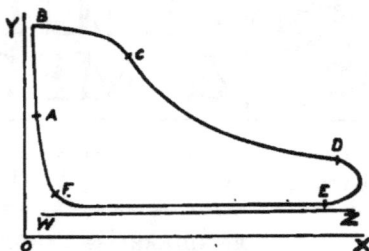

Fig. 3.

the return stroke. The point at which
the exhaust valve closes is hard to locate
definitely, but is to be found somewhere
near to point F. After the exhaust valve
has closed, the steam remaining in the
cylinder is compressed, and the pressure
rises until it reaches the pressure represented by point A.

One cycle has now been completed, and
as the admission valve again opens the
piston begins to move forward. It is
evident that the work done is represented
by the area enclosed by the curve ABCD-
EFA. If the indicator diagram is plani-
metered (or the area found in some other
way), and this area is divided by the
length of the diagram a pressure is obtained which is known as the mean effective pressure. The pressure so obtained
is equivalent to a pressure which, if allowed to act through the whole length
of the stroke, would produce the same
amount of work as does the varying pressure which really does act upon the piston.
WZ represents on the diagram the pressure of the atmosphere, and is 14.7 lbs.
above the zero line of pressure.

In computing the horse-power of an
engine from an indicator diagram the
formula, H.P.$=\dfrac{PLAN}{33,000}$, is made use of.

In this formula, P equals the mean
effective pressure, L equals the length of
the engine stroke, A equals the area of
the engine piston, and N equals the
number of revolutions per minute. P is
measured in pounds per square inch,
L in feet, and A in square inches. Thus
it is at once seen that PLAN gives us
the work done in foot-pounds, and since
a horse-power is the quantity of work
equivalent to the raising of 33,000 lbs.

through a distance of 1 ft. in one minute's time, it is possible to obtain the result in horse-power if $PLAN$ is divided by 33,000.

In Figs. 4, 5, 6, 7 are shown indicator cards obtained from actual tests. Figs. 4 and 5 are from tests made upon a high-speed single-acting engine, while Figs. 6 and 7 are from tests made upon a Harris-Corliss engine. These cards and the computations given for them are simply given to show how the horse-power of an engine may be obtained and no claim is made that the engines are run at their highest efficiencies. We see from the cards in Fig. 4 and Fig. 5 that the engine valves were not set exactly right, while the hump in the curve of Fig. 4 near release, and the reverse loop in the upper part of the curve of Fig. 5, tell us that probably a small amount of steam leaked past the piston.

Fig. 4.

$L = 3.49$ in.
$A = 0.75$ sq. in.
$P = \dfrac{A}{L} = 0.308$ lbs.
Scale of the spring $= 60$ lbs.
Therefore M.E.P. $= 18.48$ lbs.
Area of piston $= (4\frac{1}{4})^2(\pi)$sq. in.
$N = 374$
$L = 8$ in.
$$H.P. = \frac{PLAN}{33,000} = 18.48 \text{ x} \frac{8}{12} \text{x} (4\frac{1}{4})^2(\pi) \text{x} \frac{374}{33,000} = 7.92$$

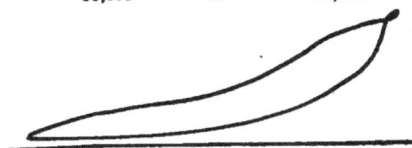

Fig. 5.

$L = 3.49$ in.
$A = 1.13$ sq. in.
$P = \dfrac{A}{L} = 0.324$ lbs.

Scale of the spring $= 60$ lbs.
Therefore M.E.P. $= 19.44$ lbs.
Area of piston $= (4\frac{1}{4})^2(\pi)$sq. in.
$N = 374$
$L = 8$ in.
$$H.P. = \frac{PLAN}{33,000} = 19.44 \text{ x} \frac{8}{12} \text{x} (4\frac{1}{4})^2(\pi) \text{x} \frac{374}{33,000} = 8.33$$

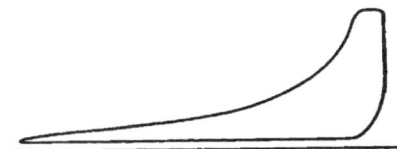

Fig. 6.

$L = 4.60$ in.
$A = 2.04$ sq. in.
$P = \dfrac{A}{L} = 0.443$ lb.
Scale of the spring $= 40$ lbs.
Therefore M.E.P. $= 17.72$ lbs.
$$\left(\frac{LA}{33,000}\right) = 0.00300$$
$N = 66$
$$H.P. = \frac{PLAN}{33,000} = 17.72 \text{ x } 66 \text{ x } 0.00300 = 3.51$$

Fig. 7.

$L = 4.63$ in.
$A = 2.03$ sq. in.
$P = \dfrac{A}{L} = 0.438$ lb.
Scale of the spring $= 40$ lbs.
Therefore M.E.P. $= 17.52$ lbs.
$$\left(\frac{LA}{33,000}\right) = 0.003103$$
$N = 66$
$$H.P. = \frac{PLAN}{33,000} = 17.52 \text{ x } 66 \text{ x } 0.003103 = 3.59$$

GLUE AND HOW TO USE IT

W. J. H.

Glue is an adhesive used chiefly for woodwork, and, being soluble in water, loses its hold if the work is kept in damp situations or is frequently wet. Glue may be purchased either in solid or liquid form. The latter is useful when only required occasionally for small work; but solid glue, which has to be melted for use, is both cheaper and better, and is most commonly used by woodworkers. Glue varies in quality; but the way in which it is used is the main factor in making it hold well. A man who knows how to use it can do better work with poor glue than a less experienced one can do with the best glue.

The joint must fit closely everywhere; the glue must be of suitable consistency, hot, applied quickly, and the joint closed and kept tightly pressed together until the glue is dry. The joint must not only be fitted perfectly before gluing, but it must be as close as possible after, the glue being in the fibers of the wood, and not in an appreciable film between the surfaces. To ensure this, the parts are always rubbed together whenever possible with heavy pressure, to force superfluous glue out at the edges. When this cannot be done conveniently, simple pressure has to suffice. In some cases even this cannot conveniently be applied except by hand, while the gluing is being done.

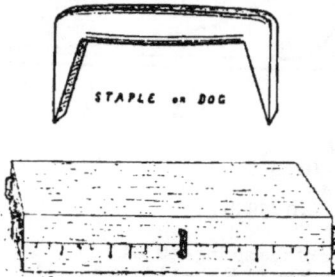

FIG. 1.

In very small joints this does not matter so much; but when there is any doubt about glue holding a joint, it is never relied on alone. Screws or nails are used in addition, and glue is regarded only as an assistance to these, and not the pri-

mary means of union. In such cases, less pains are taken to make a good glue joint. The insertion of screws or nails will spoil it as a perfect glue joint if they are put in before the glue is dry, and, generally, if screws or nails are to be used, it is very objectionable to wait a day or more for glue to dry before inserting them, in order to gain a very slight advantage in the character of the joint.

FIG. 2.

When glue alone is used, the work must be laid aside for drying, the time allowed depending on the size and character of the work and the subsequent operations on it. A small bit glued on anywhere will dry quickly; a large joint requires more time.

Glue is dissolved for use in a glue-pot, consisting of an outer pot for water only, and an inner pot for the glue. Water is added to the latter to reduce it to suitable consistency. Heated in this way, the temperature does not rise above that of boiling water, and the glue does not burn and stick to the pot as it would in a single vessel. For occasional use at home a special glue-pot is not necessary, as glue can be melted very well in an old tin placed in a saucepan of water. Before heating, glue should be allowed to soak in water for a few hours. This softens it, and it can then be melted in a few minutes, or as soon as its temperature can be raised to the usual maximum height. If not previously soaked, it takes a long time for it to melt, and lumps of imperfectly dissolved glue remain in it for hours; besides which, prolonged heating weakens glue. It is best to make small quantities at a time, and use it up without many reheatings. The process of heating is also hastened if the glue is broken into small pieces before putting it in the water. Glue must be kept as clean as possible, and especially free from grease of any kind. In workshops where glue is constantly being used, two pots or more are kept, so that while one is in use glue can be soaking in cold water in the other ready for heating when wanted. In other cases a pot is cleansed at night, and glue put in to soak ready for use next day.

Fig. 4.

Glue-brushes also required experienced treatment. While the glue is being dissolved, a stick should be used to stir it with. The brush is not to be put into it until about to be used. A new brush should be soaked for a little time in hot water before being put into the glue. Brushes should not be left in glue that is allowed to get cold. They should be taken out as soon as the glue is done with, and soaked in hot water, so that the bristles will not get stuck together. For gluing large surfaces, a large brush is necessary to spread the glue as quickly as possible. For applying glue to very small and intricate places, suitable sticks of wood pared thin at the end are better than brushes. Sometimes the bristles of a new brush are too long and flexible, and bend too easily. The brush can be stiffened by binding string around the bristles just below the handle. When the brush has worn shorter, this string is removed.

For large surfaces glue is diluted rather thin; for very small work it should be comparatively thick. It is better, if possible, to warm large surfaces before applying the glue, so that it shall not become chilled too much before the joint can be closed; in which case it fails to hold well. Sometimes hot water is applied immediately before the glue to serve the double purpose of warming

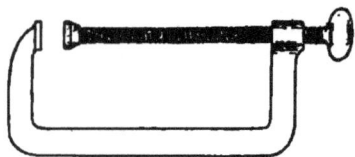

Fig. 3.

the wood and diluting the glue for a large surface. When glue is kept constantly hot, it is necessary to add water occasionally to keep it thin enough for use. In diluting it for immediate use, hot water should be added. If cold is used, a little time should be allowed for the glue to become heated up again. Thickening is slower process, requiring the addition

of more glue broken very small, or time allowed for the water to evaporate.

After being glued, joints are kept under pressure in various ways, depending on the character of the work. The staple, or dog, and the way it is used is shown in Fig. 1. As many of these as may be necessary are simply driven in with a hammer nearly as far as they will go, and, owing to the way their spikes are shaped, they draw the joint together. It will be seen in Fig. 1 that all the taper of the spikes is on the inside, the outer edges being parallel. When the glue is dry, the staples are pried out with a screw-driver. They make holes in the wood, but these are seldom objectionable. They are oftener used in end grain than in side grain.

The hand-screw, Fig 2, is in very common use for squeezing small joints together. Hand-screws are generally used in pairs; but the number required depends entirely on the size of the work. After being adjusted, the final tightening is accomplished by a turn of the screw farthest from the gripping portion of the jaws. This causes them to close at that part, and spread slightly more open at the opposite end where the screw is thrusting. It will be seen that the screw nearest the beveled or gripping end passes entirely through both jaws, but is threaded only in the farthest one. The other screw passes through one jaw, and merely pushes against the other, its end fitting in a hole to keep it in position. A great deal of squeezing power can be exercised with a hand-screw; but care must be taken to see that the jaws are gripping the work parallel. Hand-screws are made of wood and in different sizes. Their capacity from tips of jaws to the nearest screw is rather limited, but is sufficient for most work, and when a broad surface has to be clamped, hand-screws are used at intervals all round the edges. If this still leaves a considerable middle portion unaffected by them, bars of wood are placed across the exterior of the work, and gripped with it to extend the area of pressure.

Next in importance to the hand-screw comes the long bar-cramp, Fig. 3. This is used chiefly for large frames, and for boards edge to edge. It is used a great deal for other purposes besides squeezing glue-joint, as, in fact, the hand-screw and staple are alike. Fig. 4 is a metal "G" cramp, which is very useful and is obtainable in a wide range of sizes. Its action is simpler than that of the hand-screw; but it is not so commonly used for woodwork, one reason being that blocks of wood are often necessary to prevent its small metal jaws from damaging the surface of the work. Another reason is that in making great variations between the distances of the jaws the workman has a way of revolving the hand-screw which opens or closes them rapidly, but with the "G" cramp it can only be done slowly. In large sizes also the hand-screw is cheaper and lighter than a metal clamp.

Besides the regular appliances for clamping, there are numerous methods which are more or less improvised to suit the work. Sometimes weights are put on top of the glued parts, often with large pieces of wood under the weight to dis-

FIG. 6.,

in at an angle around the edges with their heads projecting sufficiently for withdrawal after the glue is hard.

The wet glue swells the surface a little, and there is always a tendency for the joint to curl at the edges. If the wood is very thick, this is impossible, and if it is extremely thin, it may be too weak to overcome the hold of the glue. When moderately thin, the outer surface is often washed with hot water to counteract the effect of the hot glue on the other side. In other cases where a small bit has to be glued on, it may be put on extra thick to prevent curling at the edges, and trimmed down after the glue is set. If pressure on the exterior can be employed, no other means of keeping a close joint is necessary.

In gluing two pieces of average size together, one of them is secured in the vise, as in Fig. 5, or is laid on the bench, as in Fig. 6, with its front end against the bench-stop, and has a staple driven into the bench behind it, or is secured behind in any other convenient way to prevent it from slipping about on the bench while the other piece is being adjusted on top. Then the other piece is held by its side in a tilted position, as shown, and glue applied to both surfaces as quickly as possible, and the piece held by the hand is turned over onto the fixed piece. Then it is slid backwards and forwards a few times on the other, and with some amount of side movement, to rub as much glue as possible out at the edges. As the glue is forced out the sliding becomes more difficult, and exact adjustment of the parts must be made without allowing the slightest

FIG. 5.

tribute the pressure. Sometimes work can be wedged or bound with string. Sometimes small wire nails are driven

separation of the surfaces. If they are allowed to separate while the adjustment is being made the glue joint will be a very poor one. They should be separated completely and re-glued. As soon as they are adjusted, staples are driven in or clamps applied. Very often staples are driven first and clamps put on as well. Then the work is laid aside for several hours to dry. In order to accomplish the clamping as quickly as possible after the gluing, the clamps required are set to size and laid ready for use before the glue is put on. Work of the size shown in Figs. 5 and 6 would be glued by a single workman, but assistance is always obtained for gluing large joints. In such cases a man stands at each end, pressing down on the upper piece of wood and doing his share in sliding it backwards and forwards, and making the final adjustment as exact as possible. There is no objection to using a mallet or hammer if the parts hold too tight to be adjusted a small amount at the finish; but it is avoided as a rule, because there is more risk of breaking the joint by a blow than there is by pushing and pulling. Joints glued in this way (assuming that the surfaces fit each other) are the most perfect glue joints possible, with the exception of veneering joints, to be spoken of further on, and the methods adopted in veneering are impossible in gluing thick wood.

Glue is used in many cases where a strong joint is impossible unless screwed or nailed as well as glued. Under ordinary circumstances glue cannot be trusted to hold end-grain or crossed-grain surfaces. The grain of both parts must run parallel. End-grain does not unite well at all to another surface. The glue sinks into the pores and the joint is easily broken. By applying two coats of glue the absorption is minimized, but there is no strength in such a union. In crossed grain the shrinkage of each piece is at right angles to the other, and this breaks the glue joint. Screws or nails are nearly always used in such cases. Plywood is an exception, and veneering is often done with crossed grain, but the wood in both cases is perfectly seasoned, and is never intended to stand even a damp atmosphere, leave alone exposure to the weather. Moreover, the gluing is exceptionally well done. Joints with parallel grain

can shrink or swell in unison without breaking the hold of the glue. Mortise and tenon joints are nearly always glued, although the surfaces in contact are entirely crossed and end-grained, but it is only in very light cabinetwork that the glue is trusted alone. There is, moreover, the fact to be remembered that such joints, and dovetails, fit with some degree of tightness independently of the glue. Another point about them is that the surfaces are too small to shrink much. In dovetails, especially, glue is quite satisfactory, because the dovetailed framework is invariably made secure by the attachment of the bottom piece or back. In putting these joints together, and in doweled joints also, there can, of course, be no rubbing of the parts to work out the superfluous glue. They are pressed together and clamped, if the character of the work permits clamping. In gluing dovetails, the form of these insures a tight connection in one direction, and in the other way they are simply driven with a mallet or hammer on a block of wood until the parts are as close together as they will go. In all glued work glue is applied to both the surfaces to be united. An exception is in gluing such things as cloth or paper to wood, when the glue should be applied to one only.

In cabinetwork and joinery, glue blocks, as shown in Fig. 7, are very frequently stuck in interior angles to assist in holding

Fig. 7.

parts together and keeping them at right angles with each other. They are used only in positions where they will not be visible when the work is finished and in place. The blocks are planed in long pieces measuring, perhaps, $1\frac{1}{2}$ in. in cross section, and are cut off in blocks 2 or 3 in. long. They are simply glued into the angles, being rubbed backwards and forwards slightly, to make a good joint, but are not clamped or secured in any way except by the hold of the glue.

They are, of course, used only for indoor work. They are common in the plinths and cornices of wardrobes and similar carcass furniture, and in staircases.

In veneering, or gluing, very thin wood on the surface of thicker, there are two methods peculiar to that class of work, neither of which are applicable to the gluing of thick pieces together. In one the wood is treated almost as if it were paper, and is squeezed to the thicker material with an appliance called a veneering hammer, superfluous glue being forced out by the pressure and sliding action of the thin smooth edge of the hammer. This makes a close union, and the veneer is so thin that it remains as it is pressed, and does not require prolonged pressure. The other method is to clamp hot plates of wood or zinc, known as cauls, on the veneer, and leave them there till the glue is dry. This method is adopted for thick veneer. The

heat of the cauls penetrates the veneer in a few moments, and causes the chilled glue to run freely again, and, combined with the pressure, this makes a perfect glue joint. For small work by this method hand-screws are used to apply the pressure. On a large scale entire boards are veneered and put in presses specially for the purpose. Veneering also differs from ordinary work in the method of preparing the joints. For uniting two pieces of thick wood it is necessary to fit the surfaces to each other, generally by planing them perfectly true. Veneer, being thin, will accommodate itself to an untrue surface, and though the veneered surfaces are approximately flat or curved, as the case may be, they do not have to be carefully fitted. An absolutely smoothly-planed surface is unsuitable for this work, and roughness is generally imparted with a toothing-plane to make a better glue joint.

TESTING WIRELESS TELEPHONE RECEIVERS

NORMAN M. DRYSDALE

It often happens that a fault in the receiver of a wireless set may be traced directly to a disconnection, either in the telephones themselves, or in the leads connecting them.

The most common method of ascertaining whether this is the case, is to connect them to a dry cell; a very loud click heard in the phones denoting a complete circuit. Now this method is very detrimental to their sensitiveness, as the comparatively large current passed through the magnet windings tends to upset the delicate degree of magnetism already existing between the ferrotype plate and the magnet poles. This degree is such that it requires but a very small additional magnetism to actuate the diaphragm. If, therefore, this balanced condition is upset, it will require a much larger addition of magnetism, thus the sensitiveness of the instruments will be reduced.

The following test provides just sufficient current to effect the diaphragm, and besides indicating a complete circuit or otherwise, also gives one some idea of their sensitiveness.

Procure a dime and cent piece, and between them sandwich a piece of blotting-paper, moistened with water. It

will be found that on applying one terminal of the telephone leads to the dime, and the other to the cent, quite a loud click will be heard in the receivers, pro-

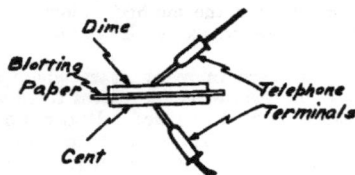

duced by the contact of the two dissimilar metals, brought about by the damp blotting-paper, generating a minute current. It should be noted that the coins should not be allowed to touch each other or the little cell will be short-circuited and fail to produce any current.

"So you heard the bullet whiz past you?" asked the lawyer of the darkey.

"Yes, sah, heard it twict."

"How's that?"

"Heard it whiz when it passed me, and heard it again when I passed it."— *American Boy.*

THE LATEST TYPES OF MERCURY CONVERTER*

DR. W. HECHLER

The theory of the converter is well known, so that the present article is confined principally to its technical aspects, though a few preliminary words may be advisable with regard to some of the physical points of the problem. Between the electrodes of a vacuum tube, the negative of which consists of mercury, a direct current at 1,000 volts is necessary to start the arc. The resistance to the passage of the current lies entirely at the mercury cathode, but as soon as the arc is formed it falls to a low value so long as current passes to the cathode. This change in the resistance is caused by the ionization of the mercury vapor. The permanent voltage-fall between the electrodes is made up of losses at the cathode, anode and in the mercury column, and amounts generally to something between 13 and 25 volts. If, now, the current falls below a certain definite value for the hundred-thousandth of a second the arc is extinguished; consequently, with ordinary single-phase current, a continuous arc is impossible, because the current periodically passes through the zero value. An auxiliary direct current of some kind is therefore necessary, and this is obtained by means of choking coils, which cause a slight overlapping of the currents in the circuits into which the whole is divided. This is, of course, unnecessary with polyphase current, but with single-phase current, there are two ways in which the choking coil may thus be arranged. In the one case the coil is arranged in parallel with the anodes of the vacuum tube, and therefore in parallel with the secondary terminals of the transformer, while its middle point is connected through the direct-current circuit to the cathode. In the other arrangement the choking coil is placed between the middle point of the transformer winding and the cathode, and is therefore in series with the direct-current circuit. The author gives, further, some oscillographic records of the currents and voltages in the different parts of the circuit, which are not, however, reproduced here. It suffices to say that the amount of overlap is a very important factor, determining, as it does, the nature of the direct-current wave which is produced. With polyphase current the overlap is already provided by the nature of the current itself, and the greater the number of the phases the less will the resulting direct current tend to pulsate.

The vacuum tubes, used as converters, differ in the number of the electrodes, which depends on the number of phases; there are also differences in size, both of the tubes and of the electrodes, the anodes depending on the maximum value of the direct current to be delivered; finally, the distance between the electrodes depends on the voltage of the direct current. As already stated, the distance between the electrodes, particularly that between cathode and anode, determines the working voltage. Up to 200 volts the internal drop is about 15, and in the shapes in which the tubes are usually constructed, it is independent of the current which these are intended to carry. The efficiency of the converter depends on the fall of voltage in the tube, and on the losses in transformer and choking coil, and it increases as the direct-current pressure rises.

The complete apparatus includes the tube with its transformer and choking coil, which can be mounted on the switchboard together with the switches and instruments. The shape which it eventually takes depends partly on the purpose to which it is to be applied, and one of the most important is that of charging accumulators. For this purpose the Allgemeine Elektricitats Gesellschaft builds units for 5, 10, 20 and 30 amperes for single-phase current, and for 30 amperes for polyphase current. The starting is effected by hand, a knob being mounted on the front of the switchboard. If this knob is turned, the tube is rotated sufficiently to bring the mercury of the cathode into contact with that forming an auxiliary anode; and on turning the knob back into its original position, the mercury connection is broken, though it has in the meantime completed a circuit from the cathode through the auxiliary anode and its series resistance to one of the main anodes. The breakage of this circuit causes a spark, which is sufficient to start the main arc. The converter obviously cannot work without load, neither can it at once be put in series with the battery; consequently it is allowed to work on a special resistance, which is put in parallel with the battery and then gradually cut out of the circuit. The whole operation of starting up requires at the outside 10 seconds, though in the smaller sizes, after starting the arc, the circuit through the battery can, if necessary, be closed at once. The regulation of the voltage is effected by varying the ratio of transformation; in the 5-ampere size, it is, however, effected by varying a resistance in series with the battery. Other methods of regulation are also adopted to meet special cases. Thus, with a converter having a range from 55 to 235 volts, the changes are effected in three stages, each of which has 21 finer stages of adjustment, by which it is possible to charge 22, 44, 66 or 88 cells in series. The only external difference in this case consists in the mounting of two simple switches in addition to the usual apparatus.

Another use to which these converters are put is for lecture lanterns and for small searchlights, and they are then constructed in 30-ampere sizes, giving 50 or 60 volts direct current. The tipping of the tube is effected automatically by a relay, which is then cut out of circuit along with the resistance of the auxiliary anode. There are also all kinds of other types. The battery type is made without any device for regulating the pressure and with automatic starting relays. Converters of this kind, as, for instance, in Hahb's magnets for ophthalmic use, and in Finsen's light treatment. They are also used for operating electromagnetic brakes on locomotives in colliery work; the starting being effected automatically, the maximum output being 10 amperes at 220 volts. This piece of apparatus

shows conclusively that it is not sensitive to shocks of a mechanical nature, the converter being mounted on the locomotive. They are also used for driving direct-current motors, which are sometimes to be preferred to those of the alternate-current type; and similarly, they are not unknown in electrochemical laboratories.

At present in Germany the largest size gives 40 amperes on polyphase circuits, and 30 amperes with single-phase, though in America it has lately been found possible to give a permanent output of 50 amperes under the latter conditions. The heating at the fused joints, carrying the leading-in wires, is the point on which the maximum limit depends. In America experiments have been made in the use of metal instead of glass for the main body of the converter; but these require the use of certain subsidiary apparatus, which can only present advantages in the very large sizes, suitable for use with large batteries and searchlights. But the simplest solution would probably be to use several converters in parallel, which would necessitate the use of something of the nature of series resistances to ensure stable working. For purposes of this nature, special types have also been devised by which some of the minor losses incidental to such a combination can be avoided.

———

The first aerial service is shortly to be initiated in Great Britain under the direct control of the Postmaster-General. King George is personally interested in the scheme, and has given permission for the use of Windsor Park as the terminus of the service, which is to have London for its starting-point. In many large London business houses special "aerial" letter-boxes are to be placed for the reception of the "aerial" missives. When collected by postmen these will be placed in Government sealed letter-bags and, if the negotiations which are now proceeding are satisfactorily completed, will be rushed across to the Hendon aerodome. There the safely-sealed bags will be securely strapped to the waiting aeroplane, which will then be piloted across to Windsor.—*English Mechanic.*

AUTO DEPARTMENT

POINTS TO BE CONSIDERED IN PURCHASING AN AUTOMOBILE

"CHASSIS"

Unfortunately, perhaps, the first point to be considered by most of us in purchasing a car is the cost. Of course it is a foregone conclusion that we want the best we can get for our money, and it is to aid the uninitiated in the selection that this article is written.

Granting that we have not to consider the cost, the most expensive car is not necessarily the best, yet it is a pretty general proposition that quality and price, with some notable exceptions, are approximately proportional. There are concerns who have built up a name for themselves in the early days, and who still build excellent cars, but who get a price for their product higher than a newer concern could get for a car of equal quality. In other words, people of means are willing to pay for a name. This applies to makes commanding a high price, since there is greater competition in the lower-priced field. Hence the writer believes that, generally speaking, one gets more for his money if he buys a medium or comparatively low-priced car than when he buys a high-priced car; but he does not advocate a cheap car in any case. Doubtless there are many who would take issue on this point and present plausible arguments in their defence. Be this as it may, the arguments the other way seem more plausible.

There are many good values to be had in used cars, but purchases of this character must be made with extreme caution, and should not be undertaken without expert advice. It is better to pay an engineer, or even a good mechanic, a reasonable fee to pick out a good car, than to be "penny wise and pound foolish" in trying to save such a fee only to find that the car you select is faulty, both in design and workmanship, and the same advice applies equally whether you are buying a new car or an old one.

After deciding what price you will pay for your car, consider well what duties it must perform, and, if possible, have it put through similar duties during the demonstration which should invariably precede a purchase. Remember that a low-powered car is very practical for city use, or where there are good roads and slight grades, but that such a car will not be very suitable for extended tours over rough and hilly roads. If you wish to take all hills excepting very heavy grades on high gear, only a car with a high-powered motor will answer your purpose and you will have to pay well for the luxury. Many advocate cars of this character in the belief that a large motor run, as it will be ninety percent of the time, at low speed, will stand up better than a small motor run more constantly at a comparatively high speed. This is undoubtedly true to some extent, but it is also true that a motor of this type run at low speed is very inefficient in point of fuel consumption. However, a much larger factor than motor depreciation is tire expense, and this will vary in almost direct proportion to the weight of the car, and will increase very rapidly with increase of average running speed. Cars with large motors must have all their parts in proportion in order to withstand the strains imposed when the motor is run at its maximum power. Hence we see that weight increases with power, and tire expense, with weight; and we reach the inevitable conclusion, that other things being equal, the lighter car has a marked advantage.

Safety is, of course, a prime factor, and reliability is next in importance, hence weight must not be decreased so as to interfere with these. Light weight without sacrifice to strength, however, is particularly important in axles and those parts between them and the road. These parts are dead weights on the tires and are not cushioned from them by springs. Hence they pound the tires, and in consequence decrease their |life. One designer whom the writer could mention considers this point so important that he has offered a dollar per pound per car for a design which will decrease the weight of these parts on his product without sacrifice to strength and durability or unduly increased cost of manufacture.

When considering the question of safety, it is well to look closely at the running gear and the steering gear. See that these parts are strong and so put together that they cannot be loosened by the severe vibration to which they are constantly subjected. Remember that a broken steering gear may easily result in a serious accident.

It is perhaps natural that men who have used horse-drawn vehicles all their lives should compare their first automobile with these vehicles. They will find a car much more heavily built than a carriage, for the reason that the former carries its own means of propulsion and can travel, in normal use, three times as fast as a horse. These conditions require heavier parts, and the result is a heavier vehicle. The heavier the vehicles, the larger the motive power necessary to propel it at any given speed, hence, with other things being equal, the lighter vehicle again has an advantage in this particular.

Simplicity in design and accessibility of parts are points well worth consideration. A car having a motor which is cluttered with a mass of piping, operating levers, rods, etc., is certainly harder to care for than a simple appearing motor. In fact, the latter type is considered much better design, and the former is likely to have inferior design in more essential parts as well as in the parts which appear on the surface.

Accessibility is a point which receives far greater consideration today than it did in the early days when cars frequently had the motor under the body and in such a position that the latter had to be lifted whenever the motor required much attention. It is still well to consider this point, however, particularly with respect to such parts as are likely to require inspection, adjustment, frequent cleaning or lubrication not automatically cared for, such as magneto, oil and water

pumps, grease cups, oil level gauge, carburetors, etc. It is well, also, to see that all parts subject to wear are covered to prevent penetration of grit.

Lubrication of all surfaces where one part slides or rolls upon another is a matter of prime importance. For pistons and bearings within the crank case, splash lubrication is usually sufficient, although other methods, not so much employed, are considered more scientific by some capable judges. In order to maintain a constant level in the troughs under the connecting rods in the "splash" system and oil pressure on the bearing in the forced system, a pump is necessary. Same should be conveniently located to enable removal for cleaning when necessary. Gear pumps are very generally used for this purpose, but piston pumps operated from eccentrics on the cam shaft are considered rather better practice. But the motor is only one of the many parts which should be lubricated. The transmission, differential, wheel bearings and steering gear worm should all be enclosed and so packed as to hold grease, or, better yet, oil, although the latter is hard to retain except by expensive construction, and therefore is seldom provided for on cars of moderate price.

The selective type of change speed gear has come to be generally accepted as standard, and is to be preferred above any other gear speed reduction., The friction drive is not to be overlooked, however, and has some inherent advantages, including simplicity and ease of operation. This type of drive requires comparatively frequent renewal of the friction surface of the wheel, however, and hence should be so designed as to enable replacements without difficulty.

Planetary transmissions have but two points in their favor: small first cost and ease of operation. They are seldom constructed for more than two speeds forward, and because of this are not often applied to large cars. In fact, they are used almost exclusively on runabouts and very light cars and are as a rule to be avoided by purchasers who are looking for durability and efficiency in transmission.

A majority of cars today are shaft-driven. This form of drive has the advantage of quiet running and greater efficiency than a chain drive, but the latter permits a strong rear axle, and is probably more dependable for very rough or hilly work and hard usage.

No car should be without brakes of ample proportions. In fact, two independent brakes are very desirable and are usually provided on the larger cars. Both the foot brake and the emergency brake should be capable of locking the rear wheels.

The frame of the car is usually of steel, but some makers still use wood reinforced with steel. Undoubtedly the latter construction is better suited for absorbing shocks and is more resilient. On the other hand, a steel frame is likely to be stronger.

For easy riding, full elliptic springs are most desirable. They require the double universal joint construction for the drive shaft, however,—a requirement which has both advantages and disadvantages— and are not so much used as three-quarter elliptic and semi-elliptic springs.

All of these points and many more should be considered by a prospective purchaser and can really be given their proper weight only by an engineer of long experience in the line. The purpose of these suggestions will, however, be realized if they serve to point out the way in a few particulars to the man who is about to invest in an automobile.

PROFIT AND LOSS

Big Returns in Electric Circuit

F. WEBSTER

It has been said that anyone, not excluding a deaf mute, will prick up his ears and take notice when offered a "sure thing proposition" for doubling his money. As the electrical engineer can offer better than double, it might be of interest to some of the present-day investors to study one of the old problems relating to capacity and inductance of circuits, this particular one giving over 5 to 1.

Thus, suppose that a certain circuit is arranged as shown in Fig. 1. This circuit is laid out so that it leaves the line at A, and has an incandescent lamp at B. Immediately below the lamp B, the circuit is divided at C, one branch having an incandescent lamp D, and an inductive coil E, and joins the other line wire at F. The other branch leads from C and has an incandescent lamp G, and a condenser H, and reaches the line wire at I. The main line voltage is 550, alternating with a frequency of 133. The resistance of the incandescent lamps is 100 ohms each, the inductance of the coil E is 0.597 henry, and the capacity of the condenser is 2.49 microfarads. The arrangement is such that when 0.4 ampere passes through lamp B, there is approximately 1 ampere passing through the lamps D and G in the branches; that is, the

sum of the amperes in the branches, or 2 amperes, is 5 times the current through the lamp B.

Several terms have been used in describing Fig. 1 that may be new to some operating engineers. While a thorough understanding of these terms as to their values and their relations to each other would require a special training in electrical engineering, yet the application of the principles involved is easily understood. A brief explanation of the meaning of each term is given in simple language below, and the reader, if he does not have an electrical training, should take it for granted that the applications are correct. The article is written for the purpose of showing the unexpected conclusions which are sometimes arrived at when using alternating currents, and not for the purpose of giving a dictionary of electrical words.

The pressure of an electrical current is called its electromotive force or voltage, and it is represented by the letters e.m.f., and the unit as measured by switchboard instruments is called the volt. The voltage corresponds with the steam pressure of a boiler, while the electric current corresponds with the quality of steam flowing from it. The pressure does not depend upon the size of the boiler, likewise the voltage does not depend upon the size of the generator. The unit of electric current is called the ampere. Any wire or device for transmitting an electrical current has a resistance which tends to cause a loss of pressure; that is, the voltage grows less as the length of line increases. The unit of resistance is called the ohm. A piece of copper wire 1,000 ft. long, and having a diameter of 0.1 in., has a resistance of about 1 ohm. In the formulas for working electrical problems, the resistance of the circuit is represented by the capital letter R.

Besides the resistance of the wire, it may be so arranged that when an alternating current is passed through it, the current will be choked back; that is, the current wave will lag behind the voltage wave and what would be its normal flow. This lag of current takes place when the wire is wound in a coil, and the effect is greatly increased by placing an iron core in the coil. The choking effect of a coil is called its inductance and the unit for measuring it is called the henry. The ordinary transformer has considerable inductance and causes the current to lag in the line. The henry is represented in formulas by the capital letter L.

For an alternating current the strength is continually varying and also its direction of flow alternates. Its strength starts at zero and rises to its greatest value, declines to zero and builds up to its greatest value in the opposite direction and then diminishes to zero. This series of wave-like operations is called a cycle, which is repeated continually. The method of designating the cycles of the current from an alternating current generator is to give their number per second or frequency and the term is represented by the small letter n.

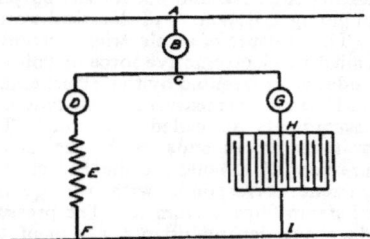

FIG. 1. CIRCUIT CONTAINING INDUCTANCE AND CAPACITY

A condenser is a device made up of flat sheets of tinfoil, every sheet in the stack being connected to opposite terminals. The sheets are separated from each other by insulating material, as shown at H, Fig. 1. A condenser can be charged with a quality of electricity and then made to discharge itself. The farad is the name of the unit of capacity of a condenser, but as this unit is very large as compared with the capacity of the ordinary condenser, it is customary to state the capacity in millionths of a farad, called a microfarad, the prefix "micro" meaning a millionth part. The farad is represented by J. The condenser causes the current wave to go ahead of the voltage wave; that is, it gives lead to the current. This lead effect is just opposite to the lag effect caused by induction and can be made to neutralize induction. A rotary convertor or a synchronous motor acts like a condenser in giving lead to the current in the line.

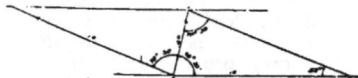

FIG. 2. PHASE RELATIONS OF CURRENTS

The method of computing the current that would flow through the circuit in Fig. 1 is by means of two formulas or rules. These formulas have been worked out in text-books on electricity, and all that is to be done in the present case is to insert the values given in the problem and complete the arithmetical work. When the resistance of a lamp and the amount of current flowing through it are known, the amount of loss of pressure in volts is found by multiplying the resistance by the current. Thus, the loss through the lamp B will be $0.4 \times 100 = 40$ volts, and the pressure left to force the current through either of the branches will be $550 - 40 = 510$ volts. The formula for finding the amount of current that flows through the left branch $C F$, where the current lags or is choked back by the inductance of the coil E, is as follows:

Square the inductance in henrys; square the frequency and multiply these squares together and the result by 39.48; to the result add the square of the resistance in ohms and take the square root of the sum; divide the e.m.f. by this root and the quotient will be the current in amperes.*

By putting the values of the terms in the problem in place of the names in the rule, it appears as follows:

$$\text{Current} = \frac{510}{\sqrt{(100^2 + 39.48 \times 133^2 \times 0.597^2)}} = \frac{510}{508}$$

equals 1 ampere, approximately, which is found by performing the arithmetical operations as indicated.

To find the current flowing in the right hand branch, where the current gets ahead of the pressure caused by the condenser, use the following rule:

Multiply the square of the capacity in microfarads by the square of the frequency and divide 25,350,000,000 by the result; to the quotient add the square of the resistance in ohms and take the square root of this sum; divide the e.m.f. by this root and the quotient will be the current in amperes.†

And when the data of the problem is inserted in place of the letters, the formula has this form:

$$\text{Current} = \frac{510}{\sqrt{\left(100^2 + \dfrac{25,350,000,000}{133^2 \times 2.49^2}\right)}} = \frac{510}{491}$$

Lag and the lead of electric currents can be expressed in terms of an angle, just as the eccentric on the shaft of an engine is set a certain angle either ahead or behind the crank position. Thus, in Fig. 2, suppose the line AC represents the direction of the current at a particular instant in the main line and through lamp B; then the current in the branch line CF of Fig. 1 will be behind CA, and that in CI will be ahead of CA. These angles between the directions of the currents are easily computed by the use

*In short form:

$$\text{Current} = \frac{e.\,m.\,f.}{\sqrt{(R^2 + 39.48 \; n^2 \; L^2)}}$$

Where letters represent quantities as stated in the text.
†The short form is:

$$\text{Current} = \frac{e.\,m.\,f.}{\sqrt{\left(R^2 + \dfrac{25,350,000,000}{n_2 \; J^2}\right)}}$$

of trigonometry. Those readers not familar with this branch of mathematics will have to consider the results only, without reference to the method of getting them. It might be stated here, however, that trigonometry is a very easy as well as a useful subject to learn.

The cause of the current in the two branches being much larger than that in the line is due to the lag caused by the inductance in the coil E, and to the lead caused by the capacity of the condenser H. The current in the branch with inductance E lags behind that through the branch with the lamp B by an angle whose tangent equals 6.28 x frequency x inductance ÷ resistance = 6.28 x 133 x 0.597 ÷ 100 = 4.989, corresponding to an angle of 78 degrees, 30 minutes. The values of angles corresponding to different tangents are given in handbooks. The current through the condenser H leads that through the lamp B by an angle whose tangent equals 1,000,000 ÷ (6.28 x frequency x capacity x resistance) = 1,000,000 ÷ (6.28 x 133 x 2.49 x 100) = 4.8077, corresponding to an angle of 78 degrees, 30 minutes. The differece in phase between the currents in the two branches is equal to the sum of the two angles computed above, or 78 degrees, 30 minutes plus 78 degrees, 30 minutes equals 157 degrees, and this angle is laid off as shown in Fig. 2. The current flowing through the lamp B and the line equals the geometrical sum of those in the two branches; that is, its value will be represented by the length of the diagonal CA of the parallelogram constructed on the lines CF and CI in Fig. 2. The lines CF and CI should be of equal length and to any convenient scale to represent 1 ampere. The length of CA will measure 0.4 of CI, or CF.

Therefore, when the get-rich-quick scheme is worked backwards, as it always is sooner or later, generally sooner, the investor who has put in a lot, always gets in return what the combination happens to deliver, which is very little.—*Practical Engineer.*

The rural delivery system costs $35,-000,000 a year; the rural carrier's daily load is absurdly small—a pitiful 25 lbs.; it could be 500 lbs. without adding materially to the cost of the service.

WIRELESS TELEGRAPHY

In this department will be published original, practical articles pertaining to
Wireless Telegraphy and Wireless Telephony

REGULATIONS GOVERNING WIRELESS EQUIPMENT ON OCEAN PASSENGER STEAMERS

The following is a copy of a circular issued by the United States Department of Commerce and Labor June 15, 1911, to the collector of customs and others concerned in an Act approved June 24, 1910:

Be it enacted by the Senate and House of Representatives of the United States of America in Congress assembled, That from and after the first day of July, nineteen hundred and eleven, it shall be unlawful for any ocean-going steamer of the United States, or of any foreign country, carrying passengers and carrying fifty or more persons, including passengers and crew, to leave or attempt to leave any port of the United States unless such steamer shall be equipped with an efficient apparatus for radio-communication, in good working order, in charge of a person skilled in the use of such apparatus, which apparatus shall be capable of transmitting and receiving messages over a distance of at least one hundred miles, night or day: *Provided,* That the provisions of this Act shall not apply to steamers plying only between ports less than two hundred miles apart.

Sec. 2. That for the purpose of this Act apparatus for radio-communication shall not be deemed to be efficient unless the company installing it shall contract in writing to exchange, and shall, in fact, exchange, as far as may be physically practicable, to be determined by the master of the vessel, messages with shore or ship stations using other systems of radio-communication.

Sec. 3. That the master or other person being in charge of any such vessel which leaves or attempts to leave any port of the United States in violation of any of the provisions of this Act shall, upon conviction, be fined in a sum not more than five thousand dollars, and any such fine shall be a lien upon such vessel, and such vessel may be libeled therefor in any district court of the United States within the jurisdiction of which such vessel shall arrive or depart, and the leaving or attempting to leave each and every port of the United States shall constitute a separate offense.

Sec. 4. That the Secretary of Commerce and Labor shall make such regulations as may be necessary to secure the proper execution of this Act by collectors of customs and other officers of the Government.

REGULATIONS
I.—*Administration*

1. The Department will appoint three wireless ship inspectors, whose districts shall be:
North Atlantic, from New York to the Canadian boundary;
Middle Atlantic and Gulf, from Philadelphia to Galveston, including Porto Rico;
Pacific, from Puget Sound to San Diego, including Alaska and Hawaii.

2. These inspectors are authorized to communicate directly in their respective districts with collectors of customs, and to co-operate with them in the enforcement of the law.

3. Collectors of customs and wireless ship inspectors, as far as practicable, shall visit ocean passenger steamers subject to the Act, before they leave port, and ascertain if they are equipped with the apparatus in charge of the operator prescribed by the first section of the Act.

4. Where an ocean passenger steamer subject to the Act is without the apparatus and the operator prescribed, or either of them, and is about to attempt to leave port, the customs officer or wireless ship inspector visiting the vessel shall—
(a) Notify the master of the fine to which he will be liable, and of the particulars in respect of which the law has not been complied with;
(b) Notify at once the collector of customs, if necessary, by telephone;
(c) Prepare in writing a report of his action, stating particulars as in (a), to be transmitted to the collector of customs. The collector will transmit a copy to the United States attorney for the district in which the port is situated.

5. The Act does not authorize the refusal of clearance in case of violation of its provisions, but specifically provides for the imposition of a fine in a sum not more than five thousand dollars upon conviction by the court. The collector of customs, accordingly, when advised that an ocean passenger steamer subject to the Act is attempting to leave port in violation of its requirements, shall at once notify the United States attorney. Subsequently he shall report the case briefly to the Secretary of Commerce and Labor.

6. The Act does not apply to a vessel at the time of entering a port of the United States. Customs officers and wireless ship inspectors may, however, accept as evidence of the effi-

ciency of the apparatus and the skill of the operator wireless messages shown to have been transmitted and received by him over a distance of at least one hundred miles, by night or day, during the voyage to the United States.

7. In cases of violations of the Act the efficiency of the apparatus and the skill of the operator will be determined by the court (see Section 3 of the Act). Collectors of customs and wireless ship inspectors, accordingly, are enjoined that the reports required by paragraph 4 (c) of these regulations must be precise statements of the facts as the basis for proceedings by the United States attorney.

II.—Operators

1. Paragraphs 3 and 4 of Article VI of the Service Regulations, annexed to the Berlin International Radio-telegraphic Convention, provide:

3. The service of the ship station must be carried on by a telegraphist holding a certificate issued by the Government to whose authority the ship is subject. This certificate testifies to the technical proficiency of the telegraphist as regards:

(a) The adjustment of apparatus;
(b) Transmission and sound-reading at a speed which must not fall short of twenty words a minute;
(c) Knowledge of the regulations applicable to the exchange of radio-telegraphic traffic.

4. In addition, the certificate testifies that the Government has bound the telegraphist to the obligation of preserving the secrecy of correspondence.

The Berlin Convention has been ratified by the following nations, dominions, and provinces: Great Britain, Canada, Australia, British South Africa, India and New Zealand, Germany and all German protectorates, France, Norway, Japan, the Netherlands and Dutch Indies, Russia, Sweden, Austria-Hungary, Spain, Denmark, Belgium, Brazil, Turkey, Portugal, Roumania, Mexico, Bulgaria, Persia and Tunis.

Wireless operators holding valid certificates issued by the Governments named above will be recognized by this Department as persons "skilled in the use of such apparatus" within the meaning of the Act unless in the case of a specific individual there may be special reason to doubt the operator's skill and reliability. Such certificates should be ready at hand for the inspection of customs or other officers before the steamer departs from the United States.

2. (a) The Commissioner of Navigation will issue operators' certificates of skill (see Appendix A) in radio-communication and operators holding them will be recognized as persons "skilled in the use of such apparatus" within the meaning of the Act, unless in the case of a specific individual there may be special reason to doubt the operator's skill and reliability. Such certificates should be ready at hand for the inspection of customs or other officers before the ship departs from the United States.

(b) To secure a certificate an operator will pass an examination in the adjustment of apparatus, correction of faults, change from one wavelength to another, transmission and sound-reading at a speed of not less than fifteen words a minute American Morse, or twelve words Continental, as the operator may elect. Operators are advised to learn as soon as practicable the Continental system, recognized by the Berlin Convention and employed by the United States Navy.

(c) The examinations will be held at the United States navy yards at Boston, Mass., Brooklyn, N.Y., Philadelphia, Pa., Washington, D.C., Norfork, Va., Charleston, S.C., New Orleans, La., Mare Island (San Francisco), Cal., Puget Sound, Wash.; at the naval stations at Key West, Fla., San Juan, P.R., and Honolulu, Hawaii, and also at the Bureau of Standards, Washington, D.C. Applicants for certificates should communicate in writing with the commandants of·the navy yards or stations named, or with the Director of the Bureau of Standards, to ascertain the day and hour when they can be examined. The certificates will be delivered at the places named.

(d) After an applicant has secured a certificate he should go before a notary public to take the usual oath for the preservation of secrecy of messages received in the line of duty.

(e) These examinations for the present will be open to—
(1) Operators actually employed as such by a wireless or steamship company, including shore operators;
(2) Operators seeking employment as such by a wireless or steamship company, including shore operators; and such applicants shall present letters from the company with which they seek employment;
(3) Applications for examination of operators of either class may be made by the wireless or steamship company in behalf of a number of operators by name.

3. Additional provision will be made later for the examination of operators by wireless ship inspectors at the New York and San Francisco customhouses and at other customhouses hereafter to be designated.

4. A wireless ship operator not possessing a certificate of skill as provided herein may present for the consideration of the visiting customs officer or wireless inspector other competent evidence of skill, or the wireless inspector may examine him, if practicable. If such examination be satisfactory, the wireless inspector will issue a certificate.

III.—Apparatus

1. When the efficiency of the wireless apparatus is certified by a foreign government, such certificate will be recognized by this Department, but the customs officer or wireless ship inspector may, if he deem it necessary or desirable, satisfy himself that the apparatus is in good working order.

2. Whenever practicable, the customs officer or wireless ship inspector shall satisfy himself on his visit before the departure of a passenger steamer subject to the Act that the apparatus is efficient and in good working order within the meaning of the Act, and, if satisfied, he shall issue a certificate in the form in Appendix B. Duplicates of such certificates shall be retained in the files of the collector of customs.

3. When inspection of the apparatus by a customs officer or wireless inspector is not practicable, the master of the steamer may furnish to the visiting customs officer a certificate in the form in Appendix C. Such certificate shall be retained in the files of the collector of customs.

4. The current necessary to transmit and receive messages shall at all times while the steamer is under way be available for the wireless operator's use.

5. A storage battery or some other auxiliary which will produce sufficient power to operate the transmitting apparatus for four hours, ordinary sending, should be suitably installed and ready for use in case of accident disabling the electric plant of the vessel. After January 1, 1912, vessels will be required to carry such battery or auxiliary.

6. One extra pair of head telephones and three sets of extra cords and one extra detector should always be kept on hand.

IV.—*Additions or Amendments*

Additional or amendatory regulations will be issued from time to time as they may appear necessary.

                                BENJ. S. CABLE,
                                Acting Secretary.

APPENDIX A

NAVIGATION SERVICE FORM 751

Operator's Certificate of Skill in Radio-Communication

This is to certify that, under the provisions of the Act of June 24, 1910, —— —— has been examined in radio-communication and has passed in:

(*a*) The adjustment of apparatus, correction of faults, and change from one wave-length to another;

(*b*) Transmission and sound-reading at a speed of not less than fifteen words a minute, American Morse, twelve words, Continental, five letters counting as one word.

The candidate's practical knowledge of adjustment was tested on a —— set of apparatus.* His knowledge of other systems and of international radio-telegraphic regulations and American naval wireless regulations is shown below:

—— —— ——.

(Signature of examining officer) —— ——

Place —— ——,   Date —— ——, 191-.

By direction of the Secretary of Commerce and Labor:
                                —— ——,

Commissioner of Navigation, Washington, D.C.

I, —— ——, do solemnly swear that I will faithfully preserve the secrecy of all messages coming to my knowledge through my employment under this certificate; that this obligation is taken freely, without mental reservation or purpose of evasion; and that I will well and faithfully discharge the duties of the office: So help me God.

            (Signature of holder) —— ——.

Date of birth, —— ——, ——.

Place of birth, ——, ——.

Sworn to and subscribed before me this — day of ——, A.D. 191-.

    (*Seal*)   —— ——, *Notary Public.*

            (On back of Form 751)
            Service Record

This is to certify that the holder of this certificate has served satisfactorily as wireless operator under my command during the period named.

Name of Steamer	Period	Master
...........From......19.., to......, 19.........		
...........From......19.., to......, 19.........		
...........From......19.., to......, 19.........		
...........From......19.., to......, 19.........		
...........From......19.., to......, 19.........		
...........From......19.., to......, 19.........		
...........From......19.., to......, 19.........		
...........From......19.., to......, 19.........		
...........From......19.., to......, 19.........		

APPENDIX B

NAVIGATION SERVICE FORM 752

Certificate of Wireless Inspection
            Port of —— ——
            —— ——, 191-.

This is to certify that I have today examined the apparatus for radio-communication on the S.S. ——, of which —— —— is master, about to leave this port for ——, and I have found the same efficient and in good working order, as prescribed by the Act of June 24, 1910.

        (Signed)   —— ——,
                Wireless Ship Inspector.
        (Or)
                Customs Inspector.

APPENDIX C

NAVIGATION SERVICE FORM 753

Master's Certificate of Wireless Apparatus
            Port of —— ——,
            —— ——, 191-.

This is to certify that I have today examined the apparatus for radio-communication on the S.S. ——, of which I am master, about to leave this port for ——, and I have found the same efficient and in good working order, as prescribed by the Act of June 24, 1910.

        (Signed) —— ——, *Master.*

---

* It is not intended to limit the employment of the holder to a particular system, but merely to indicate the particular system in which he was tested for adjustment of apparatus.
   This certificate is valid for two years, subject to suspension or revocation by the Secretary of Commerce and Labor for cause. It should be kept where it can be shown to officers of the customs or other officers of the Government just before the ship leaves port.

DESIGNING AND DRAWING OF A SCREW PROPELLER

Considered from the Standpoint of Descriptive Geometry; also from the Shop or Pattern-Maker's Method

GEORGE JEPSON

Editor's Note:—The directions given in this article apply to the design of both ship and aeroplane propellers; the latter, however, being equipped with but two blades instead of four, in accordance with modern practice.

The drawings here may be considered part of a third year course given in a technical school or an evening school. Therefore the person who attempts to make these drawings should have a good knowledge of the principles of Projection.

We will first consider the drawing from the descriptive method. The principle employed in making the drawing for a ship's screw or propeller, each blade of which ·is practically a warped or helicoidal surface, is precisely the same as that in making the drawing for any other warped surface, such as a V or square-threaded screw.

Such surfaces, viewed from any standpoint, are always seen in oblique projection. The propeller may have two, three or four blades. In an ordinary screw, whose center lines are equidistant apart, all of the helical surface is used, while in the propeller only part of such a surface is used. To make a drawing of a propeller (Fig. 1) having the following dimensions:

Pitch, 4 ft. 6 in., or 54 in.;

Diameter 3 ft., or 36 in.;

Diameter of shaft is 4 in.; diameter of hub, 9 in.; the length of the hub is 10.3 in.; the taper of the shank is .75 in. to a foot; the thread of the nut that holds the propeller upon the shaft has 2.5 threads per inch, regardless of its other dimensions, and should be left-handed when the screw propeller is right-handed. The nut should be securely fixed to keep it from backing off.

The width of the key equals 0.22 times the diameter of the shaft, plus 0.25 in. The thickness of the key equals 0.55 times the width. The thickness of the blade is shown in section on Fig. 1, and is made 0.5 in. for every foot of diameter of the propeller, measured at the center of the shaft, as shown, and the point is made one-sixth that of the root. In addition to the thickness of the blade shown on blade *D* (Fig. 1) it also shows in section

the different widths and thicknesses of the blade upon each of the different concentric circles.

Then, to draw the warped surfaces of the blades, first draw two perpendicular diameters, and make a circle which represents the disc view of the imaginary cylinder, which circumscribes the greatest diameter of the propeller. Then divide the radius of the circle, beginning at the circumference, into four or more equal divisions, and through those points (2, 3, 4 and 5) draw concentric circles. The fourth circle may coincide with the hub.

Now divide the circumference of the disc view into any convenient number of equal divisions, just as the end view of the design of a common bolt would be divided.

Suppose the propeller to have four blades, and the number of divisions of the diameters to be 36 (9 for each arc of 90 degrees). From these points of divisions on the outer circle draw radial lines, thus dividing the inner concentric circles into the same number of similar and proportional parts.

Now draw the hub and the outline of the four blades about the four center lines, according to your best general idea of their shape (about which experts differ very much). They should be of such a dimension that they will be contained within about 16 of the radial divisions—four divisions, more or less, for each blade.

The lines governing their shape will intersect the concentric circles, which are really end views of helices, and the radial lines in various points.

To draw the edge view of the propeller (Fig. 2), wherein the propeller shaft is seen parallel to the projecting plane, first project the imaginary cylinder containing the propeller, together with its axis, and draw the projecting lines of every point found in the disc view.

Begin with the points x and x', the middle points upon the ends of each blade, locating these points by means of a line drawn perpendicularly across the line just protected from the points x and x' in the disc view.

Now, having located the two points x and x', lay off from those points one-half of the pitch on each side of the points, and divide the pitch into 36 equal divisions already used in the disc view, and draw by means of these divisions, all or only that part to be used of the four helical curves which form the ends of the blades.

As we are to make the blades rake away from the ship, the position of x' at the middle of each blade is on Fig. 2 placed 2 of the 36 divisions to the right of x. (It may be placed a greater or less number of divisions.)

Now draw the helical curves Nos. 1 and 5 containing the points x and x', respectively, and making x and x' the middle of its own pitch. It should be clearly understood that the helical curves forming the bottoms of the threads of any bolt, V or square, travel just as far along the bolt as the helical curves forming the point of such threads. Therefore, beginning at x' (Fig. 2) the division already laid off may be used.

Now connect by straight lines the points x and x', also the points a, b and c, as shown upon the drawing in the larger curve No. 1, with the similar points, a', b' and c' in the smaller curve numbered 5. These points have already been located by projection from the disc view. These lines represent

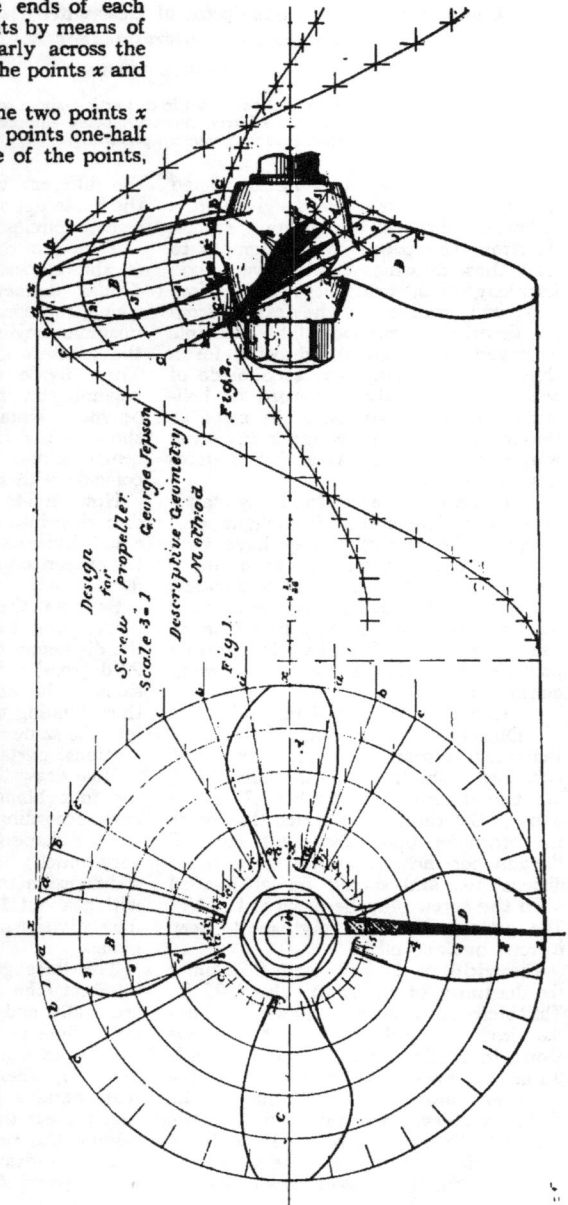

Design for Screw Propeller. Scale 3 = 1 George Jepson Descriptive Geometry Method

Fig. 2.

Fig. 1.

disc view, as seen in the edge view in oblique projection.

Draw the other helical curves, 2, 3 and 4; then locate in similar lines in the edge view all other points of the disc view, thus completing the edge view minus the thickness of the blade.

The thickness is best shown upon the blade extending toward the draughtsman, which is at right angles to the projecting plane, and which is marked "blade A" in the drawing (Fig. 2).

In the edge view the concentric circles of the disc view are represented by helical curves, all of which are shown as being drawn upon the front surface of the blade.

The section of the blades shown upon "blade D" (Fig. 1), shows the different thicknesses at the different concentric circles.

Having already obtained the widths in projection upon A and B in both Fig. 1 and Fig. 2, we will now proceed to show as much of the thickness as is possible in both figures, more especially in Fig. 2.

Referring to the section drawn upon "blade D," a line is seen passing through the center of the shaft and intersecting the different concentric circles at the points 1, 2, 3, etc., and the thickness is shown extending from this line toward the stern of the ship, viz., it extends away from the center line, as shown in the section, and is measured back at right angles to the lines drawn upon the face of "blade A" (Fig. 2), beginning at point x' and similarly from all other points upon the oblique center line x' and x. Then through the two points showing the apparent width, and the point measured back from the center of those points showing the thickness, we shall have three points through which we will draw an arc of a circle, which represents the thickness in section.

As stated, the thickness nearest the hub is .5 in. for each foot of the diameter of the screw, and the point thickness equals one-sixth that of the root.

Suppose that in Fig. 2 on "blade A" we have drawn all the sections in their respective positions, they will appear to overlap each other, as shown in the drawing.

If we now draw a line tangential to the backs of those sections, we shall have shown a line representing the thickness

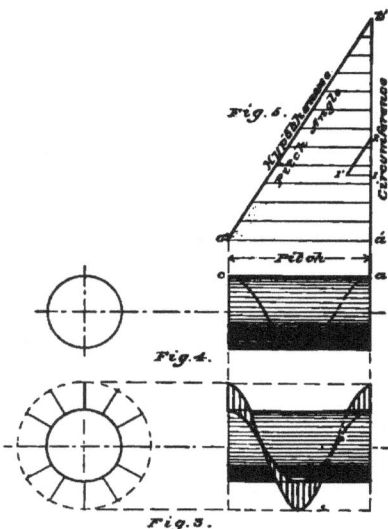

Fig. 6.

Fig. 4.

Fig. 3.

of the blade in that particular position. Any change of position, of course, would change such a line. Now it is for the draughtsman to transfer this thickness to the disc view.

It is almost impossible to represent by lines the merging of the blade into the hub without a model.

WORKSHOP OR PATTERN-MAKER'S METHOD

If a straight line, shown in the end view of Fig. 3, placed at right angles to the axis of a cylinder, be moved along, and at the same time around, the cylinder, as shown in the other view of the same figure, the line will generate a surface called a helicoidal or warped surface, part of which is used to form the blade of a propeller. The path traced upon the cylinder by the end of this line is known as the "helix or helical curve," and is again shown in Fig. 4. The top element of the cylinder in Fig. 4 is marked a c, and represents the pitch.

At some convenient distance from ac and parallel to it, draw the line a'c' (Fig. 5), which again represents the pitch of the screw. At a' erect a perpendicular a'b', whose length equals the circumference of the cylinder of Fig. 4. Join b' and c', and we have the hypothenuse of a right angle triangle. This represents the developed helical curve. Then the angle formed by the lines a'b' and b'c represents the "pitch angle" of the screw propeller.

If upon the line $a'b'$ in the right angle triangle we assume two points 1 and 2, any distance apart, say one-sixth of the line $a'b'$ (which is the circumference), and erect a perpendicular at point 1, whose length is also one-sixth of the pitch, and draw the smaller triangle 1'21, we shall find that its hypothenuse 1'2 is parallel to the hypothenuse $b'c'$, and all angles of the triangle are equal. Hence, in the drawing and designing of a screw propeller, instead of using all of the helical curve, as in the Descriptive Geometry method, we shall use only a certain portion of the pitch and a similar portion of the circumference.

We will now make a drawing of a propeller having the following dimensions:

Pitch, 4 ft. 6 in. equals 54 in.
Diameter, 3 ft. equals 36 in.
Radius, 1 ft. 6 in. equals 18 in.
Circumference, 9.42 ft.

Then, according to our text, we can obtain the pitch angle of a screw propeller by constructing a right angle triangle, whose long side will equal the circumference of the circle at the blade tip, and whose short side will equal the pitch of the propeller, the angle formed by the long side and the hypothenuse being the pitch angle.

For convenience in drawing we will make a similar but smaller right angle triangle, whose sides are obtained by dividing the sides of the larger triangle by 6.28 rather than by 6 as illustrated above.

The convenience of this choice is readily seen by noting that the *long* side becomes $\dfrac{\text{Circumference}}{6.28}$ = radius, which can be laid off at once. This, of course, requires the *short* side to be $\dfrac{\text{Pitch}}{6.28}$ = in this case $\dfrac{54''}{6.28}$ = 8.59''.

This gives for the present design the right angle triangle having these dimensions: 18 in. radius of screw for long side of triangle, and 8.58 in. for short side of triangle.

First draw the outer circle (marked *xo*) of the disc view (Fig. 7) to any scale. The scale of drawing here is 3 in. to 1 ft.

This is to be a four-bladed screw. Draw two diameters of the circle, as in Fig. 7, at right angles to each other. The four radii of this circle are the center lines of the four blades. The outline of the developed blade of the Standard Admiralty or Government screw is a true ellipse, the major axis of which equals the radius of the screw (in this case 18 in.) and the minor axis is made five-tenths that of the major (in this case 9 in.). A successful builder of steamships in this country makes the disc area of the screw, for single screws, 35 to 36 percent of the immersed midship section of the vessel, and that of double screws 42 to 46 percent.

Now draw the true ellipse, marked B', in Fig. 7, by trammel or otherwise, on the two dimensions given, 18 and 9 in. The blades in that position will occupy an arc of about 60 degrees. Therefore on either side of the radial line of blade B lay off an angle of 30 degrees, as shown by Fig. 7. Now draw a series of equidistant concentric circles (in this case 4), making one of those circles coincide with the hub. (For dimensions, see Appendix A). These circles are numbered 0, 1, 2, 3, 4. (0 is the circle of the greatest diameter), the fourth circle being part of the hub, as stated. This network of lines formed by the two radial lines and concentric circles partially contain the elliptical blade B in its parallel position.

To draw the edge view of this blade, assume any point Z on the horizontal center line, and at some convenient distance to the right, as shown on Fig. 8. As this screw is to be a right-hand one, measure from point Z to the left, as shown (8.59 in., as per formula), calling this line ZY, and from Y erect a perpendicular (YX) whose length equals the radius of the screw (18 in.). Draw a line from X through Z, say, to Z'; the line XZ is the hypothenuse of right angle triangle XYZ. This hypothenuse represents the pitch angle of the screw for this part of the blade. Now project the points 1, 2, 3 and 4 in the radial line in Fig. 7, into the line XY in Fig. 8, which also represents the radius of the screw, and draw from those points through Z indefinitely. Those lines will represent the pitch angles of those particular parts of the blade; *i.e.*, as the diameters of the different concentric circles become smaller, the angles necessarily become larger, because the pitch remains the same.

Z and O represent the same point in blade A; Z being the tip of the screw blade, practically has no dimensions (see section on Fig. 7). Now from Z in Fig. 8, we will lay off (on their respective pitch angles) the different true widths, taking those widths from the true shape of the ellipse in Fig. 7 (from the blade marked B). Through these points we will draw a smooth curve when we shall have an outline (minus the thickness) of blade A in projection, and when this blade is extending towards the draughtsman and at right angles to a vertical plane. From blade A, Fig. 8, we may now project the apparent widths to the blade A in the other view in Fig. 7. These apparent widths may now be transferred to the other three blades, B, C and D in Fig. 7. The other view of B, Fig. 8, is found by the intersections of vertical lines, drawn from the points in blade A, with horizontal lines from the same points in blade B, Fig. 7.

This will complete the outlines of all the blades in both views. The thickness is best shown upon the blade A, Fig. 8, whereupon we may imagine and draw sections taken parallel to a vertical plane, hence at right angles to blade A, and at the distances from the center of the screw which coincide with the concentric circles, and shown overlapping each other on blade A, Fig. 8. One method of determining the different thicknesses at the different concentric circles is shown in Fig. 7, blade D, whereon a section is taken through the center of this blade, passing through the center of the shaft. The thickness is shown measured from this line back towards the stern of the ship. The greatest thickness is shown at the center of the shaft, Fig. 7, and is made .5 in. for every foot of diameter of the screw.

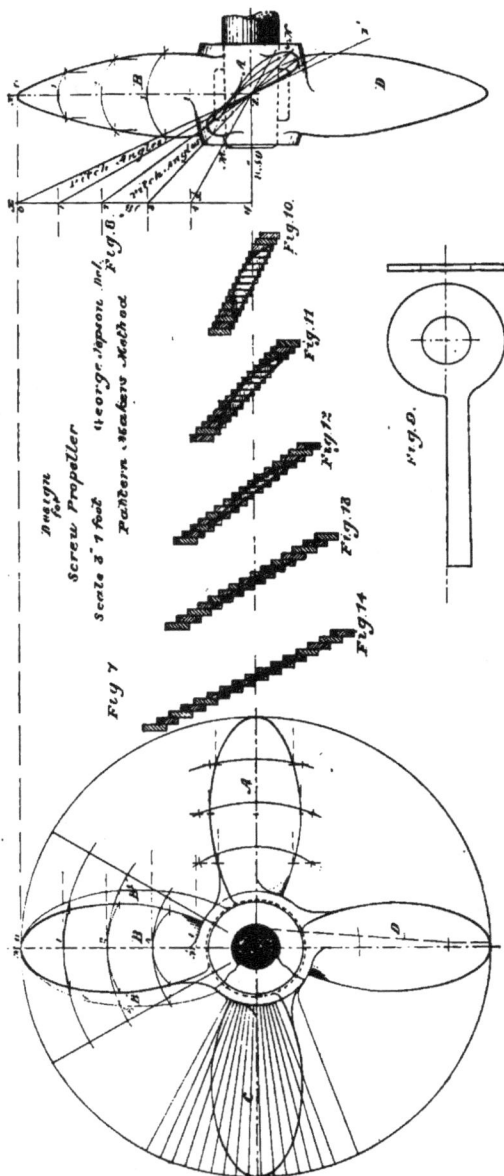

This, being 3 ft. in diameter, would make the thickness $1\frac{1}{2}$ in. So from this point, $1\frac{1}{2}$ in. back from the line, a line is drawn to the tip of the screw which, as previously stated, has no dimensions. Yet it is customary to give some small dimension and make it run parallel for a short distance from the tip, as is plainly shown in Fig. 7. Now, referring to the line nearest to the center of the screw $Z4$, Fig. 8, having already shown upon this line the true width, and point Z being the center, we will set off from Z, and at right angles to the line, the thickness shown by the cross section on the concentric circle, Fig. 7, blade D.

We shall then have three points through which we will draw an arc of a circle. That section will then represent the true section of this blade, taken at that particular place. The other sections will be similarly drawn upon their respective lines—$Z3$, $Z2$, $Z1$. A line drawn tangential to these sections on blade A, will represent a back line; that will be the nearest we can come to showing the thickness, as it merges into the hub of the screw. We are now ready to get out the stock from which to make the model.

Suppose that on Fig. 8 the horizontal distance between M and N is 7 in., and across the hub it measures 8 in., and that we use stock $\frac{1}{2}$ in. thick, it will take 16 pieces $\frac{1}{2}$ in. thick to make the screw, including the hub, which, of course, is part of the screw. The shape of the stock is shown in Fig. 9; the face of the stock is in a plane passing through the center of the shaft.

The cross sections of the assembled 16 pieces of stock for a blade are shown in Figs. 10, 11, 12, 13 and 14, wherein each section is represented isolated from, and parallel to, its own section in Fig. 8.

From the center point of Fig. 10, and upon the center line, we will measure eight $\frac{1}{2}$ in. each side, and draw through those points vertical lines, as shown in Fig. 10. The depths of these pieces of stock will of course be limited by the thickness of the section, as shown. On blade C, Fig. 7, we have shown the lengths of these 16 pieces of stock; also the widths as they are placed, one over the other, to occupy an arc of about 60 degrees. The pieces are held in that position by glue and nailed temporarily until the glue sets.

Appendix A

A usual proportion for hubs equals the diameter of the screw divided by 3

to 4), making the dimensions for this hub 9 in. to 36 in. divided by 4 equals 9 in. diameter.

The length of the hub equals the diameter of the shaft, multiplied by 2.5, making the length of this hub 10.3 in.

The following formula for finding the diameter of the shaft may be used:

Diameter of the shaft equals the cube root of (70 multiplied by the horse-power, divided by the revolutions per minute.)

Assume the revolutions to be 100 per minute, and assume the horse-power to be 100. Then:

$$D = \sqrt[3]{\frac{70 \times 100}{100}} \qquad D = 4.121 \text{ diam of shaft.}$$

The formula for finding the horsepower, when diameter of shaft and revolutions are given, is the following:

The horse-power equals the cube of the diameter multiplied by the revolutions, the product divided by 70.

$$H.P. = \frac{D^3 \times rev.}{70} = H.P. = \frac{70 \times 100}{70} = H.P. = 100$$

To find width of the key:
Width of key = 0.22 x the diameter of the shaft + 0.25.

Thickness of key = 0.55 x its width = 0.55 x 1.156 = 0.6358 = $\frac{5}{8}$ in.

THE GROWTH OF ELECTRICITY ON RAILROADS

AUSTIN C. LESCARBOURA

A problem claiming the attention of the foremost engineers and authorities is before us. Shall steam locomotives be replaced by electrically-operated cars, and to what extent? The question is interesting, the opinions contradicting, and while the writer is not in a position of voicing an opinion, he will endeavor to review the subject from a popular standpoint.

Fig. 1
Salt Lake and Ogden Railway Cars

The idea of electrically-operated railways is not a new one, and the first attempt at electrical cars dates back to 1835, when Stratingh and Becker of Gronigen, and Botto in Turin, constructed, respectively, their magneto-electric carriages. These experimenters are not the only inventors who were working on the problem, but their names stand forth as successful workers. The first practical electric railroad came much later, at the electrical demonstration of the Industrial Exposition of Berlin, in 1879. The tracks on which the car traveled were of a closed oval form, and about 900 ft. long.

The car consisted of a platform mounted on four wheels. Upon this platform, and covered with a wooden top, reposed a large Siemens motor, with its shaft revolving in parallel direction to that of the rails. By means of suitable gearing, the mechanical power was conveyed to the wheels. The current was available space on the platform to accommodate the motor; whereas it is the practice at the present time to have the powerful motors placed below the car, and all the space above utilized for other purposes. Other exhibition railroads were built, but it was not until 1881 that the first permanent electric railroad was inaugurated at Lichterfelde by Siemens and Halske.

The installation of an industrial railroad at Breuil-en-Auge, at about the same time, created wide interest in the electrification of railways, and its possibilities. This system was designed by Clovis Dupuy, and proved a success in every respect.

The rails extended 2,040 meters over swampy ground, and owing to the difficulty of insulating the current under such conditions, it was decided to use accumulators in a tender hauled by the locomotive. The battery was composed of six cells of the Faure type, weighing 8 kg. each, and requiring seven to eight hours to charge. The locomotive proper contained a Siemens machine, occupying approximately half of the available space in the car. The motor not only performed the mission of propelling the car, but also served to rotate rollers which hauled long linen strips into the car.

Fig. 2
The Hudson and Manhattan Railway Cars

and the haulage power consisted in pulling the accumulator car, weighing 700 kg., and six other carriages, each weighing 800 kg., making the total weight of 6,400 kg., at the rate of a few miles per hour. As for the gathering of the linen, it did this work at the rate of 125 meters per 48 seconds, or the work formerly requiring seven men four or five hours, in one-half hour with but two men.

The Lichterfelde railroad was but 2½ km. in length. The motor was,

Fig. 3
The New York Central and Hudson River Locomotive

contrary to the previous types described, placed under the car, and between the two sets of wheels. It was a four-pole opposed affair, similar to two bipolar Edison motors placed facing each other with a common armature for both. Flexible belts conveyed the mechanical energy to both sets of wheels. The regulation of the speed was controlled from the platform, by means of inserting resistance as desired. This car represented many points embodied in regular street cars today, and the general appearance was identical but on a smaller scale. The car was capable of developing a speed of 40 km. per hour with 26 passengers. Copper strips were used between the ends of the rails, similar to the "bonding" process used today.

In 1884 the Molding-Bruhl line was lengthened, and overhead feed system employed. Instead of wire, hollow tubes were used, with a slot at the lower side. "Boats," or sliding parts fitting within these tubes and protruding through the slot, collected the current which traveled down the flexible wire to the car. The return current was sent through the rails. Two pipes were connected on each cross piece, the entire length of the track, for these "boats" could not pass each other, and the single track was used for cars

traveling in opposite directions at the same time. It is noteworthy that this was the first railroad on which electric cars with more than one operating at a time were introduced as it was previously thought that such a performance would be impossible. These sliding contacts even today would possibly solve the difficulty of the trolley systems, when the small wheel leaves the wire. However, there are many devices which have been brought forward within the last few years, and which are simpler and just as effective, but strangely are not used in regular practice.

Coming nearer home, we find the first practical railroad of any size built in Richmond, Va., by Frank J. Sprague. The total rails covered 12 miles, with many curves and some 10 percent grades. The initial equipment consisted of 40 motor cars in 1888, and after conquering many difficulties, success was final, giving to America the embryo of its present day network of electric street railways.

The next ten years witnessed a continual battle being waged by various inventors with unique and queer ideas. Various interests were urging municipalities to place the ban on overhead trolley systems, claiming them "dangerous to human life, and a menace to property;" and so the battle kept raging, but the various interests were gradually being eliminated, with the most worthy remaining. Then the roads began to grow with remarkable rapidity, and the

Fig. 4
The Great Northern Electric Locomotive

public was gladly induced to furnish the necessary money to carry on the work. Today, we find the electric car in every city of reasonable size, and, in fact, for inter-urban service, competing with the

Fig. 5
The New York Central Power House

steam railroads, both for rates and for speed. Fig. 1 shows a model type of car, in this case being the cars used in the electrification of the Salt Lake and Ogden Railway. The total length of the road is 41 miles, and it has been electrically operated during the last few years only. Each car is equipped with four 100 h.p. Sprague G.E. motors, with the Sprague system of multiple unit control.

The first example of the electric current invading the steam railroad field was in 1895, when the General Electric Company undertook the electrifying of the Nantasket Branch of the New York, New Haven & Hartford R.R. Following this successful displacement of steam came the electrification of the Baltimore tunnel on the Baltimore & Ohio R.R. Since that time, there has been a gradual increase in the electrification of steam railroads, until today the question arises, "Shall steam locomotives be abolished?"

Various railroads have electrified their systems in whole or in part, and have different ideas as to the change. It is alleged that one road, after electrifying this 100 miles of double track, finds same a failure. Then again, some roads are very enthusiastic over their substitu-

tion, and prepare for more to follow. The New York Central & Hudson River R.R., probably the second railroad in the East, has electrified its system to High Bridge, N.Y., and contemplates extending this system within the immediate future to Croton, N.Y., a distance of 34 miles from the terminus. It is the understanding that the system will eventually be electrified to Buffalo, and current obtained for the best part, if not for all the system, from Niagara Falls water power.

Before giving just brief data on the electric locomotives we will roughly cover the right of way. For supplying current, there are two methods generally used, with the various systems falling under one or the other of these two methods. The overhead system usually is employed with either a trolley wheel making the contact or a flexible diamond-shaped mechanical affair, which raises or lowers by means of compressed air at the will of the motorman. This arrangement is far more practical than trolley wheels for fast locomotives. The third rail system consists of having an additional rail laid on the outside of the two regular traffic rails. The current is collected by means of a contact "shoe." It is used in various forms, but the generally accepted and most improved type is where the rail hangs downwards, and the contact is made from underneath, allowing the three other sides to be completely encased in wooden covers. This makes accidental contact impossible, and only deliberate intention to incur personal injury can cause accident.

The electric locomotives are employed for hauling freight or passenger trains. For suburban traffic, unit motor cars are used; and each car has its own motor equipment, and is controlled from the master controller by a series of wires and magnetic switches. The Southern Pacific R.R. operates the Oakland and Alameda (Cal.) division with such cars. The trolley wire carries 1200 volts direct current, which is collected by a special collapsible arrangement. Each car is equipped with four 125 h.p. motors and the Sprague G.E. control. The Hudson & Manhattan Railroad, operating the cars through the Hudson River tunnels, uses similar cars, which are constructed entirely of steel, and have cement floors. Each car is equipped with two 160 h.p. motors and the Sprague multiple control. The New York Subway cars are similar in general design. An illustration of the tunnel cars is given in Fig. 2.

Electric locomotives vary considerably in design, but more or less the details are universal. In Fig. 3, we have a photograph of the type used on the New York Central & Hudson River R.R. The equipment of that road consists of 137 unit motor cars, and 47 100 to 115-ton gearless locomotives. Contrary to the general practice, no gears are used in these locomotives, but the armature is built directly on the shaft of the wheels, and, in fact, becomes part of same. The current is collected from a third rail by means of a contact shoe which catches the rail from underneath. In cross-overs or confused track switches, overhead trolley wires are used, and the collectors on top of the locomotive brought into play. The following data gives the complete electrical and mechanical details of interest:

Electrical

Voltage, 600 direct current.
Rated Amperes, 3,000.

Rated tractive effort, 20,000 lbs.
Maximum tractive effort, 35,000 lbs.
Total rated horse-power, 2,200.
Number of motors, 4.
Type of motor, G.E. bipolar.
Speed at rated amperes, 40 miles per hour.

Mechanical

Diameter of drivers, 44 in.
Number of driving wheels, 8.
Diameter of guiding wheels, 36½ in.
Total wheel base, 36 ft.
Rigid wheel base, 13 ft.
Width overall, 10 ft. 1 in.
Length, 43 ft. ½ in.
Height over cab, 13 ft. 9 in.

Weights

On drivers, 142,000 lbs.
Per driving axle, total 35,500.
Per driving axle, dead weight, 12,900.
Per guiding axle, total, 22,000.
Electrical equipment, 60,000.
Mechanical equipment, 170,000.
Total 230,000 lbs.

Another example of a successful type, is that of the Great Northern R.R. used on the electrified Cascade Tunnel division, which is noteworthy, inasmuch as it is our first three-phase railroad in America. The current is collected by trolley wheels from the overhead wires, at a potential of 6,600 volts. There is a continuous 1.7 percent grade on the system, and in some sections this even rises to 2.2 percent. Trains of 2,000 tons are hauled up the grades at a speed of 15 miles per hour. The entire installation and equipment was furnished by the General Electric Co., and Fig. 4 illustrates the locomotive used, of which there are but four at the present time in use on the Great Northern R.R., comprising the initial equipment. The data on this type is as follows:

Electrical (Continuous Rating)

Voltage, 6,600, three-phase A.C. 25 cycles.
Rated amperes, 91.
Rated tractive effort, 25,000 lbs.
Maximum tractive effort, 57,000 lbs.
Speed at rated amperes, 15.2 miles per hour.
Total horse-power, 1,000.
Number of motors, 4.
Type of motors, G.E.I. 506.

Fig. 6
Atlantic City Power Station

Mechanical

Diameter of driving wheels, 60 in.
Number of driving wheels, 8.
Diameter of guiding wheels, none.
Total wheel base, 31 ft. 9 in.
Rigid wheel base, 11 ft.
Width overall, 10 ft.
Length, 44 ft. 2 in.
Height over cab, 14 ft. 3 in.

Weights

On drivers, 230,000 lbs.
Per driving axle, total, 57,500 lbs.
Per driving axle, dead, 18,300 lbs.
Electrical equipment, 108,000 lbs.
Mechanical equipment, 122,000 lbs.
Total, 230,000 lbs.

Among the many electrified systems, the following are of note, and a passing word on each would be of interest. The Detroit River Tunnel, on the New York Central Lines, connects the Michigan Central tracks in the States with those in Canada. The tunnel is double-tracked with long grade approaches at each end. The current is supplied at 600 volts D.C. from the Detroit Edison Co. The total equipment consists of six 100-ton locomotives, similar to the type described in the preceding paragraphs of the New York

The Baltimore & Ohio R.R. has much improved its original electrification, and at the present time has four 80 ton, two 90 ton geared locomotives, and the original three 90 ton gearless locomotives.

The West Jersey & Seashore R.R. operates between Camden and Atlantic City, N.J., and is a branch of the Pennsylvania system. The road was converted from steam to electricity in 1906. The cars are of the multiple control unit type, each car having two 200 h.p. motors and the Sprague G.E. control. The entire mileage includes about 150 miles of track, and the power equipment of one power house of 6,000 kw. capacity, with eight sub-stations located along the lines. Sixty-eight cars are in operation at present, and maintain the schedule time of over 60 miles per hour.

The West Shore R.R. operates one of its divisions between Syracuse and Utica, which was electrified in 1907. The total mileage is 44 miles between terminals, and is equipped with third rail. The main power plant is located at Utica, and consists of Curtis steam turbine generators, transmitting 60,000 volts over the line to the four sub-stations. Each car is equipped with four 75 h.p. motors

Fig. 7
The Motor Equipment of the Pennsylvania Locomotive

The Pennsylvania Railroad after exhaustive studies selected Westinghouse locomotives (600 volts direct current) for its New York terminal installation, which is one of the greatest engineering accomplishments of the age. These, the most powerful electric locomotives in existence, haul heavy passenger trains from Manhattan Transfer station near Newark, N.J., to the magnificent new Pennsylvania terminal in the heart of New York City. In addition to the locomotives, the entire equipment is of the Westinghouse type.

One typically progressive feature of the Pennsylvania locomotives is the use of Westinghouse field control for speed regulation. With the field control the locomotives can run at very high speeds, when necessary, and at the same time they can start the heavy through Pull-

are connected by driving rods to the wheels, as shown in Fig. 7. A complete view of the locomotive is given in Fig. 8.

In Fig. 9 we have an illustration of a N.Y., N.H. & H. electric locomotive drawing a train. The line is of the single phase alternating current type, and supplied by overhead feeders of a substantial construction. The railroad has 35 miles of double track main line electrified, leading out of New York to Stanford, Conn. The locomotive easily handles an 800 ton passenger train at 50 miles per hour, or a 1500 ton train at 40 miles per hour. The trolley voltage is 11,000 volts, and the highest insulation used throughout.

The power plants comprise one of the largest expense factors in the change from steam to electricity. Two methods are available for transmitting the current to be used, one of which is to supply the high potential current (alternating) directly through the trolley wire or third rail to the locomotive, and the other, which is more common, to transmit high voltage alternating current to sub-stations, where it is passed through transformers and rotary converters, then re-transmitted in the form of lower voltage and direct current to the third rail or trolley. The New York Central R.R. employs the latter method, having two power houses of 20,000 kw. at Port Morris, N.Y. Curtis steam-turbine-driven generators of 5,000 kw. per unit are used. The alternating current is transmitted to the numerous rotary converter stations where it is changed to direct current as well as stepped down to 600 volts. Fig. 5 shows one of the power stations at Port Morris. Fig. 6 illustrates a typical rotary station, in this particular case being that of the West Jersey & Seashore R.R. at Atlantic City, N.J. The transformers are noticed at the left-hand of the picture, while the rotary converter is noticed to the right of these.

The Great Northern R.R. has a power plant consisting of waterwheel-driven generators. The current of 33,000 volts is transmitted 33 miles to the mouth of the Cascade Tunnels, where a transformer station steps it down to 6,600 volts, the working voltage of the electric loco-motives.

Whether electricity can prove superior to steam on the trunk railroads, is a problem upon which only the highest

Fig. 9
N.Y., N.H. & H. Electric Locomotive

authorities can verse an opinion. A capable engineer has stated that while electricity will eventually replace steam for both local and long-distance passenger traffic,. it will never replace the time-honored freight locomotive with its long string of cars. The writer has before him at the present moment an old engi-neering book which states in part: "At some future day, we may publish a special work on the history of the development of the electric railroads in America, and elsewhere, but at the present time (1891), there are so many electric railroad enter-prises before the public, that it would be taking a flying shot to describe them further than to mention some of the chief in success and excellence." It illustrates that from the confused mass of ideas and inventions existing in 1891, the final perfected street railroad, employing elec-tricity as motive power, was evolved. Why is it not logical, in consequence, to expect that the electric railroad will in as many years hence have been simplified and universally adopted?

A SAFETY GATE FOR THE STAIRS
H. JARVIS

A small gate, either at the top or bot-   In Fig. 1 we show such a gate fixed at

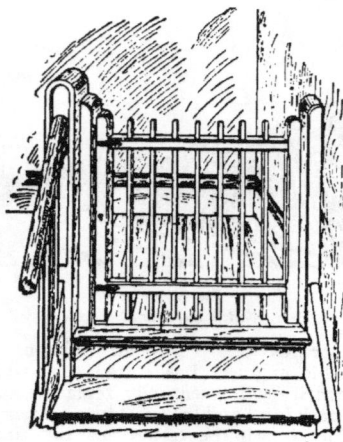

Safety Gate for the Stairs

and the width will, of course, depend on the space to be filled. The hanging and shutting pieces should be fixed first—a suitable size is 3 in. wide by 2 in. thick. The former must be fitted over the base-board, as in Fig. 1, and a screw or two into this will fix the bottom end. The top end, however, is what requires to be fixed most firmly, and to ensure this we must insert a wood plug in the wall, screwing into this, and covering over the screw-head with a plug of wood.

This method of fixing is shown sectionally in Fig. 3, where I shows the joints in the brickwork, and the plaster on the walls, K the wood plug, with the screw in it, fixing the hanging piece A, and L the plug covering the screw head.

To find the joints in the wall, if of brick, probe through the plaster with a brad-awl or wire nail, and then clear out all the mortar to the depth of some 2 in., and to the width of $1\frac{1}{2}$ in. Make a wood plug as Fig. 4, and drive in very tightly. Cut off level with the wall, or slightly under. Now place the hanging piece in position, and mark the height of the plug. Bore in $\frac{1}{2}$ in. from the face, so that the head of the screw will go in, continue through for the shank of the screw, and then insert the latter, screwing up as tightly as possible, but do not insert the plug L until later, in case the hanging piece has to be removed during the time the work is

The fixing of the shutting piece is more simple, it having to be screwed to the wood newel only. It will no doubt be fully understood that all shaping and chamfering of the hanging and shutting pieces must be done before they are fixed. They must also be kept both to the same height, and also immediately opposite each other, as well as perfectly vertical. Should the wall or the newel be out in this respect, the new pieces must be so scribed as to put matters right; the opening for the gate will then be parallel.

The gate should be made some quarter of an inch narrower than the opening, so that no planing will be required to make it fit. The sizes of the various parts are as follows: stiles and rails 2 in.; square bars, 1 in. x $\frac{3}{4}$ in.

The making of the gate, after the wood is planed up truly, is very easy. The stiles are mortised to take the rails, as Fig. 5, being afterwards rounded at the top ends and chamfered, as dotted lines.

The rails must be tenoned to fit into the stiles; also mortised for the bars, as Fig. 6, and then the corners chamfered off as dotted lines.

The bars simply require planing to size, the tops rounding, and the sharp corners slightly broken, just the extreme edge only taken off, not to show any bevel at all.

The bars should be a fairly tight fit in the rails, and should be inserted in the mortises before the rails are placed in the stiles, as if left till after, they have to be driven through the top rail a considerable distance, which is not likely to improve the finish.

The rails should be fixed to the stiles with pins. The bars should be tight enough to hold themselves, but to make certain they will not move, a small brad may be driven into each from the under side of the bottom rail.

The complete gate is shown in Fig. 7 ready for hinging. This should be done with a pair of brass or stout steel butts, as shown sectionally at M, Fig. 8. The hinging is a comparatively easy matter, and need not be detailed here. We would, however, remind our readers that the right method is to let one half of the hinge into each member, as shown, and not the whole thickness in one, and simply screwing to the other.

Fig. 1.

Fig. 8.

Fig. 4.

Fig. 3.

Fig. 11.

Fig. 2.

Fig. 6.

Fig 12

Fig. 10.

Fig. 7.

Fig. 9.

Fig. 5.

Fig. 8, which consists of the latch, fitting into a mortise in the stile as shown, and pivoted on the pin, O. The knob P is connected to the latch, forming a means of working it, from the stair side only.

The latch projects far enough beyond the stile to slide up the striking plate, and drop in the notch of the same, the closing being automatic. The front and side views of the latch arrangement are shown in Fig. 9, and the same of the striking plate in Fig. 10.

Although the latch is a secret one there is a possible chance of baby fingers finding out the working of it, and it is best to supplement the fastening with a small brass bolt fixed on each of the rails of the gate, on the stair side, as shown in Fig. 11. These cannot be reached except by hanging over, so that they are absolutely safe as regards the young ones. An alternative method of making the gate, as far as the bars are concerned, consists of substituting round rods for the square bars. Thus instead of mortises, holes bored through the rails only would be required, as in Fig. 12.

The gate should be made from wood to match the hand-rail and ballusters, or if these latter are painted it may be made in deal or whitewood, and painted to match. If fixed as described, it is a tenant's fixture, and may be taken away in case of removal at any time.

Using a Life-Preserver

"The worst trouble about a life-preserver," said an old sailor, "is that few people know what to do with one when it's thrown to them. Many a man would drown in trying to get a life preserver over his head. The average person struggling about in the water would try to lift up the big life-ring and put it over his head. That only causes the man to sink deeper and take more water into his lungs.

"The proper way to approach a life-preserver in the water is to take hold of the side nearest you and press upon it with all your weight. That causes the other side to fly up in the air and down over your head, 'ringing' you as neatly as a man ringing a cane at a country fair. After that the drowning man can be rescued."

Life Insurance Solicitor Got a Hearing by Novel Method

A sale is made in the preparation.

Tom Lowry, the late traction magnate of Milwaukee, was one of the wealthiest men of the Northwest. He had been solicited by all the best life insurance salesmen in the country—or rather he had been approached by them; for as fast as they came within reach he threw them out of his office.

He became the talk of the entire life insurance fraternity, and obviously some of this got back to Tom, and he continued his attitude towards insurance just as a matter of pride.

Whenever a local general agent hired a new solicitor he first sent him over to Lowry as a courage test and to toughen up his hide a bit.

Now, anybody with any acquaintance with Tom Lowry at all knew his propensity for betting. He would bet on anything that contained any element of chance. He used to sit in the old Sherman House, Chicago, with companions and bet that out of the first twenty men next to pass five of them would have whiskers; then he would go to a ball game and bet that out of the next five men to bat three would fly out.

One day a young, rather cadaverous looking individual called at Lowry's office and asked that his card be presented to him—the card, by the way, merely contained the man's name.

Lowry sent back to know what the man wanted.

The cadaverous one replied that he wanted to make a bet.

He was admitted.

When the caller was fairly seated be said: "Mr. Lowry, I want to wager $1,800 to $100,000 that you will die within the next year."

"I'll take the bet," said Tom.

"All right," returned the man, "just sign this." Here the salesman presented an insurance application blank that had been previously made out.

Lowry signed it.

Time of sale, three minutes.

Time of preparation, as long as it took to think of it.—*Gibson's Magazine.*

The oiler's suggestion is as good as the manager's if it helps along the work.

THE EFFICIENCY TEST OF A LEAD STORAGE BATTERY

A. SPRUNG, E.E.

THEORETICAL DISCUSSION

In testing a battery for its efficiency it is necessary to determine the relation between the watt-hours output and the watt-hours input required to store up the battery in the same condition as it was previously (at the beginning of discharge.)

To make a very accurate determination, it is necessary before making the test to take the dimensions of the plate and vessels, weight of plates, electrolyte and containing vessel, the battery being fully charged. After making the test the same process should be gone over in order to determine the rate of disintegration and probable life of the battery.

When the battery has been fully charged the following will be noticed:

1. *The terminal e.m.f.* becomes approximately 2.5 volts, depending upon the rate of charge, *i.e.*, for a time more or less than eight hours there would exist a slightly smaller or greater value.

2. *The density of the electrolyte* measured by the hydrometer should gradually increase until it becomes constant at 1.8 specific gravity, unless the charge rate is so great that the water present becomes rapidly decomposed. It is advisable to use lead-weighted hydrometers. To keep a uniform density the electrolyte should be agitated when hydrometer readings are taken.

3. To distinguish the positive plate from the negative:

Positive	Negative
1. Formation of gases at the plates showing complete oxidation.	1. A complete reduction to spongy lead.
2. Dark brown or chocolate color, blackening indicates overcharge.	2. Dark slate color.

4. *Cadmium Test.*—Cadmium, when immersed in the electrolyte of a lead storage battery, gives reliable indications of the potential of the positive or negative plates with respect to itself. Insert the cadmium stick (connected to one terminal of a voltmeter) into the electrolyte, and connect the other terminal of the voltmeter first to the positive and then to the negative plates. If the

running and charging conditions are correct, the voltmeter reading, as caused by the respective potential differences, will be nearly 2.5 volts for the positive plate and zero for the negative plate.

Thus, in making a test, the cell should be first charged to normal rate until the above stated "full charge" conditions are obtained. It is then necessary to *discharge* the cell at constant current rate until the terminal volts=1.8. During the *discharge* make note of the *temperature, specific gravity of electrolyte* and *cadmium voltage* about every 15 minutes. If the potential difference between the negative plate and cadmium stick=.25 volts before cell voltage=1.8, the discharging should be considered finished. If this should occur more than once, either local action or small capacity of negative plates are the cause. If the rate of discharge equals twice the normal rate, the terminal volts on discharge should equal 1.7. If the rate is four times the normal, the voltage should be 1.6. After discharging, the cell should be recharged with constant current, then note the terminal volts, cadmium volts, specific gravity and temperature every 15 minutes. If the readings indicate a complete discharge, the charging current should be reduced to one-half the normal. When the voltage, specific gravity, temperature, etc., become constant the charging should be discontinued.

A lead storage battery should never be completely discharged, as it is thereby likely to become "sulphated."

The chemical reactions which are believed to take place in a lead storage battery may be represented as follows:

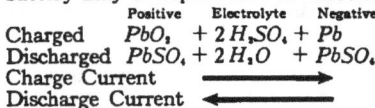

	Positive	Electrolyte	Negative
Charged	PbO_2	$+ 2 H_2SO_4 +$	Pb
Discharged	$PbSO_4$	$+ 2 H_2O$	$+ PbSO_4$

Charge Current ⟶

Discharge Current ⟵

Chemical analysis shows that lead sulphate exists in the discharged plate. The density of the electrolyte decreases during the discharge of the cell, corresponding to the consumption of the sulphuric acid and the formation of water as shown in the above reactions. From a thermochemical standpoint the energy

produced from the formation of lead
sulphate from metallic lead and the
peroxide corresponds to the voltage obtained.

EFFICIENCY TEST

In a storage battery the *efficiency* is
the ratio of the amount of discharge to
what is required to bring the battery
back to its original condition. *Efficiency*
can be expressed in ampere-hours, *i.e.*,
ampere-hours efficiency equals ampere-
hour output divided by the ampere-hour
input. The efficiency can be expressed
in watt-hour, *i.e.*, watt-hours on discharge
divided by watt-hours on charge. This
is the real efficiency, because it considers
the energy and includes the voltage as
well as the ampere-hours. The ampere-
hour efficiency will show the action of the
battery but is useless commercially. It
is generally about 15 percent higher than
the watt efficiency. It will possibly give
an apparent efficiency of over 100 percent
at times; this is due to the fact that

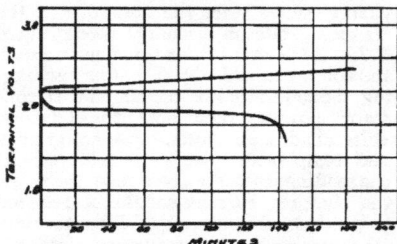

owing to a residual charge in the battery,
it is possible to draw out more than is
put in.

In general practice, it has been found
that the efficiency of a storage battery
plant when in good condition varies from
75 to 80 percent.

The results should be expressed in the
form of curves drawn between time and
terminal volts, cadmium volts, specific
gravity, etc.

To obtain the watt-hours, multiply
the average voltage by the ampere-hours.

EXPERIMENTAL SOLUTION

It is required to find the commercial
efficiency and characteristic curves of a
lead storage battery by means of an
experimental solution. To begin with,
the connections for the test are made
in the usual manner. The method
of operation of test was as follows:

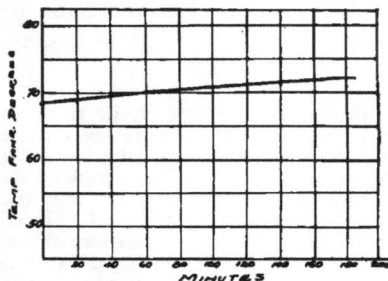

The battery is charged by throwing
the switch on the charging side. In order
to show the sudden jump in voltage,
readings are taken in rapid succession
at the start and later on as the curve
becomes more straightened out readings
were taken at longer intervals. The
readings taken were, *first*, the time;
second, with switch closed, the voltage
of the battery, *i.e.*, charging voltage at
closed circuit; *third*, with switch open,
the voltage of the battery, *i.e.*, terminal
voltage on open circuit. Next, the plus
and minus cadmium readings were taken.
Then the specific gravity and temperature readings were taken. In the meanwhile the current was held continually
constant by means of a carbon pile
rheostat. The same holds true for discharging the battery. In the test the
charging was started at 2.2 volts (high
due to a residual charge) and became
fully charged at 2.47. These values
are within the limits of 1.7 to 2.5 volts.

As seen from the curves the *rise in
voltage* is gradual, excepting at the
very beginning and at the end. When
the cell is fully charged it is seen how the
voltage becomes approximately constant
from the curves. After a value of 2.47 volts
was reached, the battery was considered
fully charged, since the reading was constant throughout the last 25 minutes,
and more charging would merely waste

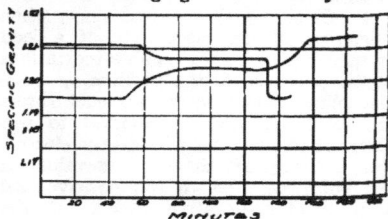

RESULTS OF TEST

RESULTS TAKEN DURING CHARGING BATTERY

Charging

Time hr.min.	Volts Closed	Volts Open	Cadmium Volts Plus	Minus	Specific Gravity	Temp. Fo	R. Cell Ohms
0.00	2.2	2.15	2.32	.10	1195	68.2	.0023
0.01	2.225	2.175	2.325	.10	1195	68.2	.00165
0.02	2.225	2.175	2.35	.10	1195	68.2	.00165
0.03	2.25	2.18	2.35	.10	1195	68.2	.0023
0.04	2.25	2.18	2.35	.10	1195	68.2	.0023
0.05	2.25	2.18	2.35	.10	1195	68.2	.0023
0.06	2.25	2.18	2.35	.10	1195	68.2	.0023
0.07	2.25	2.18	2.35	.10	1195	68.2	.0023
0.08	2.25	2.18	2.35	.10	1195	68.2	.0023
0.09	2.25	2.18	2.35	.10	1195	68.5	.0023
0.19	2.25	2.18	2.37	.09	1195	68.5	.0023
0.29	2.25	2.18	2.37	.08	1195	69.5	.0023
0.39	2.25	2.18	2.38	.08	1195	69.5	.0023
0.49	2.29	2.20	2.39	.08	1200	69.5	.0023
0.59	2.30	2.21	2.40	.08	1200	70.	.003
1.14	2.35	2.25	2.42	.07	1205	70.5	.0033
1.29	2.35	2.25	2.43	.07	1205	70.5	.0033
1.34	2.38	2.28	2.45	.07	1205	70.5	.0033
1.41	2.38	2.28	2.48	.07	1205	70.5	.0037
1.46	2.39	2.28	2.48	.07	1205	71.	.0033
1.52	2.40	2.30	2.46	.07	1205	71.	.0033
1.59	2.40	2.30	2.49	.07	1207	71.5	.0033
2.04	2.41	2.31	2.50	.07	1207	71.5	.003
2.11	2.41	2.32	2.50	.07	1207	72.	.004
2.20	2.42	2.30	2.50	.07	1207	72.5	.0037
2.29	2.43	2.32	2.51	.08	1207	72.5	.0037
2.36	2.44	2.33	2.51	.08	1212	73.	.004
2.40	2.45	2.33	2.51	.07	1212	73.	.0043
2.44	2.46	2.33	2.52	.07	1212	73.5	.0043
2.52	2.47	2.34	2.52	.06	1213	73.5	.004
3.00	2.47	2.35	2.52	.06	1213	73.5	.004
3.05	2.47	2.35	2.52	.06	1213	73.5	

RESULTS TAKEN DURING DISCHARGE OF BATTERY

Discharging

Time hr.min.	Volts Closed	Volts Open	Cadmium Volts Plus	Minus	Temp. Fo	Specific Gravity	R. Cell Ohms
0.00	2.19	2.23	2.36	.16	73.5	1211	.0013
0.02	2.05	2.12	2.30	.16	73.5	1211	.0023
0.03	2.03	2.10	2.23	.16	73.5	1211	.00255
0.04	2.02	2.10	2.22	.16	73.5	1211	.003
0.05	2.00	2.09	2.21	.18	73.5	1211	.003
0.07	2.00	2.08	2.21	.18	73.5	1211	.00255
0.08	2.00	2.08	2.20	.18	73.5	1211	.00255
0.09	2.00	2.07	2.20	.18	73.5	1211	.0023
0.10	2.00	2.07	2.20	.18	73.5	1211	.0023
0.12	2.00	2.07	2.20	.18	73.5	1211	.0023
0.15	2.00	2.05	2.19	.18	73.5	1211	.00165
0.17	1.99	2.04	2.19	.18	73.5	1211	.00165
0.23	1.99	2.04	2.19	.18	73.5	1211	.00165
0.33	1.98	2.04	2.19	.18	73.5	1211	.002
0.46	1.97	2.02	2.18	.18	73.5	1211	.00165
0.57	1.96	2.01	2.18	.18	73.5	1211	.00165
1.02	1.95	2.01	2.18	.19	73.5	1208	.002
1.07	1.95	2.01	2.18	.19	73.5	1208	.002
1.18	1.95	2.01	2.17	.20	73.2	1207	.002
1.33	1.93	2.00	2.16	.20	73.2	1207	.0023
1.48	1.91	2.00	2.13	.22	73.2	1207	.003
2.03	1.88	1.98	2.12	.22	73.0	1207	.0033
2.09	1.84	1.95	2.12	.22	73.0	1207	.0036
2.13	1.82	1.95	2.11	.26	73.0	1207	.0043
2.16	1.81	1.95	2.11	.28	73.0	1196	.00465
2.20	1.79	1.94	2.10	.30	73.0	1196	.005
2.21	1.78	1.93	2.09	.32	73.0	1196	.005
2.23	1.72	1.92	2.09	.35	73.0	1195	.0066
2.24	1.61	1.91	2.09	.50	73.0	1195	.01

Average Voltage:
  Discharge = 1.945
  Charge = 2.32
Average Amperes = 30
Ampere-hour:
  Charge = 92.5
  Discharge = 72.5
Ampere-hour Efficiency:
  Efficiency = 72.5

$$\frac{92.5}{} = 78.4 \text{ percent}$$

Watt-hours:
  Charge = 214.5
  Discharge = 141.
  Efficiency = 141.

$$\frac{214.5}{} = 65.7 \text{ percent}$$

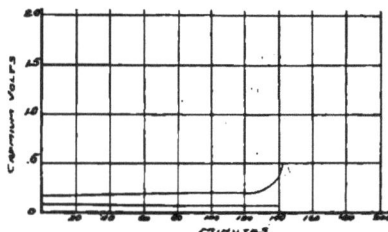

the energy in producing gases. It was noticed that the external voltage was higher in charging than in discharging, because of the internal resistances of the battery and the consequent voltage drop, which must be overcome in charging.

From the curves the exact amount of watt-hours can be determined at any time during the run.

It is seen how the *specific gravity* of the liquid rose from 1.195 to 1.215 from discharge to charged. This is due to the electrochemical actions taking place during the charging.

When the battery was fully charged bubbles of gas were given off.

The cadmium readings were taken to show their relation with the external voltage, *i.e.*, external voltage plus negative cadmium should equal at all time the plus cadmium. The positive cadmium readings ranged from 2.32 to 2.52, the latter being a value denoting full charge. The negative cadmium kept very low as would be expected, ranging from .1 to .06.

The resistance decreased on charging, due to the fact that the formation of a stronger solution is taking place. Since the resistance decreases with the strength of the solution, hence a decrease of *R*. The resistance of the solution would always naturally decrease when the density increases.

The temperature of electrolyte increases with charging, therefore *R* increases. The effect is small as compared to increase in specific gravity as affecting resistance.

In taking readings on open circuit we could bring out the values of the internal resistance, since the action is just like that in a motor,

or $e = P - IR$

where e = c.e.m.f. in this case charging voltage

$$\text{Commercial Eff.} = \frac{\text{Watt-hour discharge}}{\text{Watt-hour charge}} \times 100 = \frac{\text{Average Voltage Discharge} \times \text{Ampere-hours}}{\text{Average Voltage Charge} \times \text{Ampere-hours}}$$

$P=$ Impressed e.m.f. in this case terminal voltage

$I=$ current throughout

$R=$ internal resistance

or $R=\dfrac{\text{Voltage on open circuit less voltage on closed circuit}}{\text{current}}$

The reverse reasoning holds for discharging. The operation of discharging is naturally the reverse of charging. The curves are relatively the same but they incline downwards. At the end of discharging the current fell so rapidly that the battery was considered discharged about one-half hour earlier than the predicted three hours.

Efficiency.—Since the test was started with a residual charge it was necessary to abandon the test with a residual charge in order to obtain a true efficiency.

A value of 65.5 percent commercial, or watt-hour, efficiency was obtained. This value is considered fair in storage battery plant, 75 to 80 percent being a good value. A value of 72.5 percent ampere-hour efficiency was obtained. This latter does not show the value of a storage battery. It was stated before that in a test of this kind it is possible to obtain more ampere-hours output than ampere-hours input.

SOME PATTERN-SHOP WRINKLES

J. A. S.

There is always something for the practical man to learn, and it does not follow that because a pattern-maker is efficient in his work, he knows everything that there is to know. Therefore, it may be of use to mention one or two little details in pattern-shop practice, which, although in themselves not very important, make for better and easier work. It is so easy to do things the hardest way. This is not an Irish bull, but a bit of philosophy.

For example, there is a right and wrong way of using a brad-awl, and the wrong way will often lead to trouble when thin, slender strips of wood have to be pierced. The method adopted by the junior craftsman is to pick up any old brad-awl and wriggle it through the wood with a wrist motion. Sometimes the edge is parallel to the grain of the wood; but it does not matter to the novice. But when the nail is put into the wood, the wood splits and there is a lot more work to do. The right way is to see that the awl has a sharp chisel edge on it, then to place that edge crosswise to the grain and to press it down through the wood without any twist. The fibers of the wood are then cut, not sprung apart, and then the introduction of the nail will not lead to a split. This seems almost too simple to write about, but it is the simple things that count.

Here is another wrinkle that saves worry: when driving a nail into a very narrow strip of wood, even when properly pierced, there is a chance of splitting the wood unless the nail, which of course has a point on it, is held with its point on a hard surface and given a tap with the hammer. This flattens the point a little, and it does not seem quite in accordance with theory that a blunted nail should cause less splitting than a sharp one. It is a matter of experience, however, that this is so. Possibly the explanation is that the flattening causes a small cutting edge to be formed at the end of the nail which shears its way through the wood instead of wedging it apart. Whatever the reason may be, the fact remains, and it is a fact worth knowing.

Every pattern-maker is acquainted with the fillets which have to be placed in the corners of a pattern in order to correct a sharp corner into a nicely-rounded curve. These fillets, if made of wood, have to be specially planed or shaped up, and are often more or less tedious and difficult to make. There are in the market leather fillets cut to the required rounded surfaces which only require to be cut off and nailed into position, and this method possesses the advantage of increased speed in working. Both methods are, however, rather costly, and an idea on this point which has been found very useful in practice may not come amiss. The central idea is in the use of beeswax, which is plastic, easy to put in place and mould, and when varnished over with the wood presents no joint or seam as is sometimes the case with wood or leather.

A very good way of preparing and applying beeswax is as follows: A piece of iron or brass tube, say 1½ in. in bore and 12 in. long is taken, and fitted with a wooden plunger a bit longer than the tube, as shown in the sketch. One end of the tube is closed with a piece of metal, as shown, in which are arranged a series of round holes ranging from ⁵⁄₁₆ in. to ½ in. in diameter. The tube is first stood upright on the bench and the melted beeswax poured into it. When this has cooled off a little, so that it is plastic but not liquid, the plunger is put into the open end of the tube and the whole arrangement placed in the bench vise in such a way that the outer end of the plunger bears against the outside jaw, while the end of the tube with the holes in it rests against the inside jaw in such a way that

all the holes but one are closed up. The vice is then screwed up steadily, putting a pressure on the plunger, which forces the wax out of the exposed hole in a long circular band which is coiled up on the bench and then stored away ready for use. In this way bands of the different diameters are prepared.

When it is required to make a fillet, all that it is necessary to do is to take a length of a suitable diameter of beeswax band and lay it along the corner in the pattern which has to be filled up by the fillet. Then a steel ball mounted on a handle is heated in a flame (say the burner used for the glue-pot) and run along the band, softening it and at the same time pressing it firmly into the corner and giving it the required curve on the outer surface. In this way the fillet is finished in a quarter of the time, or less, that it would take to make the old-fashioned wooden fillet. The steel balls above

mentioned are easily prepared. Ordinary steel balls are taken and the temper drawn, and are then drilled and tapped for a piece of steel wire which is tapped and screwed into the ball. Each wire has a ball at each end, and the balls range in diameter from ⁵⁄₁₆ in. up to ¾ in. They are kept handy on a small rack near to the bench, as shown in the second figure.

These time-saving wrinkles are only types of a great many more which have been devised to expedite the work of the pattern shop, but which cannot be mentioned in detail here. Enough has however been said to show that a little thought can at times save a great deal of work.

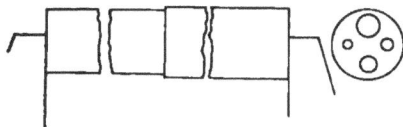

Aeroplane Chases Liner with Parcel for Voyager

One of the most novel purposes for which an aeroplane has ever been chartered was to carry a pair of eyeglasses to a passenger on the *Olympic*, just as she was leaving New York on her first eastward run. W. A. Burpee, the millionaire seedman of Philadelphia, broke his glasses as he was about to board the *Olympic*, and had a wireless message flashed to his optician to have a new pair delivered to his London office by the next boat.

The optician, however, saw a better way of delivery, and in five minutes had Thomas Sopwith, the English airman, then flying in New York, on the wire. The glasses were delivered to him by automobile, and he started in pursuit of the *Olympic*, reaching her as she was passing Fort Hamilton on the way out. He swooped down within a couple of hundred feet of the deck and dropped the padded package which contained the glasses.

SHOW-CARD WRITING

The show-card has become so important a factor in window display that a mastery of the art of show-card writing is a most valuable accomplishment for the window trimmer. A writer in the *Apparel Gazette* gives the following elementary lesson on the art, which will be found valuable by the beginner.

Before attempting to practise at all the beginner should supply himself with the right tools, which does not mean the buying of a very elaborate or expensive outfit, brushes, inks, paints, rulers, erasers, etc., but simply supplying himself with one good brush and a bottle of show-card ink.

For practice paper use common manilla wrapping paper with a smooth surface. Rule as indicated on the chart given herewith.

Before you make a stroke see that the brush is dipped several times into the ink, each time drawing the brush across the neck of the bottle the entire length of the hair, letting the ink run back into the bottle and reversing the brush or turning it over, so that as the ink flows from the brush it reduces the end of the brush to a flat, square or chisel point. This is important—next to having good utensils is the proper manipulation of the same. So very important is the part

It is not likely that your local paint supply store can furnish you with the brush necessary, but they are in communication with the manufacturers of them and can place a special order for a red sable pencil or quill with a square point, No. 11 or 12, ⅞ or 1 in. stock, or for a red sable rigger, No. 11 or 12. This brush will probably cost fifty cents. Your local stationer can, no doubt, supply the show-card ink in 2 oz. desk bottles, which sell for about ten cents. A foot ruler and a lead pencil complete the outfit.

These utensils, especially the brush and ink, are imperative and greatly facilitate the progress of the beginner.

The accompanying sketch illustrates the parts used in making both full-face and shaded letters of both upper and lower case alphabets. The assembling of these parts, spacing and slanting the same in harmonizing curves and straight strokes, produces alphabets that are very pleasing to the eye and that are easy to read.

of brush manipulation to the beginner that very little progress can be made if he is careless about it.

Having the brush in good working order, begin the practice with Fig. A, the downward stroke. Holding the brush is the next important part. Take the brush as you would your pen, between the first finger and thumb, but not resting on the second finger, holding the brush so that you can roll it freely back and forward between them. The hand should rest upon the table, only the fourth and little fingers touching it. The brush should be held almost perpendicular, the handle in line with the second joint of the forefinger, not the third, as in writing.

Downward strokes in Figs. A and B are made with the finger movement, bringing the brush down, bending the fingers. Do not touch the pencil ruling, either top or bottom, but have the strokes just within. Fig. C is made with the arm movement, holding the fingers stiff and moving the hand backward, resting the fleshy part of the forearm upon the table. Fig. D is made with the arm movement, turning or rolling the brush, as directed above, between the forefinger and thumb, so that the flat part of the brush faces across the paper, not downward. Move the hand to the right, using the fourth and little fingers as a rest and guide.

The same idea is followed in making all the strokes contained in the chart, beginning and finishing the strokes, as indicated by the arrows. The beginner may have some trouble with the strokes that have pointed extremities, but if the brush is touched lightly at the start and lifted quickly at the finish of the stroke, this difficulty can be overcome.

LATHE SCREW-CUTTING
H. W. H. STILLWELL

Much valuable information has been written upon lathe work, some in a plain, practical and common sense manner, easily understood by the mechanic of average intelligence, and still more in technical language, which is almost impossible for the young mechanic to figure out for himself unaided.

Many of our best mechanical authorities seem to lose sight of the fact that perhaps the greater part of the readers of their articles are non-technical men and not familiar with much of the technology which their articles contain, nor with the higher mathematics they often employ.

It is the purpose of this article to place before the young mechanic some useful and valuable information in as simple and non-technical a manner as possible.

Almost all of the modern lathes are indexed; the index is figured or found from one common number, by which the teeth in each of the gears may be divided exactly, and the additional number of teeth in every next larger gear will be the same as the common number. For example: If the smallest gear of a lathe is of 24 teeth, the next 28 teeth, the next 32 teeth and so on, then the common number is 4. If the teeth increase by 5, then the common number is 5. If by 6, then 6, etc.

In lathes having no common number and with irregular gears, any number may be used to multiply by, which will then be the common number. This rule also holds good for lathes that have a common number in case it is desired to cut a thread that the index does not show.

DRIVING SCREW

The driving screw of a lathe is the next thing of importance that the operator must thoroughly acquaint himself with,

to be able to correctly figure threads and gearing for that particular lathe.

In most lathes the true relation of the driving screw to the lathe spindle is maintained. In all such cases the driving screw is correctly represented and figured by the number of its threads per inch.

On a lathe where the true relation of the driving screw is changed by reason of a different sized gear on the feed spindle to that on the lathe spindle, the number of threads per inch on the driving screw does not correctly represent the same, and the correct driving screw must be found, which is done as follows:

Rule.—Take two gears of equal size, use one as driver, the other as driven, with any convenient size intermediate gear. Cut a thread, and the number per inch thus found will be the drive screw.

RULE 1
HOW TO FIND TWO GEARS

Rule.—Take the driving screw as a numerator and the required threads per inch as denominator; multiply each by the common number. The new numerator thus found will be the driver, and the new denominator will be the driven. In other words: the result of multiplying of the driving screw (the new numerator) must be placed on the lathe spindle and is the driver.

And the result of the multiplying of the required threads per inch (the new denominator) must be placed on the driving screw and is the driven.

Example

D.S. 4	Thds. 8	C.N. 10
$\therefore \frac{4}{8} \times 10 =$	$\frac{40}{80}$	Driver
		Driven

NOTE.—This rule is the fundamental principle of all rules for threading or thread-cutting.

If a right-hand thread is to be cut, use one intermediate gear. If the thread is to be left-hand, use two intermediate gears.

NOTE.—When the driver or spindle gear is non-changeable or assumed, use this rule to find the driven.

⌐ *Rule.*—Multiply the number of teeth of the spindle gear or the driver by the desired number of threads, then divide that product by the number of threads per inch on the driving screw. The result will be the gear to put on the driving screw or the driven.

Example

Spindle gear or
D.S. 4   Thds. 8   driver 32
∴ 8 x 32 = 256, then 256÷4=64, the driven

RULE 2

HOW TO FIND FOUR OR COMPOUND GEARS

Two gears will cut many plain and many fractional threads, except when very fine or very coarse. To cut very fine threads with two gears, that required on the lathe or feed spindle would be inconveniently small, and that required on the driving screw would be inconveniently large. It would be difficult, if not impossible, to find such small or large gears in the shop.

In all such cases it will be more convenient to use four gears which will be found as follows:

Rule.—First find two gears as shown in Rule 1; then find two numbers to multiply together into the driver without a remainder; and two numbers to multiply together in the driven without a remainder; then multiply each of the two numbers so found by the common number. The result will be the two drivers and the two driven to use.

Example

D.S. 4   Thds. 20   C.N. 6
4   24
— x 6 =   —
20   120

We now have 24 as the driver and 120 as the driven. Take 24 and find two numbers that will multiply together and make the same, as, 4 x 6=24; then take 120 and do likewise, as, 10 x 12=120. Then multiply 4 and 6, and 10 and 12 by the common number. The result will be the two drivers and the two driven

to use, and the full example will read as follows:

D.S. 4   Thds. 20   C.N. 6
4   24 . 4 . 6.   24 . 36 drivers
— x 6 = ———————— x 6 = ——————
20   120 . 10 . 12   60 . 72 driven

To prove the gears so found, use this rule.

Rule.—Multiply the first driver 24 by the desired number of threads 20; divide that product by the first driven 60; then multiply that quotient 8 by the second driver 36, and divide that product by the number of threads per inch on the driving screw 4. The result will be the second driver 72.

Example

24 first driver
20 desired threads
——
First driven   60)480(8 quotient

36 second driver
8 quotient
——
D.S.   4)288(72 second driven
28
——

RULE 3

HOW TO CUT FRACTIONAL THREADS
PER INCH

Fractional threads per inch means: a certain number of whole threads and a fraction of another contained in one inch.

The easiest way to measure fractional threads is as follows: take any number of threads until they measure even inches, then count the number of threads in that number of inches.

When the pitch of a screw is given in the form of a fraction, as ⅜ pitch, then the bottom figure indicates the number of threads, and the top figure the number of inches. ⅜ pitch would be one thread in ⅜ of an inch, or 8 threads in 3 in., or 2⅗ threads per inch.

Rule.—Find the number of whole threads in even inches, then find the number of threads on the driving screw in the same number of inches and multiply by any common number, same as Rule 1.

For instance, the desired thread is 4½ per inch, and the driving screw is 6 threads per inch. It will in that case be seen that 9 whole threads are found in two inches, and 12 threads in two inches of the driving screw.

The example will therefore read as follows:

Example

D.S. 6	Thds. 4½	C.N. 4
D.S. instead of 6	use 12	48 driver
	x 4=	—
Thds. instead of 4½	use 9	36 driven

D.S. 4	Thds. 3¼	C.N. 6
D.S. instead of 4	use 16	96 driver
	x 6=	—
Thds. instead of 3¼	use 13	78 driven

D.S. 4	Thds. 11½	C.N. 3
D.S. instead of 4	use 8	24 driver
	x 3=	—
Thds. instead of 11½	use 23	69 driven

D.S. 4	Thds. 2⅓	C.N. 5
D.S. instead of 4	use 12	60 driver
	x 5=	—
Thds. instead of 2⅓	use 7	35 driven

RULE 4
HOW TO CUT THREADS PER PITCH

Rule.—Multiply the top figure by the number of threads per inch on the driving screw, use that product as the numerator and the bottom figure as the denominator, then multiply each by the common number.

Example

D.S. 4	Pitch ⅜ in.	C.N. 6
$\frac{4 \times 3}{8}$ =	$\frac{12}{8}$ x 6 =	72 driver
		48 driven

When the pitch is measured by whole inches use this rule:

Rule.—Multiply the number of whole inches by the number of threads per inch of the driving screw, and use that product for the numerator, and for the denominator use 1 and proceed as per Rule 1.

Example

D.S. 2	Pitch, 3 in.	C.N. 16
3 x 2 =	$\frac{6}{1}$ x 16 =	96 driver
		16 driven

When the pitch of a thread is given in one or several whole and part of another inch, as: 1 thread in 2¾ in. proceed as follows:

Rule.—Find the number of even threads in even inches, which will be 4 even threads in 11 in., then multiply the number of inches (11) containing the even threads, by the number of

threads in one inch of the driving screw. Use that product as numerator and the even threads, which is 4, as the denominator.

2¾ pitch we find contains 4 even threads in 11 in. We find driving screw at 2 threads per inch contains 22 threads in 11 in. hence the example will read as follows:

Example

D.S. 2	Pitch, 2¾ in.	C.N. 5
Numerator	D.S. use 22	110 driver
	$\frac{}{}$ x 5 =	—
Denominator, Pitch, 2¾ use 4		20 driven

RULE 5
PITCH OF DRIVING SCREW—HOW TO FIND GEARS FOR SAME

When the driving screw of a lathe is measured and designated by its pitch, as for instance ⅜ pitch, use the following rule to find the gearing for any desired thread.

As shown above, the top figure of the fraction always designates the inch or inches, and the bottom figure always designates the number of whole threads contained in the same; consequently a ⅜ pitch driving screw would have 8 threads in 3 in.

Rule.—To find the driver simply take the bottom figure of the fraction and multiply the same by the common number, as Rule 1. To find the driven take the top figure of the fraction and multiply the same by the desired number of threads to be cut, and then multiply that product by the common number.

Example

D.C. ⅜ pitch	Thds. 6	C.N. 4
$\frac{3 \times 6}{8}$ = $\frac{18}{8}$	$\frac{8}{18}$ Reverse x 4=	32 driver
		72 driven

In order to avoid becoming confused in the use of this rule, another way to use the same is as follows: Place the figures of the pitch ahead of the example, following that reverse the same and proceed to figure as per rule. Your example would read like the following:

Example

D.S. ⅜ pitch	Thds. 6	C.N. 4
$\frac{3}{8}$ $\frac{8}{3 \times 6}$ =	$\frac{8}{18}$ x 4 =	32 driver
		72 driven

THE HOME CRAFTSMAN

Ralph F. Windoes

ETCHED PIPE RACK

In the pipe rack illustrated we have a very attractive piece of etched metal work. The material needed to construct the rack is as follows:

1 piece No. 20 gauge copper or brass, 5 x 12 in.

1 piece No. 20 gauge copper or brass, 2⅛ x 12 in.

Rivets, asphaltum, acid (nitric), lacquer, fine steel wool.

Make a full-sized drawing on a piece of paper of the shape and design of the back. The photograph illustrated here is only a suggestion of a number that can be used, and it is strongly advised that the craftsman originate his own. When the drawing has been satisfactorily completed, clean the metal and transfer the design to it with carbon paper.

Cut the shape out with snips and the jeweller's saw, and cover the parts that are not to be etched with asphaltum. Place in a weak solution of the acid and leave until the etching is completed. Remove and wash in clear water and color as has been directed before in this series.

The narrow strip has six holes cut into it that have a diameter of ¾ in. The center for these holes is ¾ in. from one edge. Lay the holes out with a pair of dividers, drill a hole in each and saw out the circles. Bend a right angle ⅝ in. from the back edge, and rivet this piece to the back.

A very finished appearance is given the edges if they are laid on a piece of iron and pounded with the ball pein of the hammer. Polish and finish the rack with a coat of lacquer.

HUMIDOR

The humidor has proven itself indispensable to the man who smokes. It is a companion piece to the pipe rack and should be designed with that end in view, though we will admit that the article illustrated does not show this point. The material list for this object includes:

1 piece No. 20 gauge copper or brass, 5 x 14 in.

1 piece No. 20 gauge copper or brass, 4 x 4 in.

1 piece No. 20 gauge copper or brass, 3⅞ x 3⅞ in.

2 pieces No. 20 gauge copper or brass, ½ x 3½ in.

Asphaltum, acid, fine steel wool, lacquer and small blotter.

Make a full-sized drawing of one side, 3½ x 5 in., and transfer it successively to the large piece four times. This leaves a ¼ in. strip for fastening.

Etch it, as has been explained before, while it is flat. Finish it up and bend on the dotted lines shown. The bottom, which is 3½ in. square, is cut and bent, making the four laps at right angles to the piece.

The cover is 4 in. square and is held in place by the two strips bent at right angles and soldered securely to it. It should be etched with an appropriate design, and after the soldering is completed, should be raised in the middle.

Beside the strips being soldered to the cover, the sides are soldered and the bottom held in the same way, as explained below.

When all processes have been completed, fit a piece of thick blotting paper in the cover and always keep it moist while the humidor is in use. Take it to a tinsmith and have him fit it with a square box of light zinc, in which the tobacco must be kept.

It is a good plan to glue a piece of felt or leather on the bottom of the humidor to prevent its scratching any article upon which it is placed.

ETCHED PIPE RACK –

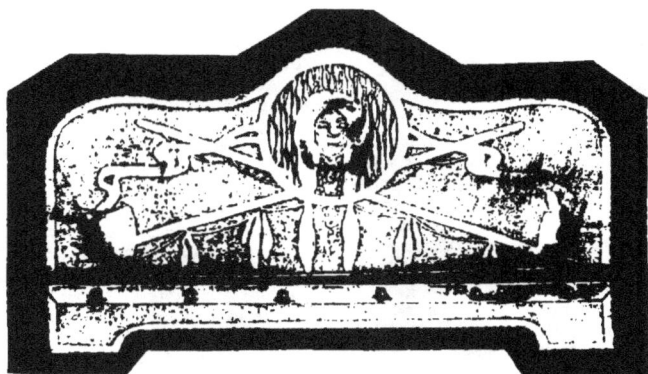

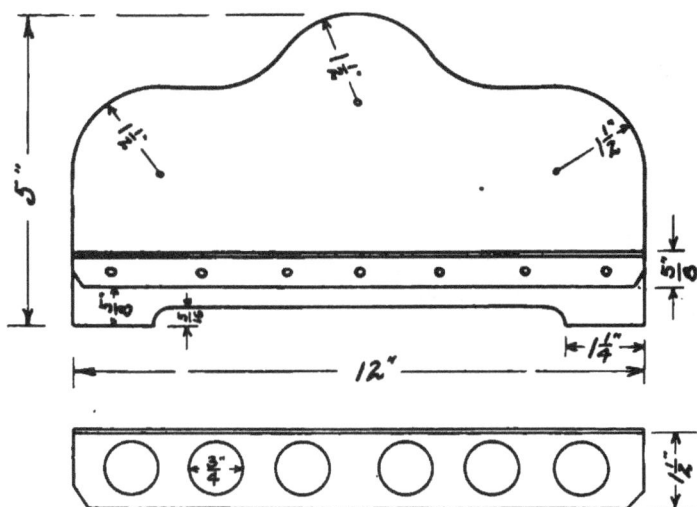

SOLDERING

In addition to the tools and materials already at hand, the following will be necessary for the process of soldering:

Annealing tray.
Mouth blowpipe.
Bunsen burner or alcohol lamp.
Borax slate.
Camel's hair brush.
Powdered or lump borax.
No. 24 iron wire.
Silver solder.

The annealing tray may be purchased from craftsmen's supply houses, or a tray made of sheet iron, about 20 in. square and 4 in. deep filled with slag will answer the purpose very well. It must have riveted corners, as soldered corners would be liable to melt.

The bunsen burner that uses gas is a better source of heat for the work than an alcohol lamp, though the latter may be used if it is of large size.

A slate slab, such as is used to grind ink, makes a very good borax slate. The silver solder may be used any gauge, but about No. 20 works the best.

The surfaces that are to be soldered must be absolutely clean. This can be accomplished by scraping them with a sharp instrument, filing them or rubbing them with steel wool. After they are clean, be very careful about handling them, as solder will not flow over grease.

Next grind up a little borax on the slate with a little water, until it is of the consistency of thick cream. If the powdered borax is used, a stick will serve for the stirring. Cut a number of pieces of the silver solder into the borax, making them about ⅛ in. long.

Coat the cleaned surfaces with the borax where they are to be joined, and bind them together with the iron wire. Lift the pieces of solder from the borax with the brush and place them along the edges to be joined, putting them about

Coal-Dust Danger

It being generally agreed that great colliery explosions, however they may originate, are spread by the progressive ignition of fine coal-dust, Prof. W. M. Thornton, of the Armstrong College, Newcastle-on-Tyne, has recently been experimenting with a view to getting rid of the dust danger in mines. His conclusions he indicated to a meeting of the members of the North of England Mining Institute at Newcastle.

After carrying out very exhaustive experiments, he had hit upon a mixture of ten parts of water-glass with one part of resinous liquid soap and one of commercial carbolic acid. This mixture, he averred, could not be improved upon from the standpoint of cost and efficiency as a wetter or binder of fine dust. He had conducted experiments not only in the laboratory, but in the gallery of the E Pit, Udpeth Colliery Co., Durham.

He summed up his researches by giving, as a counsel of perfection, the advice that a dusty road in a mine should have the timbers and sides thoroughly, though quickly, swept by a brush, that being immediately followed by hand-spraying, especially on the tops of timbers, or fine spraying into the dusty air.

Dr. Thornton's conclusions were rather warmly contested by Dr. Bedson (Professor of Chemistry at the same college). Dr. Bedson declared that most of the substances with which Dr. Thornton had been dealing were dangerous, and he (Dr. Bedson) should classify them down a mine as he would classify matches. Soap dust was nearly as inflammable as coal dust, and, when the soap solution had dried, would result in a larger quantity of volatile matter being present than before.

———

Double Glazing for Sound-Proofing

HUMIDOR.

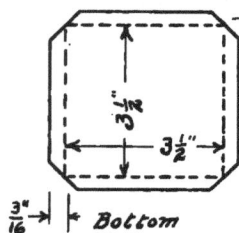

$3\frac{1}{2}$

$3\frac{1}{2}''$

$\frac{3''}{16}$ Bottom

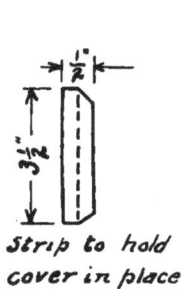

$\frac{1''}{2}$

$3\frac{1}{2}''$

Strip to hold
cover in place

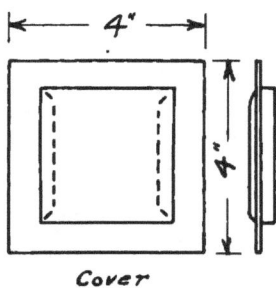

4"

4"

Cover

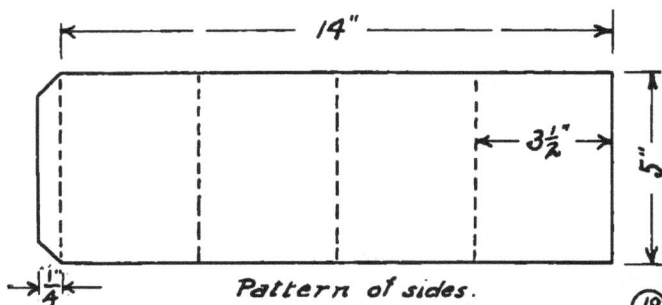

14"

$3\frac{1}{2}''$

5"

$\frac{1''}{4}$

Pattern of sides.

(19)

R.F.W. 1911

DRY CELL TESTING*

W. B. PRITZ

There are today at least 100 brands of dry cells upon the market, varying in quality from very efficient and reliable cells to those which must be considered very inefficient; and from these the consumer must select that brand which in his opinion will give best service. Before any conclusion can be reached regarding the adaptability of a cell to a particular service, the consumer must either accept the guarantee of the manufacturer or make for himself, as best he may, a service comparison of the cells in question.

The present diversity in the methods of testing cells is very troublesome for both the manufacturer and consumer. The former is often called upon to guarantee the service of his product when subjected to certain tests which have little or no relation to any which he may have adopted, and although it is not impossible, it does become quite difficult for a manufacturer of dry cells to calculate from the results of his regular tests just what may be expected of his product when subjected to a particular test required in the specifications of his customer. The adoption of some standard tests which would hold between manufacturer and consumer alike would, therefore, be very advantageous to both. The establishment of such tests and the publicity which would thereby be given to the whole subject of testing would decrease the tendency of the small consumers (who, taken collectively, must use a considerable percentage of the output) to place reliance in the so-called tests which now form the basis for their judgment of a brand of cell, such as amperage or voltage readings.

It is easily shown that there is in no sense a relationship between service and short-circuit current. It is true that the amperage reading of a cell, coupled with a familiarity with the particular brand, does serve as a good indication of the age of the cell or the presence of any serious defect; however, the customer who judges solely from the short-circuit current is very apt to obtain inferior quality, and yet we are informed by dealers that at least 90 percent of the customers who buy cells off the shelves demand that they be so read and in most cases select that brand giving the higher current.

Let us consider, then, the requirements of a satisfactory test for dry cells and the conditions which have the greater influence upon the results obtained. In general (other than for purposes of research) there are but two reasons for desiring a test upon dry cells: (1) To ascertain what life may be obtained from a brand of cells in a certain service; (2) to ascertain which one of several brands will give the longest life in that particular service.

With the former object in view the knowledge is best obtained by actual use of the cells in connection with the appliance. In some cases this is the only feasible way in which the definite information sought can be obtained. The great majority of tests are carried on, however, with the second object in view—viz., the comparison of two or more brands of cells for use in a particular service. Where the amount of testing is large, it is impossible, even were it expedient, to use the actual appliances for testing cells, and it becomes necessary to devise special testing methods and apparatus such that results obtained therefrom shall be comparable to the results obtained from the cells when placed in actual service. This is, we take it, the one necessary condition which dry cell tests must fulfil.

There have been tests devised which seek to go further and make the operating conditions not only comparable, but as similar as may be to the operating conditions of the service for which the test is intended. Upon this point there is some diversity of opinion. Some authorities claim that a test is of greater value and is more reliable the more nearly the conditions of test approach those of service, and, following out these claims, have devised certain tests which are rendered quite complicated, requiring much attention and apparatus for their continuance, by the introduction into the method of some of the irregularities to be expected in service. It is questionable, however,

*Abstract of a Paper read before the American Electrochemical Society.

if results of greater meaning are obtained from such strict adherence to service conditions. At best, such a test is but an approach to actual service, which must be continually varying from time to time and from locality to locality. Again, the apparatus necessary to carry on an irregular, intermittent test is very complicated and requires much careful attention. This feature limits its use to the large consumers and manufacturers.

Questions are often asked regarding the advisability of testing various brands of cells by connecting them in series and discharging them simultaneously. In general, we would advise against this method, though there are occasions when it is the only one that can be employed. Especially is this true when test conditions, such as length of contact and recuperation periods, strength of current and temperature of the battery, cannot be held constant for the separate testing of various brands. It is much preferable, for instance, to use the series method of test upon an automobile than to test the various brands upon different machines or upon the same machine at different times.

Temperature is a most important factor in dry cell testing, and a large part of the non-uniformity in the results of tests may, we think, be traced to temperature variations. For instance, the current increases as the temperature rises, and the differences in the current obtainable increase considerably as the temperature falls. The influence of temperature upon service is greater than it is supposed. Tests show great variations are obtainable between 60 and 95 degrees, between which values practically all testing is done. This makes the importance of temperature regulation very evident. The effect of temperature also varies with the nature of the service; high temperatures giving longer service where the drain is heavy, while low temperatures are favorable for light drain service lasting over a considerable period. Of prime importance in dry cell testing is a knowledge of the effect of temperature upon the rate of deterioration of cells when left on open circuit. To determine just what this effect might be, a number of cells were stored at seven different temperatures. The following table gives the

after 10 weeks, expressed as a percentage of the initial values:

Temperature	Initial Amp.	Percent Drop in Amp. in 10 Weeks
41°F	18.1	4.4
77°F	22.0	10.0
95°F	21.0	19.0
113°F	22.8	25.0
131°F	23.0	52.0
149°F	20.5	71.0
167°F	21.0	98.0

The greatly increased rate of deterioration on open circuit at high temperatures no doubt accounts for the poor service obtainable at such temperatures over long periods of time. A knowledge of this effect of temperature would undoubtedly explain the cases, which are at times brought to our attention, of cells rapidly deteriorating upon a dealer's shelves, from "no cause whatever." The cells should be stored in a dry, cool place, and not in the corner behind the stove.

In interpreting the results obtained from a test of various grades of cells a caution must be given against drawing definite conclusions from the outcome of a single or a small number of tests. When the matter of choosing a brand is of much importance it is necessary to run a series of tests over a period of six months or a year. In this way a very good idea may be obtained of the average service results which may be expected.

In regard to the terms in which the results of dry cell tests should be reported, there is some difference of opinion, some authorities contending that the ratings should be expressed as the number of ampere-hours given by a cell under specified conditions to a certain working voltage value. Others claim that more practical meaning is attached to a statement of the length of time during which the cell is able to maintain its working voltage above the specified limiting value. It is the author's opinion that that method of rating should be used which gives to the consumer the exact information which he desires; hence, the rating of cells by the length of service of a given kind which they are capable of giving is favored.

It is perfectly evident that the consumer is interested in the length of service which he is able to obtain from a battery

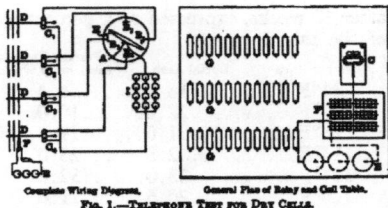

Complete Wiring Diagram. General Plan of Relay and Coil Table.

Fig. 1.—Telephone Test for Dry Cells.

At first thought it might appear that the ampere-hour capacity of a dry cell bears such a relation to length of service, that either method of rating would give the same information. Such is not the case, however. Differences in discharge rate cause vast differences in the number of ampere-hours obtainable from the cell. From the nature of the discharge curve, it is quite difficult to calculate the length of service obtainable under the conditions limiting the ampere-hour rating, while to go further and deduce from the rated ampere-hour capacity under one set of conditions, the output to be expected under different conditions becomes a hopeless task. Equally hopeless is the final interpreting of the result in terms of length of service. Again, a dry cell does not always give length of service which is proportional to ampere-hour output. It is easily possible to produce two dry cells which will give equal lengths of service, the ampere-hour output of which will show marked differences. Hence a statement of the ampere-hour capacity of two brands of cells is very apt to be misleading, as it does not settle definitely which is the better cell to use upon a given service. For research purposes the energy or ampere-hour output of a cell is very useful, but as a practical rating for dry cells it is not satisfactory, especially from the standpoint of the consumer.

The author then proposes for consideration certain tests covering the two most important services, namely, telephone, and gas-engine ignition. The nature of these services was fully outlined in Papers delivered before the Society by Dr. J. W. Brown[*] and Mr. D. L. Ordway,[†] and therefore will be touched upon only in so far as is necessary to establish the relationship of the service to the tests which we propose.

It is recommended by one of the important telephone companies that the battery should at all times give a current of more than 0.14 amperes at the end of one minute after it has been disconnected from the transmitter and connected in circuit with an ammeter and a resistance of 20 ohms. This specification necessitates that the working voltage be constantly 2.8 volts or more. The author has been using for some time a test which requires a minimum of care and supervision and which is quite comparable to the more complicated tests. It subjects three cells in series to a discharge through 20 ohms resistance for a period of two minutes each hour, during 24 hours per day and 7 days per week, until the working voltage reaches the limiting value of 2.8 volts, the results being reported as the number of hours' service to this cut-off point. The energy drawn from the battery in this test is approximately equivalent to that consumed in the more complicated methods; the batteries give practically equivalent periods of service, and the number of cells, resistance in circuit and cut-off point are identical. The regularity of this test which we propose eliminates the mechanical appliances necessary to obtain decreased night and Sunday service. The left-hand portion of Fig. 1 shows diagrammatically the apparatus necessary and its arrangement. The hand of the clock A revolves once per hour, closing by means of the contact H the circuit of the battery I in turn through the contacts B, B_2, B_3 and B_4. This current magnetizes the cores of the telegraph relays C, C_2, C_3, C_4, causing the extended armature arms D to fall, bringing the inverted U-shaped fingers F into mercury cups, through which the test battery circuits E are closed. Each contact B is of such length that two minutes are required for the passage of the contact H. At the right of the figure is shown the arrangement of the relay and coil table. The test batteries are stored underneath the table and are read from the mercury cup F.

Fig. 2 shows a representative discharge curve obtained from a battery of three $2\frac{1}{2}$ in. x 6 in. dry cells of a well-known brand. The curve passes through the

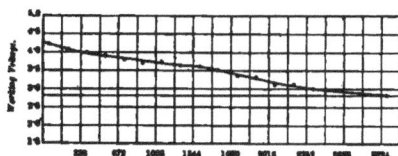

Fig. 2.—Hours on Test.

values of the working voltage at the end of the discharge periods.

Ignition practice also presents great diversity in operating conditions, and formulating a satisfactory test to meet the demands requires much consideration of the various systems used. In developing a test suitable for this service, a 4-cylinder automobile engine, equipped with a standard make of spark coil was used. With different numbers of cells in series, resistance was cut into the circuit until the engine failed to operate satisfactorily. Limits of satisfactory ignition for these particular conditions were thus reached. Readings at this point were taken of the working voltage, average drain on the battery and the value of the impulse which the battery was capable of forcing through the primary winding of the spark coil.* The work was then duplicated, using in turn all the leading makes of coils. Thus it was possible to formulate what may be considered as average limiting conditions of actual ignition service. It was found that in the majority of cases five or six cells in series gave ample voltage. Efficient service was obtained in some cases with as low a drain on the battery as 0.2 or 0.3 amperes. It is very probable, however, that 0.5 ampere more nearly represents the average drain which could be maintained by the adjustment of coils without proper instruments.† A battery was found to give inefficient service when the impulse of current which it was able to send through the primary winding of the spark coil fell below 2 or 3 amperes, Fig. 3 showing a diagram of the arrangement.

From these findings the following test has been developed and adopted as standard for ignition service. A battery of six cells is connected in series with a

16-ohm coil, which permits an average drain on the battery throughout its life of approximately 0.5 ampere. Readings of the working voltage are taken at intervals throughout the test, the most important reading, however, being the impulse of current which the battery is able to force through a ½-ohm coil connected in series with the ammeter and in parallel with the 16-ohm resistance. When this value at the end of a period of contact falls below 4 amperes, the battery is considered unfit for service and discarded. This value is taken, rather than 3 amperes, in order to be conservative. All conditions adopted from investigation as representing the average ignition service have thus been duplicated. In settling upon the length of time during which the battery should be discharged, many users of automobiles were interviewed, and the consensus of opinion was that, after the novelty of an automobile wore off, the length of time during which a car would be in operation would average not far from two hours a day. The test batteries are therefore discharged for this length

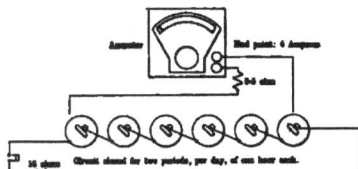

Fig. 3.—Ignition Test.

of time each day. The time is divided, however, into two periods, one hour in the morning and one in the afternoon. As a test for lighter service, the battery is sometimes subjected to two drainperiods of half an hour each. If the amount of testing be not too great, the circuits may be closed by hand, in which case no apparatus is needed other than instruments for reading, a number of 16-ohm resistance coils and a 5.0-ohm coil for use in taking the impulse readings.

This test has been used satisfactorily for several years. By slight variations, chiefly in the time of discharge, it may be made representative of any system of various types of ignition service.

---

*A full discussion of the impulse value and its importance in ignition service is contained in: D. L. Ordway, "Trans." Amer. Electrochem. Soc., XVII, 361 (1910).

†The value of the approximate drain, viz., 0.5 amperes, does not apply to the various single-spark appliances now to be obtained, which are much more economical of current.

Since it is often the case that quite an interval of time may elapse between the manufacture and installation of a cell in service, a knowledge of what may be expected during this period becomes of no little importance, and hence an open circuit or "shelf life" test is quite essential. The ideal method for such a test would be the determination of the decrease of service capacity due to storage over definite periods. This practice, however, would entail much labor and expense where the amount of testing to be done is large.

The method generally employed to determine shelf life consists of reading the initial voltage and short-circuit current of a representative sample of cells, followed by current readings at intervals of one or two months, depending upon the nature of the cells in question. The ammeter for reading short-circuit current should be dead beat, and with its leads should have a resistance of 0.01 ohm. Two 30 in. lengths of No. 12 lamp cord make very convenient leads. The results are merely indicative of increase in internal resistance, and bear no definite relation to the service which the cells may give. However, this information, coupled with familiarity with a brand of cells, becomes a very good indication of its quality. It also serves to indicate any serious defects of manufacture. The cells are kept on the shelf until the short-circuit current has fallen below 10 amperes. This point is arbitrarily chosen, as it represents a point below which it would be difficult to market the cell. For practical purposes, the results are expressed as the number of months during which the short-circuit current remains above this cut-off point. Much more meaning, however, is attached to the rate at which the current falls, generally reported as the drop in amperage expressed as a percentage of the initial amperage. This is especially true when investigation of the quality of cells is the object. For practical purposes, however, the first rating given, i.e., months to 10 amperes, is perhaps preferable.

There are, of course, many miscellaneous services in which dry cells are used, tests for which must be formulated as occasion demands and to suit the particular needs of the case. For the more important uses, however, viz., telephone and ignition services, we trust uniform methods of testing may soon be adopted which will be satisfactory to all.

A CORNER CUPBOARD

Such a fitting as the one illustrated herewith, can always be used to fill an odd corner, and will usually be found to occupy only space which would otherwise be entirely vacant. It can be made of any variety of selected straight-grained pine, and presents no especial difficulty in construction. Owing to its shape on plan, it will be easily understood upon reference to the drawing that while the side view shows the true shape of the sides and back, but foreshortens the door, which is at an angle of 45 degrees to the sides which are 2 ft. 6 in. x 5 in. and about ⅝ in. thick, are cut to the curved outline, shown on the side view, and the shaded portion fretted out entirely; then the two back pieces are prepared, and they will also have to be in two widths glued together to make the 1 ft. 4 in. required, and are 2 ft. 10½ in. high, with a square piece cut out of the bottom, and curved and fretted at the top to match the sides. These six parts can then be glued and neatly bradded together, thus forming the carcass of the cupboard.

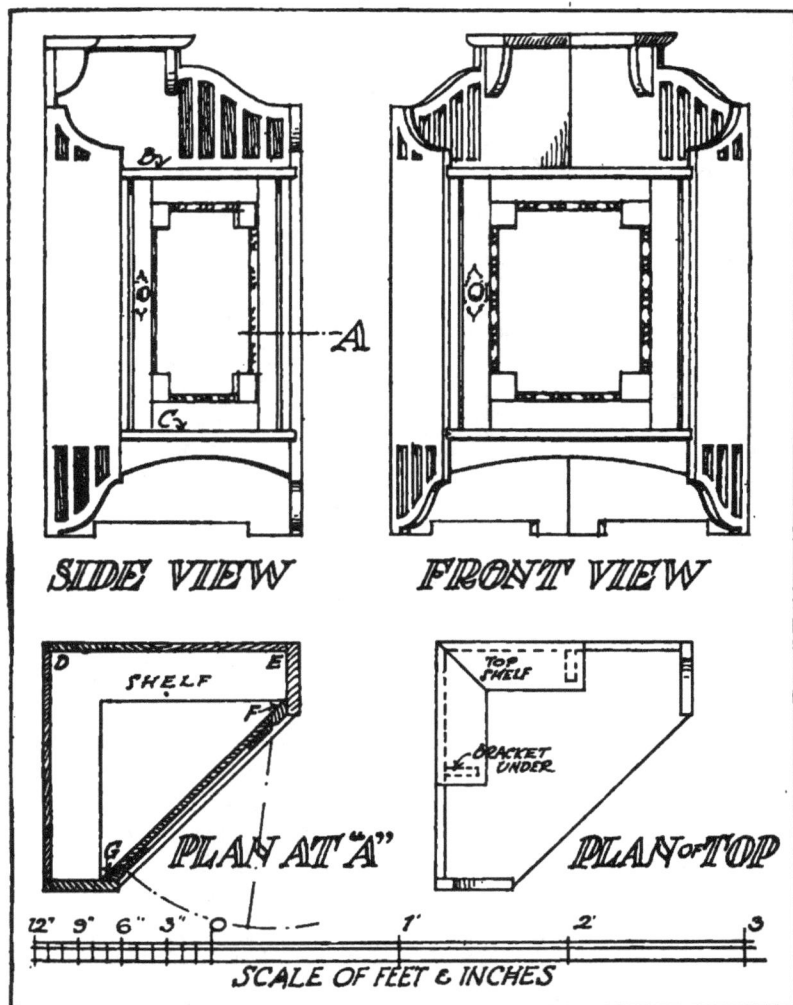

SIDE VIEW   FRONT VIEW

PLAN AT "A"   PLAN OF TOP

SCALE OF FEET & INCHES

plan at *A*); the outer edge of these pieces might be slightly chamfered.

The only part now remaining to be described is the door, which will need careful fitting and hanging. It may be framed up with a panel in the orthodox way, if the craftsman cares to undertake the task, or a piece of three-ply of the full size of the door may have thin strips 2 in. wide glued along its edges to represent the styles of a door, and if this possible to detect the difference without opening the cupboard. The door is decorated with 1½ in. square pieces about ¼ in. thick in the corners, and a small ornamental bead strip is filled in between as shown. A fillet should be fixed along *G*, projecting a little to stop the door, and neat hinges and a suitable spring catch or cupboard lock should be chosen.

The interior can easily be fitted with light shelves or pigeon holes to suit indi-

DEFLOCCULATION *
EDWARD G. ACHESON, SC.D.

It is with much diffidence I speak on a subject that has not yet emerged from the embryonic state. My latest experimental researches had to do with it; I believe it will rapidly grow in importance in the scientific and industrial world; and finally, much work of a strictly scientific character remains to be done to reduce the fragmentary knowledge we now have of it to an exact science.

In my labors devoted to working out and developing industrial and commercial projects, I have upon several occasions found reactions, conditions and results that did not harmonize with the accepted theories and formulae of scientific men. Being an earnest believer in publicity, in order that any possible benefits that might accrue to the common welfare may the more quickly be enjoyed, I shall lay before you a detailed account of my experiments on the deflocculation of inorganic bodies. It will become very evident as my story unfolds, that throughout the series of experiments described and the working hypothesis employed, I was wholly disregarding the prevailing theories, and that I unconsciously entered the field of colloids.

Having worked out the problems involved in the manufacture of graphite from coal and other carbonaceous materials, I undertook, in the summer of 1901, the introduction of this artificially-made graphite into the crucible trade. My first efforts were devoted to the making of a satisfactory crucible of my graphite, using as a binding material, American clays. Many failures were met with, and I found it difficult to locate the cause of the failure, whether with the graphite or with the clay. I soon learned the manufacturers of crucibles in the United States invariably used as a binding agent for the graphite in the crucible body, clays imported from Europe. I secured samples of these imported clays, and found them much superior to the American ones in plasticity and tensile strength.

Chemical analysis failed to disclose the cause of the physical differences existing in the clays. The question involved interested me greatly, and I decided to endeavor to determine what produced the variations. I found it generally stated in the books that residual clays were non-plastic, and sedimentary clays were more or less plastic. Here was the starting point. Plasticity was developed by or during the act of transportation from the point of formation to the final resting-place of the clay. I did not believe there was anything in the simple act of the suspension in water that would produce the effect noted, and therefore looked for the cause of the foreign matter carried by the water. It seemed the most likely agents were the organic substances washed from the forests into the running waters. With this idea in view, I made a few experiments with those substances I thought likely to be found in the washings of vegetation. One of my early experiments was to treat kaolin with a solution of tannin, and I at once noticed less water was needed to produce a given degree of fluidity; also that the tensile strength and plasticity were increased.

Tests were made on the increased tensile strength of clay, as the result of treatment with organic matter, and it was found that briquettes made of Harris kaolin and dried at 120°C. would break with a load of 5.73 kg. per square centimeter, while the same clay, after treatment with two percent of catechu for a period of ten days, then formed into briquettes and dried at 120°C., would not break until the load was increased to 19.75 kg. per square centimeter,—an increase of more than 244 percent.

I now began to wonder whether or not the effect I had discovered was known, as it might have much value to an industry of such colossal dimensions and antiquity as clay-working. Moreover, it would be amazing if it should not be known, in view of the tremendous amount of experimental work that had been done on that art. I searched for some record of the addition of organic matter to clay during its working, and only one instance could I find, that of the Egyptians in brick-making, as recorded in the fifth chapter of Exodus. The accepted theory of using straw fiber as a mechanical bind-

*The text of an address delivered by Dr. E. G. Acheson of Niagara Falls, inventor of Carborundum, Acheson-Graphite, etc., before London Society of Chemical Industry, November 7, 1911.

ing agent had never appealed to me. Straw, however, contains no tannin, and the effect I had found had always been produced with tannin, or a substance containing tannin. I procured some straw, boiled it with water, decanted the resultant reddish brown liquid, and mixed it with clay. The result was like that produced with tannin, and equal to the best I had obtained. It now seemed likely that the Egyptians were familiar with the effect I had discovered, and believing this was why they used straw in making brick and were successful in substituting stubble for the straw, I called clay so treated "Egyptianized Clay."

The effect of organic matter, as typified by tannin, in producing deflocculation and a resultant colloidal state of clay is very readily shown; for instance, I have here some powdered kaolin, a small quantity of which I will place in a test tube, add water, and after shaking, set aside. Another portion of the kaolin I will put into a beaker, and moisten with a water solution of tannin, to which a small amount of ammonia has been added. After a thorough mixture has been made, using a glass rod to stir with, to eliminate as much as possible any grinding action, I will add more water and divide the contents of the beaker between two test tubes. To one of the tubes containing tannin-treated clay I will add a little common table salt. The three tubes I will place here before you, that we may examine them later.

In the summer of 1906 I succeeded in making artificially a high grade graphite which I wished to make applicable to all kinds of lubrication. To meet the various demands, it would be necessary to have it remain diffused in liquids lighter than itself; for instance, water and petroleum oils. Recalling the effect of tannin on clay, which caused it to remain diffused

duce a small quantity of it into a test tube, add water, and after shaking, set aside. Another sample I will place in a beaker and moisten with a solution of organic matter, and after thoroughly stirring with a glass rod, I will add water and divide it between two test tubes, to one of which I will add table salt. These three tubes I will place beside those holding the clay, to be examined later.

The actual amount of deflocculating effect produced on the graphite in the beaker is very small indeed. In commercial work considerable mastication and time are required. I have here a bottle containing water having two to three tenths of one percent of its weight in deflocculated graphite, the deflocculation having been produced by a treatment similar to that I have applied to the graphite in the beaker, and a little of it being poured on a filter, you see the black liquid running into the test tube below the filter. The paper retained none of the graphite on its upper surface, all of it having passed into and through the paper. I will now add two or three drops of acid to the black liquid in the tube, and after warming over a spirit lamp, will throw it on another filter paper, and you now see a clear, colorless liquid descending into the tube below the filter. This is the water in which the graphite in a deflocculated condition was diffused; the graphite having been flocculated the acid is now retained on the upper surface of the filter paper. The effect I have produced with the acid could have been produced with a solution of salt, lime water, or any one of that large list of substances known as electrolytes, even so weak an acid as carbonic acid, if caused to bubble up through water carrying deflocculated graphite, will cause flocculation and sedimentation.

Upon being deflocculated, the graphite is diffused through the water in a colloidal

Si_4C_2O. The effect can be produced with a long list of organic bodies; for instance, tannin, or organic substances containing tannin, also with solutions containing the gum of the peach and cherry tree, or extracts from straw and grass. The drainage from the barnyard proved to be very efficient. I speak of these organic substances as agents when used to produce deflocculation, and they act as protective colloids to the deflocculated body.

Some minutes have now passed by, and we will examine the tubes containing the clay and the graphite. We find the clay that had been mixed with plain water has settled. The mixture of clay, water, organic matter and salt has also cleared, while the tube containing the clay, water and organic matter remains muddy. In like manner, the tube containing the disintegrated graphite in water has cleared; the second one containing water, graphite and organic matter remains black; while the third tube, which was set up the same as the second, but had a little salt added to it, has cleared. Apparently a great affinity existed between the organic and the inorganic substances introduced into the water. The inorganic body abstracted the organic from the water, and in doing so, was deflocculated. Each particle as it was thrown off was enveloped in an aqueous jelly of the organic agent, or at least such was the working hypothesis I followed to arrive at my results, and I find it difficult to think of this breaking up stopping short of the final subdivision with the resultant separation into individual molecules, or the smallest particles into which a body can be subdivided without loss of identity.

As I have already stated, I deflocculated clay in the year 1901 and graphite in 1906, and immediately afterwards a number of other bodies. I early understood I was producing colloidal conditions of these bodies, but not until the summer of the present year, 1911, did I read any treatise on colloids, being familiar with this state of matter only in a very general way. During the summer I procured a copy of the book, "Colloids and the Ultramicroscope," as written by Dr. Richard Zsigmondy and translated into English by Jerome Alexander of New York. I found the book extremely interesting,

my deflocculated graphite subjected to ultramicroscopic examination. Mr. Alexander kindly undertook the examination. He found the graphite in the deflocculated condition to be in a true colloidal state, the particles being in rapid motion, and he estimated their average size in linear dimensions to be 75 millimicrons. Seventy-five millimicrons are seventy-five millionths of a millimeter, and it would require slightly more than 13,000 of the particles to extend one millimeter. Now, the particles of disintegrated graphite used as the crude material from which to produce deflocculated graphite, pass, as I have stated, through a sieve having 40,000 meshes to the square inch and their maximum linear dimensions is such that it would require thirteen of them to extend one millimeter. Hence, the particle of disintegrated graphite is one thousand times greater in linear dimension than the deflocculated one. These are figures that certainly test our powers of appreciation.

I have been asked, "Why don't you speak of the graphite as colloidal?" Knowing now that it is in a colloidal state, I speak of it as being colloidal, but when I am speaking of my process I am talking of a method of producing deflocculation. When does that process of deflocculation stop? Is it short of the final subdivision and the throwing off of the molecule? I think not. I believe we are here dealing with molecules. Their size may not agree with what they should be as computed in accordance with accepted theories, but, nevertheless, I cannot conceive the subdivision once started, in the presence of sufficient deflocculating agent, will stop short of the final, with the freeing of the molecule and the creation of the colloidal state.

How did all this I have been telling you come to happen? The following quotation aptly tells how:

"It's generally the fellow who doesn't know any better who does the thing that can't be done. You see the blamed fool doesn't know it can't be done, so he goes ahead and does it."

I honor the man anywhere who in the conscientious discharge of what he believes to be his duty dares to stand alone.

AN ASTRONOMICAL TELESCOPE

The photograph illustrates an astronomical telescope of the reflecting type; with a speculum or reflector of 70 in. focus and 7 in. diameter.

Fig. 1 (not to scale) illustrates the principle of the reflecting telescope. A is the speculum, which receives the light from a star or other object. The image is transmitted to the eyepiece B by the small elliptical mirror C suspended in center of tube.

The telescope was designed with a view to simplicity of parts, ease of construction, and low cost. All the machining can be done on a small lathe (the writer used a Drummond 3½ in.). The instrument is mounted equatorially, which, briefly explained, is that the main a is inclined so as to be parallel to the axis of the earth; and by mounting in this way, it obviates the necessity of having to follow an object in the field of view in a vertical and horizontal direction, as with a simply-mounted telescope.

The first thing was to make rough drawings of each part, then a set of scale drawings with all measurements. Wooden patterns were made for all parts constructed from castings. The stand is made from No. 16 gauge planished steel plates,

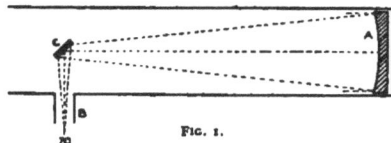

FIG. 1.

riveted with flush joints and a flange at foot, for screwing down to wood base. The top of stand is sheet iron, ⅛ in. thick. The tube, 6 ft. 3 in. x 9 in. inside diameter, is made from No. 20 gauge steel plates, in three pieces, riveted with flush joints. The declination a is a hollow steel shaft of seamless steel tubing, 1½ in. outside diameter. The polar a is the same material. The triangular frame, which carries the whole mount, is made from iron castings, bolted together. On the top end of the polar axis is a forging, on which is mounted the declination axis in cast iron bearings bushed with brass.

The cradle which carries the main tube is secured to the end of the declination axis. The tube is held in the cradle by steel straps, 1⁄16 in. thick, as shown in sketch. The eyepiece requires no special description. For the speculum, I obtained two circles of ordinary plate glass, 7½ in. in diameter, and 1½ in. thick, with top edges beveled. They are ground together entirely by hand with various grades of carborundum powder and finally with jewellers' rouge, the result being, one is ground concave and one convex. The concave one is the speculum, which is finally silvered on the surface. The process of grinding is extremely interesting and really not difficult.

The small elliptical mirror, shown in drawing, is mounted in a small brass fitting and suspended in the tube at 45 degrees. The theoretical diameter of the minor axis of small mirror is 1 in., but is made slightly larger in practice. The reason why it is elliptical is that a cone cut at 45 degrees gives an eclipse for a section. Hence the smallest shape that will transmit all the rays of light at a right angle would be an elliptical mirror of this section size. Viewed

(Continued to page 59)

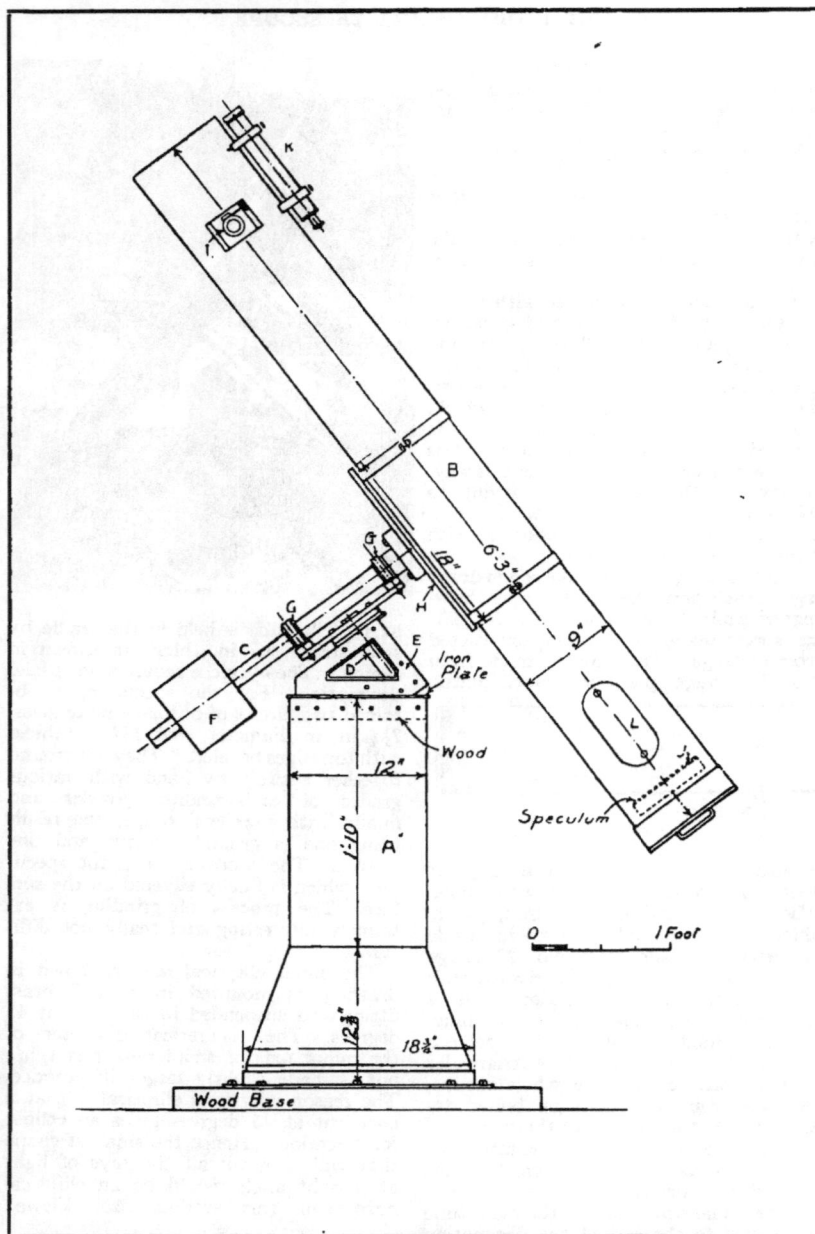

MAN-CARRYING KITES IN WIRELESS SERVICE

FRANK C. PERKINS

The young Boston kitor, Samuel F. Perkins, is seen, in the accompanying illustration, at an altitude of 350 ft., on his man-carrying kites at St. Louis. It is held that in future wars, or war maneuvers in times of peace, the man-carrying kite is likely to play an important part. The recent experiments were made at Boston of the man-carrying kites, whereby a one-wire aerial was carried aloft and proved a most successful medium for the receiving and sending of wireless messages. In the wireless experiment the aerial was carried aloft several hundred feet, and it was a revelation to know how easily and quickly an emergency wireless station could be established in time of war. Messages were sent to the office of the Los Angeles Examiner; and stations such as those at San Diego, Catalina and San Pedro wondered what station was interfering with their calls. Another thing that the man-flying kites have done and may do further, is to furnish a target for the practice of sharpshooters carried in aeroplanes. The kites can be made to bob about in a manner to keep the sharpshooter on the alert to make a hit, and they can have attached life-sized figures as further targets.

It is stated that the altitude record made in California was 385 ft., this being also a duration kite record, the designer of these man-carrying kites staying up in the air for 90 minutes. The young Boston kitor owes his life to the fact that several of the many kites by which he was suspended at Los Angeles, parachuted and prevented him from dashing to death on the earth. He was 200 ft. in the air when the accident occurred in which aviator Chas. Willard collided with and cut the cable with his bi-plane, thus severing all connections of the kites with the earth. Aviator Willard injured his front control, but was able to land immediately and safely. Although three kites were wrecked the kitor Perkins in dropping 200 ft. landed without serious injury, the remaining kite acting as parachute.

An Astronomical Telescope

(*Continued from page 57*)

through the eyepiece, the mirror appears circular.

There are many improvements which might be added, such as worm wheel and rack for rotating the instrument, or clockwork might be substituted for the same purpose. Then graduated circles could be mounted on the axes—the one on the declination axis divided into degrees; and the one on the polar axis divided into 24 hours. Then the position of a star being known in right ascension and declination, or in other words, latitude and longitude, it is only necessary to adjust the instrument till the pointers on the circles indicate these positions, when the object will appear in the field of view. Then again, a camera may be fixed to the eyepiece and many interesting photographs taken, especially if the telescope has clockwork motion to enable it to automatically follow the object being photographed, though this is not absolutely necessary.

An instrument of this size, though fairly portable, would be better mounted in a small wooden observatory. The amateur will find the construction and assembling considerably easier than the delicate work entailed in model work.—*Model Engineer and Electrician.*

THE LARGEST ELECTRIC SIGN IN THE WORLD

L. E. ZEH

The largest electric sign in the world with marvelous motion effects and most elaborate in construction is one of the night wonders to be seen along the "Great White Way," of Broadway, New York. This monster picture sign is 100 ft. long by 70 ft. wide, and shows in brilliant and flaming colors the arena of a Roman amphitheatre as bright as day, with a

The great race takes place for about half a minute, then it is in darkness for an equal time, this being repeated during the entire evening.

Directly over the race is suspended a great steel curtain that is 20 ft. high by nearly 100 ft. long, about 2,000 sq. ft. of surface. Mounted on the top of this curtain is the title of this great display:

The Largest Electric Sign in the World, Roman Chariot Race in Action

chariot race in headlong progress. In the center of the picture is the leading chariot, flying along to victory and showing brilliantly against the dark background of the arena wall, the top of which appears crowded with spectators. Behind on the arena path appear other chariots, striving to overtake the leader. The horses all appear to be running at break-neck speed, yet never overtaking the main chariot, which represents the "Leaders."

"Leaders of the World"—made entirely of electric bulbs, which remains illuminated continually. All during the evening there appear continuously on this curtain the announcements of the different business concerns; the one leading concern in various standard lines of business is shown as the "Recognized Leader of the World." Every one of the 40 announcements appearing on the display is repeated every 9 to 10 minutes during the entire evening. Over 95 miles of wire was used to complete this sign. Altogether there are 20,000 electric bulbs in the sign, which consume about 600 h.p. There are 70,000 separate electrical connections. About 2,750 electric switches are employed, and the lamps are flashed at the rate of 2,500 flashes per minute. A number of Broadway night pedestrians have acquired the habit of setting their watches by the big electric clock. The minute hand of the electric clock alone weighs 140 lbs., and the hour hand weighs only a few pounds less. The full size of the sign is 62 x 50 ft. Diameter of the dial is 20 ft. The numerals are each 4 ft. long. The minute hand is 9 ft. long, and the hour hand 7½ ft. long. The clock is connected with the hands by a shaft 40 ft. long. The weight on the drum of the clock is 125 lbs., and the counter balance on the minute hand weighs 62 lbs. At night the great clock, one of the largest in Manhattan, is lighted by 3,368 electric lamps.

The Electric "Scissor-Grinder"
F. J. KOCH

Vail, forever, to the old familiar tocsin of the scissor-grinder's bell and his harmonicals: "Scissors to grind; butcher-knives to sharpen!"—for an enterprising Frenchman has devised an automatic device, whereby electricity drives the cart, or what you would, on which the sharpening apparatus is mounted, gives the tender a ride, and works the stone as well. According to informant, as soon as the motor is set in motion, the grindstone revolves.

Clip from the newspapers interesting articles you want to read but have no time for at the present. Slip them into a large envelope, and when you make a journey put this in your bag. You can then enjoy them leisurely.

Tufts Wireless Club
Installation of New Outfit Increases Membership of Society

With plans completed for the installation of a portable wireless outfit, similar to those used by the United States Signal Corps, interest in the Tufts Wireless Club is on the increase. The organization has been offered the services of such an outfit by Captain Harry G. Chase of the Massachusetts Volunteer Militia Signal Corps, the outfit being used successfully in the recent maneuvers. Captain Chase is professor of physics at the college. The outfit will be placed at once in Robinson Hall, and in a few weeks the members are planning to set in active operation a station enjoying a sending radius of from 400 to 500 miles. Already nearly fifty applications for membership in the club have been received.

THE NEW MILITARY ANTOINETTE MONOPLANE

The Antoinette Company have recently produced a monoplane designed solely for military purposes. For this reason all vulnerable parts had to be eliminated. Accordingly, the new monoplane has no stay-wires, and the wings consequently have been greatly increased in thickness to take the additional strain. The wings are flat on their under-surface, and inclined at a considerable dihedral angle. The wheels are encased in sheet metal shields, designed to diminish head-resistance and to act as a "keel-plane" at the same time. This double device for maintaining lateral stability—dihedral angle combined with keel-plane—is by no means recent and has lately come in for strong condemnation on the score of its bad effect in disturbed air. There is a pronounced fore-and-aft dihedral angle. The body is entirely encased in metal, the forward portion (containing the motor and pilot) in steel plates. The pilot obtains his view through the loop-holes at the side and through the floor which is transparent. The following are the dimensions: span, 52 ft.; length over all, 38½ ft.; plane area, 602 sq. ft.; weight net (without fuel or passengers), 1,870 lbs.; net loading, 3 lbs.; 100 h.p. motor.—*Aeronautics*.

DOWEL JOINTS AND HOW TO MAKE THEM

The system of fixing joints together by means of dowels, or small round pieces of wood fitting tightly into each of the two pieces of wood to be joined, is a very good one, being strong, easy to do, and adaptable to almost any kind of joint in use. At the same time, easy as the method of doweling appears to be, and perfect as a good doweled joint is when finished, unless the several operations are set about in the right way, and carried out correctly, the result will probably prove a greater failure than any other form of joint would be under the same conditions.

In this article it is our purpose to show how to properly carry out the operations necessary to success, as well as to show how and why failures occur.

In Fig. 1 we show an upright standing on a flat piece, level at the back, a condition which often occurs to anyone making an article of furniture. To make this joint successfully with dowels, the best plan is to clamp the two pieces firmly together in their correct positions (as in Fig. 2) and bore a series of holes through the flat piece into the other, and drive in the dowels, gluing them, of course. When dry, this joint will be found very strong, and if the holes are bored so that the dowels spread both ways (as in Fig. 3) it will be stronger still.

The disadvantage of the above is, the ends of the dowels show in the finished work, and in many cases this will rule it out of order, as it would have to be in a case as Fig. 4, where the flat piece rests on the thin upright. To set out for this doweling the positions should be reversed, so that the upright stands on the flat, it can then be marked closely round, a middle line made, and the positions of the dowels marked on this line, as Fig. 5. The end of the upright must be marked in the same way as Fig. 6, and where the lines cross is the spot for the point of the boring bit, while Fig. 7 shows the section of the finished joint. In dealing with this thin wood in doweling, it is important that the holes should not be bored through, and equally important

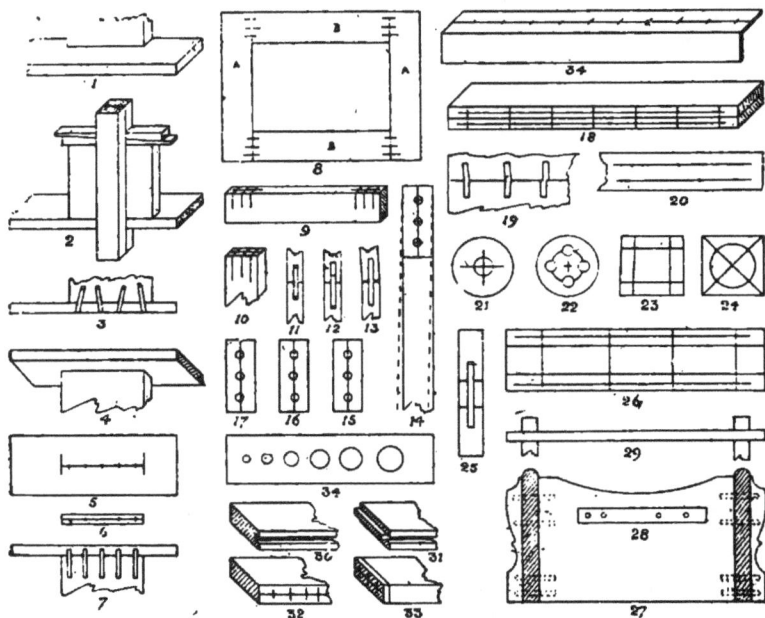

that they should be bored as deeply as the thickness of the wood will allow, and in order to comply with both of these conditions, it is best to place a block or gauge on the boring bit, so that it can go so deep and no deeper.

INSERTING THE DOWEL PINS

In putting joints such as these together, the dowels should be driven into the *thin* part first; if the attempt is made to work the other way difficulties will be met with.

In doweling a rectangular frame together, it is best to lay the four parts on the bench in the position they have to occupy when finished, as, in Fig. 8, and then mark the position of the dowels across the joints, as shown. These marks are then squared across the edges of each of the stiles, A, as in Fig. 9, and the gauge mark crossing these marks, gives the spots from which to start the holes. The rails, B, must be squared over at the end, as Fig. 10, and the gauge mark made as before.

In boring holes for dowels in such work as this, great care must be taken to get them quite straight and parallel with the sides of the wood, also to make them an equal depth, so that there is just clearance room for the dowels, as in Fig. 11. Should the holes be too shallow, the joints will not come up at all. If too deep the dowels are buried in one part, leaving very little length in the other, as in Fig. 12; and if not bored straight, the effect is as in Fig. 13, and although this to a certain extent can be put right by the application of force, it is far better if it comes right without it.

ACCURACY OF POSITION

Another very important matter in this connection is that the holes be bored exactly in the right places. In Fig. 14 the middle hole is correct, the top one slightly to the right, the bottom one slightly to the left, the effect being to make the stile run as dotted lines, instead of as the solid lines, straight with the rail. The three holes should line exactly, as Fig. 15; if irregular, as Fig. 16, the result will not be what is required, and if bored all too much one way, as Fig. 17, the stile may be straight with the rail, but the joint will not be flush, consequently it means planing off the wood at both sides, making it thinner than it

should be, and not only wasting time but timber as well.

In jointing two long pieces together they should be set out as in Fig. 18, laying them face to face on the bench, and squaring the marks across both edges, making the gauge marks along the middle. The holes must be made as true as possible both ways; or there may be a difficulty in making the joint come up close. Fig. 19 shows three very common faults in boring, all of which should be guarded against.

In jointing up thick wood, a double row of dowels may be used, as in Fig. 20, but the gauge lines should both be made from the face side of the wood, in case the two pieces are not exact in thickness.

In jointing timber endways dowels are indispensable, the general rule being to use one large one in the center, as Fig. 21. This, however, is not the best way, four small dowels, as in Fig. 22, being much more satisfactory. If the circle is struck from the center spot, as shown, and the four holes set out at equal distances on it, the two parts will come together correctly.

Should the timber to be joined be square in section, do not gauge round the four sides, as in Fig. 23, but work with a circle, as in Fig. 24. The latter method will come right in any case, the former only if the timber is at exact right angles every way, which is hardly to be expected.

In doweling a narrow piece between two wider ones, as in Fig. 25, it is best to let the dowels run completely through the narrow piece, as shown, and to set them out by laying the narrow piece edgeways between the two wide ones laying flat, as in Fig. 26, the marks can then be squared across the edges of the middle piece, and the dowel holes *must* be bored from each edge. In putting together, the dowels should be inserted in the narrow piece first, leaving them to project at each edge, the other pieces will then go on easily.

The same method may be adopted in doweling together ornamental shaped boxes, such as would in fretwork be put together with a half-cut-through joint, the dowels passing through the sides of the box into the ends, and the continuation pieces fixing onto the end of the dowels, as in Fig. 27. In this case the dowels should be glued into the ends of the box first, the sides driven on, then the ornamental pieces. Fig. 28 is the edge of one of these latter, with the holes bored, and Fig. 29 shows the plan as seen from above.

In doweling together panelled framing, where the ends of the stiles and rails are as Figs. 30 and 31, the groove in the one and the tongue on the other make it difficult to bore the holes truly. This may be obviated by boring the holes before the tongue or the groove is made, and Figs. 32 and 33 show them set out for this purpose.

Ready-made dowels should always be used, but these do not always prove true to size, therefore a dowel plate is necessary. This is simply a plate of steel about $\frac{1}{4}$ in. thick, with a series of holes in it, as Fig. 34, corresponding to the various sizes of bits used. The dowels, after they are cut to length, are driven through the proper hole; there will then be no fear of their proving too tight when putting the work together.

The arrangement shown in Fig. 35 is sometimes used to set out the dowel holes. It is made from brass or iron, and a small hole is drilled at each of the cross lines. The template is laid on the wood and the holes marked through the perforations.— *Hobbies.*

Inflating a French Military Balloon

QUESTIONS AND ANSWERS

1701. Electric Locomotive Engineer. R. H. H., Lancaster, Pa., asks where a person can learn to be a first-class electric locomotive engineer. Ans.—It is not so much or alone a matter of education, but of training and employment. You should of course have at least a high school education, and no defects in eye-sight, especially in the accuracy of detecting colors. You should have some facility with machine tools. It would be well if you obtained employment in the railway shops of either the Westinghouse or General Electric Company. It would do no harm when making application to state just what your ambition is. You should afterwards seek employment with a railway company, but this will not be difficult to obtain once you have the factory experience. We do not know what the pay is, but your aim should actually be not to remain in the engineer's grade, but to secure an executive position.

1702. Induction Motor. A. B. S., Chicago, Ill., has an induction motor, in which the rotor has a distributed winding with its three terminals connected to three slip rings. The name plate reads: "Westinghouse, Type F, 200 volts, 3-phase, 60 cycles, 30.5 amperes per terminal, 10 h.p. rotor circuit, 140 volts between rings, 32.6 amperes per ring." The correspondent wishes to make a three-arm rheostat, Y connected of proper resistances to fit the motor. Cast-iron grids will be the material. Ans.—The 140 volts between rings will be the case at starting only, when the exterior resistance is all "in." As the speed increases, you cut the resistance out, just as in the case of a direct-current motor. Finally it is all cut out, the voltage between ring approaches zero. It would be quite zero were it not for the resistance of the brushes and connections to the rheostat. This does not mean that 140 volts are used in driving the rotor current through its windings, for when the rotor is at full speed there is very little "slip," and the voltage induced in the coils is very much less. It will be a variable amount, however, dependent upon the load upon the motor. If the exterior resistance is to be regarded as non-inductive,—but with iron grids it will not be entirely ohmic,—the maximum resistance in each leg should be closely 140÷32.6, or 4.3 ohms. At the first step you should cut out nearly half of this, say let the remainder be 2.3 or 2.5 ohms. Have about 1.5 ohms left after the next step: then have .8 or .9, then .5 or .6; next .2 or .3, and finally zero.

1703. Magneto. H. B., Lyndenham, Ont., uses a low-tension magneto to light two 6-volt lamps, and to operate the induction coil on his boat. In trying to measure the output of the machine the voltmeter and ammeter are so unsteady as to prevent reading them. Lamps burn fairly bright, but rather unsteady. How can it be determined if any overloading is taking place? Ans.—We should say that if the lamps have a fairly long life, you are probably not running the machine at over voltage. The coil is supposed to endure 6 volts from a battery, and this ordinarily means a greater current than from a dynamo that gives the same voltage. The flickering may be due to the vibration of the engine that momentarily removes the driving power.

1704. Chemical Rectifier. E. S. B., San Francisco, Cal., sends a blue-print of a transformer and rectifier which he intends to build and asks the following questions regarding it: (1) Is the transformer correctly designed? (2) Is the rectifier correctly designed and the correct solution chosen? (3) What is a good book that it would be possible to get and in which it is explained how chemical rectifiers are made? Ans.—(1) With the one exception of the number of primary turns, we should say that your transformer is correctly designed. From the blue-print we assume that you propose to use the "Ferranti" type of construction, which is very popular among amateurs on account of its simplicity. The number of primary turns you specify would mean that the iron be worked at a rather high density, and we suggest that you increase the number of turns to at least 750 in the primary, with a proportionate increase in the number of secondary turns in order to bring secondary voltage to proper value. This would allow a working density of 30,000 lines for the core. (2) The rectifier will operate satisfactorily if made according to the specifications you give, although we have a preference for the four-cell type, as described in Mr. Thos. C. Stanleigh's article in our January, 1911, issue. The same article contains a description of a transformer similar to the one you propose to build. (3) With the exception of the article mentioned, we have no record of a work covering the subject.

1705. Quenched Spark Gap. W. J. H., Jr., Hartford, Conn., intends to make a quenched spark gap for use with a Clapp-Eastham ¼ k.w. transformer. He asks: (1) How many pair of 6-in. plates will be required? (2) Will ¼ in. brass be suitable, and if so, how deep should the grooves be cut? (3) What is the best method of clamping the gaps together? Ans.—(1) The best results will be obtained by varying the number from ten to fifteen. (2) The ¼ in. brass is perfectly suitable and the grooves should be made ⁵⁄₃₂ in. deep. (3) Take two oak sticks, 1 x 2 x 9 in., and, placing the gaps between them, bolt the sticks together.

1706. **Transformer.** E. F. W., Chicago, Ill., wishes to know if the *Electrician and Mechanic* has at any time in the past published an article covering the construction of a small transformer of about 125 to 150 watts capacity, transforming 110 volts a.c., 60 cycles, to lower voltages ranging from 26 to 5 volts. Ans.—In our January, 1911, issue you will find an article by Mr. Thos. C. Stanleigh, covering the construction of such a transformer. We shall be glad to supply you with a copy of this issue on receipt of fifteen cents.

1707. **Arc Lamp.** E. C. B., Madison, Ill., asks: (1) How much resistance is needed in a 110-volt a.c. circuit, to work a simple arc lamp of hand-feed style? (2) Would it be advisable to make such a resistance from No. 14 galvanized iron wire, and, if so, how much wire would be required? Ans.—(1) The resistance required will depend upon the amount of current you wish to use; in other words, upon the intensity of light you desire. As you do not state this, we shall assume that you wish to use a current of 30 amperes, which is about the smallest it is practicable to use on an alternating current circuit, and still obtain a fairly steady arc. Placing the resistance of the arc itself at a low figure, as an alternating arc of this type should be very short, you would require a resistance of approximately 3 ohms in your rheostat. (2) Galvanized iron wire of No. 14 gauge would heat very quickly under such a load, and unless it were doubled, we do not advise the use of it. As we do not know the particular grade of wire you refer to, we are at a loss to state its resistance, and in lieu of this we give the resistance of the three grades known as "E.B.B.," having a resistance of 29.08 ohms per mile at 68 degrees F.; "B.B." having a resistance of 57.44 ohms per mile, and "Steel," with a resistance of 67.88 ohms per mile. If you double the wire in order to get the greater carrying load capacity, you must obviously use twice the length in order to obtain the same resistance. For the purpose you mention, a choking coil or step-down transformer would prove far more economical in current consumption, and you could get practically twice the light for the same amount of current used.

1708. **Condensers.** W. W., Montello, Mass.,

1709. **Transparent Cement.** F. J. K., Rochester, N.Y., asks for a formula of a transparent cement to be used for wood and cloth. Ans.—There are several cements and glues on the market which would satisfy these requirements, but one of the easiest and simplest forms to use would be white shellac.

1710. **Batteries.** G. A., Burlington, Wis., writes that he intends to install an electric light on his bicycle, and which will be operated by dry batteries about 1½ x 1½ x 4 in. in size. He asks: (1) for a formula for making dry batteries. (2) Whether home-made cells would equal standard cells in length of life, etc. (3) An opinion on the use of a sal-ammoniac solution in a Gladstone battery to raise its voltage. Ans.—(1) The following formula is said to yield a serviceable filling for dry batteries: charcoal, 3 oz.; graphite, 1 oz.; manganese dioxide, 3 oz.; calcium hydrate, 1 oz.; arsenic acid, 1 oz.; glucose mixed with dextrine or starch, 1 oz. Intimately mix, and then work into a paste of proper consistency with a saturated solution of sodium and ammonium chlorides containing one-tenth of its volume of a mercury bi-chloride solution and an equal volume of hydrochloric acid. Add the fluid gradually, and well work up the mass. (2) Home-made batteries should equal standard cells in length of life, provided the home-made batteries were carefully made. However, as standard cells are sold at such reasonable prices it would probably be more satisfactory to buy them. (3) It would be inadvisable to use a sal-ammoniac solution in a Gladstone battery.

1711. **Transformer.** R. W. L., Detroit, Mich., asks: (1) How many layers of empire tape to be put over bare core? (2) How many pounds of primary wire would be fully sufficient to buy? (3) Transformer is to be operated on 110 volts d.c. with electrolytic interrupter. Would this alter specifications, and, if so, how many layers and yards of empire cloth would be required for insulation between primary and secondary? Ans.—(1) The wire core should be insulated with eight layers of empire cloth. (2) Between 3½ and 4 lbs. of No. 14 d.c.c. wire will be required. (3) For operation on 110 volt circuits with the electrolytic interrupter, we would suggest that you immerse the entire coil in a tank of transformer oil, as the insulation would not be sufficient to withstand high voltage otherwise. At least 20 layers of empire cloth should be used for insulation between primary and secondary. You can readily compute the number of yards required yourself.

1712. **Eight-inch Spark Coil.** H. M. G., Spencer, Mass., asks: Where can I get definite

with two layers of the larger wire for primary; an insulating tube of mica; secondary sections not larger in diameter than 5 in. nor thicker than ⅛ in.; number of sections dependent upon your skill as winder; insulating discs of four layers of empire cloth between each section and its neighbor.

1713. **Telegraph.** E. H., St. Paul, Minn., has a telegraph line connecting three stations on the American Morse, or closed circuit, plan. Stations *A* and *B* are 1,150 ft. apart, and each has three gravity cells. *C* is 200 ft. from *B*, and has two gravity cells. Grounds at Stations *A* and *C* consist each of plates 1¼ in. x 2 ft., in the soil near stone walls. The cells at each station work the home instruments properly, but the line as a whole is not operative. What is the reason? Ans.—Without further data we cannot more than guess at the difficulty, and that is, that you do not have enough battery power. It is certain that however high the line or ground resistance is, and how insensitive are the instruments, you can overcome these troubles with more batteries. A better way may be to reduce the ground resistance. Certainly this would be the first and cheapest step to take. We do not quite catch the construction of your ground plates, but presume it is merely thin copper. Thinness is no objection, but, unless it actually rests in water, it is of rather meager dimensions. Can you not connect to a water pipe? That makes the most effective "ground." You did not state the size of the line wire you used, nor how it was insulated. Further, you may have only "sounders," in circuit, and these wound with too few turns. A resistance in each of at least 20 ohms will be needed, and "learner's" sets, having only 4 ohms resistance, would not work on very small currents. Your line is too short to require relays, but sensitive sounders will suffice. Let us hear if you are able to remedy the defect.

1714. **Motor-boat.** F. W. F., Brooklyn, N.Y., asks if it is possible to control a 2 h.p. gasoline engine in a boat from the bow,—the starting and variation of speed being accomplished without going to the engine itself? Can it be done without recourse to use of compressed air? Ans.—Perhaps you have in mind more than is conveniently possible. Since such engines are ordinarily of the two-cycle type, you will have difficulty and complication in trying to start other than by hand. The reversal of direction of rotation, too, is a matter for manual attention. Mere variation of speed can of course be accomplished in any one of several ways, and by simple means. You can put a sprocket wheel on the carburetor valve, use a section of plumber's chain to fit into it, and connect it to bow of boat by means of a brass wire or cable. Reverse pull may be applied through a tension spring, or a return wire. Several turns around a wooden cylinder will give the wire sufficient grip, or a second sprocket chain may be employed.

1715. **½ H.P. Motor for Vacuum Cleaner.** J. D. C., Cleveland, Ohio, asks: How to wind the motor, as described in the September issue for 3-phase 60-cycle, a.c. Ans.—In a three-phase motor there are three sets of coils, one coil per phase per pole. Each set must take the full voltage of the line. Therefore the number of conductors in each must be equal to the total number on the corresponding single-phase machine. However, each set needs only to carry one-third the equivalent single-phase current, so the wire has one-third the area and can be gotten in the same space. The current per wire

$$= \frac{5 \text{ amp.}}{3} = 1⅔; \text{ No. 24 wire is suitable. Since}$$

there are 960 conductors and 8 coils per phase, there will be 60 turns per coil. Wind 24 coils and insert them as described. The direction of current is reversed between each coil of a phase. The six terminals can be combined in delta to 3, as described at various times in *Electrician and Mechanic.* The relation between the three terminals determines the direction of rotation. Reversing any pair reverses the motor.

CORRESPONDENCE

EDITORIAL

It is always the desire of the publishers of this magazine to supply to their readers such articles as are most desired, and will be of the greatest general interest. Since it is almost impossible to know what the readers really care for, unless they will make their wants known, we wish that any who may have ideas or suggestions to offer would send them in. Perhaps there may be a particular subject in which you are greatly interested, or some piece of machinery or apparatus which you would like to see described. If such is the case, write us about it, and provided the subject is one which would prove of general interest we will be glad to publish such an article at an early date. How would some of the following subjects appeal to you personally? A series of articles on Mechanical Drawing, starting with a description of the drawing instruments and some of the more elementary forms of work and continuing the subject through details, assemblies and isometric drawing. A department on Telephone Engineering, one on Wood-Working, one on Automobiles, and one on Boat Building. Articles on Marine Engines; Steam, Gas and Water Turbines and Oil-Consuming Engines. These are but a few of the departments and articles which we are considering publishing during the coming year, and if you would be particularly interested in any of them, be sure and write us a letter saying so.

We find that we can use to advantage good clean copies of July, 1908, issue. If any of our readers can send us this number in good condition we will be pleased to extend or enter a subscription for three coming months in payment for same.

Complaint has been made by certain readers that the articles on wireless telegraphy, published in recent numbers, have not included enough of an elementary character. A reference to the past files of the magazine will show that *Electrician and Mechanic* was the first popular magazine in the United States to satisfactorily treat of wireless telegraphy

from the amateur's standpoint, and that our constructional articles have included practically every piece of apparatus which the average amateur might find useful and could construct.

As many of these articles can no longer be supplied by us, we are willing, if a sufficient demand exists, to treat subjects of general interest in the shape of new and up-to-date articles. We particularly request, therefore, that every reader interested in wireless telegraphy will write us fully his views on the subject, telling us what we could publish that would interest him. *Don't put this off and let the other fellow do it. Do it yourself, and now!*

Many of our readers are sending us subscriptions for other publications and cannot always understand that deliveries are not made promptly by many of the larger publishing houses, owing to the fact that the mailing lists are closed several days in advance of publication, and the new names are often entered to begin with the following number. For instance, many of our orders now being transferred to other magazines it will be impossible to make deliveries on issues earlier than the January number and in some cases it may be well into the first of the year before the first number due in the year 1912 is sent out. We offer this explanation to protect ourselves, as the delay is not occasioned at this end of the transaction, but in the enormous volume of business that is turned at this time of the year, by some of the larger magazines, it is several weeks before the orders are reached in turn.

We would be glad to receive manuscripts which treat of subjects of general interest. Although it is not absolutely necessary that such articles be illustrated, yet articles which are accompanied by drawings, cuts or photographs are much preferred. Stamps must be enclosed if the return of unavailable manuscript is desired.

TRADE NOTES

"The Blitzen ¼ K.W. Transmitting Set," and "The Precision Potentiometer," are the titles of two circulars which have been recently issued by the Clapp-Eastham Company, of Cambridge, Mass. The transmitting set is quite inexpensive, costing $35.00; but the apparatus is of standard quality in material and workmanship, while being both efficient and of pleasing appearance. The set consists of the following pieces of apparatus: a Blitzen transformer, a plate-glass condenser, a helix of new design, a special zinc spark gap and a key. The entire set, with the exception of the key, is mounted in and on a very attractive mahogany cabinet, with binding posts provided for line and ground wires. The potentiometer has a resistance of 1,000 ohms, a sufficient resistance to prevent rapid depletion of batteries. The winding is of bare wire, spaced apart on a metal core from which it is insulated. The cost of the potentiometer is $4.00.

To drill a hole in a steel conduit box with an ordinary bit brace has been unsatisfactory to every electrician that has undertaken the job. It has always meant hard work, and unsatisfactory progress. No pressure, vise, or clamp is available, and all he has recourse to is to simply grind, grind, grind, till a hole of some sort was bored in the metal.

Most electricians still have a bit brace in use today, not knowing that a very simple appliance known as the Red Devil Chain Drill is being put on the market by the Smith & Hemenway Co., 150-152 Chambers St., New York City, the makers of the well and favorably known Red Devil electricians' tools, etc.

The Red Devil chain drill, an illustration of which is shown herewith, consists of an attachment for use with an ordinary bit brace and is practically a portable drill press. It is made of steel throughout and the operating principle is very simple. It is fitted with a Universal chuck which will accommodate a round or square shank drill, has an automatic ball-bearing feed and requires absolutely no pressure on the part of the operator to cut through the hardest metal, marble or slate.

The Red Devil chain drill has but recently been brought to our attention, and we find upon investigation that it is a very serviceable and practical appliance that the electrical worker will appreciate. It is small and compact, very light in weight, and can be easily carried in a tool bag.

The attention of our readers is called to the advertisement in this issue, and further information with regard to it can be had by addressing the makers.

We are in receipt of a pamphlet on the subject of the "Selection and Proportion of Aggregates for Concrete." The pamphlet is written in an attractive form and treats of a subject which is of great importance in connection with the manufacture of concrete. Copies of the pamphlet may be obtained by writing the Vulcanite Portland Cement Co., Fifth Avenue Building, New York City.

The Excello Arc Lamp Company of New York City have sent us a pamphlet on the subject of Flaming Arc Illumination. The story of how the "flaming arc" came to be developed, a presentation of the elementary principles of the "flaming arc" and a comparison of the efficiency of this type of arc with the "enclosed arc" are all included in the pamphlet. Readers who are interested can obtain one of the pamphlets from the Excello Arc Lamp Company.

Department of the Interior
Bureau of Mines announces New Publications

(List 6—October, 1911). Bulletin 13. "Résumé of producer-gas investigations, by R. H. Fernald and C. D. Smith. 1911.

Miners' Circular 5.—"Electrical accidents in mines; their prevention and treatment," by H. H. Clark. 1911.

Bulletin 24.—"Binders for coal briquets," by J. E. Mills. Reprint of United States Geological Survey, Bulletin 343.

Bulletin 26.—"Experimental work conducted in the chemical laboratory of the United States fuel-testing plant, St. Louis, Mo., January 1, 1905, to July 31, 1906." Reprint of United States Geological Survey Bulletin 323.

Bulletin 27.—"Tests of coal and briquets as fuel for house-heating boilers," by D. T. Randall. 45 pp., 3 pls. Reprint of United States Geological Survey Bulletin 366.

Bulletin 35. "The Utilization of fuel in locomotive practice," by W. F. M. Goss. 28 pp. Reprint of United States Geological Survey Bulletin 402. Copies will not be sent to persons who received Bulletin 343, 323, 366 and 402.

The Bureau of Mines has copies of these publications for free distribution, but cannot give more than one copy of the same bulletin to one person. Requests for all papers cannot be granted without satisfactory reason. In asking for publications, please order them by number and title. Applications should be addressed to the Director of the Bureau of Mines, Washington, D.C.

BOOK REVIEWS

The Copper Handbook: A Manual of the Copper Industry of the World. Vol. X, 1910–1911. Compiled and published by Horace J. Stevens, Houghton, Mich. Price, $5.00.

This massive volume of 1,902 closely-printed pages is one of the standard works of reference of the world, enjoying an authority in its field which is beyond question. The work is a monument to the industry and ability of its compiler, and is the result of many years' assiduous labor, the present edition containing about eight times as much matter as the first one. The first part of the book is devoted to chapters on the mining, refining and uses of copper, containing a complete summary of our present knowledge of the metal. Then follow a series of chapters on the copper fields of the world. The bulk of the volume is occupied by detailed descriptions of the copper mines of the world, in which no less than 8,130 mines and mining properties are described, the prospects and financial situation of each company being fairly discussed. The author does not hesitate to say harsh things of rascally promoters and some of the descriptions form very spicy reading. The book closes with exhaustive statistical tables, and is an indispensable reference book to all interested in its field. The publishers will send the book without advance payment on a week's approval to anyone ordering it.

Motion Study: A Method for Increasing the Efficiency of the Workman. By Frank B. Gilbreth. New York, D. Van Nostrand Co., 1911. Price, $2.00 net.

The author of this book is enthusiastic as to the value to the world of a well-directed study of the movements involved in any routine work, with a view to so adjust the work to be done and the means of doing it to each other as to produce the maximum amount of production with the minimum of fatigue. He analyzes all the variables of the case, drawing his illustrations and examples mainly from the work of the bricklayer, and draws the conclusion that the standardizing of the trades is a great problem for the government to undertake for the benefit of its people. An interesting and stimulating book on a vital topic.

How to Grow and Market Fruit: Practical Explanations and Directions for Making Fruit Trees Produce Profit. Published by Harrison's Nurseries, Berlin, Md. Price, 50 cents.

Out of the fullness of knowledge given by many years experience in propagating, growing and marketing fruit of all kinds, the publishers of this book have produced a most valuable manual for everyone who owns a home. It contains thoroughly practical descriptions of every phase of fruit-growing, simply written and fully illustrated. With the present prices of fruit, there is a greatly increased interest in this phase of agriculture, especially in the East, and the book is consequently most timely and useful.

The Boy's Book of Warships. By J. R. Howden. With over 100 illustrations from photographs.

New York, Frederick A. Stokes Co., 1911. Price, $2.00.

Equally as fascinating as its predecessors on steamships, railways, and locomotives, this book is one which every mechanically-minded boy should possess and enjoy, and the elders will find it as valuable as the younger generation. The author sketches briefly the development of the fighting ship from the earliest times to the introduction of the steamer and the iron-clad, and then pursues in detail the development of naval types down to the present time. Every detail of the construction and equipment of fighting ships is fully explained and adequately illustrated, and the book is well calculated to impress the mind of a landsman with the importance and value of naval defence preparations.

The Law of the Air. By Harold D. Hazeltine. New York, George H. Doran Company, 1911. Price, $1.50.

It is at the present time quite generally believed that a state or nation should have a certain amount of control over the air-space lying above its territory, but as to how complete this control shall be and in what manner exercised authorities differ. Mr. Hazeltine treats this subject under three heads: The Rights of States in the Air-Space; The Principles and Problems of National Law and The Principles and Problems of International Law. A careful review of the best existing opinions is presented, together with some original ones held by the author, and the volume is really a most interesting one.

The "Mechanical World" Pocket Diary and Year Book for 1912. Manchester, England, Emmott & Co., Ltd., 1912. Price, 25 cents.

A most convenient, compact and useful little handbook on Mechanical Engineering has just been issued by Emmott & Co., Ltd., of Manchester, England. This is the twenty-fifth issue of this handbook and many new and interesting features have been introduced. The entire book has been thoroughly revised, and as the tables, illustrations and descriptions are excellent this work should prove of value to all who are in any way connected with or interested in mechanical subjects. The book is well bound in cloth and the price is remarkably low.

The "Mechanical World" Electrical Pocket Book for 1912. Manchester, England, Emmott & Co., Ltd., 1912. Price, 25 cents.

This little handbook on Electrical Engineering is similar in style to the "Mechanical World" Pocket Diary, and it is fully illustrated, well printed and strongly bound. The matter devoted to lighting has been entirely re-written and a number of new sections added. Among the more interesting of the sections may be mentioned the following: One, treating the principal defects of dynamos and motors, with suggestions as to remedies; one, describing the construction, rating and testing of high-tension apparatus, and one describing electrical measuring instruments. Numerous tables and a diary for 1912 are included in the book.

NEW MAP OF NORTH AMERICA

Forty-two Colors Shown on Map of the Continent Issued by United States Geological Survey. Work of Great Value to Scientists and Schools. Sold by Government at Nominal Price

The most notable map publication of the year is the large geologic map of North America just issued by the United States Geological Survey. It represents an exceptional type of engraving and lithographic color work, and is printed in four sheets, which fitted together and mounted make a map 6 ft. 5 in. high by 5 ft. wide, the largest piece of work ever issued by the Survey. The scale is 1 to 5,000,000, or 80 miles to the inch, and the plan of projection is in harmony with the universal world map on a scale of 1 to 1,000,000, in that it shows the units of publication of the world map, each of which embraces four degrees of latitude and six degrees of longitude.

Each Sheet Printed Fourteen Times

The color scheme of the map is a striking one. In all there are 42 color distinctions, varying from a brilliant red to pale tints approaching white. These were produced by 14 separate printings from lithographic stones, requiring in many places two or three combinations of color to produce the desired effects. If the weight of paper and heavy stones lifted back and forth in the printing of this job were to be computed it would run into the hundreds of tons. The accuracy of the "register," or fitting together of the color blocks in small areas throughout the map, is remarkable. The work was done in the Survey's own engraving and printing plant, and it is believed that there are few if any other establishments in the United States capable of turning out such a production. The 42 color distinctions represent as many divisions of rock strata. Thus the rocks of seven divisions of the Paleozoic era are each represented by a color, besides three separate colors for undifferentiated rocks, and there are other colors for the divisions of the Mesozoic, the Tertiary, and the Quaternary.

The coloring of the map is both effective and pleasing. The scheme is systematic in that the colors range in prismatic order from yellow in the upper portion of the geologic column through greens, blues and purples to pinks and browns at the base. The colors for the igneous rocks, both plutonic and volcanic, are mostly bright red. Viewed as a wall map, the map of North America shows only the larger geologic units, as the smaller divisions are represented by different shades and tints of the same or closely allied colors which are indistinguishable at a moderate distance.

Valuable for Detailed Study

Viewed close at hand these minor distinctions can be read and the map can be used for detailed study limited only by the scale. When it is used as a wall map the regions illustrating different types of geology stand out boldly. The great Canadian shield of pre-Cambrian rocks is represented by a subdued color in a pattern simulating crystalline texture. Parallel bands of darker colors from New Brunswick to Alabama mark the trend of the Appalachians, while the broad area of blue and gray colors to the west represent the coal fields of the interior, and a fringe of yellow colors to the east and south represent the Coastal Plain sediments. A brilliant vermillion coloring over much of the western part of the continent from Alaska to Central America strikingly portrays the volcanic activity in this region during the Tertiary period, and the broad area of green and yellow in the Middle West marks the last stages of deposition of sediments in the interior sea which covered that part of North America in Cretaceous time and in the continental depressions in Tertiary time, including many of the great coal deposits of the public domain.

The map embodies all the available published data and unpublished manuscript maps in the offices of the Survey and corrections from geologists in all parts of the country, based on a former geologic map of North America, published by the Survey in 1906, in co-operation with the Canadian and Mexican geological surveys, for the International Geologic Congress which assembled in the city of Mexico in that year. As an example of the interest taken in the publication of the present map, it may be stated that important corrections to the map of 1906 were received by the Survey from a leading geologist of France.

Sold at Cost of Paper and Printing

Not only will the geologic map of North America be indispensable as a wall map in colleges and schools where geology is taught, but each student will desire a copy for the study of broader problems in a real and regional geology and will wish to carry a folded copy on railroad trips across the continent.

This map is now on sale by the United States Geological Survey at the nominal price of 75 cents each. It is safe to assume that any private map-publishing house would charge $5 to $10 a copy for such a map.

Listing of Wireless Stations by M.I.T. Wireless Society

At the regular monthly meeting of the Massachusetts Institute of Technology Wireless Society it was decided that the society shall make a chart on which will be recorded all of the amateur wireless stations in the vicinity of Boston. The chart will also show the location, "call letters" and power of each of the stations. The reason for doing this is that at the present time no complete and accurate list of amateur operators is available, and it is important that such a listing of amateur stations should be made. Any amateur wireless operator whose station is located within one hundred miles of Boston is urged to send his "call letters," location, and the power of his station to Mr. E. M. Mason, Secretary of Mass. Institute of Technology Wireless Society, Boston, Mass.

ELECTRICIAN & MECHANIC

| VOLUME XXIV | FEBRUARY, 1912 | NUMBER 2 |

A 100-WATT STEPDOWN TRANSFORMER

HOWARD S. MILLER

The accompanying drawings give all the data necessary for the construction of this transformer, but the following notes may prove useful to the novice.

To make the spools on which the coils are wound, cut out two pieces No. 3 from ⅛ in. fiber, soak them in hot water, and bend into square tubes over a piece of hard wood 1¼ in. square. Make the joints in the tubes come on the middle of a side, and not on an edge. Cut out the spool ends, No. 4, and place one on each end of the square tubes.

Put the fiber tube on a 1¼ in. square wooden arbor, place it between the lathe centers, and wind on 7 layers, 70 turns per layer, of No. 20 B. & S. d.c.c. copper

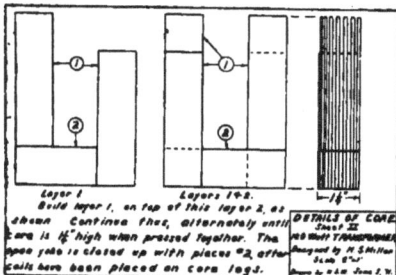

Layer 1
Build layer 1, on top of this layer 2, as shown. Continue thus, alternately until core is ⅛" high when pressed together. The spool jab is closed up with pieces ☰ after coils have been placed on core legs.

Layers 1+2.

DETAILS OF CORE
Sheet II
100 Watt Transformer
Designed by H S Miller
Scale 6"
Drawn by H.S.M. June 2, '11

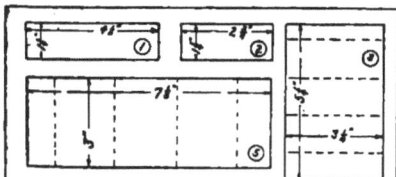

wire. Over this place the fiber tube No. 5, holding it in place with a few turns of string until the secondary winding is started.

On top of this insulating tube wind the secondary, putting on 2 layers, 25 turns per layer of No. 10 B. & S. d.c.c. wire. Bring the ends of these windings through holes drilled in the heads of the spools. These holes should all be on the side of the spool which is to be *outside* when the spools are in position on the core legs, to avoid interference with the clamping bars and the core. The two spools are both exactly alike, and when assembled, the primary coils, and the secondary coils on each are connected in series, so that the current flows in opposite directions on the two legs.

If various voltages are desired from the transformer, taps may be brought out from the secondary winding at the proper points. Every ten turns on the secondary give an e.m.f. of about 1 volt. An arrangement of taps which gives 20 different voltages from ½ to 10 in steps of ½ volt is: taps on the following turns—20, 40, 60, 80, 85, 90, 95. These taps may be brought to a row of binding posts, or the beginning of the secondary and the first four of the taps may be connected, in order, to the contacts of a five-point switch, and the last three taps and the end of the secondary connected, in order, to the contacts of a four-point switch. A wire is then connected to each switch arm, and these form the low tension terminals. Num-

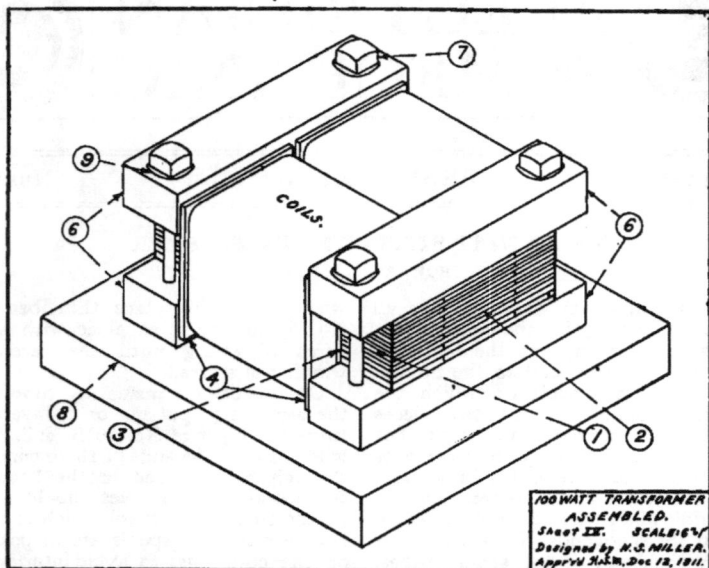

100 WATT TRANSFORMER
ASSEMBLED.
Sheet IX.   SCALE 1⁄2 f
Designed by N. S. MILLER.
Appr'd H.A.M. Dec 12, 1911.

used on the four-point switch minus the number of the one on the five-point switch equals the desired voltage. Fig. 5 shows this connection of the secondary taps.

The secondary winding is most conveniently tapped at the required points by means of strips of No. 24 B. & S. gauge copper ⅛ in. wide, and of any desired length. A hook bent on one end of this strip, as shown in Fig. 6, allows the tap to fit the wire closely. Thus when the open end of the hook is bridged over by soldering, an efficient joint results, which when filed up and taped with thin silk, increases the diameter of the wire only slightly, thus preserving the uniformity of the winding. In placing the taps, wind up to the required point, mark the location of the tap on the wire, and then unwind a few turns. This makes it easy to remove the insulation at the marked point, and solder the tap in place, with the assurance that it will come in exactly the right place when the wire is rewound.

If a constant secondary voltage other than 10 is desired, wind the primary as above; but for the secondary use a new size of wire found as follows: divide 100,000 by the desired voltage. The quotient is the size in circular mils of the wire which will safely carry 100 watts at the desired voltage, and the corresponding gauge number can be found from any wire table. Put on ten times as many turns of this wire as you wish volts, winding half the turns on each spool.

The core is built up as shown in sheet II, using two No. 1, and one No. 2 pieces in each layer. If desired, instead of using the three pieces for each layer, a piece might be cut which would have the shape of the three assembled as shown. A piece of such shape would be more difficult to cut from the sheet iron, but would have the advantage of being easier to assemble, and of allowing less magnetic leakage. If regular transformer iron is not available, these pieces may be cut from tin cans, for it has been found by experiment that the iron in cans, being fairly soft and only about .0125 in. thick, generally gives better results than the grade of sheet iron commonly obtainable. Before assembling, all these iron pieces should be given a coating of thin shellac varnish, and allowed to dry.

The core is most easily built up by

building around nails driven in a board, allowing the open yoke to project about 2 in. over one edge of the board. When the core is of the required height, another board is placed on top of the core, and the whole pressed together in a vice. The spools which have been wound may then be pushed on the projecting core legs as far as possible, and the vise released slightly to allow a little more of the board to be pulled out. The vise is then tightened and a piece of the board split off with a chisel, allowing the spool to be pushed further on the leg, and continuing thus until both spools touch the closed yoke. Then the open yoke is closed up with pieces No. 2.

The clamping bars may then be placed on the yokes, and the core drawn together with the bolts, which also pass through, and secure the whole to the base.

The primary and secondary terminals, and taps, if any are used, may be brought to binding posts, terminal blocks, switches, etc., and proper fuses inserted in primary and secondary circuits, according to the individual needs or desires of the builder.

If desired, the whole transformer may be immersed in transformer oil contained in a sheet iron box. Such a construction would allow a more ready dissipation of heat if transformer were to be used constantly, and would preserve the windings.

The following are the directions and drawings for the construction of a compact and durable relay, combining all the features of a commercial instrument:

The core is made of a piece of well-annealed Norway iron, 2⅛ x ⅜ in. At one end a No. 19 hole is drilled ½ in.

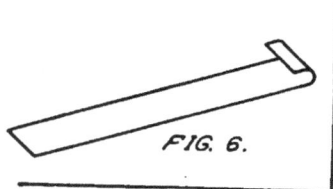
FIG. 6.

deep and tapped for a 10-32 thread. Over the core is forced two fiber "heads," 1 in. in diameter by ⅛ in. thick, placed ⁹⁄₃₂ in. from the ends. The entire space is then wound with No. 36 s.s.c. copper wire, so that the wound bobbin will have a resistance of about 500 ohms. The coil ends are then soldered to the leading out terminals at one of the "heads," as seen in the drawing. The yoke is made in an L-shape, of soft iron. The one end is drilled ¼ in. deep and tapped for a 4-36 threaded brass rod, the thread of which is ¾ in. long, for the check nuts, in order to regulate the movement of the armature. This latter piece is also made in an L-shape. The upper part has a hard rubber stud projecting ¹⁄₁₆ in. above the surface so as to insulate the armature, etc. The springs are made of brass or bronze, insulating each other by fiber washers and held fast to the yoke by two fillister screws. The contacts may be made of either platinum or German silver, though the former is superior. A piece of rod 1¹¹⁄₁₆ in. in diameter is threaded ⁹⁄₁₆ in. long with a standard 1⅝ in. pipe die for the instrument cover, which is made of brass tubing 3¾ in. long, one end having a disc of the same diameter soldered tight and the other tapped ¼ in. deep with a 1⅝ in. thread. These two parts may be made at any plumbing shop at a small cost. The attachment piece is last constructed, being made of soft iron in an L form, the upper part having three countersunk holes in order to fasten instrument where it is most suitable. The coil, yoke, etc., are held in place by a 10-32 threaded brass rod and nut to the attachment piece, but insulated by fiber washers, as seen in the drawing.

By following these directions closely the builder will have a strong and reliable instrument which may be placed on either base, behind switchboard or wherever space is available, as there are no parts to get out of order and is ab-

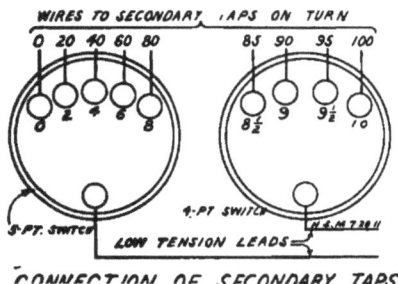

WIRES TO SECONDARY TAPS ON TURN
0 20 40 60 80   85 90 95 100
0   2   4   6   8   8½   9   9½   10
5-PT. SWITCH   4-PT SWITCH   E.&M.7.20.11
LOW TENSION LEADS

CONNECTION OF SECONDARY TAPS

MUTILATORS OF TELEGRAPHY AND TELEPHONY

FRANK M. EWING

There is nothing in this world more exasperating or nerve-racking than for a train dispatcher or an operator to struggle along, quite frequently under great difficulties, trying to receive from a mutilator of the Morse telegraph code or converse with a person with indistinct articulate speech over a telephone circuit. When the quality of Morse is good it is a great pleasure to work, and an experienced operator can answer questions or carry on a conversation while receiving without any discomfort. If the sending is mutilated then the mechanism of the ear thus combines the functions of both separator and transformer, while almost every nerve is strained in order to properly translate the mutilations.

There is really no excuse for operators remaining in ignorance of their defects, provided they are open to conviction. The bad senders endeavor to back up their claim of good Morse by mentioning one or two operators "who receive from them all day without breaking," not taking into consideration the fact that those patient and good-natured victims have evidently familiarized themselves with their combinations and put down what they mean instead of what they send. The various characteristics of bad sending noticed are as numerous as those existing between the different styles of chirography or conversation of different people. The peculiarity of their sending lies in the lengthening of the first or last dot of a letter, running the spaced dot letters together, dropping dots off some of the letters, running some letters of words together and spacing the others, making different combinations out of them and not allowing the proper interval of time for the letter and the spaces. Thus R sounds like "Ti," C like "It," figure 3 like "v," the word coat like "Is at," them like "Thw," Emporium like "Wposium," Harrisburg like "Hasspburg," President "P R & I don't." etc., and almost every word transmitted must be figured out by the receiving operator by making due allowance for the mutilation.

Accurate sending is more desirable than high speed. It is well to remember that operators are no judge of their own

Morse, and therefore should not try to see how fast they can send until they have had considerable experience. Who has ever heard a mutilator admit that his sending was poor? Now and then one may concede that his speed is below par, but as to the quality of his Morse, there can be no question, the fault is always with the receiver.

Some entertain the erroneous idea that firm transmission of the alphabet depends largely upon the pressure brought to bear on the key, and by pursuing that course do not allow the muscles of the fingers to fully relax between the formation of one dash or dot and another. The result is that a dot is lengthened into a short dash. The custom of timing for ascertaining the speed of sending should be very sparingly indulged in, for it is likely to produce careless habits.

The speed of sending should be graduated to suit the capacity of the receiver; the latter should never be crowded. Fast sending is seldom indulged in by strictly first-class operators, but fast time is made by them on account of their firm, steady, even gait.

Accept the average receiver's opinion regarding your sending before you decide for yourself that your sending is all right, for the poorest operators often think their sending is good. If the receiver tells you that you do not space properly, or calls your attention to some particular fault, do not get angry, but take the hint, and try to remedy your weak points. There should be no difficulty in correcting one's faults, as a mutilated Morse character can be detected instantly by anyone who will listen carefully to his own sending.

It was thought that the introduction of automatic transmitting machines would put a ban on the bad sender, hence his future toleration depended upon reform or a machine. Experience has shown it simply divides mutilators into two distinct classes, those using the Morse key and those using transmitting machines. Statistics show if a person is poor at handling one instrument, he rarely becomes an artist in handling another. Of machine sending there is this to say: Those employing an ordinary typewriter keyboard like the Yetman transmitter, will transmit perfectly-formed Morse characters provided the discs are clean and the electric contact is good; and the machines enable some senders who have lost their grip to do good work. All that is required is to simply touch the characters and the machine transmits them over the wire; but if the discs are allowed to accumulate dust, or become rough, the signals will be light owing to high resistance, or drop out altogether. The machines, operated by a side motion, of the Mecograph or Vibroplex type (with the exception of dot letters) require as many movements of the hand as the Morse key.

Sending machines are so trying to receivers and so unsafe when not properly adjusted that the advisability of prohibiting their use is being seriously considered. Automatic sending devices are very often so adjusted that dots are made at the rate of 80 or 100 words a minute, while the actual speed made by the operator is only 30 or 40 words a minute. Everyone, especially at repeating stations, notices that the signals from sending machines are thin and drop out when not properly adjusted and manipulated. Sending operators always know when the dots are needlessly fast and they can add to the comfort of the receivers and help to make good signals by giving careful attention to the adjustment of their machines.

Better results might be obtained from sending machines if the tinkering process were not applied to them. No machine on earth will hold up or withstand the onslaughts of a professional tinker who imagines he is thoroughly familiar with the mechanism of all kinds of machines. A great many machines of various types have been literally ruined in a short time in the hands of these artists. If a machine is properly adjusted and simply let alone it will last indefinitely, doing good work if proper care and judgment is exercised in handling it.

It is amusing to hear operators with one or two years' experience, comparatively young men or women in the point of years of service, with practically no sending worth speaking of during their assigned hours, say their arm is playing out and they must get a sending machine

out entirely. The reason they want a sending machine is because it is something new, they want to try it out and experiment with it, much to the discomfort and dissatisfaction of the receiving operator. Their sending resembles coal rattling down through a chute and the same with a poor sender on the Morse key. There are great many operators in the telegraph field who are troubled with a tired feeling, and they will surround themselves with all kinds of utensils and devices in order to see how near they can get to something to do their work without them making any effort to do it themselves. On the other hand we have men and women in the field who have done the heaviest kind of telegraphing for a period of 30 or 40 years, transmitting the prettiest Morse you would wish to listen to with the ordinary Morse key; never complaining about their arm giving out nor expressing any desire to try an automatic sending machine.

The "go as you please" sender, for whom no apology can be made, is the product of pure carelessness or indifference. Quite frequently we see him in a telegraph office in a lounging attitude with his feet elevated on the table higher than his head while sending, simply tapping the key and making no effort to send perfectly-formed Morse characters. When he is through sending he will sit upon the table with his feet upon his chair in order to accumulate as much dirt upon the chair as possible before using it to receive. It corresponds with his sending. When he is called to receive he will sit down upon the dirty chair, wrapping his legs around the top of his typewriter, which he imagines is a very graceful and easy position to receive in.

The "go as you please" sender is an operator who never sends two consecutive words or sentences at the same rate of speed or in the same style, and is never sure of a word until he hears the last letter completed and is then so surprised at his execution that he usually stumbles all over the word that follows.

The proper position for holding the key and the one adopted by the majority of the most speedy and perfect operators, is to rest the first and second fingers on the top and near the edge of the key button, with the thumb against the edge

the second fingers so as to form the quarter section of a circle. Avoid straightness or rigidity of these fingers and the thumb. Partly close the third and fourth fingers. Rest the elbow easily upon the table, allowing the wrist to be perfectly limber. When the proper "swing" is acquired, the forearm moves freely in conjunction with the wrist and fingers. The fingers and thumb should act as the end of a lever, the wrist and forearm doing the work. Let the grasp on the key be moderately firm, but not rigid. Grasping the button tightly will quickly tire the hand and destroy control of the key, causing what is termed telegraphers' cramp. Avoid too much force or too light a touch, and strive for a medium firm closing of the key in order to obtain uniform duration of the period of electric contact. It is not the heavy pressure of the key but the evenness of the stroke that constitutes good sending.

Telegraph repeaters can be adjusted for both light and heavy senders, but not for an uneven sender. A telegraph repeater adjusted for either a light or a heavy sender might be out of adjustment for a perfect sender. The motion should be directly up and down, avoiding all side pressure. Never, of course, allow the fingers or thumb to leave the key; that is, do not tap, pound, or strike the key with the fingers, or allow the elbow to leave the table. I have seen operators who were careless smash the key shut and knock the circuit breaker knob across the room. It is well to remember there are others working in the office, quite often at the same table, and they don't want to be disturbed with your efforts to demolish the key. The correct method of sending is an easy one, and when it is properly done, an operator should be able to send for twelve hours continuously without tiring.

Since the typewriter has come into general use for writing down the telegraph messages as the operator receives them from the sounder, making the receivers' work much easier, there is no danger of worrying an experienced operator by sending too fast. A good typewriter operator can write from 60 to 70 words a minute and more, but an expert telegraph operator cannot send steadily over 45 or 50 words a minute; consequently

to writing the message, to insert the "time received," the operators personal sign, etc., even when receiving at the fast rate mentioned. Every young operator should learn to operate the typewriter rapidly and accurately. It is generally conceded by expert operators that a double bank typewriter is best adapted for telegraph work on account of its simplicity and the ease by which the operator can manipulate the machine, no shifting for capital letters and other characters being necessary.

Mutilations and misunderstandings occur in telephony as well as telegraphy. Acoustics is that branch of physics which treats of the phenomena and laws of sound and sound waves. There are two distinct definitions of sound: First, sound is the sensation that is perceived when the nerves of hearing are properly excited; and, second, sound is a physical disturbance capable of producing on the auditory nerves the sensation of hearing. According to the first definition, therefore, sound is the sensation itself, while according to the second, it is the stimulus or cause of the sensation.

The physical disturbance capable of exciting the auditory nerves is a wave motion passing from some vibrating body through some material medium, which is usually air, though it may be any gas, solid or liquid. It is well established that all action between points or bodies separated by space is due to vibrations of the medium filling this space, no matter what that medium may be. In the phenomena of light, heat, or electricity, the medium is the ether; while in the case of sound, some more tangible medium such as a gas, liquid, or solid, is needed.

If these waves originate in or are communicated to the medium in which the ear is situated, then at each recurring condensation the elastic membrane, called the *tympanum*, or drum of the ear will be pressed inwards, and at each recurring rarefaction will be drawn outwards. These vibrations will be transmitted by means of a chain of bones, termed the *hammer*, *anvil* and *stirrup*, to the membranous wall that closes an internal cavity, called the *vestibule*, through it and some canal-like passages filled with a liquid and containing ramification of the auricular nerve, which the vibrations

ends in minute rods or fibers, each of which seems to vibrate at a definite frequency, and each one is excited only by a wave having the same period of vibration.

The greater the degree of condensation and rarefaction of the medium in a given time, the greater will be the motion of the drum of the ear that acts on the nerves. Hence, it follows that the function of the human ear is the mechanical transmission to the auricular nerve of each expansion and contraction that occurs in the surrounding medium, while the function of the nerve is to convey to the brain the sensations thus produced. From the above, one can understand why it is possible to make some persons who are deaf on account of an unnatural condition of some part of the ear mechanism hear by the use of apparatus that collects and transmits sound vibrations through the teeth and bones in the head to the auricular nerve. The nerve itself must, of course, in order to accomplish this, be in a natural state, free from disease.

All vibrations that set up waves in the manner already mentioned are not capable of producing the sensation of sound.

A uniform series of vibrations, a definite number of which are produced in a given time, and which are within limits capable of exciting the auricular nerves, is called a tone. Thus a simple musical tone results from a continuous, rapid, and uniformly recurring series of vibrations, provided that the number of complete vibrations per second falls within certain limits.

If, for example, the vibrations number less than 32 per second, a series of successive noises are heard, while, if their number is greater than 40,000 per second, the ear is not capable of appreciating the sound. Of course, different people have very different powers of hearing and different articulate speech. The number of vibrations of a musical tone is somewhere between 35,000 and 32 per second, and the number of vibrations produced by the human voice when talking is between 61 and 1,035 per second. In ordinary conversation, the average frequency is about 300 per second.

All sounds have three characteristics, variations in which enable us to distinguish between the different sounds we hear. They are termed loudness, pitch and timbre.

Loudness is that characteristic of sound which depends on the amplitude of the sound wave. It depends on the amount of energy in the vibrations producing the sound.

Pitch depends entirely on the number of vibrations per second, that is, on the frequency. A low rate of vibration produces what is called a low tone and a high rate a high or shrill tone.

Timbre is the quality of sound, and depends only on the form of the sound wave. A pure tone is one produced by a simple vibration. The quality of a tone may therefore be said to depend on the form of the resultant wave.

The successive vibrations set up by the vocal organs, forming distinguishable and intelligible sounds, are called articulate speech. These vibrations, which are the most complex in the whole realm of sound—so complex, in fact, as to defy mathematical analysis; but it is certain that their variations in loudness, pitch and timbre depend on the facts already outlined. By means of these variations, we are not only enabled to understand the words spoken by others, with all their various shades of intonation and corresponding shades of meaning, but we are able to distinguish between the voices of the many people with whom we are acquainted.

The letters T, V, B, P, and the words "weaver," "stever," "lever," are difficult to distinguish over a telephone on account of the similarity in the pronunciation. On some circuits they number each letter of the alphabet and give the number of the letter in order to insure accurate transmission. Quite frequently a doctor is called by telephone to see a patient and calls at the wrong house, going many blocks out of his way on account of indistinct articulate speech or similarity of the pronunciation of names. Persons with an impediment in their speech who are not a success in speaking over a telephone circuit, make good telegraph operators transmitting good Morse. When the 'phone receivers are off the hook any noise in the vicinity of the telephone passes through the telephone making it difficult to receive conversation. When telegraph keys are open no noise passes out over the circuit.

ENGINEERING LABORATORY PRACTICE—Part III
The Ericsson Hot-Air Engine
P. LE ROY FLANSBURG

In one respect at least all heat engines are similar, namely that they all receive heat from some source, transform a certain portion of this heat into work and then reject the remainder of the heat. Since we know that the efficiency of a heat engine is equal to heat transformed into work divided by the entire amount of heat supplied, it is at once apparent that maximum efficiency will be obtained when all of the heat is added at the highest practical temperature and the heat rejected is rejected at the lowest possible temperature.

$$e = \frac{Q - Q_1}{Q}$$

Where e equals efficiency.

Q equals heat applied (B.t.u.)

Q_1 equals heat rejected (B.t.u.)

The ideal type of cycle for carrying on such a heat change is known as the Carnot cycle, and if this cycle could be exactly obtained we would get maximum efficiency from the engine. However, it is not possible to exactly follow this cycle, and it therefore serves merely as an ideal type which is approached as nearly as practical conditions will allow. In obtaining the Carnot cycle, heat is supplied to and withdrawn from a constant mass of working substance, and the hot-air engine is the only engine which attempts to follow this example of the Carnot engine.

There are various makes of hot-air engines, such as the Stirling, the Ericsson and the Ryder Compression hot-air engine, but in this article I shall confine my discussion to but one type, the Ericsson. The attempts to develop hot-air engines on a large scale have been practically abandoned, but they are still used for small domestic pumping stations since they are free from danger and require but little attention. The chief difficulty with any hot-air engine is to transmit the heat to and from the working substance.

In the Ericsson hot-air engine but a single cylinder is used, and in this cylinder are two pistons, each connected to the fly-wheel by a linkage system. The duty of the lower piston (or transfer

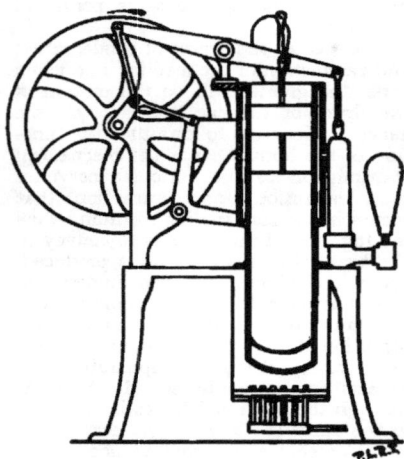

Fig. 1.

piston as it is called) being to transfer the air contained in the cylinder, alternately from one end of the cylinder to the other, and it is important that the air should be transferred at exactly the proper time. The upper piston is known as the main or air piston.

Heat is applied at the bottom of the cylinder, and the bottom of the cylinder is allowed to become as hot as is practical. The upper end of the cylinder is kept cool by means of a water jacket, and all of the water pumped is passed through this jacket.

The operation of the engine is as follows: The air is first compressed in the upper part of the cylinder, it is then transferred to the lower part of the cylinder, where it becomes heated and expands. As it expands it furnishes the power.

Fig. 1 shows a diagrammatic view of the engine.

The cycle upon which this engine works is called the Stirling cycle, and is represented in Fig. 2. The air is heated at constant volume, expands at constant temperature, is cooled at constant volume

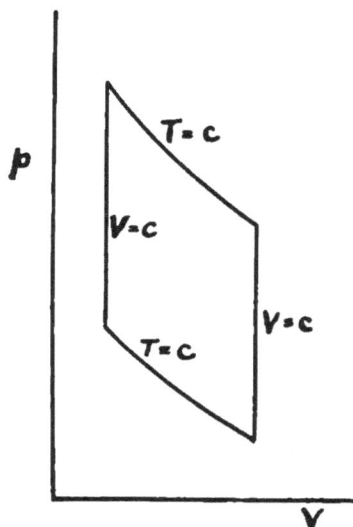

Fig. 2.

and is compressed at constant temperature.

The pump which the engine operates is direct connected by a lever arm. As the engine (like all hot-air engines) is but single acting, a large fly-wheel is used and the momentum of the fly-wheel continues the revolution until it is given an additional impulse by the engine. The engine makes but little noise and the same air is used over and over again.

The following results were obtained as a result of tests carried on by the author.

DESCRIPTION OF TEST

Three ten-minute tests were run on the Ericsson hot-air engine, to determine the horse-power developed, the cost per horse-power hour, and the thermal and mechanical efficiencies of the engine. The discharge pressures were varied from 3 to 13 lbs., and readings were taken every two minutes of the discharge pressure, the revolutions per minute, indicator and the gas-meter dials. Indicator cards were taken every four minutes during each run, and the total amount of water used per run was measured

The horse-power developed was calculated by means of the M.E.P., as given by the indicator cards, the dimensions of the engine and the speed. The cost per horse-power hour was found from the horse-power developed, the cubic feet of gas burned per minute and the cost of the gas per 1,000 cubic feet. The thermal efficiency is merely the ratio of the output in indicated work to the input which is due to the heating value of the gas. The mechanical efficiency is the ratio of the output in water work done to the indicated work done. (The water work done being equal to the product of the pounds of water pumped per minute times the total head acting on the water.)

DATA

Diameter of piston......................8 in.
Length of stroke......................$3\frac{1}{4}$ in.
Height of center of gauge above water
    level in tank.........................48 in.
Cost of gas (per 1,000 cu. ft.)...........$0.80
Heat of combustion of gas (per cu. ft.)..600 B.t.u.

OBSERVATIONS

	Test No.		
	I	II	III
Rev. per min..........	68	76	79
Dis. pres. (lb.)........	3.3	8.2	12.2
M.E.P. (lb.)............	2.69	2.97	3.71
Cu. ft. gas (per min.)..	.45	.45	.45
Lb. of water (per min.)	41.6	40.5	39.8

COMPUTATIONS

PLAN

$$\text{Indic. H.P.} = \frac{PLAN}{33,000}$$

Test No. 1.

$$\text{Indic. H.P.} = \frac{2.69 \times 3.88 \times 50.3 \times 68}{33,000 \times 12}$$
$$= 0.09$$

Test No. 2. (similarly)
Indic. H.P. = 0.11

Test No. 3. (similarly)
Indic. H.P. = 0.15

$$\text{Thermal Eff.} = \frac{\text{Indic. Work}}{\text{Heating value of gas}} \times \frac{1}{778}$$

Test No. 1.

$$\text{Thermal Eff.} = \frac{0.09 \times 33,000 \times 100}{600 \times .45 \times 778}$$
$$= 1.42\%$$

Test No. 2. (similarly)
Thermal Eff. = 1.73%

Test No. 3.   (similarly)
Thermal Eff.—2.35%

***

$$\text{Mech. Eff.} = \frac{\text{Water H.P. output}}{\text{Indicated H.P.}}$$

Test No. 1.

$$\text{Mech. Eff.} = \frac{41.6 \times 11.6 \times 100}{33,000 \times 0.09}$$
$$= 16.3\%$$

Since total head (3.3 x 2.3)+4—11.6 ft.

Test No. 2.   (similarly)
Mech. Eff.—25.6%

Test No. 3.   (similarly)
Mech. Eff.—25.8%

Cost per H.P. hour.

Test No. 1.
$$\frac{.45 \times 80 \times 60}{0.09 \times 100,000} = \$0.24$$

Test No. 2.   (similarly)
Cost=$0.20

Test No. 3.   (similarly)
Cost=$0.14

RESULTS

	Test No.		
	I	II	III
Indic. H.P......	0.09	0.11	0.15
Thermal Eff.....	1.42%	1.73%	2.35%
Mech. Eff........	16.3%	25.6%	25.8%
Cost per H.P. hour	$0.24	$0.20	$0.14

IGNITION AND IGNITION METHODS

ROGER B. WHITMAN

(Of the Bosch Magneto Company)

The state of perfection of the present-day internal combustion engine has not been reached without deep study and investigation, in the course of which it has been realized that ignition has vastly more to do with efficiency than was at first believed.

The early conception of ignition was the production of a spark some time toward the end of the compression stroke, and if this spark was successful in igniting the mixture, that was all that was desired. The character of the spark, the accuracy of its production, or the exactness of its timing were points that were disregarded by the designer, because he did not understand that these had any bearing on the power output or on the fuel consumption of the engine.

The modern designer takes an entirely different view of the subject, however, and it may be of interest to outline the problem as it is now understood.

To appreciate the fine points of this problem, the engine must be considered in its true light as a heat engine pure and simple.

The mixture that is drawn into the cylinder during the inlet stroke represents a certain heat value, and the efficiency of the engine depends upon the manner in which this heat is applied to the expansion of the gases. Any condition by which some of this heat is lost, or by which it is not applied directly to the forcing of the piston outward on the power stroke, will reduce the engine efficiency.

The first step in the securing of efficiency will be to study the points at which losses of heat may occur, and to adopt means by which these losses may be prevented.

The charge of mixture represents a certain heat value and has a certain maximum pressure. To exert the greatest possible proportion of this pressure against the piston, each particle of mixture should be made to give up its heat at the instant when the piston is at the end of the compression stroke and is ready to move outward on the power stroke.

To gain this result, it would be necessary to ignite each particle of the mixture at the same instant, and thus to have ignition and combustion occur at top dead center. The mixture would thus be compressed into a minimum space before ignition and the rise in pressure due to combustion would then be most abrupt, the piston being driven outward with maximum force.

No existing ignition system will permit the ignition of all of the particles of mixture at the same instant. The system in use, therefore, permits ignition of the mixture at one or two points from which the flame is expected to communicate itself to the remaining mixture particles.

In a perfect mixture, each particle of gasoline vapor will be surrounded by the particles of air necessary for combustion, and to ignite the mixture it will be necessary to raise the temperature of these particles to the point at which the chemical change known as combustion will occur.

Under usual motor conditions, the heat developed by the electric spark is depended upon to raise the temperature of certain of these particles to the point at which they will ignite, and the flame thus started is communicated to the particles of the mixture immediately surrounding it, thus being propagated throughout the entire charge.

To our senses, the spread of the flame from the point of ignition is instantaneous, but in comparison with the speed at which a gasoline engine operates, the time required is very considerable, and must be taken into consideration. Thus there enters into our calculations the period of time that must elapse between the instant at which ignition occurs and the instant at which the entire charge will be inflamed.

It is desired to apply to the piston the greatest pressure possible, and, obviously, the greatest possible pressure will be produced at the instant when combustion is complete. At this instant, therefore, the piston should be at the top of its stroke. We must not overlook the fact, however, that some pressure is produced at the instant when ignition occurs, and that this pressure will be constantly increasing as combustion spreads. If combustion is to be complete when the piston is at its top point, it is clear that ignition must occur while the piston is still moving upward on the compression stroke. For the last portion of its stroke the piston will therefore be subjected to this pressure, which is rising to maximum, and by which the piston will tend to be driven backwards; at the same time the momentum of the fly-wheel is urging the piston upward. Some of the power of the engine will thus be required to force the piston upward, and in this is found one of the most serious of the losses in engine efficiency. If the

momentum of the fly-wheel will force the piston against the pressure in the combustion space to top center, but the result of the conflicting pressures will be shown in abnormal wear of the wrist pin, crank pin and main bearings.

All motorists have had experience with a back-fire when cranking an engine, and known that it is the production of maximum pressure in the combustion space before the piston reaches the top of its stroke, the result being that the engine starts to run backward. This same condition in a lesser degree exists in a running engine under the normal condition of ignition occurring before top center.

The charge of mixture represents a certain heat value, and can be made to exert a certain definite pressure upon the piston. To get the best possible results, all of this pressure should be exerted against the piston when the latter is at the top of its stroke. If some of the pressure is exerted before the top center is reached, less pressure will remain to act on the piston during the power stroke. This entails a double loss, for not only is the rotation of the crank-shaft somewhat retarded, but the maximum pressure developed at top center is reduced. The effect is shown in an increase in the consumption of fuel and in a reduction of the power output.

Another loss that results from ignition earlier in the stroke is due to the absorption of heat by the cylinder walls; these surfaces being of metal are natural conductors of heat, and, of course, the longer the period during which the flame is in contact with these surfaces, the greater will be the heat absorbed and wasted in this manner.

The obvious way to reduce loss of power from these causes is to produce ignition as late in the stroke as possible, but this is limited by the necessity for having combustion complete at top center.

The remedy will therefore be to hasten ignition as much as possible, or, in other words, to reduce the time necessary for the propagation of the flame throughout the mixture.

One of the most important factors in this is the location of the spark plug, which should be so placed that the dis-spread are as short as possible. If, for instance, the plug is located in a valve pocket on one side of the cylinder, the distance through which the flame must spread will be practically maximum, and the operation will require more time than would be necessary if the spark plug were located in the cylinder head. Furthermore, the plug should be so located that its points are actually plunged in the mixture, and not set in a cavity or pocket. Engines are occasionally seen with valve caps that are solid and possibly an inch thick. If a standard plug is screwed into such a cap, the spark points will be found to set some distance up from the internal face of the cap; the spark will ignite the mixture which is in the hole or pocket, and some little time will be required for the flame to spread down through the hole and to be communicated to the mixture.

Such a construction will require considerably more advance of the spark than would be necessary if the spark points were in direct contact with the charge.

The size of the ignition is also a factor that determines the time required for combustion. The ideal ignition spark should be a mass of flame with as large a surface as possible, for this will result in the ignition of a large number of mixture particles. It should be understood that the spark must come into actual contact with the mixture particles in order to ignite them, and if the spark is thin, it will be quite possible for it to pass through a throttled mixture without actually coming into contact with any of the particles. With a spark that is in the nature of a flame this cannot take place. A large spark not only insures ignition, but makes combustion more rapid, for combustion will certainly be more rapid if, for instance, 100 mixture particles are ignited by the spark instead of but one.

Following along this line brings us to the proposition that it might be better to ignite the mixture at two widely separated points instead of but one, on the theory that this will reduce the time necessary for the propagation of the flame.

If, in a T-head cylinder where the valves are arranged opposite to each valve cap and a second one in the exhaust valve cap and sparks are caused to occur at these plugs at the same instant, the time required for the spread of the flame throughout the whole charge will be much less than would be necessary were the flame to originate at one side and be required to spread across the entire width of the combustion space.

This has been theoretically admitted for a long time, but the difficulty in its practical application lay in the securing of apparatus that would permit the production of two sparks at absolutely the same instant.

Ignition apparatus of this character has now been perfected, however, with results that are satisfactory from every point of view. It may be said at the beginning that it is essential to locate the spark plugs properly. If the two are set side by side in the inlet valve cap, for instance, there will be no gain through the use of two-spark ignition over one-spark. To secure proper results from this system, it is necessary to separate the plugs and to locate them so that the flame will have an approximately equal distance to spread in all directions from each.

A series of comparative tests was recently made at the Automobile Club of America before the Society of Automobile Engineers on an engine arranged for operation either with one-spark or with two. These tests showed that the maximum power output possible with single spark ignition was equaled by two-spark ignition at considerably less than one-half the advance, while with two-spark ignition it was possible to increase the maximum power output by 16 percent.

At first sight it seems somewhat extraordinary to claim that the power output of an engine will be increased 16 or more percent by producing ignition at two points in the cylinder instead of at but one, but the line of reasoning that we have followed makes it clear that the gain is due to the preventing of losses that follow early ignition.

The two-spark ignition system has been used on racing cars for over a year, and every car entered in the recent Gold Cup and Vanderbilt Cup Races was so equipped. By actual tests these cars

six miles per hour more than was possible for them to obtain with single-spark ignition.

The tests at the Automobile Club above referred to showed that the maximum output of 24 h.p. obtainable with single-spark ignition was reached at a speed of 1,750 revolutions per minute, while with two-spark ignition 24 h.p. was produced at a trifle less than 1,500 revolutions per minute. In other words, two-spark ignition delivered equal power at 250 less revolutions per minute; or six gallons of gasoline and two-spark ignition will do as much as seven gallons of gasoline and single-spark ignition.

It was further shown that 1,750 revolutions per minute was the maximum speed possible to obtain with single-spark ignition, while with two-spark, the maximum speed was nearly 2,000 revolutions per minute. Two-spark ignition is thus seen to give greater economy in consumption and greater flexibility than is possible with single-spark ignition, however favorably the single-spark plug may be placed.

Not the least advantage of this system is its great reliability, for one plug may become fouled without interfering in the slightest with the operation of the other. It has further been many times demonstrated that oil has far less effect on this system than it has on a single-plug system, and that over-oiling that would put a single-spark magneto completely out of business will not interfere in the slightest with the perfection of the operation of two-spark ignition.

Realizing the necessity for causing ignition to occur as late as possible in the stroke, it follows that the ignition apparatus should produce the spark at exactly this point and at no other.

If the apparatus selected does not produce this result, and if it permits the spark to occur a little earlier on one stroke and a little later on another, the result will be an unsteadiness in the operation of the engine, a reduction in power output, and an increase in gasoline consumption.

Anyone who has had experience with an automobile knows that the engine will run more steadily and more powerfully on a high-tension magneto than it will on a battery-and-coil system, but the reason for this is not always under-

the magneto produces a spark absolutely accurately and without variation, while with the coil-and-battery system the point at which the spark will occur will vary considerably.

The battery timer may make contact at the proper instant, but this does not mean that the spark is produced accordingly.

Upon the closing of the battery circuit by the timer, the battery current is permitted to flow through the primary winding of the coil, with the result that the core becomes magnetized. The effect of this is to draw the vibrator blade away from its contact, and thus to break the battery circuit, the consequent collapse of the magnetic field causing the induction of a high-tension current in the secondary winding. It is this current that furnishes the spark.

It is seen that the electric current is required to do certain work between the closing of the circuit by the timer and the production of the spark at the plug, and the lack of accuracy in the system lies in the fact that the current does not always consume the same time in performing these functions. This can be demonstrated on the apparatus that consists of a shaft that may be driven at variable speed by an electric motor. This shaft carries a pointer that travels around the inner side of a graduated ring. One end of the shaft carries a battery timer, while the other end drives a high-tension magneto, the magneto armature and the timer revolving at the same speed.

The circuit is so arranged that the spark produced by the magneto or by the coil may be caused to pass between the moving pointer and the graduated ring.

Turning the apparatus slowly by hand with the magneto thrown into the circuit will show that the spark is produced at the zero point of the graduation. By throwing in the electric motor, the speed may be increased to anything up to about 1,500 revolutions per minute, and it will be seen that the magneto spark invariably occurs at the same point.

In other words, the point in the rotation of the shaft at which the magneto spark occurs is not affected by the speed.

As the speed increases the igniting ability of the spark evidently increases, at 1,500 revolutions per minute it endures for about 30 degrees of rotation.

Throwing the magneto out of circuit and cutting in the battery, the apparatus may again be turned slowly by hand. The first battery spark will be seen to appear at the zero point, and at low speed there is an apparent sheet of flame for the entire 40 degrees during which the timer is making contact.

Running the speed up slightly it will be seen that this sheet of flame is broken up into a series of single sparks which occur very close together. Throwing in the electric motor, it will be seen that at 500 revolutions per minute the distance between the successive sparks is increased very considerably.

Each of these sparks corresponds to a single movement of the vibrator, during which the battery circuit through the primary winding of the coil is broken.

Another interesting thing is that the first spark no longer occurs at the zero point, but some 20 or 25 degrees afterwards, and this lag will immediately be recognized as representing the time required for the electric current to perform its various functions between the instant when the timer closes the circuit and the instant when the spark appears.

The delay in the production of the spark may be corrected by moving the timer so that contact is made some little time before the spark is actually required. The lag due to the work that the electric current must perform is thus overcome mechanically by moving the timer.

If the spark is observed, it will be seen that it does not always occur at the same point, but varies considerably, the total variation being 8 or 10 degrees.

At the instant when the timer closes the circuit the vibrator contact may also be closed, but, on the other hand, the vibrator contact may be open, the blade not having come to rest from the move-

or, in other words, the voltage in the battery must be raised.

If the voltage of the battery could be changed to correspond with every change in the speed of the engine, better results might be obtained, but a vibrator blade would still be needed that would be in actual and good contact every time that the timer closes the circuit. Furthermore, it would be necessary to insure the actual closing of the circuit at the timer, for when the timer contacts are covered with grease or dirt, the circuit may not be actually closed until the moving part of the timer is half-way across the timer contacts.

The timer that was used with the testing apparatus was operating under perfect conditions and the contacts were clean and uncorroded. This is not often the case with the timers that are used on automobiles, and consequently the results of the use of such apparatus on an automobile are far worse than is here indicated.

The device of doubling a wire on itself before winding it into a resistance coil reduces the inductance of the coil to a very small quantity, but unfortunately introduces a considerable capacity which is equally undesirable if the coil is to be used in alternating-current measurements. Chaperon's method of winding the coil in sections, in each of which successive layers are wound in opposite directions and the magnetic area of each layer made the same, reduces the capacity considerably, but the more recent suggestion to balance residual inductance and capacity has been taken up by Dr. E. Orlich, of the Reichsanstalt, with marked success. He winds one layer of wire on a slate slab five by twelve centimeters and three or four millimeters thick with rounded edges, then places bridges over the edges and winds the second

HOW TO MAKE A HORIZONTAL SUN-DIAL

CHAS. HEATH

As a model engineer for some years, it struck me that the application of such small abilities as I had might be directed into some original channel, and that some articles might be constructed which afterwards would be more interesting to others than the class of work I had hitherto done. The most simple which crossed my mind was a sun-dial, which, even if inaccurate, always gives a quaint appearance to a garden; and, on the other hand, if constructed properly and with due regard to the equation of time caused by the variation in the sun's velocity, it can be made a really useful article.

This old relic of an age when clocks were scarce is by no means difficult to construct, and can be made on a painted wood base to represent stone, or on a square stone, slate, or marble, all of which can be scratched sufficiently deep to make a permanent readable dial. Brass, of course, screwed on a wood base, and the lines filled up with black wax, afterwards ground off level, is equally suitable, and will probably be chosen by workers accustomed to the use of this material.

The style, or gnomon, will be of brass, and can be cast from a pattern, or built up of ½ in. square rod. The shadow of either edge of the top surface of the gnomon is cast by the sun on either side of the dial—the left side for the morning hours and the right side for the evening ones.

The angle from the horizontal of the gnomon should always be that of the latitude of the place—in London 51½ degrees; but the dial will be all right if the angle is cut equal to the latitude, wherever it is.

The usual difficulty is in marking, as, owing to the dial which receives the shadows being oblique to the plane in which the sun is supposed to move, equal angles of the sun's apparent motion become unequal upon the dial. A mathematical man can calculate these easily enough; but as all modelmakers are not mathematicians, I propose to show a method by which a dial can be scaled without any calculation whatever and quite mechanically, which is perfectly correct, and with slight adjustment can be used to indicate the divisions on a dial of any inclination or obliquity.

It is necessary first to make the gnomon (Fig. 2). This can be cast to any design or ornament which the art or plagiarism of the maker can suggest, providing the top face is straight, the bottom face is straight, and that the angle formed where their planes meet is that of the latitude. We will suppose we are in London, and it will be 51½ degrees. This can be got out by a protractor, or

(Continued on page 88)

An iceberg detector is the invention of Professor Barnes, and it takes the form of a particularly delicate electric thermometer, which records changes in temperature up to so minute a point as one-thousandth of a degree. This instrument can be carried attached to a ship's hull, but under water, and would record the temperature of the water on a dial, which may be placed in any convenient spot in the ship—the bridge, the chart-room or the captain's cabin. By watching the temperature of the water, as recorded on this dial, the ship's navigators would be able to tell when an iceberg is being approached, and also to compute with considerable accuracy its distance from the ship. So gigantic is an iceberg that it will cool the water around it for a distance of several miles, and the iceberg detector, in favorable circumstances, will begin to give its warning at a distance of ten miles.

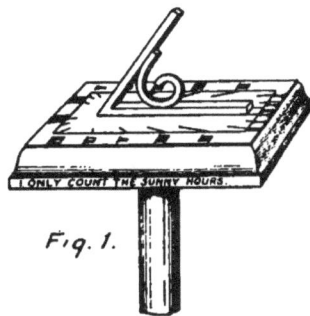

I ONLY COUNT THE SUNNY HOURS.

Fig. 1.

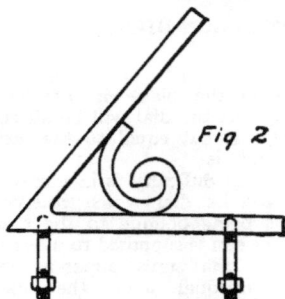

Fig 2

formed by making a triangle, whose sides are 12, 9½, and 7½, respectively. This is very near the correct tangent of 51⅛ degrees, being 1.2571. For a 12 in. dial a top edge of 7½ in. will be enough, as the point will be fixed back from the edge, in order to allow room for the shadows of morning and evening hours, before and after six o'clock. If built up from ½ in. square brass, it can be halved by filing, riveted, and soldered together (Fig. 3), making sure that the angle formed is 51⅛ degrees. Drill the bottom for two or three ¼ in. studs, long

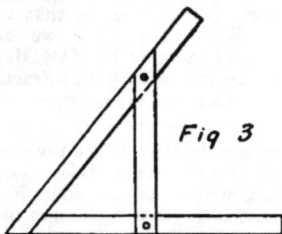

Fig 3

enough to go through the base and bolt up dead square, missing the center, which will be required to carry a socket capable of being revolved to adjust to the correct orientation last. If a casting, when filed up true and square to the angle, the same drilling, tapping and studding, will be required.

For the base, we will suppose wood, which must be either framed up 12 in. square, and a piece of good clean stuff screwed on all round, or a stout piece, 1¼, 12 in. square can be procured, and be protected from warping by fillets screwed up under it, crossway of the grain. Any edge can be worked on it, or left dead square. It must be centered on the underside, and a ½ in. back plate

to take ½ in. gas barrel screwed on. When in place, the barrel being centered in plaster or cement in any pedestal, will enable the socket to be screwed up with sufficient stiffness to remain, yet allow of slight adjustment, which need only once be made.

A square of brass ⅛ sheet to fit the base must now be procured, flattened, but not yet polished, and fastened to the

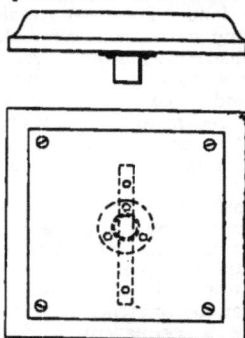

Fig 4

wood by flush countersunk screws of brass, and the holes drilled on a center line to take the studs right through the wood (Fig. 4). The gnomon being temporarily bolted in place, and its base filed till it stands square on each side, the whole affair must be leaned against a well, with the point of the gnomon resting on the table in such manner that the top face of gnomon, tried with a square, is perfectly vertical to the face of the table (Fig. 5).

Have ready a half circle of stout brown paper (as large as the table will carry) which has been divided with chalk lines into twelve equal divisions, and these

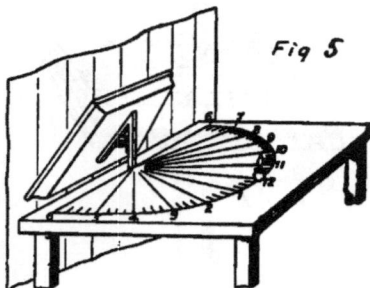

Fig 5

into quarters (Fig. 5). Adjust
this on the table in a shady
time of day or place, putting a
piece of candle, lit, on the
center line of the paper at its
curved edge, shifting the paper
round till the shadow of top
part of gnomon falls exactly
centrally on the dial. This is
the 12 o'clock line, and will
be ½ in. wide, and a slight

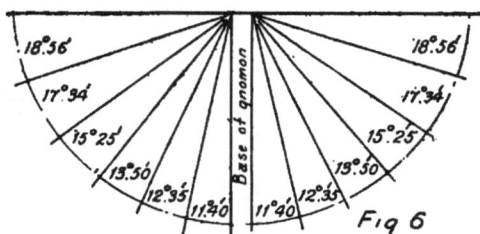

Fig 6

scratch with a needle on each side of the
shadow will mark it. Then fix the paper
by drawing pins, and shifting round the
candle to the end of next radial line,
again scratch the place of shadow without
moving the dial, repeating this operation
for each of the radial lines. When all
are marked, you will have the dial as
Fig. 6.

Unscrew the gnomon, and with a ruler
as guide and the point where the gnomon
meets 12 o'clock line scribe the lines
shown in Fig. 7. Roman figures are
best for a job of this kind, as they can
be cut in with a diamond-shaped tool
held against a steel rule. For the earlier
and later hours continue 5 night for 5
morning, 4 night for 4 morning, and the
opposite side for 7 and 8 in the evening,
which, in an open situation, will be visible
in the neighborhood of longest day.
When all is cut as shown, and the inter-

mediate lines cut, if required, by the
same method, lines can be well filled
with hot wax, the brass plate screwed
onto its foundation, and the whole ground
off flat and polished. The gnomon can
be attached and the dial screwed onto a
vertical ½ in. threaded gas barrel and
adjusted to show 12 o'clock on either
April 15th, June 14th, August 31st,
December 25th, in any year, and with
plus or minus the equation of time as
shown in any almanac for any day in the
year the correct time can be got as well
as a shadow can be read.

If it be decided to make of framed
wood, the lines can be marked in pencil
and ruled with black paint on the white
painted ground, such an inscription on
the edge written, as, "Time flies," "I only
count the sunny hours," or "Tempus
fugit," complete.

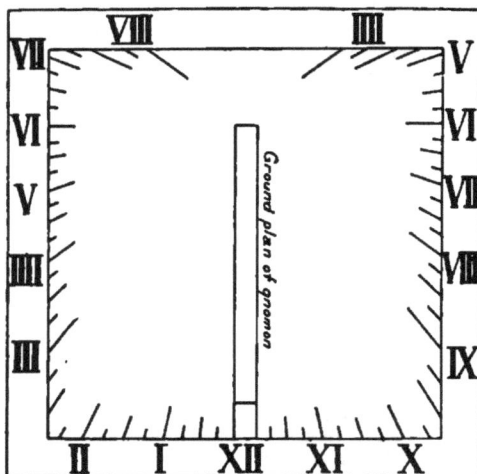

Fig 7

GARAGE OF THE EDISON ELECTRIC ILLUMINATING COMPANY
OF BOSTON

The Edison Company having entered upon an educational campaign for the purpose of stimulating the sale and use of electric vehicles in its territory, realized the advantages to be obtained by the existence of an up-to-date, thoroughly equipped and skilfully managed strictly electric garage in the heart of the business section of Boston.

The necessity of such a station was more forcibly brought to the company's attention by the complaint of certain electric vehicle owners that their trucks were not giving the high grade of service experienced in other cities. They readily argued that with proper care and scientific charging the results obtained would be far more gratifying, and the claim was made that during the early stages, at least, of electric vehicle operation this could be brought about only by the establishment of a garage under the auspices of interests willing to sacrifice present gain to the ultimate success of the larger proposition.

After various conferences and discussions the Edison Company decided to provide and operate such a garage under the auspices of the Electric Vehicle Association of America, and a committee was appointed to formulate and carry out such a project.

A one-story concrete building, owned by the Edison Company and formerly used as a garage for both electric and gasoline vehicles, was procured, and a fund was appropriated for its complete equipment. In fitting up the building no attempt was made toward artistic effect, but every effort was exerted to provide means for the best of service in the simplest manner possible, one desire being to prove to prospective garage owners the inexpensive manner in which a model garage can be provided without sacrificing quality of service or the convenience of the public.

The portion of the building fitted for this purpose covers an area approximately 100 x 50 ft., with a roof clearance averaging about 30 ft., and it is figured that this space will accommodate between 20 and 30 vehicles.

The current for charging is obtained from the Edison Company's direct current 3-wire mains, and is conducted to two General Vehicle charging panels located in the operatnig room of the northeast corner of the building. Each panel has mounted upon it six charging switches, instruments and General Electric charging rheostats from which the current is carried to Anderson charging receptacles conveniently located about the charging area.

Sufficient space has been reserved for additional equipment when required.

Charging plugs with sufficient lengths of cable are provided to accommodate vehicles in any part of the building.

The floor consists of a smooth concrete surface, and the entrance is of sufficient size to admit the largest truck when fully loaded.

The washstand, repair shop and battery space are located at one end of the building, while at the other are facilities for charging sparking batteries, also the work bench, with a full equipment of modern hand tools for light repair work.

The office, located in one corner of the room, is comfortably fitted for the convenience of operators, and provision is made for the keeping of operating and maintenance data in detail, so that exact costs of all descriptions can be figured out for the customer's information, or for commercial comparisons.

A gallery at the farther end of the room contains the stock room where supplies and spare parts are stored, also the drivers quarters provided with lockers, table and chairs for their convenience.

The entire room will be lighted with Mazda lamps equipped with Holophane reflectors, so spaced as to give adequate illumination at a minimum expense.

It is the plan of the Company to offer its garage services at a flat rate per vehicle per month, this service to include electric charging, irrespective of mileage, washing, flushing, battery testing, inspections, polishing, expert advice, adjustments and minor repairs, and will be ready at all times to undertake overhauling battery; washing, lead burning, renewal of electrolyte, extraordinary repairs, replacement of worn parts or broken parts and provide and put on tires at the regular commercial rates.

Spare vehicles will be available at the garage for towing home in case of accident or failure of power, at reasonable rates and for rental when machines are laid up for painting or repairs.

In placing this garage at the service of the public it is not the intention of the Company to entice vehicles away from other garages where they are receiving satisfactory service, but to provide a home for commercial electrics in its immediate vicinity, and to offer the best of service to those who cannot procure it elsewhere; and the Company confidently expects, in view of the increasing interest already being shown in electric vehicles, that not only will its own garage soon be taxed to its utmost capacity, but the excellent garages now existing in other parts of the city will also be filled to overflowing.

As has been stated before, the garage is to be conducted under the auspices of the Electric Vehicle Association of America, and any just criticisms or suggestions of patrons will be gladly received in the effort to prove that the electric vehicle is a power and necessity in modern transportation.

A NEW FIRE-ALARM SYSTEM

The recent test of the newly installed air-alarm fire protection equipment at Governor's Island in New York harbor, headquarters Department of the East, United States Army, furnished striking proof of the need of replacing archaic fire protection methods with the creations of modern genius.

The recently completed installation of the system at the Quartermaster's warehouse and dock gave a basis of action, and it was arranged that when the final test on the dock was made the garrison fire alarm should sound. The test was given in the presence of General Grant and staff and other army officers of note.

In just forty-two seconds from the time the test fire was started on the open dock, the alarm rang in. The garrison alarm was sounded according to the system long in vogue at the Island, the location being announced through the telephone to company headquarters.

There was a misunderstanding of the message, the run was made to the wrong place and it was full fifteen minutes from the time the alarm was sent out before the soldiers reached the dock.

The contrast between the almost instantaneous work of the air-alarm and the errors and delay following from the operations of a system that involved human agency in transmitting the fire location to company headquarters, was so marked that it is believed the old system will now be done away with at the post and modern methods substituted.

The Governor's Island test is prelimi-nary to similar tests in many other structures at United States Navy Yards and Army Posts in various sections of the country. .General Frederick D. Grant, commanding the Department of the East, before whom the test was made, believes the air-alarm solves the fire prevention as well as fire protection problem, as does Brigadier-general Torney, head of the Medical Corps, U.S.A., who has caused its installation in the Medical Supply Depot at Washington.

It is distinguished from others by the employment of an entirely new principle in automatic fire alarms. A small continuous alloy tube ($\frac{1}{8}$ in. in diameter) is distributed over the ceilings of the buildings to be protected. The ends of this tubing, on each floor, terminate in a case containing two diaphragms. Electric wires running from these cases control gongs on all the floors and an annunciator, placed at the entrance or other prominent part of the building.

In the event of a fire in any section of a building, the rapid rise of heat on the ceiling, which is the result of the exposure of even a small quantity of flame, causes a corresponding expansion of the air in the tubing. This expanded air is carried through the tubing to the diaphragms at each of its ends, and causing them to expand, closes the electric circuit, which in turn operates the gongs and indicates on the annunciator the exact location of the fire.

If the elevator to success is stopped— try the stairs.

TESTS OF DIRECT CURRENT MOTORS AND GENERATORS

Description of the Factory Tests to which Direct Current Motors and Generators are Subjected

HOWARD M. NICHOLS

Commercial motor testing practice differs, to a considerable degree, from the theoretical tests described in testing manuals. The most noticeable difference is the greater attention paid to purely mechanical features, by the testing department of a manufacturing concern.

The following remarks on direct current motor and generator testing covers the practice of one of the largest electrical manufacturing concerns in their small motor testing department, where motors and generators from $\frac{1}{4}$ to 25 h.p. are tested. Only the tests that are given to standard machines are described, as the methods of making the special tests, such as field saturation, etc., that are required to develop a new line of machines, follow closely the instructions given in testing manuals covering the testing of direct current machinery.

Usually two machines of the same horsepower rating are set up on a testing base, and the test is conducted simultaneously on both machines. Readings running light are taken on each machine. They are then belted together, one machine being run as a motor and the other as a generator; and the load being taken up by water rheostats. If the machines are rated as motors, power readings are taken on the machine running as a motor, the connections are then reversed and readings are taken on the other machine running as a motor. If the machines are rated as generators, readings are taken on each machine operating as a generator. During the heat run the connections are reversed at each reading so as to get power readings on each machine operating under rated conditions, i.e., as motor or generator, as the case may be.

GENERAL TEST

Before starting the tests on a machine it should be carefully inspected and the following points noted:

See that nameplate is properly stamped and agrees with the test card.

Check connections and see that all are tight and free from paint.

See that commutator has no high bars, is free from paint, splashes of solder, cuts, etc.

If brushes are not properly fitted, fit them with sandpaper.

See that brush holder and field wires do not strike armature, and that brush holder turns freely with setscrew loose.

See that all bolts and nuts are tight and that the pulley fits properly. Under no circumstance use a pulley that is too large.

After oil is placed in the boxes examine for leaks due to blow holes in the castings.

After connecting the machine in circuit and starting up, note the following points:

Try commutator for smoothness. This can be determined by holding the sharpened end of a lead pencil on the revolving commutator. Any rough spots or high bars will cause the pencil to vibrate. If the commutator is slightly uneven it can be smoothed up with coarse sandpaper, but if there are high bars the armature should be placed in a lathe and the commutator turned true, using a fine diamond point tool.

Check the armature for being magnetically central. Unless sure that the machine sets level on the base, remove the pulley and check for being level by placing a spirit level on the shaft. If the machine is found to be level start the motor running and with a piece of board press on first one and then the other edge of the rim of the pulley; if the armature is central, when the board is removed it will oscillate back and forth. If an armature hugs either end it will cause the bearing to get hot. In making this test on a generator run the machine as a motor.

Inspect oil rings and see that they turn and carry oil in the proper manner. If this precaution is neglected a hot-box is likely to result.

Check for balance by running light as a motor and placing the hand on the frame and feet of the machine. If it is out of balance there will be excessive vibration. A sprung shaft will cause this trouble; also armature not being properly balanced after winding.

Check the direction of rotation, facing the commutator end of the machine. If the machine rotates in the wrong direction it is due to a wrong bringing out of the terminals.

SHUNT MOTOR TESTS

Resistance.—If the mechanical inspection and direction of rotation are correct the motor is ready to be started on a heat run. Measure the resistance of the field with a Wheatstone bridge, taking several readings with different ratios in the bridge arms. Measure the resistance of the armature by the drop of potential method. Remove the carbon brushes and substitute for them copper brushes whose ends are chamfered so that they only cover a single segment. Use only one set of copper brushes regardless of the number of poles. The copper brushes are made from solid bar copper and are used only for taking resistance. Count the number of commutator segments between brushes, so as to make sure that the brush holder is adjusted to bring the brushes with the right number of segments apart. See that the copper brushes fit properly, otherwise they will arc and burn the commutator. With the brushes properly in place make a small prick punch mark in the outer end of the segments on which each copper brush rests. Do not strike too hard a blow, or a low commutator bar will be the result. These marks are used for definitely placing the sharp-pointed prick points that are attached to the voltmeter leads. They also serve to make it possible to take hot resistance of exactly the same set of coils, and thus eliminate any errors due to variations in resistance of the cross connections, etc.

Before applying current to the armature, place a wedge in the air gap between the pole piece and the armature, to prevent the armature from turning. If this precaution is not taken the armature is likely to start to turn, due to the current in the armature, inducing sufficient field in the poles of the motor to cause rotation, and the arcing at the copper brushes will burn the commutator badly. After wedging the armature connect it in series with a water rheostat, for regulation, and a circuit having a potential around 50 to 100 volts. Use an ammeter capable of reading the full load current of the machine and a low reading voltmeter. Protect the ammeter with a short-circuiting switch, which should be kept closed, except when taking readings. Remove voltmeter from circuit before opening same. Take about five readings of volts and amperes, starting in with full load amperes and reducing the current at each reading. Take the temperature of the air, armature, and field coils.

Readings Running Light.—Set the brushes at the neutral point running light. To determine this point run the motor at rated voltage, both clockwise and counter clockwise. The neutral position is that position in which the motor runs with the same speed in both directions. To cause the motor to run faster in a given direction turn the brushes in the direction in which the armature is rotating. After having set the brushes on neutral, hold the armature volts at rated value and take armature amperes, field amperes and speed.

Heat Run.—If possible secure a duplicate of the motor for load, if not, get a machine with the same horse-power rating and use pulleys that will give the proper speed on both when they are belted together. Make sure that machines are properly lined up. Check the neutral point under full load. Load the generator on water rheostats; fine adjustments of load can be obtained by putting a rheostat in the generator field. Only a small amount of resistance should be cut into the generator field, however; not enough to change the field current an appreciable amount from the field current obtained under normal operating conditions, otherwise the proper temperatures will not be obtained in the machine run as a generator. Start the machines free and throw on the load when they have come up to speed, being sure that all ammeters are short-circuited. Take readings of line volts and amperes, field amperes, speed and temperature of field coil and frame every half hour. On special machines run until temperatures are constant. Standard machines will be given a definite length of run, depending on the type and rating. Hold the load constant throughout the run. The heat run on most machines is taken at rated full load, but commutation should be observed at 50 percent over load. Never run a test with resistance in the field of the motor, unless the motor is designed to operate with resistance in the field. Short-circuit ammeter after each reading and disconnect voltmeter. Take particular note of the commutation and watch bearings for heating. At the end of the test check the neutral point under full load and mark the yoke and bearing with white paint, locating the neutral position. While the machine is still hot take readings running light as previously described. Take resistance of field and armature, and temperatures of armature, field and commutator. Record the condition of commutation throughout the run. Also test the machine for vibration, during the run, by placing the hand on various parts of the frame.

Insulation Test.—Insulation test with high potential should be made while the machine is hot. Connect all the terminals together and test between them and the frame. Test 110 and 220 volt machines at 1,000 volts a.c. for one minute, and 500 volt machines at 1,500 volts a.c. for one minute. If a ground develops test out the armature, field and brush holder separately. If the ground is in the field locate the defective coil by successive tests of each coil.

Take insulation resistance at 500 volts d.c. All machines must show at least one meg-ohm resistance. To make this test connect one side of a 500 volt d.c. generator to a 500 voltmeter, connect the voltmeter to the windings of the machine under test and connect the other side of the testing generator to the ground. Care should be exercised to use a testing generator that is free from grounds, as otherwise a short circuit is likely to result when one of the lines from the generator is grounded.

Let V represent the voltage of the testing generator, v the deflection obtained with voltmeter in series with the line and testing voltage applied to the machine, r the resistance of the voltmeter, and R the insulation resistance of the machine under test. Then

$$R = \left(\frac{V-v}{v}\right) r$$

The "pumping back" method of making heat runs is sometimes used on large machines to save power. This requires that the machines be of the same horse-power and voltage. Differences in speed can be corrected by using pulleys of different diameters. Connect both machines up to run as motors and belt them together. With the switches all open in one machine, bring the other up to rated speed. Close the field switch of the second machine, and check voltage generated, being sure that the direction of flow is the same as on the line, as otherwise a bad short circuit will result when it is attempted to throw the machine in on the line. Insert a field rheostat in the field of the driving motor, and bring the speed up slightly so as to ge full line voltage on the second machine. Now throw machine number two in on the line, and bring the speed up on number one until full load current flows. Check the load frequently throughout the run, and take readings as in the standard heat run.

SERIES MOTOR TEST

Do not attempt to take free readings on a series motor. Take cold and hot neutral points under full load. When taking speed care should be taken to hold the load constant, as the speed varies with the load. The "pumping back" method should not be used for heat runs and great care should be taken during all tests to see that the load does not fail, since if it should the motor would run away and smash things up generally. Take all other tests as described for shunt motors.

COMPOUND WOUND MOTORS

Take the same tests as for shunt motors. When getting neutral, reverse armature connections instead of field, to reverse direction of rotation. It should be noted that the speed varies with the load.

TEST OF GENERATORS

Run the generator as a motor, at rated voltage and speed, and set the brush holder on the neutral point. Take resistance of field and armature. Then belt to a motor of the proper speed and horse-power. Connect a field rheostat in the field of the motor for the purpose of varying the speed. Also connect a rheostat in the generator field. Bring the generator up to rated speed and adjust the field rheostat to give rated no load voltage. Then take no load readings of field volts and amperes and armature volts.

If the generator fails to pick-up, make sure that it is running in the right direction and that there is no break in the shunt field circuit. Also note that brushes bear properly on the commutator and that the springs are snapped down. If after noting all of these points the trouble is not located, open the shunt field circuit and put a voltmeter across the armature and note the deflection due to residual magnetism. Then close the shunt field circuit. If the deflection decreases it shows that the armature voltage is opposing that due to residual magnetism. The remedy is to reverse either the field or the armature connections. Sometimes excessive vibration will prevent a machine from picking up. This can be overcome by increasing the tension on the brushes.

The same general instructions for heat runs on motors applies to generators. In compound generators place a temporary shunt across the series field so as to get rated full load voltage at full load current. The final adjustment of this shunt should be made at the end of the heat run while the series windings are still hot.

LOCATION OF TROUBLE IN TESTING MOTORS AND GENERATORS

Motor will not start.—May be due to open circuit wiring leading to the motor or in the motor itself. If this is the trouble starting rheostat will show no flash as the handle is allowed to fly back from the first notch. Go over the connections to the motor and inspect starting rheostat for open circuit. See that springs bear properly on the brushes and that the leads are properly connected. The brush yoke may be improperly set. Try shifting the brushes. One or more field spools may be reversed. Check polarity with a pocket compass. First excite field to its normal value, and then open the circuit and check the polarity of the poles. If the field coils are properly connected the poles will alternate north and south. Do not attempt to check polarity with the field excited, as the compass needle is likely to become reversed.

Excessive Field Current.—Often due to short-circuited field spool. Can be detected by taking the voltage drop across each individual spool. May be due to improper field coils. If this is the trouble the total field resistance will be too small.

Sparking at the Commutator.—May be due to wrong lead of brushes, bad fit of brushes, high bar, or rough commutator. An open-circuited coil in the armature will cause a very viscous bright spark at the brushes, and the commutator segments to which the coil is connected will be blackened.

Flashing Over.—May be due to too weak a field, wrong lead of brushes, heavy overload, short-circuited armature, or any other trouble that causes excessive sparking.

GENERAL INSTRUCTIONS

When starting a machine for the first time make sure that the oil wells are properly filled and that the oil rings turn properly at about one-half the speed of the shaft. If the box on the pulley end starts to get warm, loosen the belt if it will still carry full load. Sometimes sand in the casting will fill the oil well with grit and cause the box to run hot. Never let a box get hot enough to smoke.

Never attempt to remove a belt from a machine with the power on. Throw off the power and wait until the machine is nearly still before attempting to throw off the belt.

Use instruments of the proper capacity to give readings within the range of their calibrations. Never leave an ammeter continuously in the circuit during a run, but short circuit it except when taking readings, as otherwise it may be injured by an excessive current due to a short circuit.

The engineering department of the University of Michigan last season graduated its second woman student, Miss Lillian Pearl McOmber. She is the first graduate from that department to take a degree of Bachelor of Science in architectural engineering. This sounds in no way compatible with "a matchless complexion and great violet eyes," and yet these are the descriptive terms of a reporter's admiring pen. Miss McOmber will specialize in steel structural work when she enters a city office.

DEVICE FOR MAKING KNOWN A SHORT CIRCUIT IN A GAS-LIGHTING SYSTEM

FRED H. HAYN, M.E.

The sketch shows a simple way in which a short circuit on a house gas-lighting system may be made known.

It often happens that the springs on the gas jet get weak, and, in extinguishing the light, the wiper is left in contact with the projecting platinum tip. This fact is not made known until the next morning when the postman tries to ring the front door-bell. It is well known that it will take several days for the battery to recover, if at all.

In the sketch, A is a set of cells, and C the ordinary coil for producing the spark at the gas jets. The wires T, T' lead to the door-bell and the push button.

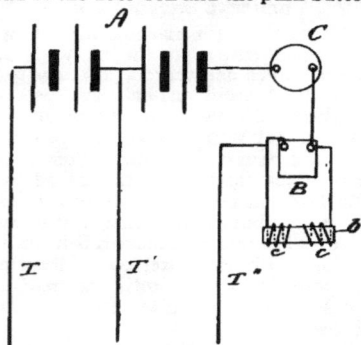

The wires TT'' lead to the gas jet. A buzzer B is placed in series with the coil as indicated., This buzzer alone, however, is not sufficient to produce the result desired, for it will be found that on attempting to light the gas, that the buzzer only will ring and the spark for lighting will be at the buzzer rather than at the jet.

If, however, two naked coils of wire be wound about a piece of carbon b, and the connections to these coils be shunted around the buzzer and adjusted toward and away from each other until the correct position is ascertained, it will be found that the spark will be at the jet and the buzzer will sound as desired.

If the lever or chain, as the case may be, be pulled down quickly, the buzzer will not sound. If, however, the wiper sticks on the platinum tip, the buzzer will be heard groaning throughout the house.

This is an exceedingly simple, easily constructed and efficient device for protecting the batteries of a house gas-lighting system.

--- --- ---

How Moving Pictures Originated

Perhaps it is safe to say that the large majority of the discoveries and inventions which have benefited and blessed as well as instructed and amused the world were the outcome of experiments conducted for altogether different results. What we know as moving pictures originated, according to the *Chicago Tribune*, in a question asked by Sir John Herschel of his friend Charles Babbage. This was in 1826, and the question asked was how both sides of a shilling could be seen at once.

Babbage replied by taking a shilling from his pocket and holding it before a mirror.

This did not satisfy Sir John, who set the shilling spinning on a large table, at the same time pointing out that if the eye is placed on a level with a rotating coin, both sides can be seen at once.

Babbage was so struck by the experiment that the next day he described it to a friend, Dr. Fitton, who immediately made a working model.

On one side of a disc was drawn a bird, on the other side an empty bird-cage. When the card was revolved on a silk thread, the bird appeared to be in the cage. This model showed the persistence of vision upon which all moving pictures depend for their effect.

The eye retains the image of the object seen for a fraction of a second after the object has been removed.

This model was called the thaumatrope.

Next came the zoetrope, or "wheel of life." A cylinder was perforated with a series of slots, and within the cylinder was placed a band of drawings of dancing men. On the apparatus being slowly rotated, the figures seen through the slotes appeared to be in motion.

The first systematic photographs of men and animals taken at regular intervals were made by Edward Maybridge in 1877.

CUTTING A SKEW GEAR FOR 2 H.P. GAS ENGINE

F. C. LEES

A few years ago I became possessed of a gas engine of about 2 b.h.p.; I got it second-hand, or possibly third or fourth-hand, for all I know! It had a number of good points, but was, even when I first had it, considerably worn. It was a "nameless" engine, and not standardized in construction, so it was not possible to obtain renewals, except as altogether special jobs. I carefully went over the brasses, and did a lot of tuning up; but there was one part which was badly worn, and obviously would require renewal before long—the screw-wheel on the crankshaft, which worked the "two-to-one" camshaft for the valves.

However, it ran for a couple of years, but at the end of that time one of the teeth was quite worn away in one spot, and it was evident that the wheel was done. I once had experience of getting a more simple wheel than this cut as a special job, and when, after much protest, I had at length paid the bill, it left me a sadder, but a wiser man! So I determined to cut a new gear myself on my 6 in. center lathe.

First, I got to work to find out what the wheel was as to pitch and angle of tooth. Those who have in any way studied skew-gearing know what an elusive term "pitch" is in this connection. This particular wheel had ten teeth, and was 2½ in. diameter over the tops of the teeth. The corresponding wheel on the camshaft naturally had twenty teeth, but its diameter was nearly the same measure, viz., 2¹¹⁄₁₆ in.

The dividers, when set to measure the distance from one tooth to another on the pitch circle round the side of the wheel, gave a "pitch" of about ¾ in.; but taken on the face of the wheel, at crown of the teeth in the line of the axis of the wheel. This is called the "divided axial pitch." When all that is said, there is still another pitch, viz., that of the complete screw (if it were completed), of which the wheel is but a short section. This last is known as the "primary pitch," says *Model Engineer and Electrician*.

Where, amidst all these considerations, shall we begin to think of how to cut a duplicate of a worn screw-wheel, with no data but what can be obtained from it? Fortunately, there are two definite points at which to attack the problem. The first is, that where a pair of skew gears are to work at right angles to one another, the angle of the teeth of the one added to the angle of teeth of the other must be equal to 90 degrees. The other point of attack is the fact that, however the tooth angle varies, if the two wheels are to gear together, the "normal pitch"—that is, the measure at right angles to teeth—must be the same in both wheels.

Bearing these two facts in mind, I began operations by dismounting the wheels—an operation easily mentioned, but not so easily carried out—and then, with a protractor, measured as carefully as possible the angles which the teeth of each wheel made with their respective axis.

It was evident at once that there was a great difference, and equally evident that the 20-toothed wheel was in the neighborhood of 30 degrees from the axis, while the 10-toothed wheel was in the neighborhood of 60 degrees, and fairly closely so. I therefore felt justified in assuming that the wheels had been cut at such angles that one was 60 degrees

appeared very like the change wheels of my lathe: these wheels are 10-diametral pitch. Taking the 20-tooth change wheel and "sighting" it against the other, it looked a fair fit. More careful measurement convinced me that the tool used to cut the skew gears had been one which would cut a 10-diametral pitch wheel of 18 or 20 teeth.

I felt that I was getting on with the problem of what to go upon, but there still remained the very big question as to how the relative movement of tool and wheel blank was to be set to produce an angle of tooth of 60 degrees to the axis on a diameter of 2½ in., or perhaps 2⁹⁄₁₆ in., as the original size of the gear. The blank was, at this point, turned up out of mild steel, that being the material chosen, though the original wheel was gun-metal.

The diameter was 2⁹⁄₁₆ in.; the measure round the circumference was 8¹⁄₁₆ in.; the width of face 1⁹⁄₁₆ in. Then I returned to the drawing-board, and striking off a rectangle $ABCD$, 8¹⁄₁₆ x 1⁹⁄₁₆ in., had before me, as it were, the surface layer of the blank unrolled upon the paper (see Fig. 1). From the bottom right-hand corner, A, I drew a line, AA', by the help of the protractor, at 60 degrees from the line AB, which corresponds with the wheel axis. Taking a pair of dividers, the line A,D was next stepped off into ten equal divisions by trial and error, and the ten positions (including the original A) marked. Lines were then drawn parallel with AA' from each point, and also from X and Y to complete the whole area.

If now this rectangle were cut out and wrapped round the gear blank, the lines would represent the tooth crowns, and the tool in making the teeth must follow them. What "pitch of screw," then, must the lathe be set to cut? We are very close to finding out, but it must be done at the drawing-board. On the line CD, it will be noticed that D, F and E represent the successive positions at which the screw threads (or wheel teeth) cut the axis of direction of advance. As there are ten threads, the one starting, say, at D will not re-appear on the same line of direction, DC, until the other nine threads have put in an appearance; so that if we again take the dividers, set them to the distance DF, and, starting

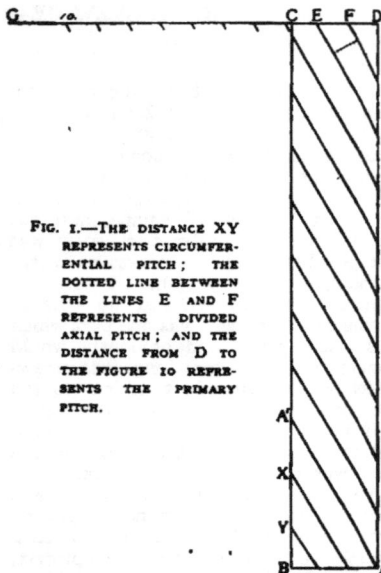

FIG. I.—THE DISTANCE XY REPRESENTS CIRCUMFERENTIAL PITCH; THE DOTTED LINE BETWEEN THE LINES E AND F REPRESENTS DIVIDED AXIAL PITCH; AND THE DISTANCE FROM D TO THE FIGURE 10 REPRESENTS THE PRIMARY PITCH.

from D, step off ten points along the line DCG, the tenth point will show where the thread which started at D re-appears on the same axis of advance. In other words, it is the pitch of the screw, a small section of which we wish to cut. This measure was so nearly 4½ in. that again I felt justified in assuming that the skew gear had been cut to a "primary pitch" of 4½ in.

Now we know what is required of the lathe setting, viz., to cut a ten-threaded screw of 4½ in. pitch. My leading screw is four threads to the inch; therefore, in order to cut a thread of 4½ in. pitch, the leading screw must make 18 revolutions to each one revolution of the mandrel. The following change wheels will satisfy the requirement:

Mandrel	Stud	Leading Screw
120	to 20	
	90	to 30

A wheel is chosen for the mandrel of a number of teeth divisible by 10, for the purpose of dividing the ten teeth to be cut. I rigged up the above; but found that, even with the back gear in, the change gearing could not be driven, so great is the proportionate speed of the lead screw, and, of course, the friction.

The only thing I could think of was to drive the train of wheels from the fast-moving end—that is, from the leading screw. The end nut of the screw was found to be tapped ½ in. gas thread. I had a pulley on a slow-speed dynamo, measuring 9 in. diameter by 2 in. on the face, and bored for 1 in. shaft. This pulley was requisitioned, a ½ in. gas socket connector obtained, and turned down to an outside diameter of 1 in. When this was screwed onto the leading screw end, in place of the nut, the 9 in. pulley was also easily fixed to it by its setscrew. I had a 4 in. pulley on the overhead shaft, and was fortunate enough to find in a box of belting scraps one or two lengths, which, when pieced together, made just the right length to go from overhead shaft pulley to leading screw pulley. I was pleased to find that the arrangement answered admirably, running quite smoothly.

The next thing to do was to grind a cutting tool. I have a tool-holder taking ⅜ in. round tool steel, a diameter which would be wide enough for the gear teeth. A templet of a tooth of change wheel 20 teeth was made, and by fixing the tool steel in a file handle, the shape of the gear tooth was reproduced very fairly.

A moment's digression may be allowable here as to the contour of skew gearing teeth. I believe I am right in saying that these teeth are bound to be a compromise between at least three opposing conditions.

The respective angles which the root, pitch circle point and the crown make with the wheel axis should theoretically be three different figures for any given position. To satisfy this a most elaborate mathematical machine would be required, and, though I believe machines have been made the designers of which claim to fulfil in execution the theoretical requirements, the fact remains that the ordinary and excellent skew gear of commerce is cut with a tool found in practice to produce a good working article, even if it has a co-efficient of friction a little higher than a theoretically correct tooth contour should possess. Therefore, I made no attempt to produce theoretically perfect curves on my tool, but copied an ordinary gear tooth as nearly as possible.

This was put in the holder, laid on a surface plate, and the straight end of the cutter turned round in the holder, till it had a rake of 60 degrees, as measured by the help of one of the very useful small sheet-steel adjustable gauges made for purposes of setting screw-cutting tools to the correct rake.

The heel of the cutter was then backed off very considerably, so as not to foul either side of the thread when cutting full depth; it was then ready for work. But first we must calculate how far the saddle must move at each cut before stopping the lathe and disengaging the nut in order to return to the loose head-stock end by the quick hand traverse. The universal rule is where the number of threads per inch to be cut will not

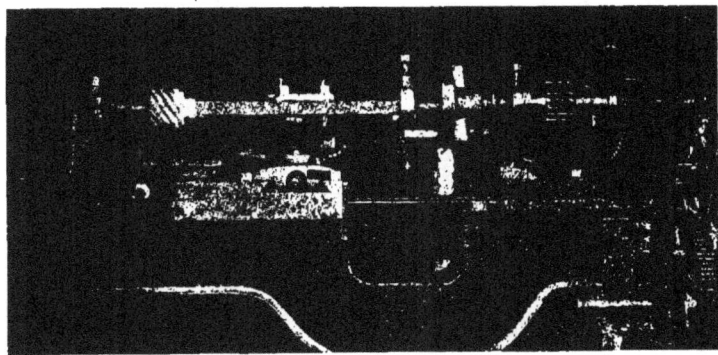

Showing the Lead Pattern Skew Gear as completed in the Lathe

divide exactly into the number per inch on the leading screw, to turn the pitch to be cut into a fraction—in this case $4\frac{1}{2}=\frac{9}{2}$—then the numerator, or figure above the line, gives the least distance the saddle must move before disengaging the nut, in order to be able to engage again when returned to the starting position. In the case of very long traverses being required according to rule, there are, of course, ways of getting round it by special tooth marking; but with the quick motion we have to deal with here, it is best to allow the 9 in. run.

The saddle then, when the tool was fixed in the slide-rest, was run up against the loose headstock, and a chalk line squared across the bed, 9 in. in front of the forward edge of the saddle, then the leading screw was turned slowly till the nut would engage easily.

At this stage we make the last of the settings (apart from the slide-rest) by marking the teeth exactly in mesh on the mandrel wheel, and on the 20-tooth wheel to which it is geared. This shows the exact position when the saddle nut will engage the leading screw correctly. Then we want to provide for dividing the ten threads; therefore, as the mandrel wheel is 120 teeth, we count round 12 from the marked tooth (or space, as it happens to be—it was space in my case) and chalk again; 12 more, and so on; we then have 10 marked teeth on the mandrel wheel. When the first tooth of the skew gear is cut, the lathe is stopped with the marked teeth in gear, the quadrant is loosened and dropped till the 20-tooth wheel is out of gear, then the mandrel wheel is turned round so as to bring the second marked tooth into gear, when the quadrant is raised, and so on for all ten positions; the teeth then are accurately spaced.

The photograph shows the arrangement, with pulley on leading screw, and the marked spaces on the mandrel wheel. The 20-tooth gear is rather too well-lighted to show the marked tooth very well. Secure the mandrel carrying gear blank against back lash by binding the carrier to the driving pin on the catchplate with wire, and we are ready to start.

I made several traverses of the saddle, starting from hard up against the loose headstock, traversing the 9 in. along the bed to the chalk line, disengaging the nut, returning by quick hand motion rack work to starting position, engaging nut again, and again traversing before putting on any cut. It worked perfectly.

I then set the tool in by cross motion of slide-rest till it was just clear of the gear blank, and chalk-marked the collar of the cross slide screw to show the position when the tool was clear, as it is, of course, very necessary to bring the slide-rest back after each cut has been taken to the clear position before returning to starting position for another cut. Then, as each cut is set in, a mark is made to show the depth to which the tool has been set as a guide for the next set in.

Accordingly, with all prepared, I set the cross slide in for the first cut and started the lathe. It was a light cut, but was a clear indication as to whether all the adjustments and reckonings had been carried through correctly. Measured with a protractor, the angle was quite correct. Putting the old wheel alongside the blank, it was also evident that the trace of the future tooth space was so nearly exactly a reproduction that no difference could be detected. This was exceedingly pleasant, and I cheerfully withdrew the tool to clearance, returned to starting position, set in the second cut, and again started the lathe. But now a series of unforeseen difficulties occurred. I have said that the blank was of mild steel, and all who have worked this material know how very much difference there is between working it with a sharp-edged tool, with plenty of top and side rake and clearance, and working with anything at all closely approximating to a tool without top and side rake. It is easy to see that a tool cutting all over its profile, as a gear cutter must, cannot have side rake, and owing to the depth of cut being full $\frac{1}{4}$ in. and the necessity of giving great backing off at the heel of the tool, in order to obviate fouling either edge of the tooth-space, only a very small amount of top rake could be given.

As the second cut came on, I perceived by the sound of the change gearing, that very great strain was being thrown on it; but the cut carried through. I brought on the third cut, with still more ominous complaints from the train of gear wheels. At the fourth cut the tool started and travelled a short distance, then a series

Showing Arrangement at Fast Headstock, with Driving Pulley on Lead Screw, and Marked Teeth on Mandrel Wheel, and Pinion for dividing the Ten Teeth to be cut

of sharp cracks and flying gear teeth told me that something had stripped.

I drew the tool out of cut, eased back the whole lathe motion, and as I did so saw the gears turning back with blank space, where teeth ought to be; moreover, on carefully examining the change wheel end, I found the stud of the reversing plate had been distinctly bent.

Here, indeed, was misfortune. However, I had started in to make a skew gear, and intended to do it. Fortunately, cast change gears in the small sizes cost but a few cents, and they were the small ones which had gone, but the stud evidently required some looking into.

I dismantled the whole lot, and found that the stud, though nearly ¾ in. in diameter, was screwed ½ in., reduced for that purpose, and entered into a ½ in. tapped hole. It was decided to fit a stronger one, and, accordingly, with ratchet brace, that ½ in. hole (after filling it hard and tight with teak wood) was drilled out to take ½ in. gas thread; turned up a new stud, with the necessary sleeve for the gears, bored out the reversing plate to take the larger stud, bored, faced and trimmed the teeth of fresh gears, and once more was ready for work. I got in several more cuts, but could hear that the strain was enormous, and

again assailed by flying gear teeth, I began to realize why machines for cutting steel gears are so very, very massively built!

If I had not seen the results that appeared when I again dismounted the change gear train, it might be difficult to convince one of their absolute truth! The gears were stripped, sheared off; no result of gears jammed too closely together; for, recognizing that all ordinary strains would be reversed when driving from the lead screw instead of from the mandrel, I had taken most particular care to have ample clearance between gears, and what portions of them were left showed that all was in thoroughly good adjustment in that way. The new stud had borne the strain without flinching, but the feather (good steel, ¼ in. x ⅛ in.), which keyed two change gears to their sleeve, was simply drawn out as though it had been wire, as the photograph shows (see Fig. 2). The two portions which were actually within the bore of the wheels remained straight and practically parallel, while the space between was drawn round and stretched by the strain!

It was, therefore, clear that though my lathe is rather massive for a 6 in., it was not sufficiently so for cutting this skew gear in steel.

I made up my mind to take the opportunity of fitting mild-steel gears on the reversing plate, and did so, then returned to the work in hand.

The original skew gear was gun-metal; the new one might then be gun-metal also; but I wished to so arrange matters that the bulk of the metal to be removed should be in some more tractable form than even gun-metal. I got a length of square mild steel, 1 in. x 1 in. in section, centered it and drilled the ends for setting between centers in the lathe. Then two short lengths of round mild steel were driven through holes drilled at right angles to the length of the bar, and a "boss" of lead was cast on to these, about 3 in. diameter and 2 in. thick. This "boss" was then turned to the size of the skew gear blank, with a little extra allowance in size, and the job put in the lathe, once more fitted up with all adjustments as at first.

Everything now went splendidly. The

cut each time. In rather over half-an-hour I had the first tooth space completed. The quadrant was slacked and dropped, number two marked space on the 120 change wheel brought into gear and tightened up. The saddle nut went in exactly as required, and before the hour was much passed the second tooth space was finished, leaving a complete tooth. I almost raised a cheer!

This first completed tooth was very carefully examined as to its profile on the edge of the blank. It was found that it had, to a certain extent, the fault of leaning over from a truly symmetrical shape—a fault very well known in this class of work. However, as the lead blank was only to be used as a pattern for casting in gun-metal, and there was sufficient material in the completed tooth to allow of correcting the shape, the next tooth was worked, carefully watching as to which side of the cutter was lagging in its work. It was very soon discovered, and by quite minute adjustments of the tool as to its horizontal angle with *the axis of the lathe centers*. The fault was corrected, and subsequent teeth worked with quite a good and satisfactory shape.

A few hours sufficed to work the whole ten teeth, and by evening the lead gear, with all the burrs smoothed by the help of a rough file, was surrounded by plaster of Paris, and next day, on removing the lead gear, a plaster mould remained in which to have the gun-metal cast.

It is very necessary to dry the plaster mould as thoroughly as possible, at a considerable heat and for some time—two days in the domestic oven is none too much.

If any moisture is left, the molten metal will explode when poured into the mould. This plaster mould was sent to a local foundry in order to have the gun-metal casting done.

The lead gear had been sent first of all with a request for a casting, but apparently it was beyond the moulder to tackle. The gear was then sent to a firm of gun-metal founders in London, but they also returned it as an impossible job. This was distinctly discouraging; but a number of other difficulties had been surmounted, and as the professionals sent the job back, the amateurs must perforce try!

repairing and engineering works was borrowed one Saturday afternoon, and about 10 lbs. of old motor brasses of good quality metal were bought for less than 15 cents per lb., and by surrounding a crucible with broken coke, held up by a few ordinary yellow bricks, some 7 or 8 lbs. of high quality metal were well melted in a few minutes over the half-hour. The crucible was lifted, the contents well stirred by a clean iron rod, and the plaster mould filled up flush. When cool, the rough bits of metal were chipped off, and a very fair casting came out.

In the lathe it was soon seen that the material had come out of the melting pot quite as good in quality as it went in. It was very nearly as tough as the mild steel, and very glad I was that most of the metal to be removed had been worked on the lead pattern.

After chucking the casting and boring out the center hole to the necessary diameter, it was mounted on a mandrel and in no very great length of time was a finished skew gear.

On being tested with the 20-tooth wheel which it was to work, it geared quite well at right angles, and with practically no backlash. So this very interesting job was brought to a successful conclusion.

A fireman was up for examination for promotion to the position of engineer. He passed a fair test on the rules and machinery, but during all of it the examiner was constantly lecturing him as to the need of economy in the use of fuel and oil, so that by the time he finished his examination it was pretty well on his nerves.

Having finished the technical part the examiner thought he would put the man in a critical position to see what he might do in an emergency. So he put to him this question: "Supposing you are the engineer of a freight train on a single track, and you are in a head-on collision with a passenger train, and you know that you could not stop your train; that a collision could not be averted. What would you do?"

The man, unstrung by the vigorous instruction he had received as to economy, replied in this way: "Why, I would grab the oil can in one hand and a lump of coal in the other and jump."—(From the

BIG FORTUNES IN LITTLE INVENTIONS

Men Who Saw the Importance of the Apparently Unimportant—Many Inventions Hit Upon by Accident—Millions of Dollars in Seemingly Trivial and Commonplace Ideas

Every time anybody in the United States pulls the cap off a beer bottle or a soda-water bottle with the intent to quench a thirst, temperately or otherwise, he puts the fraction of a cent into the pocket of one William H. Painter, of Baltimore, writes William Atherton Du Puy in the *Scientific American*. A good many people have pulled these caps in the last few years and Painter is consequently an ever-increasing millionaire. Yet the cap for bottles is a small thing, an idea crystallized and patented. The patent is the source of the millions.

Painter, however, carried his patent in his pocket for six years before he succeeded in interesting capital in its manufacture. Then a man of means advanced the necessary capital in return for a half interest in the patent, and a company was formed. At the end of the first year he and Painter each had a net $27,000 in his pocket. Now the invention has crowded all other stoppers for fizzy water off the market, and a big factory in Baltimore turns out the caps by the million every day.

A MILLION-MAKING STOPPER

Before the time of Painter there was a man by the name of De Quillfeldt, who lived in New Jersey, and who invented a stopper that took the trade away from the corks of our youth. This stopper was of rubber and was tightened by a wire attachment which was pulled down as a lever on the outside of the bottle. A decade ago they were generally used on milk bottles. De Quillfeldt is said to have made $15,000,000 out of his patent. He might have amassed a competence had it not been for William Painter and another equally clever person who fitted a piece of pasteboard into the neck of a milk bottle and took the business away from him.

An idea that is perhaps simpler than the pasteboard stopper is the "hump" on the hooks that furnish so much employment for married men just before theater time. Women had been fastening their dresses up with hooks and eyes for a generation, and it is probable that some one had made a lot of money out of the original invention. But hooks had a way of coming unfastened much to the chagrin of the neat and fussy. Then came the genius of the hook and eye. A man who was wide awake despite his residence in Philadelphia, bent one of these hooks so as to make a hump in it. He tried hooking it up and found that it remained hooked. He patented it and has monopolized the business through his "see that hump" advertisements ever since.

One day a man stood behind his wife while she put up her hair. The hairpins of those days were straight pieces of wire. They did not "stay put" very effectually. The woman in this case bent her hairpins before putting them in. Her husband saw her do it. The result was the invention of the crinkly hairpin which is today used in carload lots by the women of the world.

INVENTION OF THE TELEPHONE

So important an invention as the telephone was made by turning a screw one-fourth of one revolution. All the millions that have resulted from the invention of the Bell telephone depended upon this slight twist of the wrist of Dr. Alexander Graham Bell. There had been men before Doctor Bell who had come near finding a way to make female gossip and masculine commercial intercourse easier. The Reis patents came nearest success. But in the Reis patents the current was intermittent. It had to leap a gap. Doctor Bell closed that gap when he turned the screw.

But Doctor Bell was not trying to invent a telephone when he incidentally stumbled upon his secret. He was working on a method of making speech visible, for his wife was deaf and dumb, and he was seeking an easy method of conversing with her. Instead, he found the method of talking over a wire to people at a distance. He did not patent the idea,

however, and it knocked about the house
for months. Finally, he demonstrated
it to some friends and they saw the possibility of its application. Upon their
advice he patented the invention. His
patent was filed at 10 o'clock in the morning, and at 3 in the afternoon another
man applied for a patent on the same
thing and lost a hundred million dollars
by a nose.

THE SELDEN CLUTCH

Such are the stories that the veterans
of the patent office gossip about in the
moments of their leisure. They tell you,
for instance, of the Selden clutch, which
is one of the vital patents that has much
to do with the control of the automobile
business of the country. It is this clutch
that enables the operator of the machine
to stop and start without having to get
out and crank his machine—sometimes.
It is interposed between the running gear
and the motor, where it keeps the car
marking time while the crossing is block-
aded.

This clutch was invented before automobiles were. For a decade after its
invention there was no opportunity of
applying it to any good purpose. Then
the automobile was invented. In fact,
George B. Selden was one of the early
builders of automobiles, and it is logical
to suppose that he built them that he
might make an opportunity to use his
clutch. Certain it is that he long had a
clutch on the automobile business. Before his patent was declared invalid about
$2,000,000 had been paid by nearly ninety
automobile makers, who found it cheaper
to pay than to engage in extensive litigation.

THE FAIRBANKS SCALE

Thaddeus Fairbanks was a New England farmer with long whiskers and much
Yankee ingenuity. In his time old-
fashioned steelyards were the only accurate means of weighing the produce
of the farm. Platform scales were unknown, for nobody had ever worked out
a method of arranging the lever that
supported the platform in such a way
that an object would pull equally no
matter upon what part of the platform
it rested. Old Thaddeus Fairbanks used
to tell the story of the evolution of the
arrangement of these levers. For a long
time the problem was upon his mind.

He used to lie awake nights and attempt
to arrange those levers. It was in the
dead of night that his thinking finally
bore fruit. The arrangement unfolded
itself and the Fairbanks scale was the
result. So did a farmer practically
monopolize the scale business of the world
and so did he write his name upon platform scales wherever civilized man buys
and sells by weight.

It is a man by the name of Hyman L.
Lipman, likewise a resident of Philadelphia, who invented the rubber eraser
that throughout our generation has been
attached to the lead pencils in common
use. It was in 1858 that the invention
was made. In those times people talked
in much smaller figures than nowadays.
Lipman was, however, able to cash in
his patents for a cold $100,000 when
dollars went much farther than they do
today.

So did a man by the name of Heaton,
resident of Providence, notice that mother
was occasioned a great deal of trouble
because the buttons constantly came off
the children's shoes. Heaton devised
the little metal staple that holds on the
shoe buttons of today, and realized a
fortune for his pains. No less clever was
a man of the name of Dennison, who
pasted little rings about the hole in a
shipping tag, and thus made an "eye"
that would not pull out.

THE SEWING MACHINE

Elias Howe conceived the idea of
placing a hole near the point of a needle,
and under the encouragement of this
small thought was the sewing machine
developed. Howe was one of the Co-
lumbuses in the development of a machine
to sew seams and deserves a monument
from the women he emancipated from
needlework. When he asked Congress
to extend the term of his patent for a
short time (one extension had already
been granted) he admitted that he had
collected $1,185,000 in royalties, but con-
sidered himself entitled to $150,000,000.

Howe had many followers who improved the sewing machine. One of the
cleverest of these was the man who patented the stitch his machine made instead of the machine itself, and thus
made infringements more difficult. Another man, Allan B. Wilson, a journeyman cabinet-maker of Pittsfield, Mass.,

exhibited the first model of what has since become known as the four-motion feed. He afterward founded the firm of Wheeler & Wilson, and became immensely wealthy. In the *Scientific American* of 1849, James C. A. Gibbs saw a picture of Wilson's machine. The working of the device was clear down to the point where the needle perforated the cloth. He wondered what happened after that. Finally he decided to make the needle work. After much thinking and infinite whittling he worked out the ingenious little revolving hook which became the important feature of the Wilcox & Gibbs machine and which made that firm wealthy.

THE CHEWING-GUM BUSINESS

There is a palatial mansion up the Hudson with a private yacht moored beneath the Palisades that is a monument to the millions that Adams made in the chewing-gum business. It was in 1871 that chewing-gum was patented and millions of willing ·jaws have wagged industriously upon it ever since.

Harry Hardwick invented an ingrain carpet with the threads of it so interwoven as to prevent wrinkling, and Hardwick is now $4,000,000 better off for his pains.

Charles Edward McCarthy was a blind man and lived in South Carolina. He devised a method of attaching mule power to a cotton gin, and lived his life out in luxury and ease while the mules did the work.

R. R. Catlin, of Washington, invented a pattern cat that need but be stuffed with hay and sewed up to become a toy. Such figures as "Billiken" and such games as "Pigs in Clover" are always a fortune to the inventor if they become popular. The rubber return ball made much money, both for the inventor and likewise for an infringing manufacturer who fought him in the courts.

THE BRASS PAPER FASTENER

The brass paper fastener, which is still generally used for thick documents, was patented in 1867 by a governmental clerk by the name of G. W. McGill. Yet it was not new, for the Romans used a similar device two thousand years ago and the modern appliance was but a resurrection.

The patent for a typewriter lay dormant for half a century in France before it ever came into use. Then a man by the name of Sholes made a machine in this country and called it Remington. Another man named Brown made a different kind of typewriter and called it the Smith. The patentees immortalized other men by their work. They made millions and also made it much more pleasant for the editor who has to read copy.

The man who invented tin cans made it necessary for somebody to invent an opener. This was done and the money corraled. A can opener is not a very laborious thing in the using, but the public is always ready to pay for things that are made easier. So, just recently, an inventive genius made a can with a seam just below the top and when the owner wants it open he has but to strike it a blow where the seam breaks and the top is off. A single Chicago packer ordered ten millions of these cans as an experiment and others followed suit. The inventor has a fortune, and the thing is but just begun.

Cutting Plate Glass

It is quite a trick to cut plate or rolled glass, and the thicker the glass, the more difficult the operation. With the trade, however, the job is not so hard to do, as there are certain rules the workmen follow that nearly always lead to success. These rules are as follows:

With a common glass wheel cutter, which may be purchased at any hardware store for 25 cents, bear rather heavily on the glass in the direction in which the fracture is intended; it will leave a white line across. With a very light hammer, one with a pein weighing about an ounce, commence to lightly and rapidly tap the glass immediately under the commencement of the line. In a short time a fine crack will be observed to start. This will follow the hammer along the line to the end, when a very slight pressure will cause the glass to separate. I have seen a workman use the rivet end of a 2 ft. iron folding rule for the tapping, but I find a very small hammer more convenient.

In the case of a ribbed plate glass the cutting must, of course, be done on the smooth side.—*National Builder*.

SERVING TRAY.

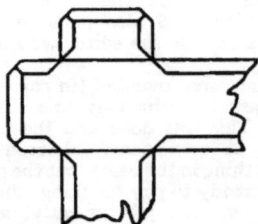

Detail of corner joint.

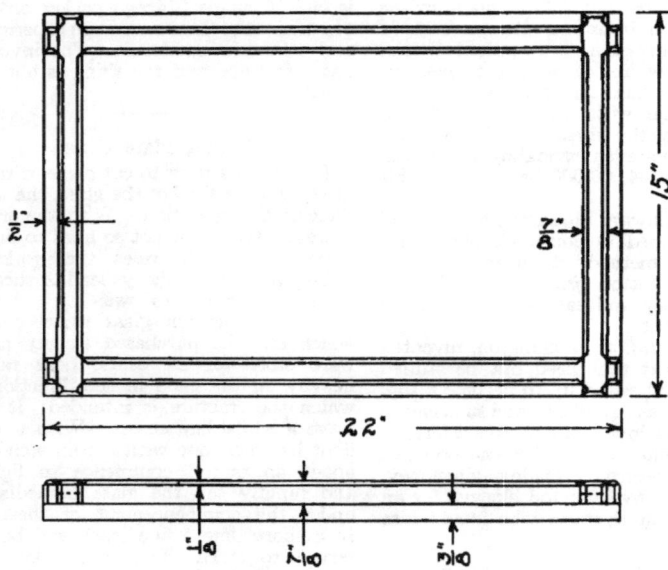

15"

22"

$\frac{1}{2}$"

$\frac{7}{8}$"

$\frac{1}{8}$"

$\frac{7}{8}$"

$\frac{3}{8}$"

20

R.F.W. 1911.

THE HOME CRAFTSMAN

RALPH F. WINDOES

SERVING TRAY

A quarter-sawed oak serving tray finishes up into one of the most beautiful pieces of craftsman furniture which is possible to produce. It is very easily constructed, and anyone with ordinary ability to use tools should have no fear about attempting it. The material needed is as follows:

1 piece $\frac{3}{8}$ x 15 x 22 in. quarter-sawed oak
2 pieces $\frac{7}{8}$ x $\frac{7}{8}$ x 15 in. quarter-sawed oak
2 pieces $\frac{7}{8}$ x $\frac{7}{8}$ x 22 in. quarter-sawed oak
Two brass handles.

The brass handles may be purchased at a hardware store. If they do not have them rigid, a pair of drawer pulls will do in their stead.

The corners of the small pieces are half-lapped. In cutting these joints be very sure that the holes cut are not wider than the finished pieces. The $\frac{1}{8}$ in. bevel is laid out with the marking gauge, and as much of it planed as possible. That which cannot be reached with a plane can be cut out with a sharp chisel.

The bottom piece needs to be sanded perfectly smooth and the edges should be planed. The frame is fastened to it with screws, well countersunk.

The finish should be applied before the handles are attached. A piece of soft felt should be glued over the bottom, or at least around the edges, so that it will not scratch any surface upon which it is placed.

WASTE BASKET

No pen sketch can do justice to the appearance which this basket offers when it is completed. And it also presents new problems to the builder which he should try to solve; hence it is to be advised that every follower of these articles try this one. The material to order is as follows, planed and sanded to finished sizes:

4 pieces 1 x $1\frac{1}{4}$ x 12 in. quarter-sawed oak
4 pieces 1 x $1\frac{1}{4}$ x 10 in. quarter-sawed oak
1 piece 1 x $7\frac{1}{2}$ x $7\frac{1}{2}$ in. quarter-sawed oak
2 pieces $\frac{1}{2}$ x $11\frac{1}{4}$ x 11 in. quarter-sawed oak
2 pieces $\frac{1}{2}$ x $10\frac{1}{4}$ x 11 in. quarter-sawed oak
8 pieces $\frac{3}{16}$ x 1 x 12 in. quarter-sawed oak
4 pieces $\frac{3}{16}$ x 1 x 8 in. quarter-sawed oak

First shape the sides and glue them up. The two wider pieces need to be $11\frac{1}{4}$ in. across the top and $8\frac{1}{4}$ in. across the bottom, taking measurements from a center line. The narrower pieces are $10\frac{1}{4}$ in. across the top and $7\frac{1}{4}$ in. across the bottom. When these are shaped, glue and clamp them together. A few small brads may help to hold the pieces together, and if the craftsman cares to use them, no objection can be raised. Placing them at an angle this way makes it necessary to plane the top and bottom edges off so that they are parallel.

The top is made up of the 12 in. pieces mitered at the corners. The bottom is made up of the $7\frac{1}{2}$ in. square piece around which the 10 in. pieces have been mitered. These two are fastened to the sides as shown.

The thin strips are used for decoration. They may be attached as shown, or in any manner which the builder so desires. They are mitered at the corner, and as this is on an angle, it presents a problem to the amateur. The most easy way to solve it is to cut the proper angle, then lay out the miter with the tee-bevel and cut it without the aid of a box.

(To be continued)

Steel tools put in a barrel of air-slaked lime will never rust. I have always kept my spades and such tools in lime.

WASTE·BASKET·

Plan of basket

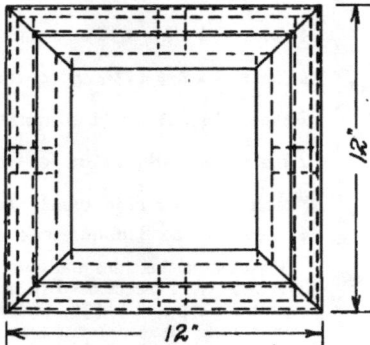

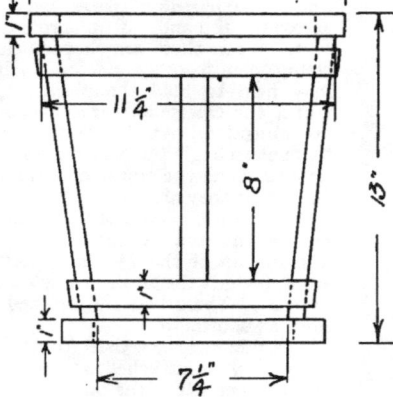

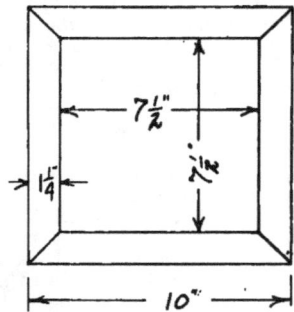

12"

12"

11¼

8"

13"

7¼"

Side elevation

7½"

7½"

1¼

10"

Plan of bottom piece

(21)

R.F.W. 1911.

PULLEYS

In the installation of any system of mechanical power transmission, vital points to be considered are the pulley arrangement, size and material. There are many points, says *Practical Engineer*, which may be brought out and enlarged upon in order to show the saving and economy of certain classes of pulleys in preference to others for the particular work at hand. Numerous experiments have been performed with the various materials, types and shapes of pulleys, and their value for different installations has been quite thoroughly worked out.

MATERIALS

In the construction of pulleys the materials now common are cast iron, steel, cast iron with steel rim, wood, cast iron with wood rim, and paper.

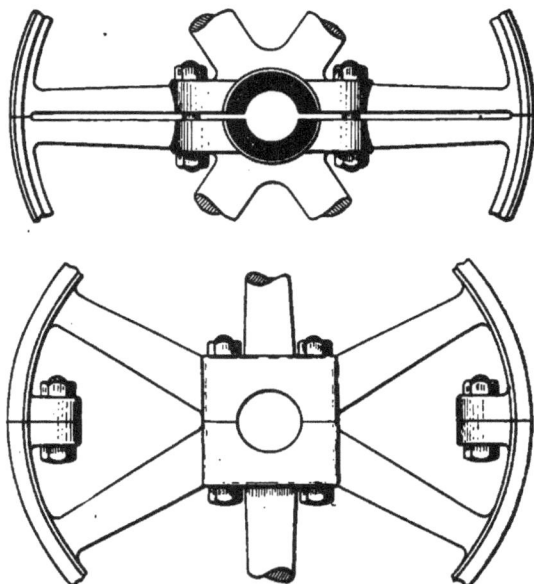

Fig. 1. Two Forms of Split Pulleys

The most common pulley now employed is made of cast iron. It presents the advantage that it is easily made, its belt friction is high, it is strong and its strength can be calculated with a certainty, the speed limit is high, balancing is easily done and the pulley is practically moisture proof. On the other hand the pulley may easily be broken by a shock or blow, and when so broken is not easily repaired. It is somewhat heavier than steel, wood or paper for the same service, the cost is high compared with wood or paper pulleys, and it is subject to internal strains due to temperature changes and improper casting.

Steel pulleys are made of either cast or forged steel in either split or solid form. The coefficient of friction is practically the same as for cast iron, but the weight and wind friction are reduced considerably from cast iron and the strength considerably greater as is also the speed limit.

Pulleys with cast iron hubs and arms with steel rims have become quite popular where high speed is a requisite. The points which recommend them for use are the strength, the light weight compared with cast iron for similar service, repairs to the rim are easily made, and, like the cast iron pulley, they are not affected by moisture of the atmosphere. On the other hand the cost of these pulleys is high, and the pulley is subject to strains due to temperature changes.

Split wood pulleys are becoming continually more popular owing to their low cost, absence of strain due to temperature, to their ability to withstand shocks and jars without breaking, their comparatively light weight, and the fact that they can be mounted upon shafts without keys or set screws. These pulleys, however, are seriously affected where moisture is present, the life of the pulley is not so long as that of the cast iron or steel pulley, nor is the strength of the pulley so great as that made of metal.

Paper and pulp pulleys are more recent developments and combine strength and durability with a high coefficient of friction.

In comparing the strength of pulleys

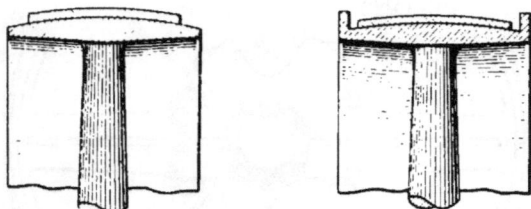

Fig. 2.   Rims of Crowned and Flanged Pulleys

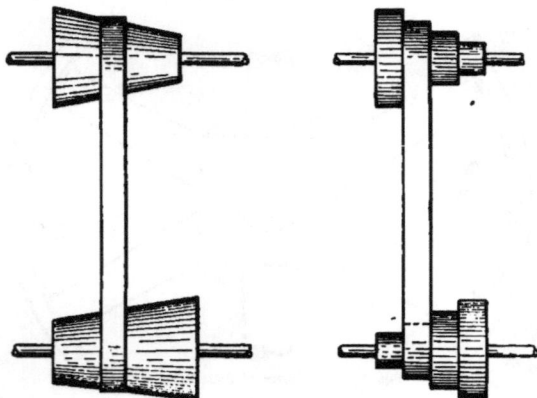

Fig. 3.   Cone and Step Pulleys

iron 16.7, paper 21.6 h.p. These maximum values occurred at about 4,500 ft. per minute speed of belt. Other tests at different belt tension showed about the same ratio.

FORMS AND PROPORTIONS

The class of work to be performed by a pulley determines largely the form and proportion which the pulley should have. A solid pulley is suitable for installing upon the end of a shaft, where it can easily be reached and taken off without interfering in any way with the alignment of the shaft or other machinery in the plant. These pulleys are set with keys or set screws, keys being used where the load to be transmitted is heavy and set screws only where the load is very light, set screws being spotted into the shaft.

When it is desired to place a pulley at the center of a shaft, a split pulley is most convenient, as it can be mounted without taking down the shaft and other pulleys placed upon it. These pulleys are generally made in two parts and may be set upon the shaft either with a key, set screw, or entirely by friction, the latter being preferable in all cases where it is possible, as no injury to the shaft is then necessary. Either the solid or split pulley may be made of nearly any material now used for pulleys, and the choice is largely a matter of opinion with the engineer in charge.

The type of rim adopted depends entirely upon the character of the work which the pulley is desired to do. The flat face, that is, where the face of the pulley is straight and parallel with the shaft, is commonly employed where a belt must frequently be shifted from one pulley to another, but for isolated work it is desirable to use a crowned pulley which keeps the belt at the center of its face, thus reducing the danger of the belt running off.

For certain classes of work where the

the results of tests made at Purdue University are of interest, showing the bursting speed of several kinds of pulleys. Two tests on solid wood pulleys gave 285 and 267 ft. per second as the bursting speed. Split wood pulleys withstood speeds of 232 and 221 ft. per second. The results recorded for bursting speeds on paper pulleys were 295 and 307 ft. per second. Two split steel pulleys gave way at 235 ft. per second.

Tests made at Sibley College, Cornell University, resulted in the following maximum horse-powers per square inch of belt cross section of belt, which were developed at 3,500 ft. per minute; when the tension on the tight plus one-half the tension on the slack side equalled 180 lbs.; on a pulp pulley 5.4 h.p. was developed, a wood pulley gave 5.5 h.p. maximum, cast iron 8.5 h.p., and paper 10.7 h.p. When the tension on the tight side plus half the tension on the slack side equalled 300 lbs., the results were as follows: Pulp 10 2, wood 10.4, cast

pulley is subject to considerable shock or jarring, the rim is flanged, thus keeping the belt from running over the edge. In some installations it is desirable that a small variation in speed is made possible during the running of a belt. In these cases a cone speed pulley is employed which consists of two cones of equal and similar dimensions, but sloping in opposite directions. A modification of this drive is the step pulley which permits a variation in speed from the line shaft to the machine operated, but these speeds are constant and cannot be changed during the operation of the machine.

The rims of pulleys are frequently perforated, which, it is claimed, increases the friction between the belt and the pulley by letting out the film of air, thus giving more efficient operation of the belt.

The arms of metal pulleys are either round or elliptical in cross section and usually extend radially from the hub, but they may be curved or double curved, which makes the pulley more elastic. Frequently in small pulleys no arms are used, the pulley being made with a solid web in the place of arms. Split wood pulleys have various forms of arms, but the most common have parallel arms on either side of the dividing line, with the addition of radiating arms when the size demands a more solid pulley.

In order to secure the best operation of a belt drive, the diameter of the pulley should be equal to or greater than 36 times the belt thickness, but in the case of link belts 30 times the thickness of the belt. The ratio between the driving and driven pulleys should not be greater than 6 to 1, and the distance betweeen them depends upon the ratio of the diameter, thus where the ratio is 2 to 1 the distance should be greater than 8 in., 3 to 1, 10 in.; 4 to 1, 12 in.; 5 to 1, 15 in.

For best operation the convexity of a

Fig. 4. Forms of Keys used for Pulleys

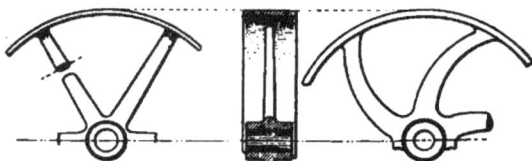

Fig. 5. Straight and Single-curved Pulley Arms

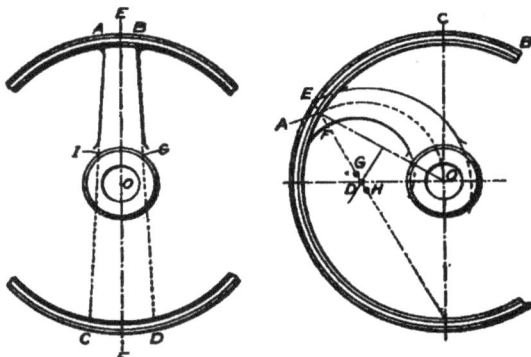

Fig. 6. Methods of laying out Arms

crowned pulley should be 1/20 the width of the belt and for all pulleys the width should be equal to 5/4 the width of the belt where an isolated pulley is employed. This may be made smaller where the belt is shifted from one to another running beside it.

The thickness at the edge of the rim for iron pulleys is usually taken at 0.01 the face width plus 0.08 in. The thickness of the rim at the center is equal to twice that at the edge, plus the amount of convexity. The rule taken for the thickness of the nave or hub is that this should be 0.4 in. plus 1/4 the diameter of the shaft, plus 1/60 the radius of the pulley, the dimensions all being in inches. The length of the nave should be equal to or greater than 2.5 times its thickness.

Keys employed for fastening pulleys to the shaft are of four kinds: the hollow key, as shown at A in Fig. 4, is used for light work and depends entirely upon

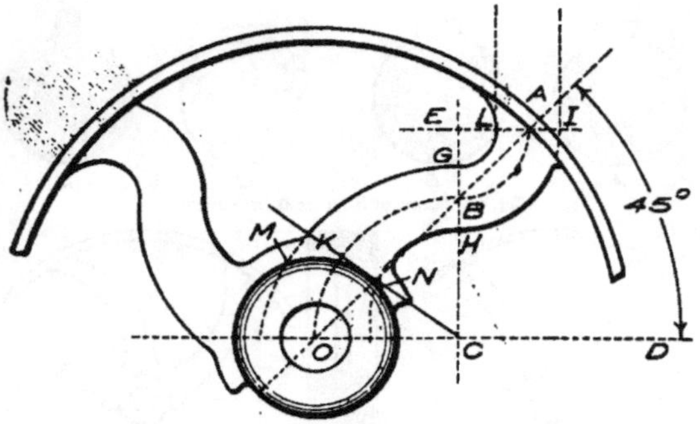

Fig. 7.   Laying out a Double Curved Arm

nave, draw circles to scale representing the rim, nave and bore of the pulley, as shown in Fig. 6. Draw the diameter EF through the center O. Lay off FD, FC and AB equal to ¾ the large diameter of the arm cross section at the nave. Draw the lines AC, BD, the parts AI and BG being the profile of one arm. The other arms are laid out in the same manner and equal to this one.

The manner of laying out a curved arm is somewhat more complicated. With a radius OA equal to the radius of the pulley scribe an arc which covers an angle of about 120 degrees. On this arc lay off AC equal to ¾ the arc between arms of the pulley. Then draw the radial line OC and perpendicular to it at O draw the line OD. From the center of the line OA erect a perpendicular ED which intersects the line OD at D. From D as a center with DO as a radius scribe an arc AO which is the axis of the curved pulley arm. Locate E and F on the line AD, each at a distance equal to ⅙ the large diameter of the cross section of the arm from A, locate G and H on the line AD, each at a distance from D equal to half AE, now with H as a center and HE as radius, draw another arc extending from the rim of the wheel to the nave. This is the outer profile of the arm, and for the inner profile use the center G with the radius GF, and draw an arc from the rim to the nave of the wheel.

One method of laying out a profile of a double curved arm is as follows: Having drawn arcs to represent the nave and rim of the pulley, draw the radial line OA at 45 degrees from the line OD, Fig. 7. Take the point B, on OA at ¾ the distance OA from O. Through B draw the line EC perpendicular to OD, then draw AE parallel to OD. The points E and C are the centers from which the axis of the arm is drawn. From E as a center and EA as a radius, scribe an arc AB and from C as a center and CB as a radius, scribe an arc BK. The line thus formed is the axis of the double curved arm. On the line EA extended, take AI equal to ⅙ the large diameter of the arm at the nave and AL of the same length, then

friction to hold the pulley from turning. The flat key, as shown at B in Fig. 4, is used for medium service and requires that the shaft be marred if this key is employed. For heavy work a countersunk key is employed, this also requires that a key seat be made in the shaft. Feather keys are used where the pulley must be shifted along the shaft. The proportions of keys of different types are given in the article on shafting in another part of this issue.

ARMS

In iron pulleys the oval-shaped arm is most common, dimensions of the cross section being, for the large diameter, twice that of the small diameter, the large diameter being in the plane of the pulley. These arms may be either radial, curved or double curved. The rule for the number of arms employed is: Divide the radius of the pulley by the width of belt, to the quotient add 5, then divide by 2, the result will be the number of arms. The long diameter of the cross section of the arm is equal to 0.24, plus ¼ the width of the belt, plus the radius of the pulley divided by ten times the number of arms; this being the diameter of the nave, the small diameter is one-half this amount. At the rim the dimensions of the arm are taken equal to ⅔ the dimensions at the nave. These proportions are admirably adapted for large pulleys, but in smaller ones they should be increased in order to take care of slight defects in castings.

To lay out the profile of a straight arm, having determined the diameter of the pulley and the dimensions of the arm and

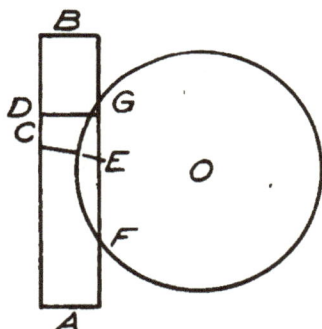

Fig. 8.   Templet used in lagging Pulleys

from *E* as a center and *EI* and *EL* as
radii, scribe arcs *IH* and *LG* respectively.
Having laid out the large diameter of
the arm upon the nave, find centers upon
the line *BC* from which arcs may be
drawn from *G* and *H* to the points upon
the nave. This gives the pròfile of one
double-curved arm and the rest of the
arms are made similar to it.

To design a set of cone or step pulleys
which will maintain an equal tension of
the belt at all speeds of the drive, two
equal and similar cones tapering in opposite ways are selected as the line of centers
for the pulleys when crossed belts are
employed. There are, however, few
cases where it is advisable to use crossed
belts for cone pulleys, owing to the severe
wear upon the belts, and to design a set
of pulleys for open belts, two similar
conoids, tapering in opposite directions
and bulging in the middle, are employed.
The line representing the centers of the
various steps is laid out according to the
following rule, given by Rankine: The
radius of the center pulley is equal to
½ the sum of the radii of the large and
small pulleys of the cone, plus the square
of the difference between the large and
small radii of the cone divided by 6.28
times the distance between the axis of
the cones. While this does not give an
exact solution it is sufficient for all practical purposes, and an exact result requires the use of higher mathematics.

LAGGING PULLEYS

It sometimes becomes necessary to
increase or decrease the speed of a line
shaft, owing to changes in the power
plant. Under these conditions it would

be necessary to secure new pulleys of
different diameters, or change the diameters of those pulleys employed under the
old conditions. The latter method is
quite readily accomplished by lagging
the line pulleys when the speed of the
line shaft is decreased or lagging the
driven pulleys in case the speed is increased.

Pulleys made of any material may be
lagged with wood, increasing the diameter even as great as 4 or 6 in. To do this,
holes are bored into the face of the pulley
into which screws are set to fasten the
lagging firmly to the pulley. A templet
for cutting the lagging is made as follows:
With a center *O*, Fig. 8, and a radius
equal to the radius of the pulley to be
lagged, scribe an arc. On the edge of
this arc, overlapping it 2½ or 3 in., tack
a thin piece of board *AB*, then draw a
radial line, *EC*, and the arc *EF* on this
board. With a band or hack saw, cut
this board along the line *FEC*. The
board may be cut off at *DG* in order to
make it easily handled. With this
templet the boards used in lagging the
pulley, are laid out and sawed accordingly.
The lagging should be placed with the
grain running around the pulley and glued

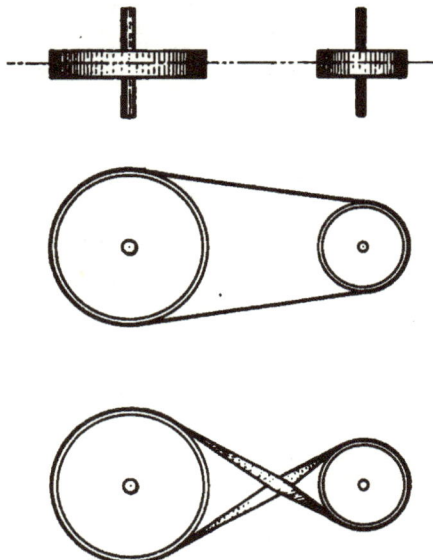

Fig. 9.   Open and Crossed Belt Arrangement

firmly together when placed in position, then fastened with screws. The face should then be turned, and finished by giving it a coat of shellac or varnish.

One of the best methods of lagging a pulley where it is not necessary to increase the diameter a great amount, is by means of leather riveted to the rim. In this manner the diameter can be increased to the extent of an inch.

A rather crude method sometimes employed is to use rope wrapped upon the face of the pulley, but it can hardly be recommended for a permanent job.

SETTING PULLEYS FOR SPECIAL TRANSMISSION

As a general law for handling all cases of pulley setting in belt transmission, it may be stated that the direction in which the belt leaves a pulley must be tangent to the center line of the pulley to which the belt goes. Keeping this in mind the following solutions of belt transmission are readily apparent.

Where it is desired to transmit power between two parallel shafts the direction of motion being the same, an open belt is used, as shown in illustration Fig. 9. This is the simplest form of belt drive and that most commonly employed.

To secure opposite direction of rotation of two parallel shafts it is necessary to cross the belt, this arrangement also being shown in Fig. 9. In both these cases the pulleys are placed in the same plane and any ratio of speed up to 1 to 6 may be obtained.

When it is desired to transmit power from one shaft to another not in the same plane, a quarter turn arrangement is resorted to. The general solution of this problem is shown in Fig. 10, where variation in the angles of the two shafts is illustrated.

(To be continued)

The gasoline engine serves a very useful purpose, but do not expect it to run the whole farm.

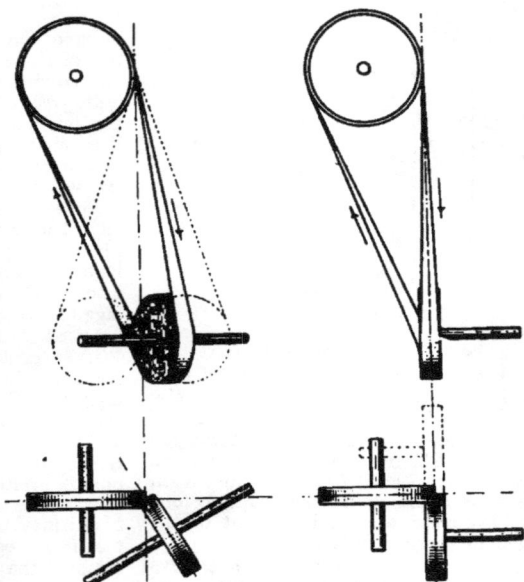

Fig. 10.   Quarter Turn Belt Arrangement

Trans-Pacific Wireless

First Communication between San Francisco and Japan

Wireless communication between San Francisco and Japan, a distance of 6,000 miles, was established recently. This was the first time that a wireless message had been received across the Pacific Ocean. When the operator at Hill Crest station caught the signals he made them out to be the call for the *Chiye Maru*, a steamer that was then due at Honolulu. He answered the signal and learned that the call came from the Japanese wireless station on Hokushu Island, in the northern part of the Japanese archipelago. The operators exchanged messages for some time.

"Mr. Cleaver, how do you account for the fact that I found a piece of rubber tire in one of the sausages I bought here last week?"

"My dear madam, that only goes to show that the motor car is replacing the horse everywhere."—*New York Times.*

VALVE SETTING

F. L. BAILEY

Almost all eccentric rods have some means of increasing or decreasing their length, and a misadjustment or unintentional change of this device is probably responsible for as many valves incorrectly set as is the slipping eccentric. If the valve is of the plain D type, it can be easily told if anything is wrong with the rod adjustment. Simply remove the valve chest cover and mark the extreme travel of the valve on both sides, then measure from each of these marks to the steam edge of the adjacent port, and if the two measurements are equal you may be sure that the rod is not at fault and the trouble lies in the eccentric. The only exception to this rule is that in case of a vertical engine the lower port is usually given a slight advantage over the upper to offset the weight of the reciprocating parts. To set the D-valve, place the engine on dead center, using the method given later, and turn the eccentric or adjust the rod according to which is at fault until the port admitting steam behind the piston is opened to an amount guessed to be equal to the lead. Then turn the engine on the other center, and if the lead opening is the same, the valve is set.

It must be remembered, however, that the D-valve is not employed so universally as formally, and the chances are it will be a piston valve with which we have to deal; and in that case we must resort to measurements alone, for it is impossible to see just what the conditions are. Most piston valves carry the live steam between the pistons, so supposing that the one in Fig. 1 does let us see the proper methods employed in setting it. First remove the valve chest head and take out the valve. Then the dimensions marked A, B, C, etc., in Fig. 1, should be carefully taken. It is well to make a rough sketch of the valves and ports and substitute the exact dimensions instead of the letters, as shown in the figure. By preserving this diagram the valve may be set at any subsequent time without removing anything but the valve chest head. After taking the measurements replace the valve exactly as before, and turn the engine over until the *valve* reaches the highest (presuming the en-

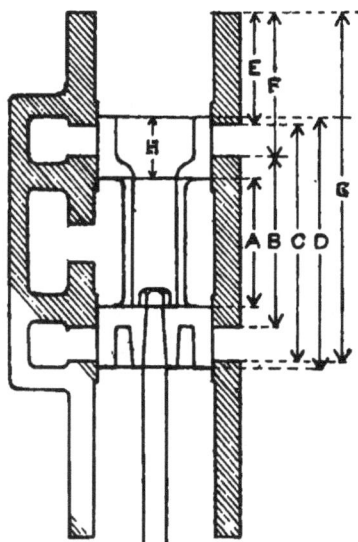

Fig. 1

gine is vertical) point of its stroke. Then carefully measure the distance from it to the top of the valve chest and add the dimension H. Subtract this sum from F, preserving the remainder. Next turn the engine until the valve reaches its lowest point, and again measure from it to the top of the valve chest adding this time the sum of H and A, and subtracting the sum of F and B. If the remainders thus obtained are equal the rod adjustment is correct, and you may expect to find the trouble in the eccentric. If, however, they are unequal (with the exception of a slight allowance, as noted above in a vertical engine) the rod adjustment is certain to be at fault. In either case the next thing to do is to place the engine on dead center. This may be done accurately by the following method:

Turn the engine about one-eighth of a revolution off center, and make a mark on the crosshead and guide in such a manner that the exact position may again be obtained. Then with a tram set on the floor or base of the engine at some finely-marked position make a mark on the flywheel. Next turn the engine back across center until the mark on the crosshead and guide again coincide, and with the tram, make another mark on the flywheel. Now bisect the space between the two marks on the flywheel and turn the engine until the bisection comes even with the tram, and the engine will be exactly centered.

Suppose the engine is centered with the crank nearest the cylinder, then, to set the valve take dimension H and subtract it from E, and then by turning the eccentric or adjusting the rod, as the case requires, make the distance from the valve to the top of the valve chest equal to the remainder. The valve will then be set "blind" or without lead. The amount of lead to gain the greatest efficiency will depend on a great many things, but in general, the higher the speed and the steadier the load the greater the lead should be. Experience alone and a close observation of the engine working under different amounts of lead will enable one to tell just how much should be given. Of course, more or less lead is given by changing the eccentric and not by adjusting the rods.

It would seem at first thought that it would be more difficult to set a valve on an engine fitted with a reversing gear, but it is really much easier, in fact it can often be done without taking any measurements or even removing the valve chest head. If, for instance, an eccentric slips on an engine with a Stephenson link gear it may be set correctly enough for all practical purposes by the following method. First, find which eccentric is at fault, and then set the reverse lever so that the valve will derive its motion entirely from the good 'eccentric. In other words if the "go-ahead" eccentric (in the case of a locomotive) has slipped, set the lever on the full backing position. Then turn the engine on dead center and make a mark on the valve stem where it enters the packing gland. Next move the lever over to the opposite side of the quadrant and turn the faulty eccentric until the mark on the valve stem is at the same position as before, and the valve is set. In the Walschaert gear one has simply to remember that the eccentric when properly set is either 90 degrees (one quarter of a turn) ahead or behind the crank, according to the kind of valve

used, and whether the motion is transmitted direct or indirect. In the Marshall gear the eccentric is either right with the crank or directly opposite it, also depending on the type of valve and the way the motion is transmitted.

There is another way to set a valve which should not be overlooked, that is, by means of an indicator. Where an indicator is available this proves a most reliable method. The indicator is put on in the regular manner and the valve adjusted until the most perfect card is obtained. Sometimes most surprising facts are revealed by an indicator on an engine apparently all right. The writer at one time out of idle curiosity put an indicator on the high-pressure cylinder of a 40 h.p. cross compound vertical engine. The card taken showed fully twice as much work being done on the lower end of the cylinder, yet no one in the engine room dreamed but what the valve was set correctly. In this case the engine was fitted with a Rites governor, making it impossible for the eccentric to be wrong. The adjusting nut on the valve stem had gradually worked loose and turned until the above resulted. With the aid of the indicator it was but the work of a few minutes to properly readjust it.

To the novice valve setting is apt to seem an accomplishment to be acquired only after years of experience, and true enough one is apt to feel a little ill at ease upon first undertaking it, especially if a piston valve is employed, but if the theory is fully understood and all measurements accurately taken little trouble should be encountered.

Industry Buys Anything

The world was made for you. All that has gone before was that you might be. If you desire wealth, it can be yours. If you desire fame, it can be yours. But you must pay the price. Industry is the only coin acceptable at the gate of success. Our Roosevelts, our Carnegies, our Edisons, have bought their way to glory by hard labor. It's "the only way." The world and all therein is—that you want—is yours, if you pay the price in the free coin of the realm—industry.—GLEN BUCK.

No lie can last a long time.—HUBBARD.

WIRELESS TELEGRAPHY

In this department will be published original, practical articles pertaining to
Wireless Telegraphy and Wireless Telephony

CAUSES AND PREVENTION OF DETERIORATION OF WIRELESS DETECTOR

ERNEST C. CROCKER

It is common experience among wireless workers that, with but few exceptions, all wireless detectors of the rectifying type become less and less sensitive with time, even though they are only occasionally used. This deterioration means continual adjustment of the detector until its entire acting surface becomes "dead" and must be renewed. In practice, there are two general types of rectifying detectors in use—the electrolytic detectors and the so-called crystal detectors. We shall consider the causes of deterioration of each type separately.

The essential active part of an electrolytic detector is a small surface of platinum in contact with an acid. The arrangement of the platinum surface makes possible two styles of detector, one in which the platinum, in the form of a wire is adjustable vertically and just dips into the liquid, and the other in which the platinum wire is sealed in a glass tube, the end of which is ground off so as to allow only the end of the wire to be exposed.

If we have an electrolytic detector "tuned" in circuit for a strong nearby station and view the detector in the dark, we can often "read" the station's message by means of the little flashes of light which occur at the fine point. This effect is more powerful than one might believe, and it is often possible to see light at the point of a delicate detector, when the sending station is only of 2 k.w. power, and is 5 miles away; and even though one uses only a small antenna.

Where there is light there is usually heat, and when we consider the powerful action which takes place at the tiny point, we can scarcely wonder that the point is often ruined. If a glass tube is used, the heat cracks the glass, and in almost any case melts the end of the platinum into a ball or even volatilizes it entirely. Apart from this violent destruction of the detector, there is a slow loss of sensitiveness which is chemical in nature and is usually due to impurities in the liquid which dissolve away the platinum. Although considered acid-proof, platinum is slowly dissolved by hydrochloric (muriatic) acid, particularly if the acid is strong, or contains nitric acid or iron. Nitric or sulphuric acids alone, when pure, scarcely attack platinum.

What has been said in regard to heating also applies to the crystal detectors, although these will usually withstand rougher treatment than will an electrolytic detector. The chemical action, however, is more noticeable, even though the air is the only source of the disturbing chemicals.

In a crystal detector we have a point sometimes made of metal, sometimes of a mineral like chalcopyrite or bornite (essentially sulphides of copper and iron), and sometimes of an artificial compound such as iron monosulphide, which is in contact with a "sensitive" surface. The most prominent sensitive materials are: silicon (an element, Si), galena (lead sulphide, PbS), iron pyrites (iron di-sulphide, FeS_2), carborundum (silicon carbide, SiC) or zincite (zinc oxide, essentially ZnO). Of these latter materials, silicon and carborundum are artificially prepared and are fairly uniform in sensitiveness throughout a given piece. This, however, is not the case with the natural minerals, the sensitiveness of which depends largely on the particular way in which they were originally formed. In the case of the minerals, pyrites and galena, one may find it necessary to examine tons of material to obtain even a pound of active substance, so unusual is the active form of these substances. Not only is this sensitive material hard to find, but it is also rather easily impaired by the chemical action of certain components of the air.

The atmosphere is composed principally of nitrogen, oxygen, carbon dioxide and water vapor, and ordinarily we consider it as having no other components. None of the above substances, except water vapor, has perceptible action upon detector materials, but some of the minor constituents of the atmosphere, such as ozone, hydrogen peroxide and nitric acid vapor, exert a powerful destructive action. At a wireless station ozone is produced in considerable quantities, and much detector deterioration is due directly to this cause. Although the amount of these substances in the air is very small, they work continuously, so that in time they have a decided action.

The action of ozone or hydrogen peroxide upon galena is to transform the conducting sensitive surface of lead sulphide (PbS) into non-conducting lead sulphate (PbSO$_4$), thus producing a "sulphating" comparable with that of a storage battery. The action upon silicon (Si) is not so noticeable, however, but this substance is oxidized to silicon dioxide (SiO$_2$), which has the same composition and about the same conductivity as common sand. Zincite is already a mixture of oxides which cannot be further oxidized, but it is nevertheless quite sensitive to water vapor and dust. Dust almost always conveys some soluble materials, which, when they come in contact with the water which is held in the crevices and on the surface of detector materials, dissolve and make a conducting solution which shunts away the current and tends to make the detector useless.

The detectors which are most sensitive to wireless currents are generally sensitive to chemical changes caused by the air. The simplest way to obtain great sensitiveness with freedom from chemical disturbance is to cover the detector with a glass jar or to enclose it in a metal box, preferably the latter, for reasons to be presently explained. Inside the enclosure should be placed a little vial half-filled with either strong sulphuric acid or dry calcium chloride. The calcium chloride is cheap and clean and can even be used on shipboard where sulphuric acid could not be used. Enclosing the detector will do away with dust and harmful gases, and the drying agent will reduce the amount of moisture in the air in the box.

Very little can be done to prevent a distant station from "breaking down" a detector, except to use a detector which will withstand the strain. It is the action of one's own sending station which must be guarded against. The best way of eliminating this trouble is to surround the detector entirely with metal, even though it be no thicker than tin-foil, and to "break" the detector circuit at the surface of the metal box, or just within it. This breaking is best performed with an electromagnet so arranged that only after it has disconnected the detector will it be possible to turn on the sending current.

One source of detector trouble that is often overlooked is vibration. Every detector should rest on an under-base of felt or some such elastic material. Moreover, the spring which makes the tension between the elements should be fairly long and flexible and the proper tension secured by much tightening upon the adjusting screw, and not by a short or stiff spring which is lightly adjusted.

In summing up, it must be said that that the only way to have really satisfactory receiving is to choose a detector which is moderately free from electrical weaknesses and to enclose it in a metal box for its protection from the sending station's current on the one hand and from atmospheric corrosion on the other. This latter precaution is almost imperative in damp or dusty places. If the detector is properly protected from vibration, as well as from electrical and chemical troubles, it will scarcely ever need readjusting, so that the disadvantages of having it "boxed in" are more than compensated by its good behavior.

U.S. NAVAL WIRELESS TELEGRAPH STATIONS AND THEIR SERVICE TO SHIPPING

The following information, furnished by the Bureau of Steam Engineering, Navy Department, is published so that shipmasters and the shipping interests may see the advantages afforded by this service.

Wireless Communications between Commercial Vessels and Naval Wireless Telegraph Stations

The facilities of the wireless telegraph stations of the Naval Coast Signal Service, including those on the Nantucket Shoals and Diamond Shoals Lightships and the one soon to be established on the Frying Pan Shoals Lightship, for collecting and disseminating information useful to mariners, and for communicating with ships at sea where not in competition with private wireless telegraph stations, are placed at the service of the public generally, and of maritime interests in particular under the rules established herein, which are subject to modifications from time to time, for the purpose of:

(a) Reporting vessels and intelligence received by wireless telegraphy in regard to maritime casualties and overdue vessels.

(b) Disseminating hydrographic information concerning menaces to navigation, etc., sent out by the Hydrographic Office, or by a branch hydrographic office.

(c) Sending out storm warnings and weather reports as received from the Weather Bureau.

(d) Sending time signals for rating chronometers on vessels at sea.

(e) Receiving wireless information from ships for the Hydrographic Office concerning serious obstructions to navigation, such as derelicts, wrecks, ice, or any information that may be useful to the Pilot Chart or Hydrographic Bulletin.

(f) Receiving weather reports for the Weather Bureau of the Department of Agriculture.

(g) Receiving wireless telegrams of a private or commercial nature from ships at sea for further transmission by telegraph or telephone lines.

(h) Transmitting wireless telegrams to ships at sea.

Information Furnished to Shipping

This includes (b), (c) and (d) above, and is furnished gratis.

Hydrographic Information and Storm Warnings

Information concerning wrecks, derelicts, ice, and other dangerous obstructions to navigation, whenever received from the Hydrographic Office, or from a branch hydrographic office, and storm warnings received from the Weather Bureau, are sent broadcast four times daily, viz., at 8 a.m., noon (immediately after the time signal, if sent), 4 p.m. and 8 p.m. Ships within range of a naval wireless station should be prepared to receive these hydrographic messages and storm warnings at the hours mentioned, and should avoid sending wireless messages at these times. One vessel sending may prevent several others receiving information necessary to their safety.

Naval wireless stations will furnish this information to passing vessels on request, whenever practicable, at other hours than those mentioned above. Should it not be practicable to send out this information on one of the hours

scheduled it will be held until the next scheduled time, except that important storm warnings, reports of lightships off stations, etc., will be treated as urgent, and sent out as soon as practicable after each hour scheduled.

Time Signals

The following wireless stations send out time signals broadcast between 11.55 a.m. and noon every day, except Sundays and holidays, for the determination of chronometer errors, and hence time and longitude at sea: Portsmouth, Boston, Cape Cod, Newport, Fire Island, New York, Cape Henlopen, Washington, Norfolk, Beaufort, Charleston, Key West, Pensacola and New Orleans on the Atlantic and Gulf coasts; Table Bluff, North Head, Mare Island and Point Loma on the Pacific coast. This service has been suspended at St. Augustine, and will be re-established as soon as practicable.

It is proposed to extend this service to the wireless stations at Guantanamo, Colon, and Tatoosh Island, if necessary arrangements can be made.

The signals are sent from the Naval Observatory, Washington, for the Atlantic coast between 11.55 a.m. and noon of the 75th meridian west of Greenwich, and from the observatory at the Mare Island Navy Yard between 11.55 a.m. and noon of the 120th meridian west of Greenwich for the Pacific coast.

The wireless sending or relay key in each wireless station is connected to the Western Union lines by a relay at about 11.50 a.m., and the signals are made automatically direct from Washington or Mare Island.

Time signals from each of the observatories mentioned begin at 11.55 a.m., standard time, and continue for five minutes. During this interval every tick of the clock is transmitted except the 29th second of each minute, the last five seconds of each of the first four minutes, and finally the last ten seconds of the last minute. The noon signal is a longer contact after this longer break.

(Note.—See diagram on back of Pilot Chart of the North Atlantic Ocean, No. 1400, of November, 1910, or North Pacific Pilot Chart for January, 1911.)

It is not necessary that an elaborate wireless telegraph installation be employed for the purpose of receiving these signals nor that a skilled operator be in attendance. Any vessel provided with a small receiving apparatus with one or two wires hoisted as high as possible and insulated from all metal fittings, or preferably stretched between the mastheads with one wire led down to the receiver, may detect these signals when within range of one of the seacoast wireless stations.

These time signals have been used successfully by vessels for rating their chronometers and have been used by surveying vessels in the accurate determination of longitudes.

Collection and Transmission of Information from Sea

All information for the Hydrographic Office and all weather reports received by any wireless station will be forwarded by wire direct to the

Hydrographic Office and the Weather Bureau, respectively, without charge.

Stations at isolated points, and other stations in important cases, will relay these messages to other wireless stations for further transmission if necessary.

Commercial Messages

All naval wireless telegraph stations with the following exceptions, viz., those at the navy yards at Boston, New York, Philadelphia, Norfolk, Puget Sound and Mare Island, and the naval stations at New Orleans and Yerba Buena, San Francisco, will handle commercial messages under the following conditions:

(1) That no commercial station is able to do the work.

(2) That no expense is incurred by the Government thereby.

(3) That no money or accounts in connection with this business is handled by any person in the employ of the Navy Department.

(4) That the handling of the commercial messages shall not interfere with Government business.

The Government handles all commercial wireless messages without charge, but assumes no financial responsibility whatever for errors, delays, or non-delivery. Every effort will be made, however, to forward all messages accepted accurately and expeditiously by the best means available. Confirmation copies of commercial messages sent through naval wireless stations will be sent only when request is made in advance, or within thirty days after messages are forwarded.

Messages of all kinds received from ships at sea will ordinarily be forwarded by land wire, the land wire charges to be collected at destination.

In case of isolated stations, such as stations on Alaskan Islands and in emergencies, these messages will be relayed to other wireless stations for further transmission if necessary.

Position reports will be forwarded to owners or agents by land wire when request is made.

Messages received by land wire at a naval wireless station for a ship at sea will be forwarded by wireless when the ship comes within range. For this reason ships should ordinarily communicate with wireless stations while passing along the coast, giving their positions.

Messages received by a wireless station for a ship which cannot be delivered for any reason will be returned to the land wire company from which it was received.

The personnel of naval wireless stations required to keep the strictest secrecy in regard to the contents of messages passing through their stations, and they are not permitted to communicate the fact that a message on any particular subject has been received.

All messages are kept on file, and senders and addressees may obtain copies of all messages as sent upon request.

Code

The Continental Code is the one used by the United States Navy, and is preferred for all wireless communication.

Changes being made in Sending Wave-Lengths of Naval Stations and Ships

All naval shore stations, except certain long-distance stations to be mentioned later, will have their apparatus adjusted for sending on a wave length of 1,000 meters as rapidly as possible. All sets will be "sharply tuned," so that it will be necessary for a vessel receiving to have her receiver very carefully adjusted for receiving a 1,000-meter wave. Otherwise the signals of a naval station may not be heard. A difference of 3 percent in wave length between the signals sent and received may be expected to cut down the strength of signals by one-half, and a receiver set for receiving on a 900-meter wave or on a 1,100-meter wave (i.e., a difference of 10 percent), may not hear the shore station at all, depending on the distance. Vessels of the Navy are having their apparatus adjusted for calling on a 600-meter wave length, and may use other wave lengths for communicating with each other. When communicating with a naval vessel she may be expected to use a 600-meter wave having the same characteristics as the long wave described above. Shore stations and ships of the Navy may be expected to receive all calls from merchant ships using those wave lengths ordinarily in use at the present time. It is only the receiving by merchant ships which will be affected by the changes now being made, as described above.

The attention of all steamship companies, ship owners, masters of vessels and operators is invited to the advantages of transmitting apparatus capable of sending with a certain wave length with one sharp crest only. Signals from such apparatus can be readily "tuned out" if the desired signals differ sufficiently in wave length and the interfering ship is not too near. A change from direct to inductive coupling between the closed or oscillating (condenser) circuit and the open or radiating (aerial) circuit will accomplish this without loss of efficiency if two circuits are carefully adjusted by a wave meter and the proper coupling between the two inductances is used. A few experiments in tuning out any naval ship or station, properly adjusted to new standard tunes, especially those with high-pitched sparks, will show some of the possibilities of ordinary wireless working in the future.

Sharply tuned transmitters involve attentive receiving operators, in order that no calls may be missed. It is suggested that each line select a wave length under 600 meters and carefully adjust the transmitters of all its ships to that tune, as is being done with ships of the Navy.

Certain stations referred to above may use a wave longer than the standard (1,000 meter) for ship communications, and these exceptions will be published from time to time in "Notices to Mariners" and in the "List of Wireless Telegraph Stations of the World."

Working Rules for most Satisfactory Wireless Communication

A vessel wishing to communicate with a coast station should commence calling when about 100 miles from the station, having first "listened in" to ascertain that she is not interfering with messages being exchanged within her range.

The power and range of many stations, however, are being rapidly increased, and vessels should note at what distances they hear certain stations working with merchant ships in order that communication may be held over the maximum distance if necessary.

Calls should not be prolonged beyond fifteen seconds and should be followed by the letters of the station calling. Reasonable time should be given for an acknowledgment before repeating the call. A number of complaints have been received that vessels frequently call for long periods without pausing to hear whether or not their call is heard, or they are interfering with other communications going on. If, after making the call, a ship hears the signal "BK" or "XXXX" made, she should take it to mean that one station communicating with another is being interfered with by her calls and that she should wait.

As the use of longer wave lengths for avoiding local interference and for long distance and overland communication will be used considerably in the future, a vessel should listen on the longer wave lengths as well as those around 1,000 meters. Otherwise she may not understand why her call is not acknowledged immediately. While intercommunication is going on between two shore stations or between shore stations and naval ships with long wave lengths no ship calls will be heard.

After the station acknowledges the call the vessel should report her position. The following manner of reporting position, etc., is preferred:

(a) Distance, of the vessel from the coast station in nautical miles.

(b) Her true bearing from coast station in degrees, counted from 0 to 360.

(c) Her true course in degrees, counted from 0 to 360.

(d) Her speed in nautical miles per hour.

(e) The number of messages she desires to transmit.

This will enable the coast station receiving a number of calls from various vessels to determine which one will pass out of range first in order that that vessel may be permitted to finish her business. When a coast station acknowledges she may state whether or not she has messages for the ship, and if she cannot communicate further with the ship at that time the ship will be informed of the length of the time it will be necessary to wait.

On receiving word to "go ahead" the vessel should send a message as follows:

(a) "HR" or "MSG."

(b) Number of message.

(c) Ship's call.

(d) Operator's sign.

line the fact should be indicated in item (h) by the initials "WU" or "PT," or other designation necessary. In an original message sent from a ship to a wireless station items (f) and (g) may be omitted.

In case of long messages the sending ship should get an acknowledgment after every twenty words, or thereabouts, before proceeding.

Communication may be interrupted at any time and the right of way given to a Government station or vessel, if necessary, or to any vessel in distress, or to send broadcast any important information.

All stations may be expected to be familiar with the methods of communication adopted by the International Wireless Conference of Berlin, of 1906, with special regard to the international signal of distress "SOS," and the signal "PRB," expressing the desire to communicate by means of the international signal code by wireless. Ships are requested not to use the letters "OS" preceding a position report, as the letters "OS" made rapidly and continuously might be mistaken for the signal of distress "SOS."

Shore stations in designating the order in which messages will be received from the vessels within range will be guided exclusively by the necessity of permitting each station concerned to exchange the greatest possible number of wireless telegrams. At all times business may be expected to be handled in the following order:

(a) Government business, viz., telegrams from any Government Department to its agent aboard ship.

(b) Business concerning the vessel with which communication has been established, viz., telegrams from owner to master.

(c) Urgent private dispatches, limited.

(d) Press dispatches.

(e) Other dispatches.

Reports to Navy Department

In order that the efficiency and reliability of the service may be steadily increased, it is requested that merchant vessels unable to communicate with any station open for public business report the matter in full to the Secretary of the Navy, Washington, D.C. The statements should be specific, giving date and hour, local conditions as regards atmospheric disturbances and wireless communications, distance from the shore station, and the statement that the wireless apparatus of the ship was in good condition, as evidenced by other communications effected at or about the same time, and that the receiver was adjusted approximately for the sending wave length of the shore station. All reports will be investigated and the cause of the trouble

Portsmouth, N.H. (*Navy Yard*)—Uses standard time. New high-frequency 2 k.w. set recently installed. Handles commercial messages.

Boston, Mass. (*Navy Yard*).—New high-frequency 5 k.w. set being installed.

Cape Cod, Mass.—Post-office address, North Truro, Mass. Telegraphic address, Navy Wireless, Highland Light, Mass.

Newport, R.I. (*Torpedo Station*).—New high-frequency 5 k.w. set being installed. Handles commercial messages.

Nantucket Shoals Lightship.—Post-office address, Care of Babbitt & Wood, New Bedford, Mass. Telegraphic address, via Torpedo Station Newport, R.I. Uses short-wave length. Communicates with ships and Newport only. Ships passing are requested to communicate by wireless or by international signals in order that they may be reported via Newport. Ships whose wireless apparatus permits should report to Newport direct.

Fire Island, N. Y.—Post-office address, Bayshore, Long Island, N.Y. Telegraphic address, Wireless, Fire Island, N.Y. New high-frequency set to be installed. Handles commercial messages.

Philadelphia, Pa. (*Navy Yard*).—New high-frequency 2 k.w. set being installed.

Cape Henlopen, Del.—Post-office and telegraphic address, Lewes, Del. Handles commercial messages.

Washington, D.C. (*Navy Yard*).—Handles commercial messages. High-power station being erected.

Norfolk, Va. (*Navy Yard*).—New high-frequency 5 k.w. set being installed.

Diamond Shoals Lightship.—Post-office address, Care of Clyde Steamship Company, Pier 36, North River, New York, N.Y. Telegraphic address, Via Wireless Station, Beaufort, N.C. Handles commercial messages. Communicates only with Beaufort.

Beaufort, N.C.—Post-office address, Beaufort, N.C. Telegraphic address, Beaufort, N.C. (Western Union only). Handles commercial messages.

Charleston, S.C. (*Navy Yard*).—Handles commercial messages.

Frying Pan Shoals Lightship.—Installation in progress. Will communicate with Charleston.

St. Augustine, Fla.—Post-office and telegraphic address, St. Augustine, Fla. Handles commercial messages.

Jupiter Inlet, Fla.—Post-office address, Jupiter Inlet, Neptune, Fla. Telegraphic address, Jupiter, Fla. (Western Union only). Handles commercial messages.

Key West, Fla. (*Naval Station*).—Handles commercial messages. Two high-frequency sets, 25 and 2 k.w., to be installed.

Pensacola, Fla. (*Navy Yard*).—Handles commercial messages.

San Juan, P.R.—Handles commercial messages.

Guantanamo, Cuba (*Naval Station*).—New high-frequency 5 k.w. set to be installed. Handles commercial messages.

Colon, C.Z.—Post-office address, Colon, C.Z. Telegraphic address, Wireless, Colon. Twenty-five k.w. high-frequency set. Handles commercial messages.

St. Paul, Pribilof Islands, Alaska.—Established July 3, 1911. Standard sending tune.

Two operators. Hours of operation will be published later. Communicates with Nome and Unalaska by day and in addition with Kodiak and Cordova at night. Handles commercial messages.

Unalaska, Alaska.—Established August 10, 1911. On Amaknak Island. New high-frequency 5 k.w. set. Standard sending tune. Communicates with St. Paul by day and with Nome, Kodiak, and Cordova by night.

Kodiak, Alaska.—Established May 28, 1911. on Woody Island. Standard sending tune. Communicates with Cordova by day and with St. Paul and Unalaska by night.

St. Paul, Unalaska, and Kodiak transmit and receive messages to and from the U.S. Army Signal Corps station at Nome and the naval wireless station at Cordova, either direct or by relaying. The last-named stations are connected with the Signal Corps Washington-Alaskan Telegraph and Cable System, and messages to and from the United States are sent via cable to Seattle. Particular attention is invited to the necessity for providing for payment in advance, required by law, for any messages transmitted over the land lines or cables of the Washington-Alaskan Military System. Commercial concerns and ships intending to send messages to the United States or to the interior of Alaska through the naval wireless station at Cordova or the Army wireless station at Nome should make a deposit at the U.S. Army Signal Corps Office at Cordova or Nome to guarantee prepayment of charges, otherwise the messages can not be sent. For the present all naval Alaksan wireless stations are authorized to relay messages of all classes among themselves, but they are not expected to communicate with any station in the United States, except on rare occasions.

Cordova, Alaska.—Standard sending tune. On the Washington-Alaskan Military Cable. Communicates with Kodiak by day and with Unalaska and St. Paul by night.

Sitka, Alaska.—Standard sending tune. On the Washington-Alaskan Military Cable. Has not reliable communication with other Alaskan stations at present. Communicates with passing ships.

Tatoosh Island, Wash.—Post-office address, Tatoosh Island, Wash. Telegraphic address, Wireless, Tatoosh, Wash. Handles commercial messages, relaying to other stations as necessary. Communication by land wire for commercial messages will be arranged if practicable.

Bremerton, Wash. (*Navy Yard*).—New high-frequency 5 k.w. set being installed.

North Head, Wash.—Post-office and telegraphic address, Ilwaco, Wash. Handles commercial messages.

Cape Blanco, Ore.—Post-office address, Denmark, Ore. Telegraphic address, Marshfield, Ore. Handles commercial messages.

Table Bluff, Cal.—Post-office address, Loleta, Cal. Telegraphic address, Eureka, Cal. Handles commercial messages.

Farallon Islands, Cal.—Post-office address, via San Francisco, Cal. Telegraphic address, via Navy Yard, Mare Island, Cal. Handles commercial messages. Relays to Yerba Buena or Mare Island.

Mare Island, Cal. (*Navy Yard*).—New high frequency 5 k.w. set to be installed.

Point Arguello, Cal.—Post-office and telegraphic address, Surf, Cal. Handles commercial messages.

Point Loma, Cal.—Post-office and telegraphic address, San Diego, Cal. Handles commercial messages. New high-frequency 5 k.w. set to be installed.

Honolulu, T. H. (Naval Station).—Handles local commercial messages.

Guam, M.I. (Naval Station).—Handles local commercial messages.

Cavite, P.I. (Naval Station).—Handles local commercial messages.

Olongapo, P.I. (Naval Station).—Five k.w. high frequency-set to be installed.

Wireless Messages from Spitzbergen

The first aerial messages were exchanged recently between Spitzbergen, high up in the Arctic Ocean, and the European wireless stations. The new Spitzbergen installation, as the Norwegian periodical *Nordland* points out, is a creditable piece of quick construction work. It had to be, for the Spitzbergen summer is very short.

It was only in the spring that T. T. Heftye, the director of the Norwegian State Telegraph Department, obtained from the Storthing a grant of funds for the purpose. Building operations were begun about the middle of July and in September the whole station was finished, comprising dwelling houses for the operating staff, an engine house, big store-houses for provisions, motor oil, coal, etc.

Green Harbor, it was found, was the only suitable place in the archipelago for the erection of buildings, although the air waves to reach the nearest European wireless station, on the island of Ingo near Hammerfest, have to pass over a lofty mountain ridge, whereby they lose a good deal in power. This was foreseen and extra powerful machinery was installed.

Working conditions in Spitzbergen are somewhat peculiar, for the thermometer often records 45 degrees of frost. The engine room has, therefore, to be kept at an equable temperature by means of an ingenious heating system. There is also a double set of machinery, comprising two oil engines of 30 h.p., two 16 kw. dynamos, two 60-cell 725 ampere-hour accumulator batteries and two 10 k.w. motor generators.

The erection of the two 200 ft. trellis masts was a matter of considerable difficulty, as an enormous quantity of concrete for the foundations had to be shipped by sea all the way from Christiania. Spitzbergen has already contacted with the German station Norddeich and the British station of Poldhu in Cornwall, but the regular traffic is conducted by way of Ingo.

Wireless Goes 4,000 Miles

New Station at Coltano, Italy, transmits Greeting by Marconi to Glace Bay, Nova Scotia

Four thousand miles from Coltano, Italy, to Glace Bay, Nova Scotia, was covered in a greeting sent by wireless to the New York *Times* recently. This is the greatest distance a wireless message has ever traveled. It exceeds by 2,250 miles the news Marconigrams regularly sent from Clifden, Ireland. The brief message which was signed by Cavaliere Guglielmo Marconi, the inventor of wireless, marked the opening of the new station at Coltano, near Pisa, the most powerful in the world. It was transmitted over the land lines' from Glace Bay.

John Bottomley, New York manager for the Marconi Company, explained that the Coltano station, which is a short distance from Pisa, had been planned to connect Italy with Argentine. A large proportion of the population of the South American republic is Italian, and it has been the inventor's hope to connect Italy with Buenos Ayres. Mr. Bottomley said that a new station was being built at Buenos Ayres for this purpose. The distance in an air line from Coltano to Buenos Ayres is approximately 7,000 miles.

The next station at Coltano will transmit messages in all directions for distances corresponding to the 4,000 miles from Coltano to Glace Bay, but as yet, Mr. Bottomley explained, there are few long-distance stations ready to receive them. The English Marconi Company has recently taken over the Russian operations and is contemplating a number of large stations in the interior of the empire which would be in touch with Coltano and Clifden.

There are no long-distance stations in Africa, but the Italian fleet and army in Tripoli are in easy communicating distance with the home shores.

PHILIPPINE ARTESIAN WELLS
MONROE WOOLLEY

The artesian well is doing more, perhaps, than any other one agency to rid the Philippines of disease. Prior to American occupation the natives carried water in hollow bamboo poles from neighboring springs, creeks and sloughs, for drinking purposes. It was nothing uncommon in the early days to see a carabao or a pony wallowing in a pond where the lavandera did her week's washing, and from which the house muchacho carried the drinking and cooking water. This custom did not, of course, obtain in the larger towns and cities, nor among the better class of natives. Now, throughout the interior, the common people in most places have, through the medium of the artesian well, water equal in purity and quantity to any other community. The Bureau of Public Works has spent much time and money driving wells wherever needed. The well drilling crews, each having an American foreman and several native laborers, operate under the direction of Mr. Frank L. Irwin, who for many years worked in the oil fields of India. At first some of the more ignorant natives stood much in fear of the spouting fountains, not understanding how water could be forced from the bowels of the earth in a huge, steady stream. But finding the water pure, with no ill effects, and the supply never ending, they are now grateful for the accommodation.

Tower of Well Drilling Rig

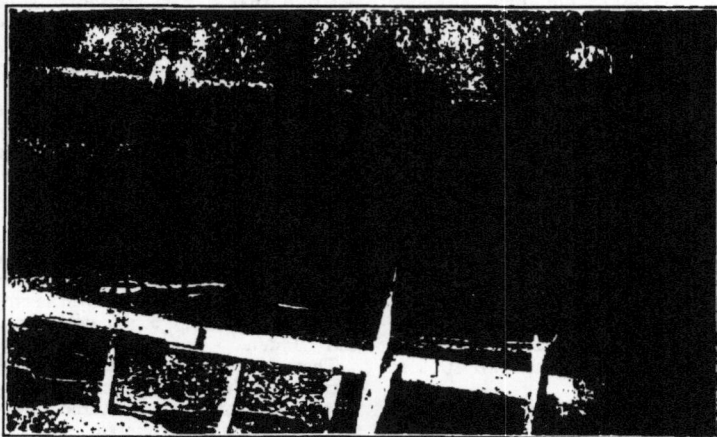

A LARGE ELECTRIC CLOCK

A large electric clock will be located 220 ft. above ground-level in a tower of the Royal Liver Insurance Company's new offices, overlooking the Princes Landing Stage, London. The clock will have four dials, each 25 ft. in diameter, while the minute-hands scale some 14 ft. in length by 3 ft. at their greatest width. Each of these huge bands is of copper, strengthened by 9 in., gun-metal ribs, and they are designed to withstand the maximum local wind pressures with an ample margin of safety. It has been calculated that an ordinary gale will entail a total pressure of no less than 11½ tons on a single dial. Each of the latter is cast in several pieces, to facilitate manufacture, transmission and erection, and the white groundwork will consist of special translucent opal glass, which, although only ⅛ in. in thickness has proved capable of withstanding a test pressure of 600 lbs. per square foot without fracture. The hours will be indicated by twelve distinctive marks, each 3 ft. x 1 ft. 6 in., and a space of 1 ft. 2 in. will separate the minute divisions. It will thus be seen that the circumference of each dial at the "minute" divisions is no less than 70 ft.

The clock will be fitted with the "B.P. Patent Waiting Train" movement. This both drives and controls the progress of the hands in such a manner that a practically continuous movement is the result. The movement will be under the control of a precision time transmitter, of Gent's design, which is already extensively employed in observatories at home and abroad. The resultant control takes effect every half minute, thus insuring absolute accuracy in timekeeping, while a distinct advantage is the separation of the timekeeping from the hand-driving element, thus eliminating errors due to

involves an ingenious automatic switching arrangement, which will not only turn on the light at dusk, and switch it off again at sunrise, but is also compensated for the gradual change in the seasons. This result is achieved by a cam, acting through a simple reducing gear, revolving once only in two years. The error due to leap years is compensated so nearly that the resultant error is only ten minutes in thirty years, at the end of which period it can be corrected and reset for a subsequent run of thirty years, in less than a minute, and without the aid of tools.

The huge clock will thus be entirely automatic, and entail no periodical attention beyond that possibly required to replace a burnt-out lamp in the lighting system.

Protection of Workmen on Buildings

It has been but a few years that the safety of employees on buildings has received legislative attention, but the list of states having laws on this subject has attained considerable length, three—Louisiana, Montana and Oklahoma—being added thereto within the period covered by a review of this matter in a bulletin of the Bureau of Labor.

The act of the Louisiana Legislature, says the *American Builder and Contractor*, calls for the installation of such devices as will protect workmen below from falling objects and requires safety rails to be placed on scaffolds, elevator shafts to be guarded, the adoption of signals for hoists, the construction of secondary scaffolds and protective floors and the determination and observance of the loading capacity of joists during the construction of buildings.

A Montana law on this subject requires scaffolds to be safe and so built as to

AN ELECTRIC WATER TANK GAUGE

JAMES P. LEWIS

It is often desirable to know the quantity of water in a tank or cistern located some distance away, without the necessity of going to look in same. As, for instance, in the case of a small water system, with a cistern several hundred feet away, on a hill, supplying water to a home.

When the owner of the device to be described desires to know when his supply is getting low, etc., he merely presses a button at his house, and the gauge shows the exact number of feet or gallons remaining.

Fig. 1

There are many other instances where such a device will prove a great convenience and time saver.

The principle of the gauge is as follows: An instrument, which is really a simple form of voltmeter, is located at the house, or any place desired. A line of common telephone wire runs from it to a small rheostat on the tank. This rheostat is controlled by the height of water, through a float and a simple system of levers, to reduce the motion. When the tank is full, all the resistance is cut out, and the pointer of the gauge reaches its maximum deflection. As the tank is emptied, the resistance is gradually introduced, the

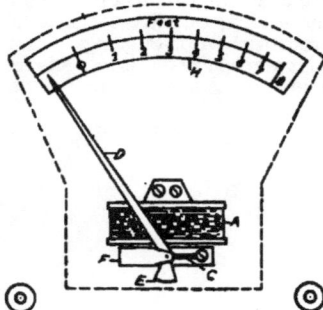

Fig. 2

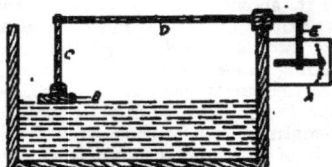

Fig. 3

gauge registering at all times an amount corresponding to the depth. The reading is taken directly from the scale, as the gauge is calibrated in feet (or gallons) instead of volts. A few dry cells are sufficient to operate same, as they are only used the short time during which the reading is being taken.

A simple and easily made form of voltmeter for the gauge is shown at Figs. 1 and 2. Fig. 1 is an end view of the coil and moving system; while Fig. 2 shows a plan view of the completed instrument. The baseboard is a piece of $\frac{1}{2}$ in. hardwood, 4 in. square, and finished as desired. On one center line, about 1 in. from the end, is fastened the coil A, which is constructed as follows: The ends are two pieces of $\frac{1}{16}$ in. brass, $\frac{3}{4}$ in. x $1\frac{1}{4}$ in. One of these has an extra lug, on one side $\frac{1}{2}$ in. long and $\frac{1}{4}$ in. wide. This is bent over at right angles, and is the means of fastening the completed coil to the base. Two or three small wood screws are used. In each of these pieces a rectangular hole $\frac{1}{4}$ x $\frac{3}{4}$ in. is cut. A strip of $\frac{1}{32}$ in. brass $\frac{3}{8}$ in. wide is bent into a closed form, so that the holes in the two end pieces just slip over the open ends of the form, where they are neatly soldered. The spool just formed is wound full of a fine gauge magnet wire, about No. 30 (single silk or enamel is best).

The moving system consists of the pivot B, which may be a portion of a large sewing needle, with both ends ground to a sharp point. Slightly below the center of this pivot and at right angles to it, is soldered a piece of soft iron, C, $\frac{1}{8}$ x $\frac{11}{16}$ in., and from $\frac{1}{32}$ to $\frac{1}{16}$ in. thick. The barest trace of solder is all that is necessary.

The pointer D is made of very thin spring brass, tapered nicely as shown.

This is soldered about $\frac{1}{8}$ in. below upper end of pivot, and so as to make an angle of about 30 degrees with armature C. Aluminum is a better material, being lighter, only it is rather difficult to fasten securely, as it can not be soldered. A small weight E is used to balance the pointer] and bring it into the correct position; that is, so C stands parallel to the hole in the coil. A U-shaped piece of brass F forms the bearings. This is $\frac{1}{16} \times \frac{1}{4}$ in. in cross section. The pivot turns in small dents, made with a smooth sharp-pointed instrument or end of a drill. Two blocks of wood are used to support the scale H, which is of heavy cardboard.

Fig. 4

A case of thin wood, or metal (not iron), should be made to protect the instrument from dirt and air currents. The outline of this case is shown in dotted lines, Fig. 2. The portion of case over scale being cut away, and a piece of thin glass or mica may be fastened on the under side of opening. The completed instrument must be fastened in a permanent upright position.

Figs. 3 and 4 show the rheostat at the tank end of the line. The baseboard A is about 6 x 8 in. Arranged in the arc of a circle, with E for a center, are a half dozen or more flat headed brass bolts D. A switch arm pivoted at E by a small bolt, slides over them successively. This arm consists of two parts, the thin springy part B, and the heavier part C, to which is fastened the operating arm.

On the under side of the base are arranged as many resistance coils as there are contacts D-1. These are merely small cylinders of wood, fastened between the two wood strips G, G, and wound with about an ounce each of about No. 32 gauge magnet wire. The method of determining the exact amount will be described later.

Fig. 3 shows the tank, float and levers which operate the switch arm. The

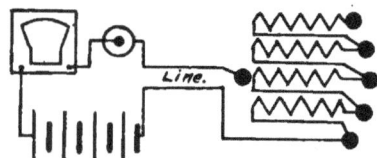

Fig. 5

rheostat being greatly enlarged out of proportion for the sake of clearness. The float B is a large block of wood. Levers C and D are light wood strips, pivoted at all connections with small bolts. D has a bearing near one end, on edge of tank. The position of this bearing will be determined by the depth of tank, and consequently the amount of motion to be reduced. A brass strip connects end of lever D, Fig. 3 and C, Fig. 4.

The rheostat should be protected by a cover of some kind.

Where this system of lever is for any reason objectionable, another method may be used to accomplish the same result. A stout cord is fastened to the float in place of the lever. This cord is run over a small pulley on edge of tank. The frame of an old clock is secured, with a train of two or three of the gears remaining. This frame is then fastened just above the rheostat, and the loose end of the cord is attached to a small drum on the fastest moving wheel of the train of gears. Another small drum is placed on the axle of the slowest moving wheel, and a cord run from it to the arm of the rheostat. With most clocks, two gears will give a reduction of 5 to 1; three gears 25 to 1. It will also be necessary to place a small weight on the rheostat arm, to keep the cords taut while they are played out by the float.

The number of batteries required will depend on the length of line, size of line wire and sensitiveness of meter. For a line about 500 ft. (each way) 6 or 8 cells will probably be sufficient. The correct number, and also the amount of resistance in the rheostat, is determined; and the calibration of the meter is accomplished in the following manner: The connections are first made as per Fig. 5 (except as the coils are not in the rheostat yet, it is short-circuited for the present). The button is now pressed to obtain a reading; with the tank full of water, the pointer of the meter should give a full scale deflection. If it does not lighten, counterbalance weight, add cells, or do both. If pointer of meter moves too far, decrease number of cells, until the desired result is obtained. The number of feet of water in tank is now marked on scale. The tank is now emptied, and sufficient resistance inserted in the circuit to give only a small deflection (say ⅛ scale or less). This position is marked 0 on scale. The resistance thus obtained is divided among the coils F, Fig. 4, in such amounts as to give fairly even scale divisions. The short circuiting wire being removed, the other positions can be marked on the scale with the water in the tank at various depths.

Although the diagram, Fig. 5, only shows four coils, there should be one or more for each foot of water to be measured.

Since for correct readings the voltage should always be practically constant, when the cells become nearly exhausted the tendency will be to give an under reading. But a correct reading can always be obtained by having an extra resistance coil, equal to the resistance of the line. Then before each reading a test can be made by switching this resistance in, in place of line. This is equivalent to the tank in full position, and the gauge should register same. If it registers less, the difference is noted. Now a reading is taken in the ordinary way, and this reading plus the difference just obtained will be the correct reading.

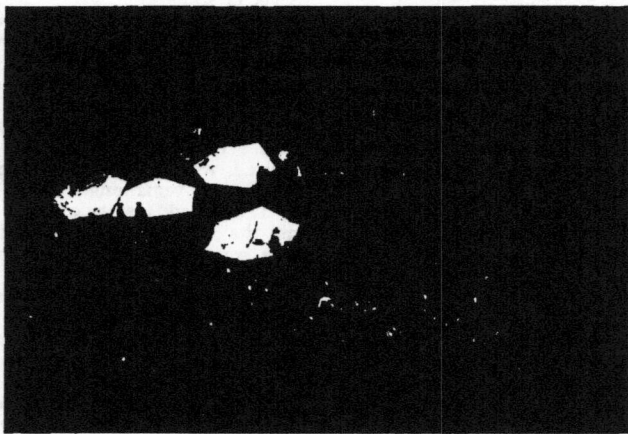

View from Man Flying Kite

Recently a practical demonstration has been made of the usefulness of the man-carrying kites to transmit wireless communications, messages having been sent from San Francisco to Los Angeles, California, and by means of the kite cable transmitted to the operator on the ground.

The altitude record made in California was 385 ft., this being also a duration kite record, the designer of these man-carrying kites staying up in the air for 90 minutes. The young Boston kitor owes his life to the fact that several of

the many kites by which he was suspended at Los Angeles parachuted and prevented him from dashing to death on the earth. He was 200 ft. in the air when the accident occurred in which aviator Chas. Willard collided with and cut the cable with his biplane, thus severing all connection of the kites with the earth. Aviator Willard injured his front control, but was able to land immediately and safely. Although three kites were wrecked, kitor Perkins in dropping 200 ft. landed without serious injury, the remaining kite acting as a parachute.

THE MOST NORTHERN WIRELESS STATION IN THE WORLD

Editor's Note—Mr. Kline is one of our most enthusiastic readers, and it is certainly very interesting to receive a description of the station in which he is located. Being, as he is, upon the outskirts of civilization, Mr. Kline would no doubt be very glad to hear from a few of our many wireless readers.

The Mail Train Leaving for Fairbanks and Eagle

WAR DEPARTMENT
SIGNAL CORPS U.S. ARMY

Circle, Alaska.

My order of September 4th just received, and I am much pleased with the books which were sent. As soon as navigation opens I will try and make a larger order.

This station was opened in September, 1908, for commercial business, and since then we have worked daily with the best of results. Are using a 3 k.w. generator and a 6 h.p. gasoline engine. I am working with the two stations, Fort Egbert (about 100 miles) and Fairbanks (130 miles); but have heard Cordova plainly (250 miles from here); they are using a 2 k.w. outfit. The telegraph stations here in Alaska are controlled and operated by the U.S. Signal Corps of the Army. From what data I can gather this is the most northern telegraph station in the world and the only one that is entirely in log cabins. We are only about 20 miles south of the Arctic Circle. We are using the Counterpoise system of wires at a height of about 7 ft. above ground, 43 wires 200 ft. long (fan-shaped), 9 aerial harps 300 ft. long, and a steel tower 80 ft. high. I enclose a few snapshots of our station.

Yours very truly,
C. E. KLINE.

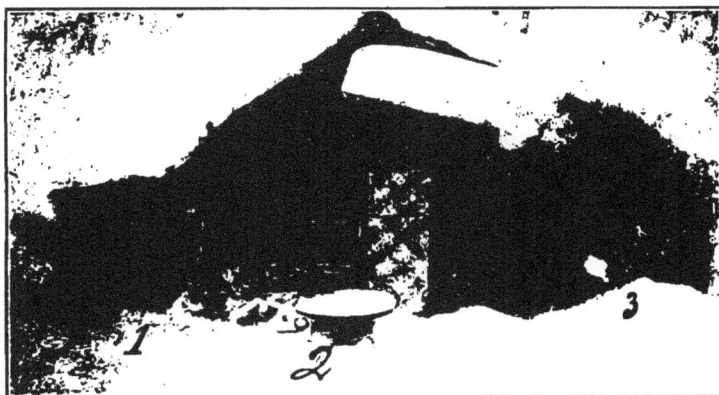

1. Engine Room   2. Telegraph Office   3. Location of Operator

AUTO DEPARTMENT

FROZEN AUTOMOBILE RADIATOR
N. M. HOWARD

The season of the year when the autoist has to look out for frozen tubes in his radiator is here. Experience is a good teacher, but it is sometimes expensive, as we learned last winter. We intended to use the car only a few times during the cold weather and thought it would be the simplest plan to use ordinary water, without any freezing compound, in the radiator, and to drain the radiator after each trip. Well, we tried this, and it worked very well for a while, until one morning we went to fill the radiator and found that one of the tubes had burst. The drain cock was open, and apparently all of the water had drained out of the radiator. For some time we were at loss to determine the cause of the trouble but finally decided that it must have been due either to the water being retained in the tube by a piece of sediment becoming lodged in the bottom of the tube, or to the tube becoming air-bound and thus preventing the water from draining out.

This experience taught us that it is never safe to assume that a radiator is completely drained, simply because the drain cock is open and apparently all of the water has run out. After having our radiator repaired we used a mixture of alcohol and water and had no further trouble. We used about one part alcohol and two parts water. Replacing the evaporation with a larger percentage of alcohol, as the alcohol boils away much faster than the water.

When a radiator is injured by having one or two tubes burst, it can be repaired without going to the expense of having the radiator pulled to pieces and the defective tubes replaced with new ones, by simply plugging the defective tubes at each end. To do this, it is necessary to cut a hole in the back of the top tank and one in the bottom of the bottom tank, near the defective tubes. Then working through these holes plug both the top and bottom ends of the burst tubes with solder, using a soldering iron to melt the solder in place. Make sure the tubes are tightly plugged and then solder patches of sheet copper or brass over the holes that were cut in the tanks to allow access to the tube ends.

To test the radiator for leaks, fill it nearly full of water, plug up the openings and connect a bicycle pump to it. A few strokes of the pump handle will give sufficient pressure to force the water out in a fine spray through any leaks if they exist. Care should be taken not to put enough pressure on the radiator to injure it.

A radiator that has been repaired in the above manner is practically as good as new, as the loss of one or two tubes does not make any appreciable difference to the cooling surface; but, of course, if a large number of tubes are defective this method cannot be used, as it would reduce the total cooling surface of the radiator to such an extent that the circulating water would not be properly cooled, and in hot weather the engine would run too warm.

COST OF HYDRO-ELECTRIC PLANTS
GEORGE E. WALSH

The harnessing of our streams and waterfalls for the generation of electricity represents one of the most important industrial developments of the day, and there are hundreds of such power plants scattered all over the country; but aside from the large commercial hydro-electrical plants there are hundreds of comparatively small streams and brooks which possess all the possibilities on a limited scale that the big ones offer. The value of comparatively small streams running through a rich farming or manufacturing region is scarcely appreciated by many, and to utilize them to economic advantage is one of the opportunities that awaits the far-seeing man of affairs.

Theoretically, size is inconsequential in hydraulic development, but as a matter of practical electrical engineering small powers require very special treatment to secure good economic results. This applies with special reference to a plant under 500 h.p., and in particular to those of 200 or 300 h.p. There are literally thousands of brooks, small waterfalls and lakes scattered over the country that could be made to yield an output of several hundred horse-power, and the only question is one of economic development and operation. Such plants could be utilized for pumping water for irrigation purposes, for operating small factories, and for lighting houses and mills. Water is the cheapest of all power when properly harnessed; but, heretofore, attention has been directed chiefly to such big affairs as Niagara. It is remarkable what tremendous power can be developed from a small stream when its location is such that a steep vertical flow can be obtained. We know that in the Rocky Mountains there are mere brooks, which are today turning generators which furnish upward of 30,000 and 40,000 h.p. This is due to their great altitude when conditions are favorable for a steep drop at a point where the power horse can utilize it.

The cheapest development of water power is where the drop is vertical, for then the penstock need be little more than the height. With heads that range from 10 up to 2,000 ft., the cost of development is subject to wide variation. In fact, a complete water power development with its electric generation station costs all the way from $50 to $300 per kilowatt of capacity. Somewhere between these two extremes will be found the cost of the average hydro-electric plants of the country. If the bed of a stream has only a moderate descent of 5 or 10 ft. per mile, it will require larger waterwheels and electric generators than another which has a sudden drop of 50 or more feet. Consequently we have the rule that the cost of development per unit of power increases as the head decreases, other factors remaining the same; and for wheels and generators the weight per unit of output decreases as the speed of revolution increases, and this speed goes up with the head of the water.

The measurement of a stream with its water flow throughout the year is the first important step in considering the work of developing it. A good many small plants have proved failures in the past through lack of exact knowledge of the annual flow of water. Although the flow was good in the winter and spring months, in the summer it was so small that there was not sufficient power to run the wheels. A small brook only a few feet deep and 20 ft. wide that flows steadily the year round is of much greater value for hydro-electric development than a river ten times as large which practically dries up in the summer months. With a comparatively high head a relatively small brook fed by springs that never dry up is one of the most valuable assets for an electrical plant. Such streams with their sources among the hills and mountains are found in many parts of the country, and they could be harnessed with the minimum of expense for electrical generation.

To take a concrete illustration a stream which has a reliable flow of 100 cu. ft. of water per second located so that a working head of 25 ft. can be used should prove a good one for hydro-electric development. Such a stream might have a maximum flow of 400 cu. ft. of water in the rainy season, and a mean flow of

200 cu. ft. and a minimum flow of 100. The latter flow must be the one on which to base calculations for reliable power output.

Now what can be done with such a stream, and how much power can be developed, and what would be the cost of installation be? These are the practical questions that many are asking today. They apply to nearly every part of the country. There are thousands of towns and villages, tens of thousands of individual farms and homes, and hundreds of mills and factories, which are interested in the development of such small hydro-electric power. The big streams and waterfalls require so much capital for their development that they cannot be utilized except by large companies and syndicates. Yet the smaller streams are of far more general importance to the whole country, and interest far more people.

The average flow of such a stream will develop just about 225 h.p. when utilized in a first-class turbine. The electrical product of this plant can be utilized for every hour of the twenty-four of each day throughout the year. For a good part of the year, say about nine months, twice this amount of power could be developed. But as few industries require continuous power throughout the day and night the plant would thus furnish far more electricity than would appear at first sight. Electric lighting power would be used only for half the time, and surplus power could be used or sold in the day time.

If the normal minimum flow was 100 cu. ft. per second or 6,000 ft. per minute, the working power for a 12-hour run could be doubled by storing some 4,000,-000 cu. ft. of water. This would mean drawing-off a 100 acre pond 1 ft. of water at night and filling it up in the day-time when the power was reduced or entirely off. In a very dry season the water stored might be reduced to 2,000,000 cu. ft. in each twenty-four hours. Six inches on a 100-acre pond would take care of this for a day, so that if by putting flash boards on the dam the water could be raised 2 or 3 ft. in a 100-acre pond the plant would be made practically safe for over 300 k.w. steady output on its lighting load.

If 1,800 k.w. hours were thus stored in the pond for 1 ft. of depth, the distribu-

tion of the power could be made to suit the demand. One might draw off 300 k.w. for three hours, and still have 900 k.w. hours left for the rest of the night or day over and above the basic 150 k.w. from the minimum flow. This would allow a total output of 450 k.w. during the hours of heavy load, and would leave 2,250 k.w. hours for the rest of the night and for a small day load. Now a plant with a peak capacity of 600 k.w. is capable of doing the lighting work of an average city of 25,000 to 30,000 inhabitants, furnishing also in the day-time a fair amount of power for light mill and factory work. But as all of this output is based on the minimum flow, it is quite apparent that for seven to nine months of the year a very much larger output can be depended upon. This larger output would come at a time when our nights are the longest, and the demand for lighting the longest. It is also the busiest part of the year, so that the sale of current could be greatly increased.

But the safe way in figuring on hydro-electric development is to consider the dry season output or the minimum flow. If this is securely taken care of the rest of the season will give no trouble. Where the mistakes have been made in the past has been to figure on the mean flow and construct a plant accordingly. Also the condition of the season must be taken into consideration. We have our wet and dry seasons. A few years ago a plant was erected in Massachusetts where hasty and insufficient data caused a great loss. The minimum flow was based on the measurements of the stream for one summer. That summer proved to be an unusually wet season, and with no previous summer records of flow on hand the plant was erected on this insufficient data. The result was that the first season after the plant was put into service unusually dry weather followed, and instead of being able to develop 225 h.p. the plant could scarcely show 175. The plant was characterized as a fraud and failure, and the engineers were blamed for their optimistic statements and claims. It gave a black eye to hydro-electric work in that whole region. Many believed that electricity was an unreliable power, and others simply said that you had to discount electrical engineers' statements by a large percentage.

Not many such mistakes are made today, or there is no reason that they should be made. Reliable data and measurements are the foundation upon which every plant should be constructed. There can be no guess work, no reliance upon the statements of the oldest inhabitants that a certain stream has never dried up in their fifty years of recollection. Science does not go upon any such proceedings.

If there is any question about the minimum flow a little extra depth in the pond will take care of the difference. Some streams have a remarkable variation in the flow, ranging all the way from 100 to 600 cu. ft. of flow per second. The Massachusetts plant referred to above was one of these streams. It was found that by increasing the depth of the pond 1 ft. the amount of water storage would prove sufficient for all purposes; but to be on the safe side an extra depth of 2 ft. was made. This solved the problem satisfactorily, so that today even more than the rated horse-power can be developed and maintained throughout the dry season. In the event of an unusually long dry summer it is more than probable that the plant could still go on serving its clients without fear of a shut-down.

The development of small streams does not usually call for large expense. For the dam a well-ballasted timber crib proves the most satisfactory, with special attention given to secure foundations and abutments. Such a dam will last indefinitely and prove very satisfactory. The power house should be placed as close to the dam as conditions will permit. The shorter the penstock the better the work will prove. If the water must be conducted any great distance to the power house site it is far more satisfactory to conduct it there in a deep open canal or flume.

In arranging the number of wheels and generator units, consideration should be given to the question of possible use and breakdowns. Enough units should be developed so that the whole plant will not be crippled in the event of temporary crippling of one. For a plant of this size four 150 k.w. machines would prove more reliable than two. More than this number would increase the cost without any material gain, and anything less might cause too severe an overload in the event of a hot bearing compelling the shutting down of one machine.

Water wheel data shows that the required maximum power per unit, about 275 to 300 h.p., could be obtained from a single horizontal wheel under 25 ft. head at about 200 revolutions per minute, or from a combined pair of smaller wheels at about 300 revolutions per minute. Either combination with a dynamo would do well, but the single wheel would be a little cheaper.

A power plant of this size is almost too small to attract promoters and financial houses interested in bonds. It must therefore be the work of private concerns. But there is no more profitable investment of funds than in such hydro-electric plants where conditions for harnessing and selling the power are satisfactory. The farmers as well as the people of small towns and villages are interested in the harnessing of the small streams. Where irrigation pumps are to be operated and farm machines to be driven by some power other than hand and horse, the hydro-electrical question should be considered. There is not a farming community or a hamlet in the United States favorably situated in regard to streams that could not be enormously benefited by the construction of a hydro-electric plant. They are being installed all over the country nearly every year, and they are proving valuable assets. They are lighting houses and running mills, factories and farm machinery cheaper than can be done by any other known power. The capitalists may not care to bother with these little brooks and streams, and the big water companies overlook them; but they are an asset to the small village and individual farmers that are of incalculable worth. Some day they will be nearly all harnessed, but at present the opportunity is awaiting the progressive men who can utilize them.

A round file may be used as a reamer by inserting it in the hole to be reamed and turning to the right instead of to the left. By doing so the file will not bind, and excellent work will result.— *Penberthy Engineer and Fireman.*

Indecision can never think well; the field of thought is never ploughed by simply turning it over in the mind.

HOW TO MAKE A DIVIDING PLATE FOR LATHE HEADSTOCK

A. HALLETT

The following method of obtaining the exact positions for the holes in a dividing plate, as fitted to so many small lathe headstocks, will be found extremely useful when the required number of holes is a large one or a prime number. With ordinary care in the drilling, a plate can be made with any number of holes desired, with a degree of accuracy unobtainable by any other method I know of in which another divided head is not used.

Let us consider, for example, that we require a plate having 74 holes in it. First construct the small drilling jig, shown in plan and elevation at A and B, Fig. 1. This consists of a piece of steel plate 1 in. square and $\frac{1}{8}$ in. thick. This is drilled for four wood screws, and also has drilled along one of its center lines two $\frac{3}{16}$ in. holes $\frac{3}{8}$ in. apart. These holes must be true to size, and their distance apart must be fairly accurate, although this is not of absolute importance.

This piece is screwed on to a piece of flat hard wood, with two strips of $\frac{1}{16}$ in. brass or steel under its two edges, so as to leave a space under the center of it $\frac{1}{2}$ in. wide and $\frac{1}{16}$ in. deep. This will be clear from the drawings. A small steel peg, C, Fig. 1, must be turned from mild steel, having the parallel end $\frac{3}{16}$ in. diameter and $\frac{1}{4}$ in. long. The knob can be anything. A short piece of rod of $\frac{3}{8}$ or $\frac{1}{2}$ in. diameter, with a spigot on one end, of the above sizes, will answer quite well. This $\frac{3}{16}$ in. peg must be a nice fit in the $\frac{3}{16}$ in. hole drilled in our jig. A strip of brass will now be required,

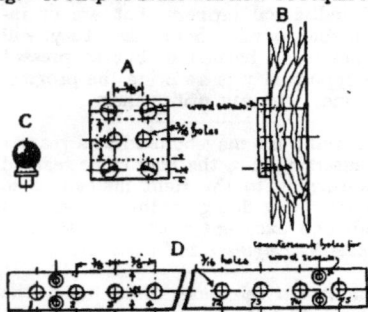

FIG. 2.

$\frac{1}{2}$ in. wide and $\frac{1}{16}$ in. thick. Hard rolled brass will be the best for the purpose. This must be a sliding fit in the space under the steel plate of the jig already made. The length of this piece of brass for a 74-hole plate will be $28\frac{3}{4}$ in. One end of this strip is pushed under the jig until the end is about $\frac{1}{4}$ in. or $\frac{5}{16}$ in. clear of the nearest $\frac{3}{16}$ in. hole. A hole is then drilled through the strip and the peg inserted in the hole. The second hole is then drilled, the peg taken out, the strip pushed along, and the peg inserted through the front hole of the jig and the second hole in the strip. A third hole is then drilled, and so on until we have the whole strip drilled with 75 holes, or one more than the number we require for our plate. The ends of the strip must then be cut off, leaving $\frac{1}{4}$ in. beyond each end hole. The ends must also be beveled off, commencing from a point $\frac{1}{4}$ in. inside each end hole to the end, so that when the strip is bent into a hoop, with the two end holes coinciding, the joint will be of the same thickness as the rest of the strip. Next take a piece of $\frac{3}{4}$ or $\frac{7}{8}$ in. pine or white-wood 9 in. square, and roughly shape it to a 9 in. diameter disc. Fix this to the lathe faceplate and turn it up on the edge until the strip, when bent round it closely, will have the two end holes exactly coinciding. The turning must be carefully done to ensure an accurate and close fit of the brass strip. The peg (C, Fig. 1) can now be pushed into the two end holes to hold the strip together, and holes drilled for wood screws in the strip at

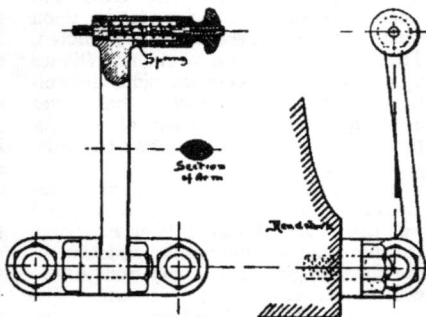

FIG. 1.

several points all round it to hold it on. Two had better be put in close to each end, as shown, to keep the ends down close together. A piece of stiff spring, having a peg of ⁵⁄₁₆ in. diameter riveted into one end, must be clamped to the lathe bed, so as to act as a stop, the peg falling into the hole of the strip. The dividing plate can then be fixed onto the front of the wood disc on the faceplate, and secured centrally, or if it be a new plate we are making it can be turned up now.

If we have a drilling attachment fitted to the slide-rest it is a simple matter to drill this plate; but if not, we must scribe a circle round the plate with a pointed tool in the rest, and mark this circle with the same pointed tool moved to and fro across the plate, moving the lathe round one hole between each mark. This latter method opens up a chance of error in the second drilling, which the use of a drilling attachment, if at all satisfactory, eliminates. Should more than one circle of holes be required on the same plate, it is necessary to make the strip, as above described, first for the largest number. This done and the plate drilled we can cut off a piece of the strip, leaving the correct number of holes for the second number plus 1, and so on.

The wood will then have to be turned down to take the shortened strip, and so on. Anyway, the strip can be kept, as it is always likely to come in for this purpose, if not required at once.

Fig. 2 shows a very serviceable spring stop for a dividing plate, and one which I have found stiffer in use than one made with a peg riveted to a piece of spring steel.

The arm and top end plunger casing is a malleable casting, or a steel forging drilled out to ¼ in. diameter to take the plunger. This is of steel (double shear or silver steel), turned at its outer end to suit the hole in the dividing plate, generally ⅛ in., and at its inner end also to ⅛ in. A knob is turned and tapped to screw onto the end of the plunger, and a fairly stiff spring bears against the shoulder left on it and the back end of the casing.

The lower end has a boss drilled and faced up true, and has a ⁵⁄₁₆ in. or ⅜ in. bolt holding it to a small bracket fastened to the headstock with studs. This allows the arm to swivel so as to bring the peg in line with the various circles of holes on the plate. The height, etc., will have to be arranged to suit the lathe under consideration, but I think the general idea is shown fairly clear.—*Model Engineer and Electrician.*

Price of Platinum

The price of platinum has lately attained to the high level of $480 per lb. troy, for refined, a figure which it has only once before touched—namely in 1906. The recent advance is attributable to causes similar to those which operated in that year. Consumption continues to expand steadily as the result of the more widespread utilization of the metal in connection with the steadily growing electric lighting industry, the jewelry manufacturing industry, and the chemical, dental and photographic trades, while new uses are continually being discovered, which though comparatively small individually have a cumulative effect. A typical example of such new uses is to be found in the motor-car manufacturing industry, where the unique qualities of platinum have gained for it a use which even high prices do not seem sufficient to discourage. The world's annual consumption of the metal has been estimated at between 350,000 oz. and 370,000 oz., of which, however, roughly one-third is represented by metal recovered from scrap and remarketed. Practically the whole of the 350,000 oz. or 370,000 oz. is consumed in the four principal markets, France absorbing about 40 percent, Germany and the United States of America each 26 percent, and Great Britain 7 percent, the small balance remaining being consumed in Russia, which is practically the sole source of supply.

New Wireless Apparatus

Professor Cerebotani, an Italian inventor, gave a private exhibit of his wireless discoveries the other day before members of the French ministers of war, posts and telegraphs, and a large number of scientists, including M. Eiffel, the constructor of the Eiffel Tower, which is now a government station. Among the apparatus employed was a pocket wireless machine, a wireless telegraph printer by means of which messages are sent as readily as writing on a typewriter, and a wireless teleautograph which enables persons to sign their signature as far as wireless waves reach.

The pocket apparatus is a little larger than a pair of field glasses and is operated by attaching its antennae to a post or tree, which, at the height of 50 ft. enables communication to be made within a radius of two or three miles. The teleprinter, a local contemporary explains, is a simple little instrument with a keyboard like a typewriter, which can be fixed to any telegraph or telephone installation. This transmits messages which appear on printed slips at the other end, but it has the advantage of being infinitely more simple than anything yet invented, and, besides, can be used with wireless. This should be interesting to railway officials in particular, since such a machine could be put at the disposal of all signalmen, pointsmen, station masters, and others, permitting them to communicate quickly and accurately with the head office. It would be also exceedingly useful for small, out-of-the-way post offices, since no special training or practice is necessary to operate it.

The teleautograph is a most simple apparatus, which can also be affixed to any telephone or telegraph line. By this a signature, a drawing or a holograph manuscript written with a pencil fixed to a flexible carriage is copied exactly on a machine at the other end. Hence a man in Paris could sign a document in Algiers or a signature in Algiers could be verified from Paris. As if these wonders were not already sufficient, we are further assured that the greater the distance the better the machine will work, although we have not been told why this should be so. The tracing of one's signature seems to be no more difficult than with a pen, and a pencil repeats it automatically wherever we want it, even should it be at the Antipodes.

Another invention of the professor is an instrument for preserving the secrecy of wireless messages. As is well known a message sent out by a wireless station is received by all stations within a certain radius, although it be only intended for one of them, because the Hertzian waves sent out affect all receivers alike. This new machine, however, allows each of a large number of stations to have its identification number, and when the Hertzian waves are set going with the transmitter at a certain number, only the station bearing the corresponding number can receive the message, all the others being cut off by a short-circuit arrangement.

QUESTIONS · AND · ANSWERS

Questions on electrical and mechanical subjects of general interest will be answered, as far as possible, in this department, free of charge. The writer must give his name and address, and the answer will be published under his initials and town; but, if he so requests, anything which may identify him will be withheld. Questions must be written only on one side of the sheet, on a sheet of paper separate from all other contents of the letter, and only three questions may be sent at one time. No attention will be given to questions which do not follow these rules.

Owing to the large number of questions received, it is rarely that a reply can be given in the first issue after receipt. Questions for which a speedy reply is desired will be answered by mail if fifty cents is enclosed. This amount is not to be considered as payment for reply, but is simply to cover clerical expenses, postage and cost of letter writing. As the time required to get a question satisfactorily answered varies, we cannot guarantee to answer within a definite time.

If a question entails an inordinate amount of research or calculation, a special charge of one dollar or more will be made, depending on the amount of labor required. Readers will, in every case, be notified if such a charge must be made, and the work will not be done unless desired and paid for.

1716. Storage Battery. W. B., West Orange, N.J., asks for directions for charging a storage battery, and if a generator made from a telephone magneto will be suitable for the purpose? *Ans.*—In our last July *Electrician and Mechanic* appeared a very complete article on this subject, the diagram given in Fig. 13 being the one you should particularly study. Instead of using a shunt field rheostat for controlling the dynamo, you will need to adjust the speed or interpose a resistance in the main line. You must allow considerable time for charging, for the case is like that of filling a large tank when you have only a small pump.

1717. The "Henry." H. P., Upper Troy, N.Y., asks (1) for an explanation, with example, of this electrical unit. (2) What is a "resonator" when applied to electrical apparatus? Ans. —(1) There is a quality connected with electrical circuits that is analogous to inertia in mechanical matters. A balance wheel objects to being started in motion, but when finally in motion it objects to being stopped. Similarly, the presence of "self-induction" in an electrical circuit hinders the growth of the current and at the next instant hinders its decay. So when contact is made, say with a spark coil for gas lighting, an appreciable time is required for the current to attain its final value as required by Ohm's law. All this time while the current is increasing, the production of the magnetism is constantly inducing a counter electromotive force in the winding. When the current reaches its final value, no increase in magnetism takes place, and this counter effect stops. Now at the break of the circuit the magnetism disappears, but in getting out of existence these lines cut through the coils and induce in them an electromotive force in the direction to prolong the current, and so vigorously as to hold a momentary flash across the gap. The energy stored up in the magnetic field is similar to that stored in a fly wheel. If you try to stop the wheel instantly there will be a crash, and it is this electric crash with the coil that you wish for the ignition. Such an abrupt break is carefully avoided, however, in case of dynamo machinery, for the momentarily great electromotive force may readily puncture the insulation. If you multiply the number of turns of wire in a given coil by the number of magnetic lines of force passing through it, and divide by the number of amperes, and then by 100,000,000, you will get the number of henries. This will usually be a fraction. (2) You will find a description in Chapter XXV, in the Engineering Series, in the July, 1908, magazine.

1718. Magneto Generator. E. K., Richmond, Va., has substituted No. 27 wire for No. 36 on the shuttle armature of a telephone magneto, and now the machine completely fails to work. Also, an induction coil, made of Nos. 18 and 31 wire, and intended to give ⅛ in. sparks, gives only ¹⁄₁₆ in. ones. What is the reason? Ans.—The winding of a shuttle armature is about the simplest of electrical arts, and you ought not to fail where thousands of amateurs have succeeded. Perhaps you are expecting too much from the present winding, possibly the brushes—if you have used a commutator—are in the wrong position, or perhaps you have failed to insulate the winding from the iron. The sharp corners that are the hardest to protect are the most important places. Use cloth over these edges and corners before covering the easy parts. See if you can get a circuit between iron and wire. Perhaps you have wound the armature for alternating currents, and therefore maintained the "ground" on one end of wire. Of course there must be a ground nowhere else.

1719. Molding. G. E., Brooklyn, N.Y., asks for directions for making molds that can be used for casting, and yet not crack as does plaster of Paris from action of the heat. Ans.— You give no hint as to the sort of metal you are casting, so we cannot advise you as directly as we wish. If you are casting lead, antimony, or any of the easily-melted metals, and are making a large number of pieces, you will find it economy to make a cast-iron mold. If only a few castings are desired, and these of brass or copper or the like, sand molds will be necessary; of course requiring new work for every casting. This labor is regularly reduced in brass foundries making large quantities of small pieces by use of "gate" patterns, whereby a number of similar parts are made in each flask.

1720. Small Dynamo. W. K. B., New York City, has the parts for an enclosed "Apple" ignition generator, and wishes data for winding it for experimental purposes, speed to be about 1,200 revolutions per minute, other specifications not being particularly definite. Ans.— We would suggest a shunt field winding, and the armature for 6 to 8 volts. With an armature core of the size you mention, 2⅝ in. long, 2⅛ in. diameter, with 12 slots ⁹⁄₁₆ in. diameter, you should be able to get fifty No. 21 wires per slot, twenty-five turns per coil for each half winding. Let the two field coils be wound with s.c.c. No. 24 wire—all you can get room for—even when bending the coils to conform to the spherical space. If you wish a compound winding, wind the coils to about three-quarters their full size with No. 25 wire, the remaining space being allowed for the series portion of No. 16 wire. It would be well, however, to wind the series portion first, perhaps two layers being sufficient. If you wish a purely series motor, let the entire space be filled with No. 15 wire.

1721. Small Dynamo. C. K., Warren, Ill., asks: (1) What would be a suitable armature winding for a small dynamo of the following general dimensions: diameter of armature core 3 in., length 2½ in., with 12 slots ⅛ in. in diameter; field magnet of Edison type of steel castings, cores being 1½ in. in diameter and 2½ in. in length; bore, 3¹⁄₁₆ in., arc, 135 degrees; winding, 11 layers 50 turns per layer, of No. 22 d.c.c. wire. A current of 5 or 6 amperes is desired, and speed not to exceed 2,500 revolutions per minute. (2) Is it practicable to generate a direct current by a method analogous to the rotary field alternator, using an alternating current to excite the field, but with a different winding on the stationary armature? The excitator alternator would, of course, be driven synchronously. Ans. —(1) With the field winding you have employed, the voltage should not exceed 10 or 12. If there is room for more wire on the field, you could then wind for a few more volts. It would have been better if you had made the magnet cores about 2 in. in diameter. Use No. 18 wire on armature, all the turns you can get. A 12-segment commutator would be much better than one with 6. (2) Your proposition is a dream that no inventor has yet realized. The induction motor driven above synchronism, and called the induction generator, is the nearest realization, and some of these in sizes above 5,000 k.w. are in operation. Of course they generate alternating currents. The "homopolar" direct current machine is suggestive of possibilities, but thus far attended with disappointment.

1722. Electro-plating. C. J. G., Lewiston, Idaho, asks: (1) What is the best solution for plating with lead, and how is it made? (2) Can a person plate with the following alloy: 9 parts of lead, 2 of antimony and 1 of bismuth? (3) Where can antimony and bismuth be obtained? Ans.—(1) Zinc and tin articles can be coated with lead by immersing them in a solution formed by dissolving litharge (one of the oxides of lead) in water and caustic potash. Similarly, iron will become coated if dipped in a solution of acetate of lead. A battery current, though very weak, can be used to encourage the deposition from the caustic potash solution, but this must be prepared by saturating the boiling hot potash and water with litharge. The process has no commercial importance. (2) Antimony and bismuth may separately be deposited upon articles, either by simple immersion, as with lead, or by the action of the electric current, the first from a tartar-emetic (potassio-tartrate of antimony) solution, the second from the double chloride of bismuth and ammonium. We cannot find, however, that the alloy can be successfully operated, and from the indifferent results

from working separately with the three constituents, think your proposition rather doubtful of success. (3) Since Missouri is the source of so much of these metals, we think you could well obtain it there. Write to the postmaster for the address of some retail dealer.

1723. Wireless. D. W. D., Montrose, Col., asks: (1) Which is better for an aerial, hard-drawn or soft-drawn copper wire; and is a six-strand aerial 50 ft. long and 50 ft. high run slanting on a slant of about 45 degrees very good? (2) I am building an open-core transformer of the following dimensions: core 1 x 10 in.; primary two layers of No. 12 d.c.c.; secondary of 6 lbs. of No. 30 enameled wire wound in 12 sections ½ in. thick and 3½ in. in diameter. What is rating in kilowatts of this coil, using proper voltage and amperage? (3) I shunted a small condenser around the spark gap, the condenser containing 120 sq. in. of tin-foil when using only three sections. The condenser emitted a hum. Why was this? No spark went across the gap. (4) Are there any large stations in Colorado? If so of what power in kilowatts? Ans.—(1) Use hard-drawn copper on account of its strength. A slope of 45 degrees is not fatal to good working, although the antenna will receive a little better in the direction opposite to which it points than in any other. (2) Such a coil would take probably 10 amperes at 110 volts, safely, but its actual output would be only about ⅛ or ¹⁄₁₀ k.w. (3) Like electrical charges repel, unlike charges attract. You have plates close together which bear opposite electrical charges, and hence attract each other. Since the attraction is periodic, it causes a hum. (4) We cannot find any station listed which is even at a communicable distance from Colorado.

1724. Quenched Spark. G. S. C., El Monte, Cal., asks: (1) In the article "The Quenched Spark System," I do not find any statement as to the number of sets of discs to use for any certain voltage, for the modification of the Telefunken gap. I note this is given for the Von Lepel generator, but I prefer the Telefunken system. (2) Using this gap on ordinary 60-cycle 110-15,000 volt wireless transformer is the note produced clear and high and penetrating, or is this high note the result of using high-frequency current (high cycle) in primary? (3) What do you think of the idea of designing and building and using a transformer, with quenched gap, with a secondary voltage of only 1,000 to 2,000 volts, and consequent high amperage? I understand the army is using a set with transformer secondary voltage of only 500 volts. Do you think this idea practical? Ans.—(1) Use as many gaps in series as is possible and still get regular sparks. (2) In regular practice, a generator of 400 to 500 cycles per second is used in order to get the high note. It is possible to get a sort of ragged high note from 60-cycle current when you have the circuit containing the secondary of the transformer in "resonance" with the high pitch. (3) Yes, one may easily use as low as 500 volts and obtain a high efficiency, but since the current thereby used is large, the gap must have excellent heat-radiating arrangements.

1725. Wireless. R. H., Springvale, Me., sends sketches and notes. Ans.—(1) The antenna you illustrate is directive, but very inefficient. The part reaching downwards from the top should be omitted. (2) The second type is much preferable, but is no more efficient in the direction you wish than if you connect to the center of the horizontal part.

1726. Electrolytic Rectifier. H. S., Santa Barbara, Cal., asks for some information as to construction of rectifying current apparatus. Ans.—An electrolytic rectifier was described in the August, 1909, number of this magazine. The electrolytic rectifier is the only kind of arrangement which can be made by an amateur. The General Electric Company will surely send you all the information about mercury-vapor rectifiers that you wish.

1727. Transformer. R. B., Oak Park, Ill., asks: The size and dimensions of core, number of "pies" size and pounds of wire in secondary; size and pounds of wire in primary winding insulation. This is for a 2 k.w. closed-core transformer to run on a 60-cycle 110-volt alternating current circuit. Ans.—You do not state what use is to be made of the transformer. If it is for a wireless station you will need a much smaller core and primary winding than if it is for continuous "line" work. We do not know of any 2 k.w. transformer designs. A ¼ k.w. transformer was described in the June and July, 1909, numbers of this magazine. It is not usual for private individuals to build transformers of such large size as 2 k.w.

1728. Telegraph. E. H., St. Paul, Minn. On page 68, January number, you recite the troubles of "E. H." at St. Paul. Ans.—I feel sure that his trouble is due to none of the causes you suggest, but is undoubtedly due to his having the poles of one of his batteries reversed. That is, in plain language, he has "copper to line" in one, and "zinc to line" in another. These two batteries then neutralize each other and the "through" line cannot work. If he will trace the current and see that it all runs one way + or − I think his trouble will disappear at once.

1729. Perpetual Motion. H. M. A., Cambridge, Mass., asks: What progress has been made with the perpetual motion picture machine up to the present time? What has been the closest that anyone has come to striking the idea? Is the machine possible and probable and what do scientists say concerning it? What is Edison's view? Ans.—The doctrine of the conservation of energy, that is, that we can never take out of any system more energy than is put into it, has been experimentally tested in hundreds of ways, and is, as far as human intellect can discern, one of the immutable laws of nature. Perpetual motion without consumption of energy is *impossible*, and no attempt comes nearer to a solution than any other. We have not consulted Mr. Edison as to his view, but have no hesitation in saying that he would speak practically as we have above.

CORRESPONDENCE

Gentlemen: I beg to hand you the following for your correspondence column, should you consider same sufficiently interesting.

The subject of transmission of thought, or telepathy, has occupied many minds for many years. I have seen and read a great deal in connection therewith, but do not remember ever to have heard even a plausible explanation of the phenomena. For many years I doubted its existence altogether, thinking that its votaries must be either victims of coincidence or that some trick like so-called Spiritualism must be responsible for their delusions. However, during recent years I have myself seen, skeptic though I am, instances of this wonderful manifestation which I could not explain on the above lines, to my own satisfaction.

Latterly an idea has come to me, call it a wild fantasy if you will, which has been growing in

always between persons who have been closely united by one means or another; for instance, between brothers or sisters, husband and wife, lovers, etc., persons, let us say, whose minds have been "tuned" in exact unison. It was recently shown by some scientist, by the use of photographic plates, that the living human body radiates electric energy of some kind, as an impression was made on the sensitive plate by it, which a corpse would not produce. If then the mind is strongly concentrated on a certain subject, it is only reasonable to suppose that this great brain activity radiates strong "thought waves" which another person whose mind is working on the same subject, and consequently tuned to receive same, should get the message.

Could we not get Messrs. Tesla, Edison or Marconi's opinion on this?

Respectfully

Wish to say I wait for your paper every month, as there are a good many articles in same which are a great help to me. Hoping you get a larger circulation and keep up your well-written articles, I remain,

MARTIN BOISEN, JR.

To the Editor of the *Electrician and Mechanic:*

Dear Sir: I will continue to read your magazine every month as I did in the past, only I will have to buy it here from the newsdealers, and just as soon as I am able to renew my subscription, I will do it without delay. Since I have started to be one of your readers I have learned more from your magazine than from a set of text-books that I have had for the last three years.

I will close now, hoping to renew my subscription soon and wishing you all possible success with your magazine, the *Electrician and Mechanic,* I remain very respectfully yours,

A. E. B.

Dear Sirs:—I am dropping you a line to congratulate you on your last issue which is one of the best I have seen. I have been looking for something on the von Lepel and Telefunken Stossesgerung systems in my three magazines for almost two years now. I experimented with the von Lepel (in the spring of 1910 I think

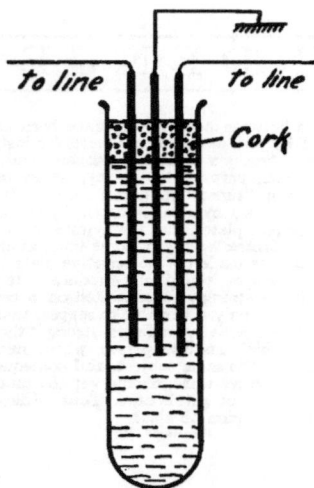

it was) and was very much impressed with its remarkable adaptability to the amateur's use. I used a transformer wound for 600 volts, 1 ampere a.c. on the primary side (110 watts that is) and an aerial 20 ft. high. I was able to work some 5 or 10 miles, talking over the city but had some trouble with raggedness of the spark. I may remark that this would be the chief difficulty of the experimenter work-

ing on a.c. It requires a remarkably exact adjustment of the condenser and the coupling between open and closed circuits to overcome it. I have noticed from time to time several weird combinations of "micrometer" gaps, condensers, choke coils, high resistances, etc., described in your magazine, as infallible cures for static on the primary of the transformer. I have found a small electrolytic lightning arrester, simply made of three pieces of aluminum wire dipping into a test tube full of sodium phosphate, to work admirably.

Yours,
R. K. FREEMAN.

Dear Sir: The *Electrician and Mechanic* is better today than it ever has been. Yes, Mr. Wanner, you are right. You *are* a crank. A broad-minded wireless operator, electrical engineer, aviator or any man of any trade, or profession, takes pleasure in reading short articles out of their general routine. An up-to-date mechanic takes pleasure in talking intelligently upon any subject in the mechanical line.

ISAAC E. WORTS.

Dear Sirs: In reply to your recent editorial, I would like to express a few of my personal views on the subject mentioned.

I disagree with Mr. Wanner in his notion that the *Electrician and Mechanic* has decreased in value, so will be one of the "five" that prefer the "new style."

I am in favor of specializing to a certain extent. I would not care to have this magazine specialize on "Wireless," as I believe that should be left to a magazine treating wireless alone, where plenty of space can be had. I would prefer it treated as before, with such articles as cannot readily be found in popular books on the subject. Every wireless enthusiast should have at least a few books for reference.

I have the same views on aviation. Its treatment in *Electrician and Mechanic* has been satisfactory to me.

Of the mechanical topics mentioned in the editorial I am especially interested in steam, gas and water turbines, oil-consuming engines, marine engines and automobiles.

Most all electrical articles are interesting, as my interest lies chiefly in that line. Aside from practical articles, an occasional one like "Some Unexplored Fields in Electrical Engineering," by Steinmetz, adds interest by treating the theoretical side. Also articles on electrical development, as, "The Growth of Electricity on Railroads," by A. C. Lescarboura, are interesting.

The "Home Craftsman," "Show-Card Writing," "Wood-Carving," and articles on the making of furniture are not so interesting to me.

Do not decrease the electrical department to make room for another. Increase the size of the magazine and raise the price first. To make things short, the magazine has been very satisfactory.

Thanking you for your time and attention, I remain,

Sincerely yours,
HOWARD A. BAXTER, Jr.

Gentlemen: That wireless department of this new January issue is certainly the thing. I have tried to find that information in others, but failed.

As to the elementary character of the past articles I wish to say that I would like to see some articles on damped and undamped oscillation and of that nature, and an historical article on the beginning of the wireless telephone.

All I can say is that you are sticking up for the name of the magazine, and as long as you don't go out of that alright.

Very truly yours,
WILLIAM BOLLES.

Dear Sirs: Complying with your request for opinions from your readers, I must say that you are treating on some subjects which are very uninteresting. Do not by any means discontinue your splendid articles on wireless, such as furnished by Mr. Getz and others. Keep on the progressive track in wireless; new ideas are what we want. Another subject which will give ginger to your magazine is a "Photographic Department." "Mechanical Drawing" is interesting, but it would be of no use to those who have taken the course at high school.

Respectfully yours,
W. P. HUSBY.

Dear Sirs: I want to voice the sentiments of Mr. Wanner, whose letter appears in the January issue of your magazine, as I have been very much disappointed several times back when I bought your magazine, expecting to find something in it worth while. I buy it principally for the wireless items, and I am no kid either, having pounded brass for about thirteen years. I think the suggestion for more wireless telegraphy and telephony a good one, but think your Auto Department is a frost and would knock it in the head while it is young. There are so many good Auto magazines published that an Auto Department in your magazine is out of place. More of the Questions and Answers would be more acceptable than plans for an "inlaid table" which only an expert could make. Drop the boys a hint that the government gets out a real Wireless Blue Book for 15 cents. I suppose it is hard for you to please everybody, and if I don't like the magazine I can quit buying it, which I will if it doesn't brace up. Your article by Marconi was bully. More of that sort would be appreciated.

Hoping for a change,
Yours, etc.,
R. L. PATCH.

Dear Sirs: I heartily agree with Mr. Wanner in the January issue. In my estimation, your magazine is not what it was one year ago. You say in your editorial that you have covered all branches of wireless telegraphy. On such a comprehensive subject as that I take it that the elementary articles must have been published at least six or seven years ago. Many of your subscriptions can hardly date back that far. May I suggest that you begin a new series for the benefit of your later subscribers.

In my opinion many of your articles are too technical for the average reader. All of your readers I am sure are not electrical or mechanical

engineers. Give us plain, simple articles on how to make wireless apparatus and woodwork.

Of course I don't mean to make your publication an entirely amateur magazine, for I realize that you must strive to please all classes.

Give us some good articles on elementary wireless and mechanical drawing, and watch your subscription list increase. Continue as you are and watch it decrease.

I would like to hear the opinions of other readers through these columns.

Yours truly,
CHAS. I. GINGRICH.

Nome, Alaska.

Gentlemen: Your magazine has just arrived here after traveling these thousands of miles over white trail, and has been duly studied and praised by our crowd of young mechanics. "It's the best yet," is the unanimous verdict. Set me down for two years' subscription.

With several members of our "Aero Club," I am working on a motor driver sled with which we hope to shorten the mail-time one-half. Will let you know the results.

Very respectfully,
PERCY L. SCHOOF.

Portland, Ore., Dec. 27, 1911.

"The Oregon State Wireless Association" has just been formed with the following officers: Charles Austin, *President*; Joyce Kelly, *Secretary*; Edward Murray, *Sergeant-at-Arms*; Clarence Bischoff, Lents, Ore., *Corresponding Secretary and Treasurer.*

They have two officers, George Schwartz and Herbert Slocum, who handle operating tests with each member in order to ascertain their efficiency at sending and receiving. The members are then given a certificate showing their ability, and then at the end of a year they are again tested to show their improvement. They desire to regulate unnecessary interference, such as testing out sparks, etc., after nine p.m.

They would like to have any of the other clubs correspond with them so that they might benefit by one another's experience.

Address all communication to the Corresponding Secretary.

CLARENCE BISCHOFF,
Corresponding Secretary of the Oregon State Wireless Association.

Gentlemen: I wish to protest against the kick of Mr. Cecil A. Wanner. I think the work of the *Electrician and Mechanic* on wireless is very satisfactory as it is. I do not know where the amateur or experimenter would go to find anything more or better.

The articles I like best are those telling how to make wireless apparatus, motors, how to solder, how to use glue, or how the experimenter may by any means adapt common materials and tools to his ends.

I am least interested in the woodwork and the engineering articles, but I do not intend this as criticism; others may like them. Naturally I would hope that the price be not increased.

Very respectfully,
H. T. VAN PATTEN.

Dear Sir: Page 68 of January number contains an article from a Mr. Wanner on the benefit of specializing in a magazine for craftsmen, on the next page is an editorial offering suggestions for future articles. As the two articles are interrelative, I will treat it as one subject.

When the average city-bred man, bookkeeper, clerk, salesman or whatever he may be, sows his wild oats, and having taken a better half and settled down to housekeeping, there comes a time when he feels the need of using his hands for the use or adornment of his home. This is due, no doubt, to heredity, and dates back to the ages when all men were craftsmen of sorts, they had to be, as they could not 'phone for a carpenter, plumber, or electrician in those days, but had to do their own jobbing.

Finally he makes up his mind, prompted by his better half, who wants a small table for the parlor, that he will make her one. He therefore sallies out to the nearest hardware store, and gets saw, plane, hammer and a gimlet or so, also some wood, and blithely sets about fabricating this table. Alas for him, he finds he can't saw a straight cut, the plane digs in, the gimlet with fiendish maliguity either goes crooked, or splits the wood, and while he nurses his blistered hands (and temper), consigns the ghastly imitation of a table to the woodbasket.

Not losing heart, however, he thinks he will try something easier, and goes about among his friends, thirsting for information. One of them introduces him to the *Electrician and Mechanic,* and in the pages of that magazine he gets a light, he sees descriptions of little things in ornamental brass work, woodwork, electrical work and others, all along the lines he wants; also a description of just such a table as his good lady desires, all described with a fullness of detail and instruction, that he feels it's a cinch to do it. Another friend enlightens him as to what tools he really needs, and shows him how to use and keep them. He gains courage and purchases more wood and tools that he finds advertised in the magazine. He learns to plane and saw, and finds a joy in seeing the wood grow to shape in his efforts; finally, comes the task of assembling the table,

he reads, "put the sides together with dowels, etc." This gives him pause. Now what on earth is a dowel? Turns up the ever ready *Electrician and Mechanic* and lo, he finds an article descriptive of dowels, and another of gluing up the aforesaid dowels and other shapes, and he is comforted. He sends for a few back numbers of the *Electrician and Mechanic* (on his friend's advice) and sees he has struck a mine of richness. Here is a scheme by means of incising, chip carving, etc., to ornament this little table. With renewed ambition, he goes on improving the work of his hands, and delighting himself and his friends by the little things he makes, and getting the sense of satisfaction that a true craftsman feels in seeing the finished article grow from the raw material.

I have only instanced the woodworker here, but the same thing applies to the worker in art metal, the worker in the mysteries of electricity, etc.

The journeyman also sees in this magazine articles of the deepest interest to him, that he does not learn in the shop and at the bench, that tends to his usefulness and education; and so, from the making and mounting of a telescope, with which to study the magnificence of the heavens, to the making of a desk set, in thin metal, all kinds and conditions of men can find something to amuse and interest them; and while one man buys the magazine for the woodworking articles and another for the art metal work, yet, they read it all through and in so doing widen their views, and see how many other things are done which, while they may never need or use them, tends to a broader vision and makes him feel in touch with his fellows in craftsmanship.

So, Mr. Editor, my opinion is, Go to it, cover everything, so long as it is done as in the past, fully and completely, so the neophyte may feel that he is watched and helped by those who know and are willing to help the learner for the sake of craftsmanship.

Let the man who wants to specialize, buy a magazine that specializes; there are plenty such.

Yours truly,

J. C. JENNINGS.

BOOK REVIEWS

Modern American Telephony. By Arthur Bessey Smith, E.E. Chicago, F. J. Drake & Co., 1911. Price, $2.00.

This interesting handbook on telephony is clearly and accurately printed on substantial paper, is well illustrated, and is strongly bound in black leather. The telephone is without doubt one of the most useful of inventions and the author has carefully described the various parts of both the telephone and the telephone system. Two of the most interesting chapters are on Wireless Telephony and the Automatic System of Switching. The Automatic switchboard has come into widespread use and since it is installed in more than eighty cities in America, many people will doubtless be interested in reading a description of the various types of automatic systems.

TRADE NOTES

The L. S. Starrett Company which has one of the largest manufacturing establishments of fine mechanical tools in the world has recently increased its capital stock to $3,500,000, of which $1,500,000 is to be six percent cumulative preferred stock, and $2,000,000 common stock. The Company was incorporated in 1900, succeeding to the business established by L. S. Starrett in 1880 and conducted by him individually up to the time of its incorporation. The capital stock of the incorporation was $100,000 in 1900, since which time the business has increased far out of proportion to the original capital. Mr. L. S. Starrett, himself, continues in active control of the business.

Henry Disston and Sons, Philadelphia, Pa., have just issued a beautifully illustrated and printed booklet describing their line of Crosscut Saws. In the booklet is described the Raker or Cleaner Tooth which is used on these saws, and reasons are advanced showing why such teeth are of value. The booklet contains some fifty pages, and there are many illustrations shown of the various types of Crosscut Saws. Copies of this most interesting booklet may be obtained by writing to the company.

The plans completed for the motor car exhibitions to be held next midwinter provide for two national shows in New York and one in Chicago. These are the twelfth annual displays made by the industry to show its progress from year to year and the schedule of dates is as follows: Jan. 6–13—Passenger Car Exhibition, Madison Square Garden, New York; Jan. 15–20—Commercial Vehicle Exhibition, Madison Square Garden, New York; Jan. 10–17—Combined Passenger and Commercial Car Exhibition, Grand Central Palace, New York; Jan. 27–Feb. 3—Passenger Car Exhibition, Coliseum and First Regiment Armory, Chicago; Feb. 5–10—Commercial Vehicle Exhibition, Coliseum and First Regiment Armory, Chicago.

More than two months before the opening date of the first of these national displays, a total of 113 different manufacturers of private passenger cars and 86 makers of commercial vehicles had been allotted space in one or more of them. Sixty-four of these are new exhibitors, having made no displays at the national shows last winter in New York and Chicago. Of the 64, 39 are builders of trucks and delivery wagons and 25 make pleasure cars.

During the two-weeks show period in New York more than 100 different makes of passenger cars and 70 makes of work vehicles will be on exhibition simultaneously. In Chicago more than 90 makes of pleasure cars will be shown during the week of Jan. 27 to Feb. 3, and the following week more than 60 different makes of business machines will be exhibited.

Exhibits will include almost every type and size of power vehicle designed for use on the public roads, from motorcycle parcel carriers and delivery wagons of 500 lbs. capacity to ponderous trucks of 10 tons capacity. Besides the more common types of trucks and wagons, there will be a number of dump trucks for contractors' use, trucks fitted with power winches for hoisting, self-emptying coal and lumber trucks, machines with special bodies for special purposes, self-propelled fire engines and combination chemical and hose wagons, police patrols, ambulances and other types for municipal and public service purposes.

In addition to complete vehicles and chasses without bodies, there will be comprehensive displays of motor car parts, equipment and supplies by more than 200 manufacturers that will fill all available space in the galleries. Large sections at Madison Square Garden and the Chicago Coliseum will be occupied by the exhibits of a score of motorcycle manufacturers.

It is only at these national shows that all the different makes of cars having more than a local sale and reputation are brought together for inspection. Without loss of time the engineer as well as the individual purchaser can study and compare the constructional characteristics of the various makes and models under practically one roof. The show promoters have recognized the importance of the time element and have segregated the exhibits of industrial, commercial and municipal vehicles, giving a separate week to them, thereby enabling the business man to look them over in a few hours and talk to exhibitors without getting into the crowds of visitors attracted by the pleasure cars or having his own and the exhibitors' attention distracted.

The two consecutive weeks' display in Madison Square Garden is under the same management that has conducted it for the last seven years. The show committee consists of Messrs. George Pope, Charles Clifton, and Alfred Reeves, with M. L. Downs acting as secretary.

The other New York show, which is to be held concurrently with the Garden exhibition, and embraces both private passenger cars and commercial motor vehicles, will open on Thursday of the first week and close Wednesday night of the second week of the show at Madison Square Garden. It is to be staged in the new Grand Central Palace, recently completed a few blocks north of the Grand Central Station now under construction. For the first time this exhibition is to be under the auspices of the National Association of Automobile Manufacturers and the management of Mr. S. A. Miles, who has conducted the Chicago show for more than a decade—ever since such shows were inaugurated, in fact.

The Coliseum show in Chicago will be, as usual, under N.A.A.M. auspices and Miles management.

Briefly, the trade situation with regard to the several exhibitions is this:

The Garden show is restricted to members of the old Association of Licensed Automobile Manufacturers and makers of electric vehicles who have been consistent exhibitors at Madison Square Garden for the last five years or more. The Grand Central Palace show is "open" to all manufacturers but will not include displays by makers who have exhibits in the Garden. All manufacturers are eligible also for the Chicago show, which is the only one that will be held in that city, and it will include exhibits by most of the makers who display at both the New York shows.

We are in receipt of a small and attractive pamphlet from the L. S. Starrett Co., Athol, Mass., describing their excellent line of vernier calipers. The pamphlet also includes directions for reading the vernier. The scale of the tool is graduated in fortieths, or 0.025 of an inch, every fourth division representing a tenth of an inch, being numbered. With the aid of the vernier it is possible to read in 1,000ths of an inch or on the calipers with metric divisions it is possible to read in 50ths of a millimeter. A copy of the pamphlet may be obtained by writing to the L. S. Starrett Co.

The State Water Supply Commission of New York State has published a pamphlet entitled "Water Power for the Farm and Country Home," by David R. Cooper. This pamphlet describes a few of the many ways in which water power may be utilized on the farm, and how by employing a water wheel to drive a dynamo both light and power can be obtained whenever needed.

The General Electric Company has just placed on the market a new piece of apparatus known as a Battery Truck Crane, which the Company describes in its Bulletin No. 4892, recently issued.

The machine is a short, heavy, storage battery vehicle, having mounted on its forward end a swinging crane, the hook of which is raised and lowered by a one-ton hoist operated from the vehicle battery.

The vehicle is used in loading, hauling, and unloading trailers, loading and unloading cars, hoisting and carrying on the hook boxes or barrels, and for stacking. The running of the vehicle, hoisting and carrying, are controlled by one man.

The crane is equipped with special attachments to suit the carrying on of the work contemplated, and the height can be made to suit local conditions.

The bulletin contains illustrations of the truck and crane in use, and describes also various pieces of apparatus used in connection with it.

In Bulletin No. 4912, recently issued by the General Electric Company, is a collection of several articles devoted to the use of this new piece of apparatus for air purification. Various applications of the apparatus are illustrated.

The Rubel School of Aviation, located in Louisville, Ky., has issued a most interesting little pamphlet, describing their Aviation Park and the course of instruction given by their school. The school was opened on December 15th by the firm of R. O. Rubel, Jr., and Co., which for more than three years has been engaged in the manufacture and sale of aeroplanes and their component parts. The equipment of the school is of good size, and they have at present three monoplanes, three biplanes, one hydroplane, one wind wagon and one glider, all of which are housed in one large hangar. The school guarantees to secure engagement to those aviators whose ability and connections are satisfactory. The school and aviation park are located just two and one-half miles from the city of Louisville, and the park comprises one hundred acres of clear, unobstructed level ground which has been completely tiled to effect perfect drainage. Surrounding the park is thousands of acres of clear, level fields.

The Tufts Wireless Club

With the installation of the antennae of the wireless outfit on the top of Robinson Hall, the Tufts Wireless Club has started active work in the field for which it was formed. The use of the outfit, which is a 1-10 k.w. set, has been tendered the club through the efforts of Professor Harry G. Chase, head of the physics department at Tufts. It is the one used during the maneuvers this last summer by the Signal Corps, M.V.M., of which Professor Chase is captain.

The club, well backed by the faculty, is in charge of three very capable and experienced men of the undergraduates. Its president, Harold J. Power of Everett, has had exceptional practice for a sophomore at college. He was operator on Colonel Astor's yacht this past summer, and on the *Harvard* the one previous. In addition to his regular college work, he is now teaching telegraphy and wireless three nights a week at the Everett High Evening School. Walter L. Kelley of Arlington, the vice-president, has been employed during the past two years in the engineering department of the Edison Company and previous to that in one of the power houses of the Boston Elevated. He now has the supervision of all electrical apparatus on the campus and in the college buildings. Joseph A. Prentiss, the secretary-treasurer, has done a large amount of research work in the last two years, and is now in charge of the installing of the outfit at Tufts.

The club intends to go into research work, and in view of that it is now testing the wireless outfit of the Signal Corps, the results of which are to be sent to the Signal Corps authorities at Washington. To add interest to the meetings, a series of twelve instructive lectures will be given by the officers of the club on the following subjects: (1) Mathematical determination of wave lengths and calibration of meters; (2) Advantages of a rotary spark gap at a wireless transmitting station; (3) Transatlantic telegraphy; (4) The work of a commercial operator; (5) Government requirements for a wireless operator; (6) Alternating current generation; (7) Armature winding; (8) Steam turbo generators; (9) Automatically controlled high-tension switches; (10 The ideal sub-station; (11) Balanced antennae; (12) Quenched spark gap.

The outfit now erected will, after testing, be replaced by a larger set which will be capable of sending from 500 to 1,000 miles and receiving from any distance.

ELECTRICIAN
& MECHANIC

| VOLUME XXIV | MARCH, 1912 | NUMBER 3 |

ELECTRIC LIGHTING IN PRIVATE INSTALLATIONS

J. A. S.

The problem of electric lighting as regards ease and convenience has been satisfactorily solved in cases where a central station service is available, as all that is required is the turning on of a switch. In country houses and private installations of a similar description, however, unexpected difficulties have been encountered, owing to the necessity of the provision of what is virtually a small power house. This involves attendance, and a choice has frequently to be made between the employment of a technical man with a technical man's wages, and the exciting and expensive recreation of trusting the plant to a

Fig. 1
General Appearance of Plant

Fig. 2
The Controller

gardener-engineer odd-jobber, whose chief virtue is an 'unfailing optimism in the midst of distressing circumstances.

Moreover, even with à good equipment of producer and gas engine, or oil engine, the times occur when the prime mover is being overhauled or has the spirit of cussedness in it, and a standby storage battery of sufficient capacity to carry a good proportion of the lighting load is then appreciated. One cannot (or should not) have a battery installation without paying for it, and apart from the initial cost, there is the fact that unless the battery is treated with due care it is apt to develop complaints the cure of which is beyond the tender mercies of the gardener aforesaid.

It is therefore interesting to note a recently devised automatic generating set which is calculated to be always on hand when required to do work. No attendant is required to start and stop it; beyond the technical knowledge required to fill the tank, and to clean and oil the engine, the man in charge may be of the usual standard of intelligence, and Messrs. R. A. Lister & Co., Ltd., of Dursley, England, appear in their Bruston system to have put their finger on a serious fault

in private lighting plants and to have found the remedy.

{ The general appearance of the plant is seen in Fig. 1. The plant is self-contained and is delivered ready to work, no special foundation being required, but the plant must be set level, and it is advisable to set a soft pad (lead or felt) at each corner for the bed plate to rest upon, in order to ensure smooth running. The circulating water tank must also be protected in frosty weather. In the case of large installations the plant may be divided into two or three different units. For instance, in the case of a 400-light installation, two units can be provided, viz., one each of 100-lights and 300 lights, or, if preferred, two each of 200 lights. These units are connected up in such a manner that a light load would start up one set (the smallest), but upon the load being increased the second set would automatically switch itself in. Any number of separate units can be controlled in this manner. It is, therefore, possible to work an installation of any size practically without attention. To start the plant it is only necessary to switch on three or four lamps or more as may be required; these lamps are lighted direct from a small storage battery. When, however, more lamps are switched on, the extra current demanded by the lamps causes a relay instrument, which is simply an automatic switch, to be put into action. The current from the battery then passes through the automatic starting switch or controller and is delivered to the dynamo, which then acts for the time being as an electric motor, and rotates in the same direction as the engine, to which it is connected by belt. The engine then revolves in exactly the same manner as if it were started by hand. The engine then commences firing, and, overcoming the motor effect of the dynamo, starts generating electricity itself, delivering the direct current to the lamps, and at the same time replacing the energy which has been taken from the battery in order to start the dynamo. Assuming the number of lamps in use being below that required to start the plant and such lamps are required for a long period, the voltage of the battery will drop; but this very fact is provided for in the system, and such a drop of voltage will also actuate the relay and cause the plant to start up and recharge the battery. When the lamps are switched off and the battery has been charged up to the standard voltage, the plant automatically stops. The engine may be driven either by gas or petrol, and the automatic arrangement, depending as it does solely on turning the lights required on or off, is remarkable. The system saves waste, as practically the whole of the current is delivered direct from the dynamo instead of being first used to charge up the accumulators and subsequently to discharge them, the loss in this operation being at least 40 percent of the power generated; and it also ensures long life to the battery, as the latter is always kept charged, and never falls below the standard voltage; it is automatically charged up, although no lamps may be switched on. The principal feature of the arrangement is its simplicity and absence of complicated mechanism, as will be seen from a brief specification of its components. The engine is a low-speed petrol or gas engine of standard design, the only addition being a simple valve lift to relieve compression for starting and stopping. The dynamo is a standard pattern shunt-wound machine coupled to the engine by means of a leather belt. The controller is similar in design to the standard type solenoid starting switch, many of which have been used for years, and a form of which is shown in Fig. 2. The switchboard consists of an enameled slate and contains an ammeter indicating the current being used; an ammeter indicating the current being generated; a voltmeter indicating the pressure; a "Lister" patent relay which controls the starting and stopping of the plant, a safety circuit breaker for protecting the battery and dynamo; a safety fuse for protecting the wiring; and a voltmeter switch. The switch-

contained and as sent out ready for use. No special foundations are required, and as the set is not fixed to the floor it can easily be moved from place to place. It only occupies a floor space of approximately 9 x 4 ft., and, owing to the efficient and automatic arrangement for oiling, the cost of lubrication with the set described is less than with some other systems.

Some comparative costs may be interesting. They are, of course, based on English prices, but they may be taken as sufficiently good for comparison. Taking an ordinary accumulator electric light installation with a low-speed engine, the cost to run 200 12-c.p. lamps would be for engine, dynamo, battery, switchboard, etc., about $1,000. An engine and battery room 20 x 12 ft. would cost about $300, to which must be added $75 for concrete foundations, and $100 for erection, starting up the plant and charging up the battery, $1,475. Taking the Bruston set the cost of engine, dynamo, battery, switchboard, etc., is slightly more, being put down at $1,100, but no battery room is required; if there

(*Continued on page* 164)

THE CONSTRUCTION AND OPERATION OF THE GASOLINE ENGINE USED ON AUTOMOBILES

A. E. KLINE

The gasoline engine and its development has been one of the wonders of the mechanical age, and it is essential that the young mechanic has at least a fair conception of its construction and mode of operation. It is with this object in view that this article is written.

Gasoline is a colorless and comparatively light-weight liquid which is distilled from petroleum. It is a mixture of hydro-carbons, chief of which are known as pentane, hexane and heptane.

In order to become usable in the gasoline engine, it is mixed with air in a proportion of one to eighteen parts, in which it is vaporized. This vapor is then introduced into the working cylinders, where it is compressed from about 80 to 90 lbs. per square inch, at which pressure it is ignited by means of an electric spark. Due to the igniting of the mixture, combustion takes place, and the pressure rises to four times that originally. This pressure acts on the movable pistons, imparting to them mechanical energy in the form of a reciprocating motion. This motion is then transformed by certain means into rotary motion in order to be made use of.

Before discussing the operation of the engine it is best that the various parts which constitute it be dealt with. A sectional view of an up-to-date engine is shown in Fig. 1. The cylinder A is a cylindrical casting of iron, at the upper portion of which, and cast integral with it, is what is termed the combustion chamber B. Two passages, C and D, communicate from the combustion chamber, and are termed the inlet and exhaust passages respectively. Each of these passages is controlled by a valve. One opens at the proper moment, and allows the mixture of gasoline to enter the combustion chamber, while the other allows the expulsion of the gases after they have been used. The spark plug E, used for the purpose of igniting the mixture, is usually placed in the combustion chamber. At the bottom end of the cylinder is what is termed the crank-case F, usually made of aluminum or cast iron, and so formed as to cover the moving parts

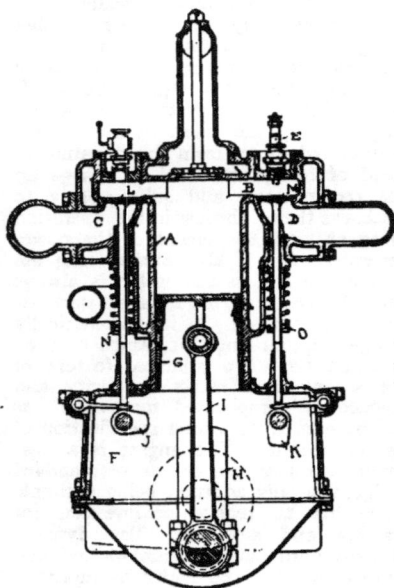

Fig. 1
Sectional View of an Up-to-date Engine

of the engine. Within the cylinder is placed the piston G, which is usually a a hollow cylindrical casting of iron, closed at its upper end, and fits very closely in the cylinder bore. The piston is the portion of the engine acted upon by the gas pressure, and transforms the energy heat imparted to it into mechanical energy. The crank-shaft H is made of a horizontal steel bar, and is supported by bearings within the crank-case. At its middle point · it is provided with an offset, which constitutes a crank. A rod I, called the connecting rod, is fastened, at its upper end, on a pin within the piston, while at its lower end it is fastened to a bearing on the crank. It will be seen that, as the piston moves down, due to the gas pressure, it imparts a rotary motion to the crank-shaft, through the intermediate action of the connecting rod.

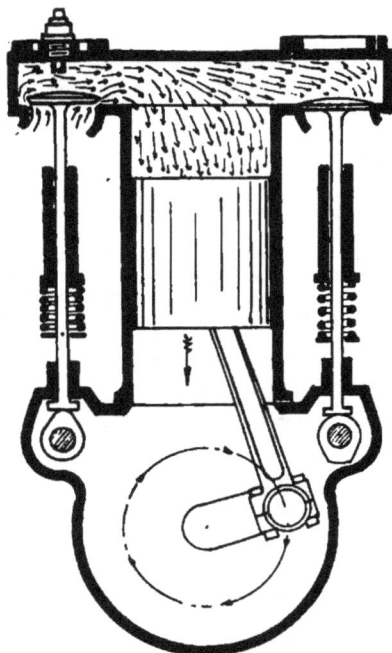

Fig. 2
Inlet Stroke

Each of these four operations occupies
the time of one piston stroke. It is
obvious that since one stroke of the piston
occupies half a revolution of the crank-
shaft, which is 180 degrees, the entire
four strokes of a cycle, occupy 4 times
180 degrees which equals 720 degrees, or
two revolutions of the crank-shaft. It
will also be noted that in but one of the
four strokes is power developed.

Figs. 2, 3, 4 and 5 show the four strokes
diagrammatically. As before stated, the
first stroke of the cycle is the admission
stroke. The piston then moves outward
in the cylinder causing a partial vacuum,
and, due to opening of the inlet valve
throughout this stroke, the gasoline
mixture enters the cylinder from the
vaporizing device known as the carburetor. At the end of the stroke, in
fact a little after, the inlet valve is closed;
the piston begins to move up, and compresses the charge in the combustion
chamber. This is the second and compression stroke. When the piston has

The valve mechanism which controls
the entering and expulsion of the gases
is clearly shown in the drawing. The
cams J and K are rotated by suitable
gears, and lower and raise the valves
L and M at the proper moments, against
the tension of the coiled springs N and O.

Having thus briefly referred to the
principal parts of the engine, let us take
up mode of operation. The great majority of engines are operated on what is
known as the Otto cycle, deriving its
name from the inventor, Dr. Otto, of
Germany, who invented it in 1876. It
is known as the 4-cycle type of operation.
By a cycle is meant a succession of operations in an engine cylinder, with a single
charge of explosive mixture. The four
operations, or strokes, succeed one another in the following order:

1. Admission of the charge into the
   cylinder.
2. Compression of the fuel.
3. Igniting and expansion of the fuel.
4. Exhaustion of the burnt gases.

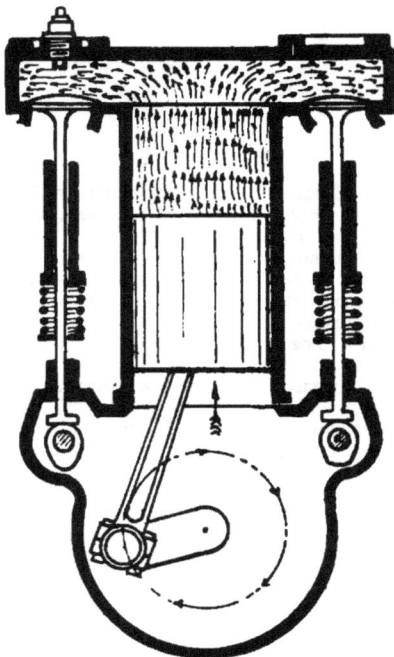

Fig. 3
Compression Stroke

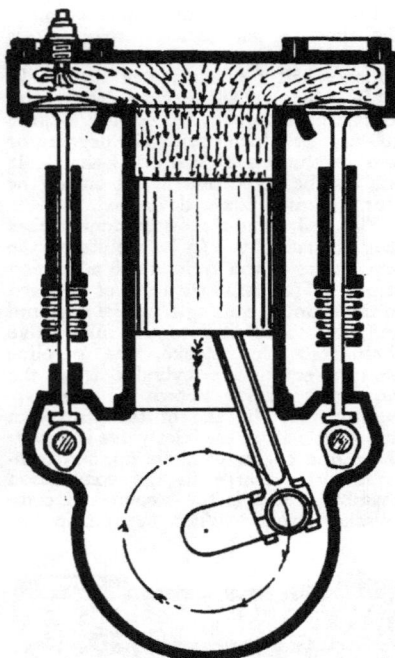

Fig. 4
Expansion Stroke

and last stroke of the cycle. On the next downward stroke a new cycle begins.

It is interesting to note the behavior of the gases in the cylinder throughout one cycle, and Fig. 6 readily shows this. This is a so-called indicator diagram, similar to those taken of a steam engine. The line 1-3-4 designates the inlet stroke, and shows the partial vacuum produced, since the line at point 3 drops below the atmospheric line. The compression stroke is shown by the line 4-5, and it can readily be seen how the pressure rises. At the instant of ignition the pressure increases enormously, as shown by the point 6, and, as the piston moves out, the pressure gradually drops again until the end of the stroke is reached. The line 1-2-4 shows the exhaust stroke, pointing out the fact that the gases are slightly compressed, due to the fact that the piston is sweeping them rapidly forward, and exerts a very slight pressure. As in a similar case of the steam engine,

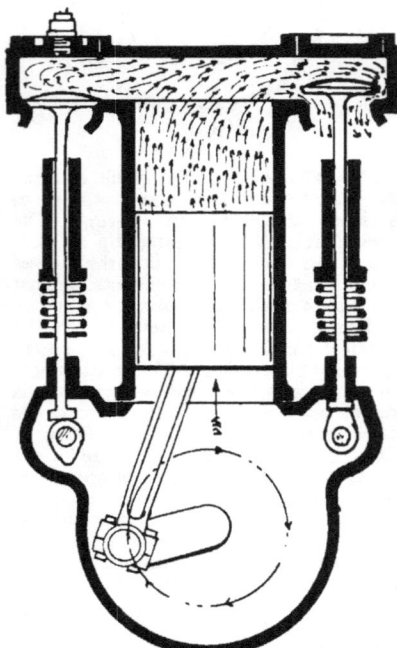

Fig. 5
Exhaust Stroke

reached the end of its return stroke, the gas has been compressed to about one-fifth the original volume.

At or near the end of the compression stroke, an electric spark, produced within the compression chamber, ignites the compressed gas, and combustion takes place. The pressure immediately rises to about four times that previous. The actual pressure in the cylinder the instant of ignition is about 250 lbs. per square inch. Under this enormous pressure, the piston is forced downward, thus transforming heat energy into mechanical energy. The gas pressure drops rapidly as the piston moves outward, due to the increase in volume, loss of heat, etc. This is third and power stroke of the cycle.

When the power stroke is almost completed the exhaust valve opens, and the expulsion of the burnt gases continues throughout the following return stroke of the piston. This completes the fourth

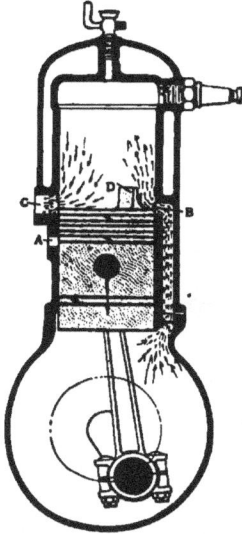

Fig. 7

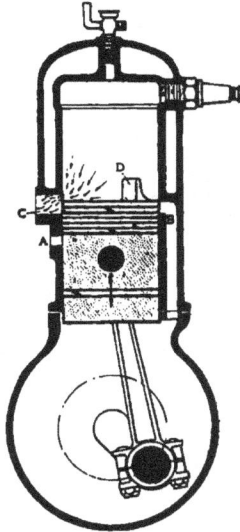

Fig. 8

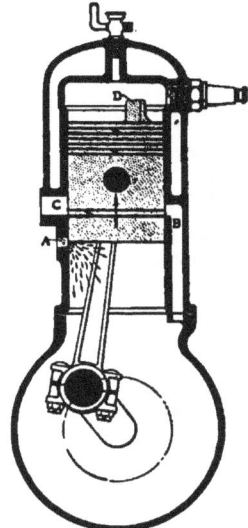

Fig. 8a

the area, 4-5-6, shows the amount of work done during the cycle.

Besides the 4-cycle engine, there is another form of engine in which the four operations, intake, compression, expansion and exhaust, are performed in but two strokes of the piston, or one revolution of the crank-shaft. This type is known as the 2-cycle engine. Figs. 7, 8

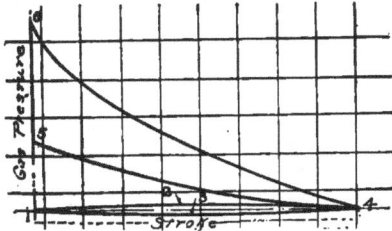

Fig. 6

and 8a show diagrammatically the operation, and also the essential parts.

The essential features of a two-cycle engine are as follows: An enclosed crank-case which is provided with a valve 'A which admits gas into it on the upward stroke of the piston. An inlet and exhaust post, B and C respectively, located

at points near the extreme outward position of the piston, and a pass or connection between the crank-case and the cylinder.

The operation is as follows: Assuming the piston to be moving upward, and is compressing a charge. As it travels up,

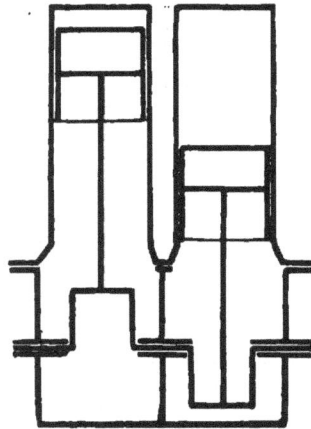

Fig. 9
Twin Cylinder Type

it uncovers the post A, Fig. 7, and gas enters the crank-case. The compressed fuel in the cylinder is ignited, expands, and the piston moves down. As it does so, it shuts off connection between the crank-case and cylinder by covering the valve B, and hence it compresses the gas in the crank-case. As it reaches the end of the downward stroke, it opens the exhaust post C, and a trifle later connection is made between the cylinder and crank-case, shown in Figs. 8 and 8A. The burned gas exhausts, and the gas in the crank, being compressed to about 35 lbs., rushes into the cylinder, as shown

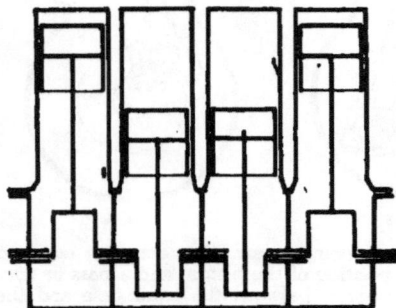

Fig. 10
Four-Cylinder Type

The piston now moves up, closes both posts, and compresses the gas above, and at the same time, a new charge enters the crank-case. The compressed gas in the cylinder is now ignited, and the piston is forced downward again.

The result is that two operations are done in every stroke; on the down stroke expansion and compression, and on the up stroke, intake and exhaust.

The chief advantage of the two-cycle engine over the four-cycle evidently lies in its simplicity, as there is a total absence of valves, cams, springs, etc. The other advantage lies in the fact, one power impulse is received for every two strokes, as compared with but one power impulse for every four strokes. Hence theoretically the two-cycle engine should develop twice the power of a four-cycle, other dimensions being equal. Such is not the case, however. In actual practice hardly twice the power is developed, due to various reasons which cannot here be dealt with.

It is an evident fact that the amount of power developed by an engine depends upon the number of cylinders, other things being equal. The average single cylinder engine develops from 6 to 12 h.p. The majority of engines employed for automobile work have more than one cylinder,

Fig. 11
Double-Opposed Type of Engine

in order to develop more power, and it may be of interest to study the methods of grouping the cylinders.

Fig. 9 shows a twin cylinder type; Fig. 10, a four-cylinder type, and Fig. 11 what is termed a two-cylinder double-opposed type. While the casting of these cylinders is quite a problem, yet there are engines manufactured at the present time which have all four cylinders cast integral, and there is also a six-cylinder type, which has just been announced, that has all six cylinders cast integral.

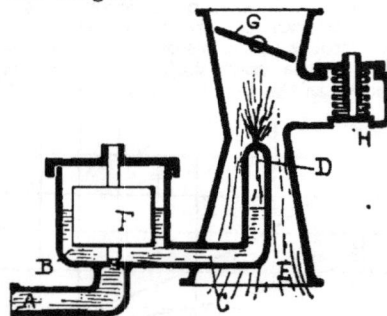

Fig. 12
Diagram Form of Carburetor

It may be well to examine the method of vaporizing the gasoline before it enters the engine cylinder. The carburetor, or vaporizer, as it is sometimes called, is used for this purpose. Fig. 12 shows a very simple type in diagram form.

The gasoline in liquid form enters at the opening A, and into a reservoir B, from which it issues through a channel C

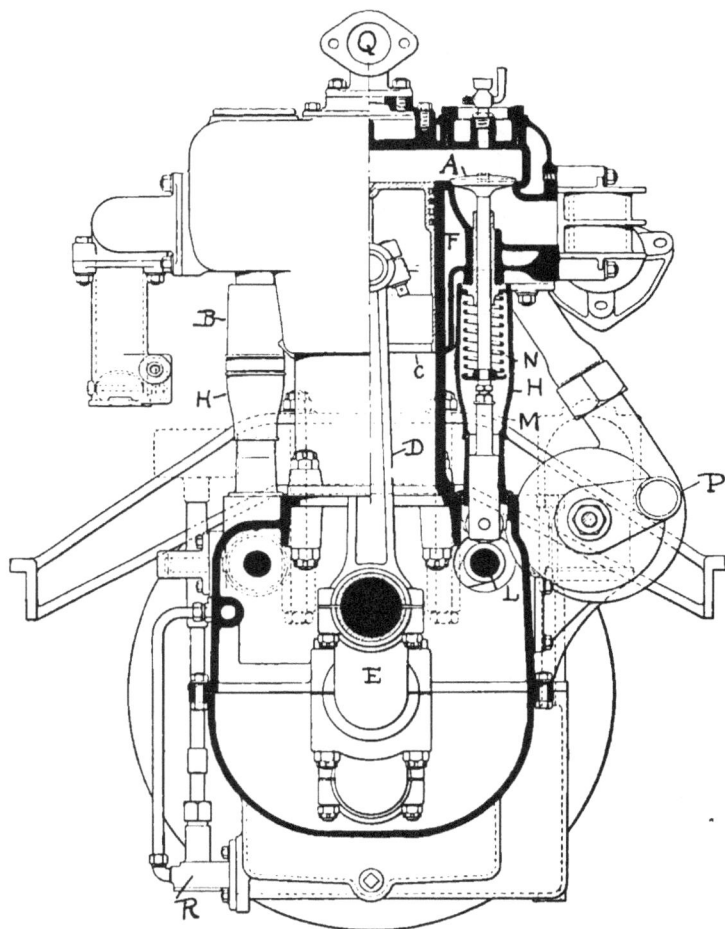

Fig. 13

into the tube *D*. As it leaves this tube, or nozzle, as it is called, it is mixed with the air, which enters at *E*, and vaporized. The vaporized mixture then passes up the chimney and into the engine cylinder. The float *F* serves the purpose of keeping the gasoline in the reservoir at a constant level. A valve *G*, called a butterfly valve, serves the purpose of regulating the amount of mixture entering the engine. The valve *H* is for the purpose of adding additional air to the mixture.

As before stated, the explosive mixture is ignited by means of an electric spark, which is produced by the well-known magneto, which is a form of dynamo generator. It generates the current, transforms it to a high pressure, or voltage, and distributes it to the various cylinders.

Having familiarized ourselves with the fundamental parts and principles of operation, let us examine a well-known type of engine, and discuss it in detail. This engine, shown in Figs. 13 and 14, will be seen to have four cylinders, cast

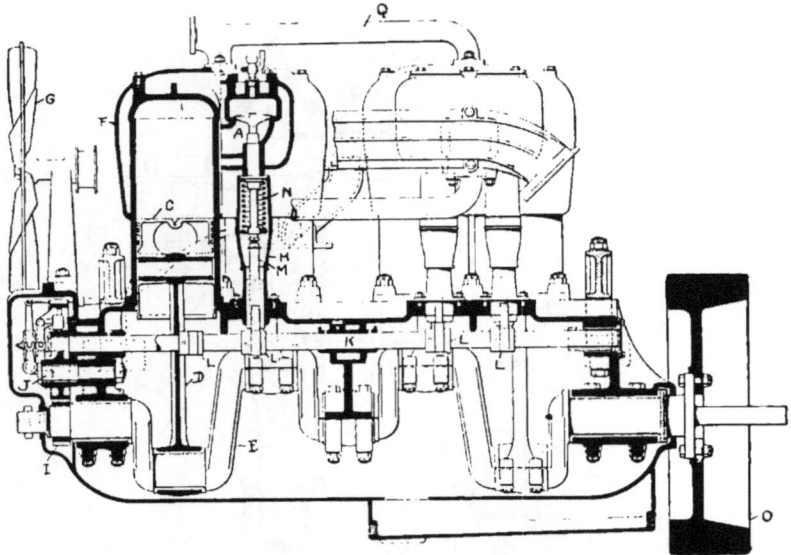

Fig. 14

in pairs, with the valves A, B on opposite sides. This type of cylinder is termed a T-type, as it resembles the letter T. But one of the valves A is shown fully. The piston C is fastened to the connecting rod D, which in turn is fastened to the crank-shaft E. There is a jacket F cast about the cylinders, through which water circulates, for the purpose of carrying off the tremendous heat generated by the explosions. A fan G also helps to cool the cylinders. The valve stems and springs are encased by the casing H, which protects them from the dust. The gear I is keyed to the crank-shaft E,

meshes with the gear J, which turns the cam-shaft K and consequently L. The cams operate the valve push rods M against the springs N, which operate the valves. The flywheel O, which is bolted to the rear end of the crank-shaft, is for the purpose of producing an even turning motion. A pump P continuously circulates the water through the water jackets and into the radiator by means of the outlet pipe Q, and an oil pump R continuously pumps the oil to the various bearings, where lubrication is required.

Figs. 15 and 16 are actual photographs

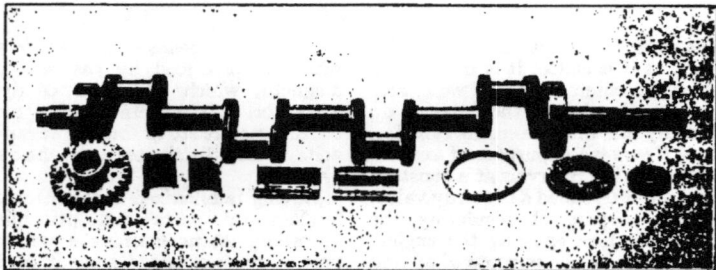

Fig. 15. Crank-Shaft and Bearings

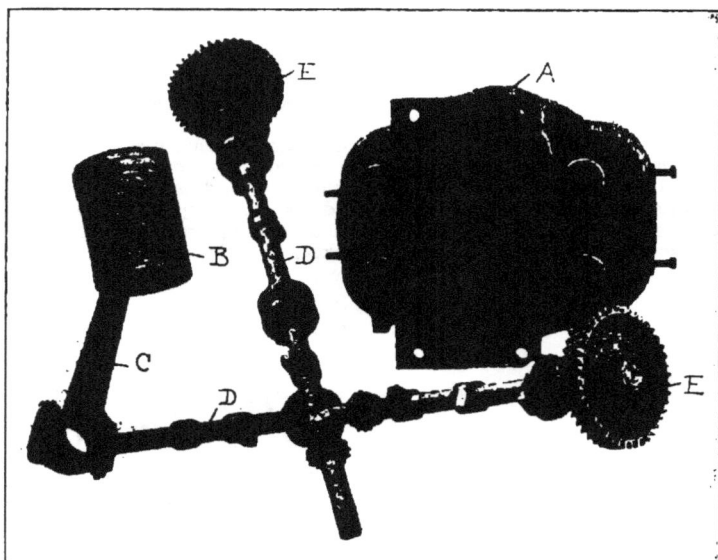

Fig. 16. Piston, Cam-Shaft and Connecting Rod

of the various parts, and are inserted so as to help the reader to more readily grasp the general appearance of them. At A, in Fig. 16, is shown the lower portion of a twin type of cylinder; at B, a piston; at C, the connecting rod; and at D, D, two cam-shafts. The helical gears E, E, which are keyed to the ends of the cam-shafts, are for the purpose of driving them from the crank-shafts. At Fig. 15 is shown the crank-shaft of a four-cylinder engine, with its bearings and the gear which meshes with that on the cam-shaft.

At Fig. 17 is shown the manner in which the pistons, connecting rods and crank-shafts are assembled in a four-cylinder engine.

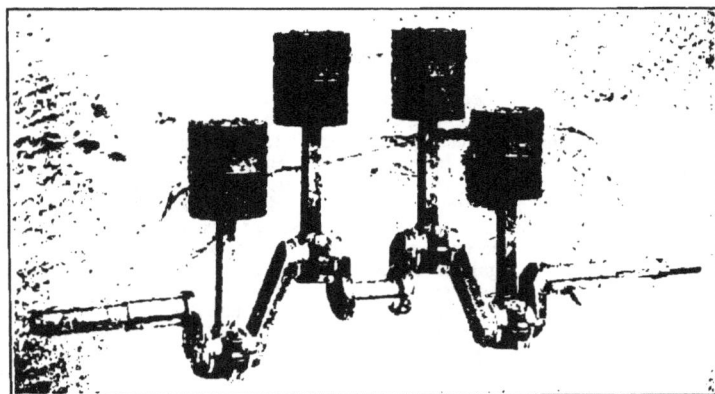

Fig. 17. Crank-Shaft and Pistons

X-RAY APPARATUS OF TODAY

H. WINFIELD SECOR

When the X-ray was first introduced as an accessory to the surgeon and physician, to aid in locating foreign bodies, diseased bones, etc., in the human body, it was usual to operate the X-ray tubes from a static machine, probably because it was a ready source of high-tension direct current at that time.

a powerful discharge, of practically unvarying intensity; but its serious disadvantages, which unsuited it to some extent for this work, were the frequent adjustment and care of the interrupter, and the secondary current, which, being an alternating one, although not harmonic, had the undesirable half wave,

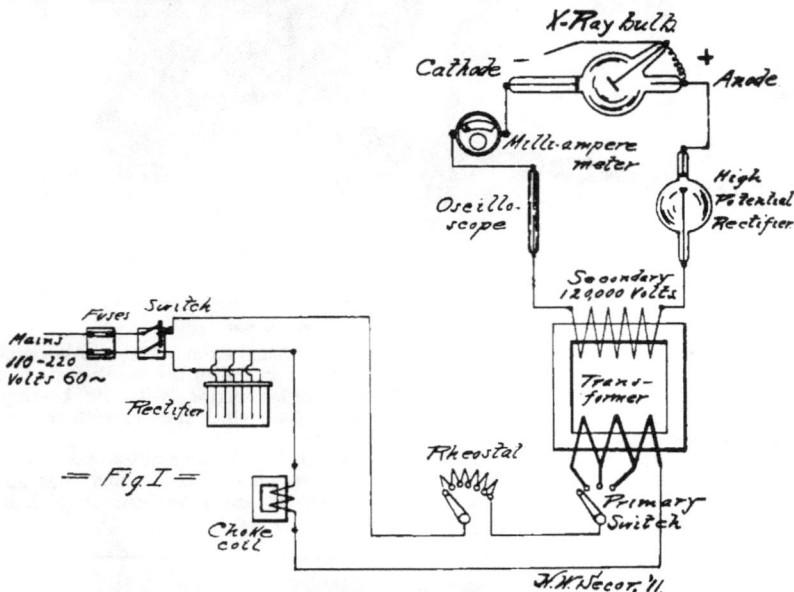

Diagram for the Production of a High Potential, Unidirectional Current

But it was not long before dissatisfaction began to be manifested among the users of these machines for exciting the X-ray tubes, and the most frequent cause given was, that in damp or warm weather it was oftentimes impossible to make the static machine generate a sufficient current, and sometimes none at all. Another point against static machines, is that their current capacity is rather limited, also, they are quite unsteady in the supply of current.

Then came the wave of popularity for the induction coil, as a source of X-ray current. This seemed to be a great boon, as it was always ready to deliver

caused at every make of the primary circuit, and known as the inverse current, which caused the X-ray tubes to have a greatly shortened life, owing to this inverse current, tending to make the anti-cathode or anode electrode the cathode, with a consequent scattering of the platinum forming the target or anode.

Thus what was wanted was a source of current which should demand practically no attention after once adjusting; absolute constancy in service, allowing the full value of the current to be had at any time, and this current should be a unidirectional or direct current, with no inverse or reverse current, whatever.

These qualifications of a satisfactory X-ray machine seem to have been fulfilled in the apparatus built by the Kny-Sheerer Co., of New York City. They build several different types of this apparatus, and as they represent the very latest in this field, they will be described in order. All of this apparatus employs a closed core, high potential, step-up transformer of high efficiency, in place of the less efficient open core induction coil.

The simplest set for the production of a high potential unidirectional current is illustrated in diagram I, and is connected directly to any 110 or 220 volt 60-cycle alternating current circuit. Use is made of a regular electrolytic rectifier, placed in series with the main line feeding the transformer, which suppresses one-half of every wave, allowing the current to flow through it in one direction only, resulting in an unidirectional current flowing through the primary circuit of the closed core transformer. The rectifier is composed of alternate iron and aluminum plates immersed in a saturated solution of bi-carbonate of soda. As shown in sketch, a variable resistance is placed in the primary circuit to vary the intensity of the secondary discharge,

also a choke coil. The primary winding is adjustable in several steps, making the outfit very flexible in its operation.

The secondary voltage is variable between 20,000 and 120,000 according to the amount of primary turns and resistance inserted into circuit. At the latter voltage the secondary discharge takes the form of a heavy spark, or rather flame, 11 to 12 in. long. To further improve the unidirectional qualities of this current, a high potential rectifier, having two electrodes sealed in a glass bulb, containing a partial vacuum is placed in series with the positive terminal of the secondary circuit and the positive electrode or anode of the X-ray bulb. The negative or cathode of the X-ray tube is connected to an oscillioscope, thence to a milliampere meter, and from this to the negative terminal of the secondary coil.

The oscillioscope, referred to above and seen in diagrams, is composed of a glass tube about 1 in. in diameter and 7 in. long, having two aluminum wires secured in it, with a short gap left between their ends. The tube is evacuated, and the presence of any inverse current in the X-ray circuit is indicated by the

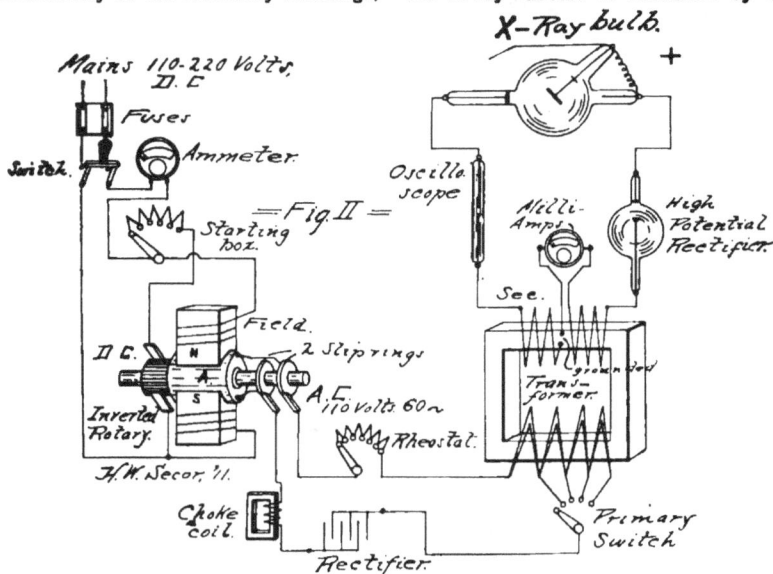

Diagram of a Set which is to be operated from a Direct Current Circuit

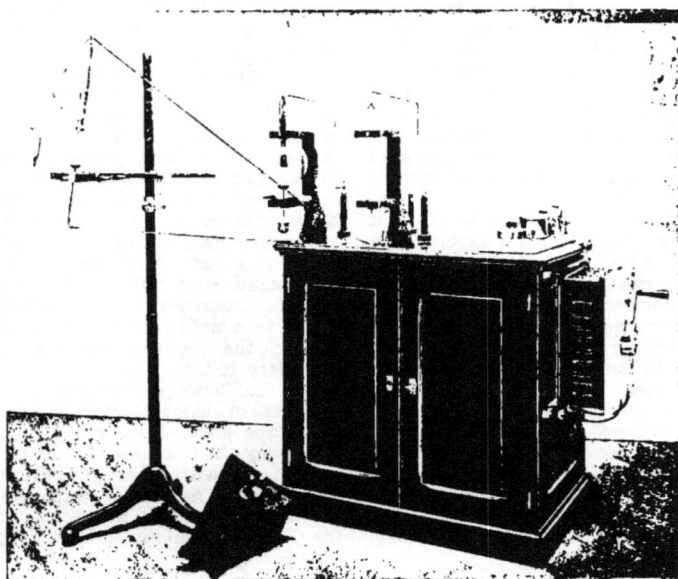

Fig. 3
X-ray Current Generator

fluorescence in the tube, extending along both wires, but if the current is flowing in one direction only (as it should), then the band of fluorescence extends only around the negative pole or electrode.

This scheme of producing a high-potential unidirectional current is quite satisfactory, but has certain disadvantages, *viz.*, the rectifier requires some attention, besides being wasteful of current, and the high-tension valve tubes, inserted in series with the anode lead, are subject to wear and tear, and must be renewed at intervals.

A set of this type designed to operate from a direct current circuit is illustrated by diagram Fig. 2. Here an inverted rotary converter (which is simply a D.C. motor with two slip rings attached to one end of the armature and connected to the regular winding) is utilized to run as a shunt motor and deliver the necessary A.C. from the two slip rings to excite the step-up transformer. The electrolytic rectifier, choke coil and variable resistance are inserted in the primary circuit, as in the set used for A.C. circuits

described above, and the rest of the diagram remains the same.

The appearance of this type of X-ray current generator is seen in Fig. 3, while Fig. 4 shows a high-potential rectifying tube, which is utilized to suppress the inverse half wave of oscillating currents. It acts the same as the iron-aluminum rectifier in low-potential circuits, *i.e.*, it allows the current to flow through it, in one direction only.

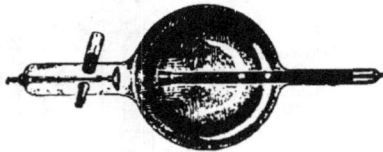

Fig. 4
High-potential Rectifying Tube

This machine, although much superior to the induction coil, with its attendant interrupter troubles and undesirable secondary alternating current, left a few points to be desired yet, and so, as "necessity is the mother of invention,"

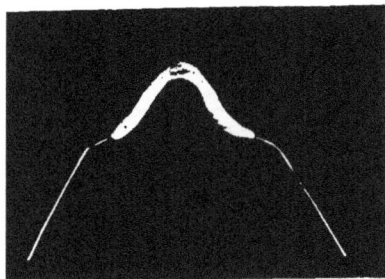

Fig. 5
An Interrupterless Set

a more improved type of generator was produced, which is certainly a remarkable machine.

As a starter, ask someone you know, if they ever saw 120,000 volts alternating current commutated into 120,000 volts direct current, and I'll wager 99 out of every 100 will look askance at you. Yet this is what this improved type of X-ray machine is capable of doing, and it is known as the unipulsator, or "interrupterless," the latter being the trade name for it.

It is regularly built in the 4.4 k.w. size, for operating either on D.C. or A.C. circuits. The high-potential unidirectional current used for the X-ray tubes may be varied from 20,000 to 120,000 volts, giving a very heavy flame discharge 12 in. or more long at the latter voltage, the appearance of the flame being illustrated in Fig. 4.

In the cut at Fig. 5 is depicted an "Interrupterless" set, complete, and designed to be operated on alternating current, single or polyphase. It is also regularly built for direct current circuits of 110 to 220 volts.

The general layout and scheme of such a set, designed for direct current circuits, is shown in Fig. 7; the scheme remaining the same for alternating current circuits, with the exception that the inverted rotary converter has its place taken by a synchronous A.C. motor and starting motor, with the necessary synchronizing devices, or an automatic self-starting

Fig. 6

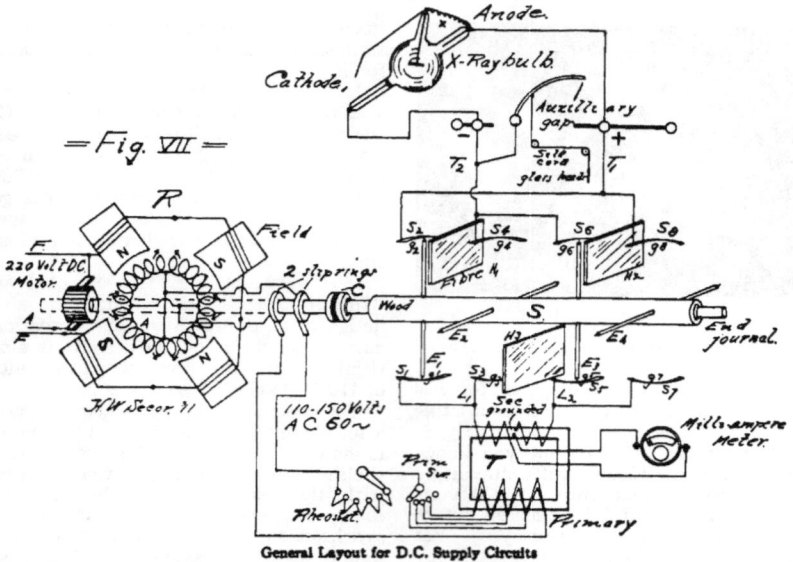

= Fig. VII =

General Layout for D.C. Supply Circuits

and synchronizing motor embodied in one machine.

With A.C. supply circuits at hand, the A.C. motor, used to drive the high-potential rectifying spindle and arms, may be a small affair, about ¼ to ½ h.p. The transformer takes its quota of A.C. direct from the mains. The position of the rectifying arms, in relation to the

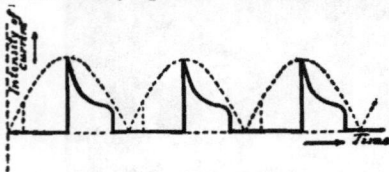

Fig. 8

reversals of the alternating current, is assured by the utilization of the synchronous motor, which runs in step with the alternators, feeding the circuit.

The general layout for D.C. supply circuits is given in Fig. 7, where R is an inverted rotary converter running as a D.C. motor and delivering A.C. from its two slip rings to the primary of the step-up transformer T. The D.C. motor circuit includes a regular starting box for shunt motors, an ammeter, and sometimes a field rheostat, although not

shown here. The primary circuit to the transformer has inserted in it, a variable resistance box (but preferably a variable inductance), and a multipoint switch to change the number of primary layers in use, and by these means, the secondary current may have its tension adjusted to any desired value up to 120,000 volts, which is sufficient to excite the largest X-ray tubes to their highest efficiency.

We now come to the ingenious device used to commutate the high-tension alternating current coming from the secondary of the transformer into a unidirectional or direct current.

This is made up of an ebonite or hardwood shaft S, coupled onto the end of

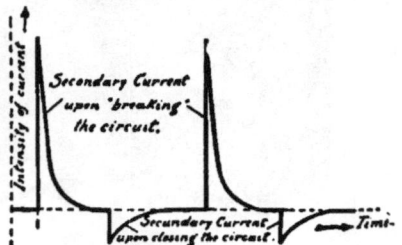

Diagram of the current in a tube supplied by an Induction coil.
Fig. 9

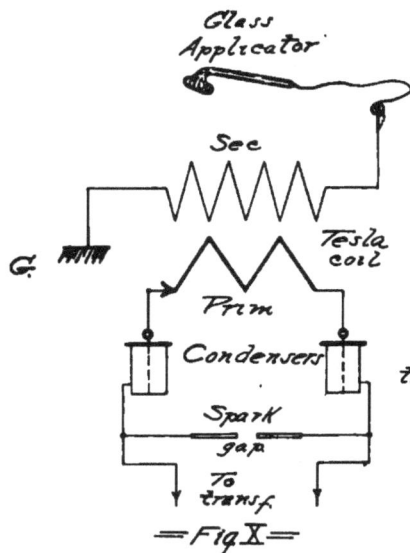

=Fig X=

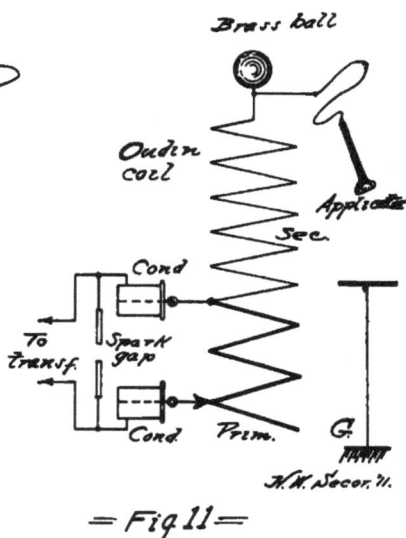

=Fig 11=

N. W. Secor. '11.

the rotary converter shaft by an insulated coupling C, and hence rotates in step or synchronism with the rotary armature. This shaft S has four ebonite rods, with conductors running through their centers, E_1, E_2, E_3 and E_4, pierced through it crosswise at 90 degrees apart, as shown.

Above and below this shaft, are placed 8 metal segments electrically connected to the transformer secondary and the X-ray tube terminals, in the manner depicted.

Taking the inverted rotary converter as a quadrupolar machine, we will have two complete cycles of alternating current, four alternations: two positive and two negative, at every revolution of the rotary armature and rectifier shaft S. Now the ebonite cross rods must be set by experiment or otherwise, so that when a positive impulse, or the positive half wave of a cycle is emitted from the secondary of the transformer, along the lead L_1, to segment S_1, that the impulse may travel a minute gap at g_1, along the ebonite covered electrode F_1, over the second gap g_2, and so to the positive terminal of the machine. The negative current flows over the secondary lead L_2, the two small gaps, g_5 and g_6 and the negative terminal wire T_2.

During the next quarter-revolution of the armature and shaft S, and while the negative impulse or half-wave of the cycle is flowing from the secondary coil over the lead L_1 (remembering that the transformer supplies an alternating current), the cross electrode E_2 has connected segments S_3 and S_4 together, through the two small gaps g_3 and g_4, leading the current to the negative terminal of the apparatus, while the positive current flows over the lead L_2, to segment S_7, gaps g_7 and g_8, segments 8, and positive terminal T_1; so it will be seen that the current is kept continually directed in one direction or it is changed from an alternating current to a unidirectional or pulsating direct one.

The pole changing parts are of necessity extremely well insulated, and have the fiber or hard rubber partitions H_1, H_2 and H_3 placed between parts of opposite polarity, where there is most liable to be a jump made by the current.

In some types of this class of apparatus the ends of the cross-wise electrodes have spring bush contacts fitted upon them, so that the current does not have to jump the small gaps, which occasion quite a little loss, as several thousand

(Continued on page 198)

A TINY WATTLESS H-CORE TRANSFORMER

How to Make a Small Step-Down Transformer for Ringing Doorbells, Lighting Small Lamps, etc.

PHILIP EDELMAN

The title of this article is perhaps a trifle misleading, since the output of the transformer is in reality about 5 watts. However, when it is remembered that this transformer is for use with 110 volts a.c. current it will readily be seen that only a very small current is required to operate the primary circuit. The current is in fact so small that it will not register on most meters,—therefore the above name. It may be left connected to the line all the time with safety, and for all purposes where only a small current is required, it is the cheapest source of current that I know of.

GENERAL PRINCIPLES

In its simplest form a transformer is very much like an ordinary induction coil. (See Fig. 1). The core is generally in the form of a ring or a modified ring, so that the magnetic lines of force can have a continuous iron path. Referring to the figure. Suppose the primary P

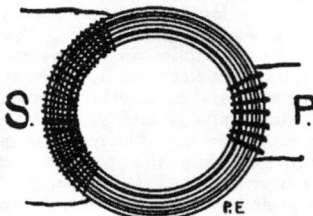

Fig. 1

has ten turns of wire, and the secondary winding S, one hundred turns. Then, if an alternating current is sent through the primary coil, an alternating current of just ten times as great a voltage and only one-tenth as great in amperage will be induced into the secondary coil. This is known as a step-up transformer. Now if the primary coil is regarded as the secondary and the secondary coil is connected to a source of alternating current, the current induced in the other coil will be only one-tenth as high in voltage as the primary coil, but ten times as strong in amperage. The transformer in this case is a step-down transformer. Of

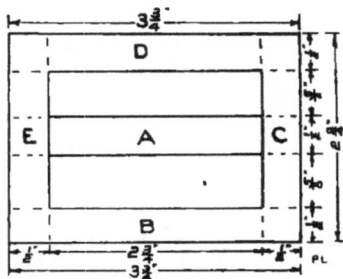

Fig. 2

course the sizes of the wires would have to be proportioned in practice to conform with the relative currents, as well as proportionate insulations and other items. Since watts equal the voltage times the amperage, it will be seen that the wattage remains the same in both the secondary and primary, if there are no losses. Among the losses the chief items are magnetic losses, and losses caused by the heating of a transformer.

CONSTRUCTION

The Core.—The foundation of the transformer is the core. It would be difficult to wind a large number of turns of wire on a round core as illustrated in Fig. 1, so a modified closed core is adopted. The core to be used in this little transformer is known as a modified H core. In this type of core the two windings are wound on the middle leg *a*, Fig. 2, and then surrounded with an iron shell, *b, c, d, e.*

From some soft sheet iron (stovepipe iron) or some soft sheet steel, cut up a pile of strips ½ in. x 3¼ in. for the middle leg *a*, until the strips form a pile ¾ in. high when compressed. Enough scrap sheet iron can usually be found around a tin shop from which to cut these strips. The cutting should be done with a foot-operated square shears. In a like manner cut up a pile with the dimensions ½ x 3¼ x ¾ in. high, for the two outside legs *b* and *d*. Only the one pile is needed for the two parts. For the parts *e* and *c* cut up one pile only ½ x 2¾ in. and ¾ in. high.

Assemble the strips for the leg *a*, as shown in Fig. 3. Before assembling the strips they should be coated on one side with a thin solution of orange shellac. It will be seen from the figure that each strip is laid so that it projects ½ in. on either side. The strips are thus arranged alternately with projections of ½ in. on each side. The pile should then be held in place by binding it with an evenly-wound layer of tape. In this taping, do not tape the ½ in. projections on each side. The other piles are left alone for the present.

Fig. 3

The Windings.—The primary for this coil will necessarily have a large number of turns of wire. In order to get them all in the small place, it is necessary to use fine wire evenly wound. Since the current which it is to carry will only be a fraction of an ampere, the wire need not be any coarser than No. 34 B.&S.

Make a spindle, Fig. 4, on which to wind the wire. Make a piece out of wood ½ x ¾ by 4½ in. long. Center it and place it in a small lathe or place it so that it may be easily rotated in some manner. On this spindle first wind a layer of common string. Then wind on five or six layers of heavy paper. Over this wind three or four layers of shellacked paper and a layer of friction tape. The shellacked paper and the friction tape should only be wound a distance of 2½ in. on the middle portion of the spindle.

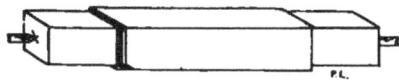

Fig. 4

Over this tape wind another two layers 2½ in. wide of shellacked paper. All this is to insulate the windings from the core. It would never do to have any grounds on the core metal. This done the winding of the wire comes next. Get 1 oz. of No. 34 enameled wire (full weight) and wind it on the central portion as evenly as possible. Be sure to keep within the 2½ in. covered by the shel-

lacked paper. To aid in doing this it is a good plan to fasten little guide blocks on the spindle. It is also well to put a layer of shellacked paper between each three or four layers of the winding. When all the wire is wound on, the primary winding should be covered by alternate layers of tape and shellacked paper, as was done before winding on the primary wire. It would ruin the transformer if a short circuit were to take place between the secondary and transformer in many cases. It would also spoil any low voltage apparatus that happened to be connected to the secondary winding. Unpleasant shocks might also result. Great care should therefore be taken to have plenty of insulation.

The secondary winding consists of three sizes of wire. The first winding is of No. 22 magnet wire. Wind on one hundred turns evenly. Leave an end

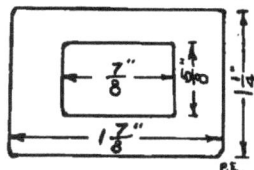

Fig. 5

of several inches for connections. Over this wind a second hundred turns of No. 24 magnet wire, leaving enough wire on the ends for connections. The third winding is of No. 26 wire, and is wound over the other two. This last winding has one hundred and fifty turns. About 40 ft. each of the No. 24 and No. 22 will be needed, and about 60 ft. of the No. 26. The windings should be arranged as evenly as possible. The ends of the three (No. 22, 24, 26) windings are attached together, the end of the first to the beginning of the second, etc. This arrangement allows of a choice of three voltages on the secondary. Test the coils for short circuits or grounds, using a telephone receiver and dry battery. If a defective coil is found it should be rewound. The whole should then be taped with some good friction tape and removed from the spindle. To remove the coil from the spindle, first pull out the string which was wound on at first. The windings will then come off easily.

Assembling.—The completed windings

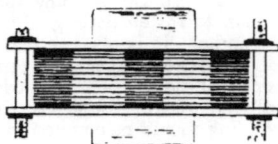

Fig. 6

are slipped over the pile *a* next. Cut out some pieces from fiber ⅛ or ¼₆ in. thick, in the shape shown in Fig. 5. One of these washers is slipped on each end of the core and winding. The pieces *e* and *c* are then fitted between the projections of the piece *a*, and then the pieces *d* and *b* are meshed in between the ends of the pieces *e* and *c*, and the whole nicely squared up. The whole shell will thus be like a grate bar. The core is held together by two clamps, one on each end. See Fig. 6. These clamps are made out of strips of iron ⅛ in. thick and are cut out ½ x 3½ in. Four pieces are needed for the two clamps. One-eighth inch from each end a ¾₆ in. hole is drilled.

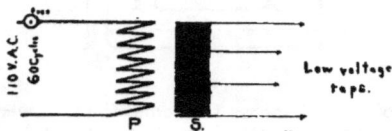

Fig. 7

The core should be clamped in a vise or clamp and then the clamp just made should be arranged on it and screwed down. The screws are made by threading four pieces of ¾₆ in. rod. The nuts should be turned as tight as is possible.

The transformer thus made may now be mounted in a suitable metal box and the proper connections made (see Fig. 7). It would be well to include a ½ ampere fuse in the primary circuit. In mounting the transformer the primary and secondary ends should be brought out to binding posts on opposite ends of the box. The secondary terminals being of low voltage need not have much insulation, but the primary terminals must be well insulated from the metal box. While the secondary may be overloaded or even short circuited without harming the transformer, it is well to use it only on normal loads.

When used as a doorbell ringing trans-former, the same wiring that is used with batteries may be used. When the secondary is on open circuit, the current consumed is practically nothing, on account of the high reactance of the primary winding.

WARNING: Never connect the secondary of this transformer direct to the line. It is only a step-down transformer and would burn up if the primary current were sent through the secondary.

Electric Lighting In Private Installations
(Continued from page 147)

are no concrete foundations, the erection costs are negligible, as the plant is self-contained and the cost of the engine room, erection and starting up can be put down as $55, giving a saving over the former method of $320. With regard to working costs, the ordinary accumulator system the approximate labor costs of running the plant, cleaning, charging batteries would work out at about $180 in England, to which must be added 12½ percent depreciation of the battery equal to $35 per annum. In the Bruston system the approximate labor and battery depreciation are insignificant; $15 would cover it. The annual saving in running cost would, therefore, work out at about $200. It will, therefore, be seen that for country locations, in particular, this system offers possibilities which render it very interesting.

Some of the steamships supplied with Marconi Wireless Equipment landing and departing to and from this country are as follows: *Rotterdam, Kronprinsessin Cecilie, Lusitania, La Bretagne, C. F. Tietgen, Baltic, George Washington, St. Paul, Berlin, Minnewaska, Cameronia, Ryndam, Finland, Kaiser Grosse, St. Paul, Celtic, La Provence, Campania.*

Messages for any of the boats are handled through any telegraph office, in a similar manner as the manipulation of a telegram.

"Why have you painted your sign upside down?"

"I carry aviation goods; I want it so that the birdmen can read it as they fly overhead."—*Washington Herald.*

ENGINEERING LABORATORY PRACTICE—Part IV
The DeLaval Steam Turbine
P. LEROY FLANSBURG

In a book written by Hero of Alexandria, more than two centuries before Christ, were described the first steam engines of which we have any knowledge. Many of these engines were very ingenious, but they were of no practical use, and it was not until 1693, that the first commercially successful steam engine was made. This engine was built by Thomas Savery, and, compared with our engines of today, it was very wasteful of steam. The pulsometer, which is still in quite common use, is a modification of this early engine. After Savery, much work was carried on by Newcomen and Watt in developing the steam engine, and in 1829, George Stephenson built the "Rocket," which was the first successful steam locomotive. While there have been many mechanical improvements made upon the steam engine since the time of Watt, there has been but one thermodynamic improvement made in the reciprocating engine, namely the use of compound expansion types of engines.

Among the earliest of the steam engines is the one known as the Hero engine. This engine consisted of a hollow sphere so mounted that the steam entered the interior of the sphere through one of the supports, and was allowed to escape through pipes bent at right angles to a diameter, as shown in Fig. 1. The reaction of the escaping steam caused the sphere to revolve upon its supports.

In 1882, Doctor Gustaf De Laval invented a turbine based on the same principle as Hero's engine. It consisted of two curved hollow arms attached to a shaft and the passage of steam through the arms caused them to revolve, thus rotating the shaft. This shaft drove a second shaft at a slower speed by means of friction wheels. De Laval invented this turbine for the direct driving of cream-separators, and many such are still in use. Seven years later De Laval patented in Great Britain a turbine wheel combined with a diverging nozzle and the steam was allowed to expand in the nozzle and was then directed against the turbine buckets. Shortly after the in-

Fig. 1.

vention of the De Laval turbine, another type of turbine was introduced by Parsons. The difference between the two lies chiefly in their methods of utilizing the expansive force of steam.

A turbine is essentially a machine in which rotary motion is obtained by a gradual change of the momentum of a fluid. The change of momentum produces, or is equivalent to a force, and it is this force which causes the turbine to rotate against the resistance of the load. In steam turbines, the expanding steam produces kinetic energy of the steam particles, and the total energy given up by the steam is thus converted into kinetic energy. Usually several steps or stages are required for this change; namely, by expanding the steam acquires velocity, this velocity being either wholly or partially utilized. After one such expansion the steam may be allowed to still further expand, again acquiring velocity, and this process may be repeated many times. The cycle of operation corresponding to each such expansion is commonly called a stage, and a turbine is spoken of as single expansion or multiple expansion, according as to whether it has one or more pressure stages. The best known and most extensively used turbine of the single stage

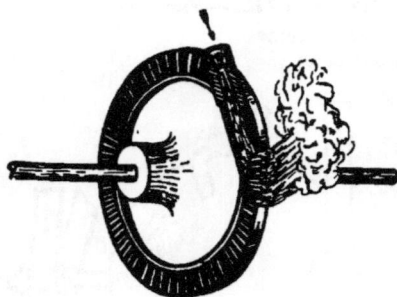

Fig. 2.

type is the De Laval. A diagrammatic view of such a turbine is shown in Fig. 2. While but one nozzle is shown in this figure, still it is common to employ several, some of the larger ones having four or five nozzles. In this turbine there are a number of turbine buckets or vanes mounted upon the periphery of a wheel. The steam is admitted to the nozzles where it expands and enters the turbine buckets with a high velocity, causing the wheel to rotate. Since in this type of turbine the wheel rotates in a chamber full of steam at exhaust pressure, there is but little resistance offered to the rotation of the moving parts. Turbines of this class are very simple, because all of the available heat energy is converted in a single step into kinetic energy. As the wheel rotates at a very high speed, it is necessary to employ a flexible shaft which can deflect enough to take care of any eccentricity of its center of gravity. If this were not done, the slightest unbalancing of the rotor would cause serious vibration. A 5 h.p. turbine revolves 30,000 times per minute, while the diameter of its wheel is about 4 in. A 30 h.p. turbine makes 20,000 revolutions per minute, with a wheel diameter of 8.9 in.; and a 100 h.p. turbine has a speed of 13,000 revolutions per minute, and a wheel diameter of 19.7 in. It is thus seen that the high speed of the turbine precludes its use for the direct driving of machinery, unless some form of reducing gear is used. Therefore it is customary to employ helical gears, which will reduce the speed 10 to 1. Gears with this speed ratio are customarily employed on all sizes of De Laval turbines. The reason for the use of this type of gears is on account of its smooth-running qualities. While the machine is speeding up, it reaches a point at about one-sixth of its rated speed at which the machine commences to vibrate. This speed is known as the critical speed. However, the flexibility of the shaft allows the wheel to choose its own axis of rotation, and after the critical speed is passed, the wheel runs with perfect smoothness, since the axis thus chosen passes through the center of mass of the wheel. It is quite easy to make the shaft spindle flexible, since the high rotary speed allows it to be made quite slender, and even on a 150 h.p. turbine, the spindle is but 1 in. in diameter.

The high speed of rotation of turbines renders them suitable for direct connection to both blowers and fans, and we find the De Laval turbine much used for these purposes.

The De Laval blades stand the impact of the high-velocity steam-jet very well, and these turbines may be run many years without replacement of blades, while, during this time, there is but little falling off in economy.

Upon first starting, after erecting, or after a long shut-down, the bearings should be flooded with oil and the nozzle valve opened about half a turn. The steam is then turned on and the governor-valve and wheel-case allowed thoroughly to heat. In order to give the bearings time to heat thoroughly, the turbine should be started gradually. If the turbine is running condensing, the condenser should be started before starting the turbine. When starting with no load, it is best to have a low vacuum of from 24 to 25 in. of mercury, and, as the load is applied, the vacuum should be raised to its maximum. As far as possible, changes in the power output of the turbine are obtained by shutting off one or more of the nozzles.

The advantages claimed for the turbine over the best of the reciprocating engines are high economy under varying loads, less weight and smaller floor space, uniform angular velocity and close speed regulation, less cost and inexpensive foundations, freedom from all vibration, adapted for highly superheated steam,

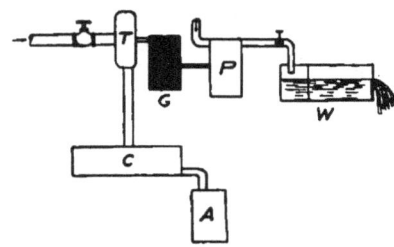

P—pump C—condenser A—tank for weighing water
W—weir T—turbine G—Gear

no need of internal lubrication, ease of erection and quickness of starting, small cost of maintenance and attendance, water of condensation free from oil and the steam economy is but little affected by wear or the lack of adjustment.

The following results were obtained as a result from tests carried on by the author.

DESCRIPTION OF TEST

The turbine was directly connected to a pump through 10 to 1 gears, as shown in the diagram. The discharge pressure of the pump was regulated by a valve placed in the pipe near the weir. (The weir was used for the purpose of measuring the water discharged by the pump.) Three tests were run, each of twenty minutes' duration and with different discharge pressures, the necessary measurements being taken for each test. The pressure of the steam entering the turbine, the weight of water from the condenser (which was equal to the weight of steam supplied to the turbine), the vacuum of the condenser, the suction and discharge pressures of the pump, the height of water flowing over the weir, (found by means of a hook gauge), were the measurements taken.

The results obtained were the work done in horse-power, the B.t.u. per minute per horse-power of water work done, and also of turbine work done, the thermal efficiency of turbine and pump and also of turbine alone, the ideal efficiency of the turbine, the percent of ideal efficiency attained by the turbine, the capacity of the pump in gallons per minute, and the steam used per hour per horse-power of both water work and turbine work done.

DATA—STEAM TURBINE

Width of weir	.8 ft.
Weight of water	62.3 lb./cu. ft.
Quality of steam at throttle	99 percent
Diam. of discharge pipe	4.03 in.
Area of discharge pipe	12.76 in.
Nozzle for 150 lb. and condensing	

OBSERVATIONS

Barometer, 30 in., 14.74 lb./sq. in.

Run No.	I	II	III
Boiler pressure, gauge in lb.	141.7	141.7	141.6
Steam pressure under gov. valve	141.8	142.0	135.3
Vacuum in condenser (inches)	26.0	25.9	26.2
Back pressure (lb.)	1.97	2.02	1.87
Steam used in 20 minutes (lb.)	88.0	83.5	77.0
Revolutions per minute	29550	29060	29700

CENTRIFUGAL PUMP

Revolutions per minute	2955	2906	2977
Discharge pressure (lb.)	22.6	29.9	40.0
Discharge pressure (ft.)	52.2	69.1	92.5
Suction pressure (in. Hg.)	14.0	12.1	10.0
Suction pressure (ft.)	15.9	13.7	11.3
Rdg. Hook gauge (in.)	6.25	5.52	4.44
Zero rdg. hook gauge (in.)	2.375	2.375	2.375
Head of water on weir (ft.)	.322	.262	.172
Water over weir per min. (lb.)	1770	1310	707
B.t.u. available per min.	4810	4570	4210
Pump efficiency (percent)	44	50	44
C (value from tables)	.601	.607	.619

Value of "velocity of approach" is small, so we may disregard it.

$$q = C \times \tfrac{2}{3}(b) \sqrt{2g}\, H^{\frac{3}{2}}$$

Run No. 1

$q = .601 \times \tfrac{2}{3}(.8) \sqrt{64.4}(.322)^{\frac{3}{2}} = .472$ cu. ft./sec.

$$v = \frac{q}{12.76} \times 144 = 11.3q$$

Run No. 1 $v = 11.3 \times .472 = 5.34$ ft. per sec.

$$h = \frac{v^2}{2g} = .0155\, v^2$$

Run No. 1 $h = .441$ ft.

Total lift = total press. hd. + vel. head

Run No. 1 $66.8 + .4 = 67.2$ ft.

Water work done

Run No. 1

$$\frac{62.3 \times 66.8 \times .472 \times 60}{33,000} = 3.58 \text{ h.p.}$$

Turbine work done

Run No. 1 $$\frac{3.58}{.44} = 8.14 \text{ h.p.}$$

Capacity in gal. per min. over weir

$$q \times 60 \times 7.48 = \text{Cap.}$$

Run No. 1 .472 x 449=212 gal. per min.

IDEAL EFF.

$$e = \frac{H_1 - H_2}{H_1 - q_2}$$

Run No. 1

$$e = \frac{1185.8 - 904.4}{1185.8 - 93.6} \times 100 = 25.7\%$$

Run No. 1
p_1=boiler pressure (abs.)   156.4
p_2=vacuum of condenser   1.97
x=99%

B.t.u. per lb. per min. = $H_1 - q_2$

Run No. 1   1185.9—93.6=1092
p_1=gov. press. (abs.)

Lbs. of steam used per min.

Run No. 1   4.4 lbs.

B.t.u. per min. used

Run No. 1   4.4 x 1092=4810

B.t.u. per h.p. of water work done (per min.)

Run No. 1 $\dfrac{4810}{3.58}$ =1340 B.t.u.

B.t.u. per h.p. of turbine work done per (min.)

Run No. 1 $\dfrac{4810}{8.14}$ =592 B.t.u.

Thermal eff. of turbine and pump

Run No. 1 $\dfrac{42.42}{1340}$ =3.16 percent

Thermal eff. of turbine

Run No. 1 $\dfrac{3.16}{.44}$ =7.19 percent

Percentage of ideal attained by turbine

Run No. 1 $\dfrac{7.19}{25.7}$ =28.0 percent

Stm. per hr. per h.p. of water work done

Run No. 1 $\dfrac{60 \times 4.4}{3.58}$ =73.8

Stm. per hr. per h.p. of turbine work done

Run No. 1 $\dfrac{60 \times 4.4}{8.14}$ =32.4

B.t.u. available per min.

Run No. 1 $\dfrac{1092 \times 88.0}{20}$ =4810

RESULTS

	I	II	III
Total lift (total press. hd.+hd. due to vel.)	67.2	80.5	103.4
Work done in h.p.	3.58	3.18	2.77
B.t.u. per min. per h.p. water work done	1340	1440	1900
B.t.u. per min. per h.p. turbine work done	592	720	835
(assuming eff. of pump under head of—	66.8	80.3	103.3
Percent as—	44	50	44
Percent of thermal eff. of turbine and pump	3.16	2.94	2.35
Percent of thermal eff. of turbine alone	7.19	5.88	5.07
Percent of ideal eff. of turbine	25.7	25.6	25.9
Percent of ideal eff. attained by turbine	28.0	23.0	19.5
Capacity in gal. per min.	212	157	84.4
Steam per hr. per h.p. work done (water)	73.8	79.0	104.5
Stm. per hr. per h.p. work done (turbine)	32.4	39.4	45.9

In Mr. A. Sprung's article on page 36 of the January issue, reference is made to the necessary connections used in finding the commercial efficiency and characteristic curves of a lead storage battery

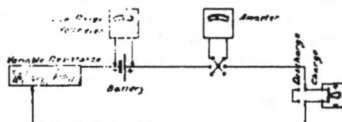

by means of an experimental solution. The cut showing these connections was omitted from the body of the article and is given here.

Common garden hose attached to an independent condenser in the power station, is a fine extractor of dirt from the inside of an armature or other electrical apparatus, where it is undesirable to raise dust. The hose should be void of any metallic substance on the end applied to the armature. The most desirable time to apply this cleaner is when the engines are off.

PULLEYS—Part II

USE OF GUIDE PULLEYS

There is no essential difference between a guide pulley and one used for transmitting power, except that its strength

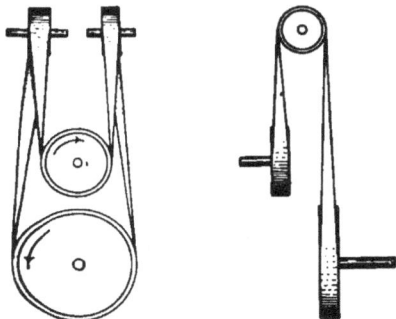

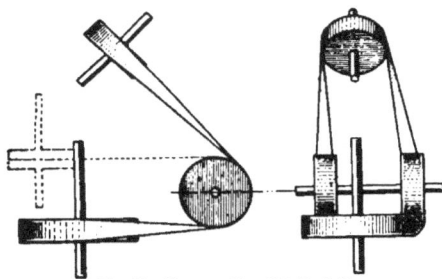

Fig. 11.   Common Use of Guide Pulley

need not be so great and that it runs loosely upon the shaft. Guide pulleys are used to direct a belt to the proper place on the main pulleys in drives that are so complicated that direct means will not suffice. Illustrations of their varied uses are shown in Figs. 11 to 16. The arrangement of two shafts so close together that a direct belt cannot be used is shown in Figs. 11, where it is necessary to employ two guide pulleys, which are placed upon the same arbor or may be separated with axes at an angle with each other.

It sometimes occurs that pulleys cannot be placed in the same plane upon parallel shafts, but must be offset, as shown in Fig. 12, and in this case guide pulleys must be employed, these being placed upon the same arbor and run in

the same direction, the diameter of the guide pulley in this case being equal to the offset of the main pulleys.

The arrangement of two shafts in the same plane but at an angle with each other is shown in Fig. 13.   In this case also, two guide pulleys are necessary to secure best operation of the belts. A solution of the quarter turn arrangement is that shown in Fig. 14, where one guide pulley is used.   The quarter turn arrangement is frequently bothersome and the use of a guide pulley is advisable in a great many cases.   Another solution of the , same problem is shown in Fig. 15, where two guide pulleys are employed.

It frequently occurs that by use of a guide pulley which is adjustable in its position, the belt can be shifted from the loose to the tight pulley, as shown in Fig. 16, without the use of a belt shifter.

BELT TIGHTENERS

To secure the best operation and highest economy from belts it is essential that a constant, uniform tension be employed while the belt is running.   Owing to the stretch and shrinkage of belts due ، to long use or changing conditions of the atmosphere, it frequently becomes necessary to employ some means of taking up the slack which is thus caused.   This

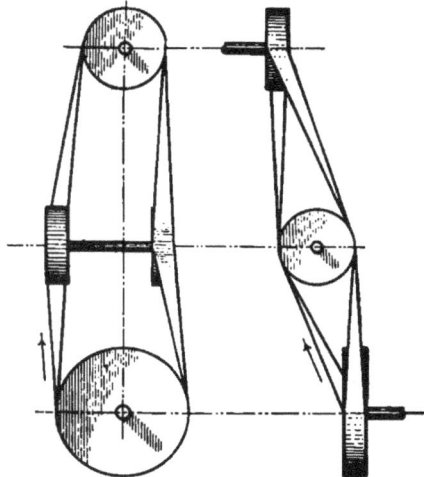

Fig. 12.   Pulleys Offset

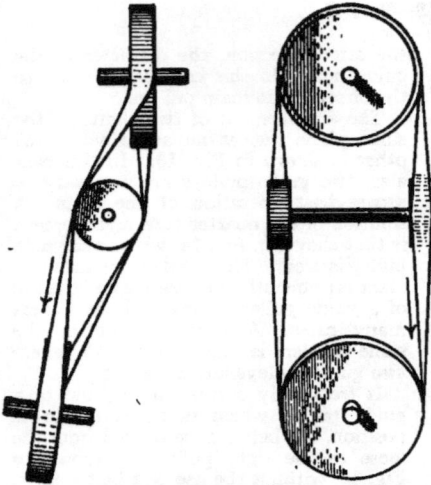

Fig. 13.   Transmission between Shafts at an Angle

is particularly the case where belts run between line shafts, or between the engine shaft and a line shaft. It is seldom necessary to employ a belt tightener between the line shaft and a motor or other small apparatus, inasmuch as these pieces of machinery are usually provided with means for shifting the entire machine upon its bed, thus taking up any slack or relieving an undue tension upon the belts.

As a general rule the belt tightener consists of a loose belt pulley running upon a shaft parallel to the shaft upon which the succeeding pulley is placed. This pulley is arranged so that it can be shifted at right angles to its shaft and held in any desired position, either by springs, weights or screws, according to the size and power transmitted by the belt.

For small belts the tightener is usually placed upon an arm and presses against the loose side of the belt a short distance from the smaller pulley, thus giving the pulley the advantage of a longer arc of contact with the belt. In this case the tightener is usually held in position by a weight or

spring of proper size, thus keeping the belt at the proper tension automatically.

For heavy duty, a large belt tightener is employed, and the tension is regulated by means of a hand wheel and screw, thus making it possible to obtain any desired belt tension: In many cases of this kind the belt tightener is employed for no other purpose than to give greater arc of belt contact on the pulleys.

Another use for belt tighteners is on vertical belts where, by increasing the tension, the upper shaft is put into service; and by decreasing it the belt will hang loose around the driving wheel, thus stopping the upper shaft. This use, however, is not recommended as good practice, for the reason that the driven shaft cannot be stopped instantly, and the belt hanging loose upon the driving pulley is subject to considerable wear at the point where it happens to come in contact with the wheel, and is liable to cause accidents by slipping to one side of the driver and catching on a key or set screw.

In design, belt tightener pulleys are similar to other belt pulleys, but where they run loose upon a shaft the hub is lined with bearing metal or otherwise highly polished. The pulley is also provided with some kind of lubricating arrangement, and should be kept thor-

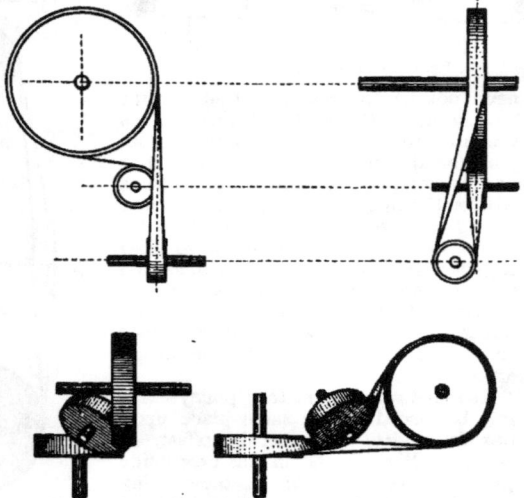

Fig. 14.   Quarter Turn Arrangements Using One Guide Pulley

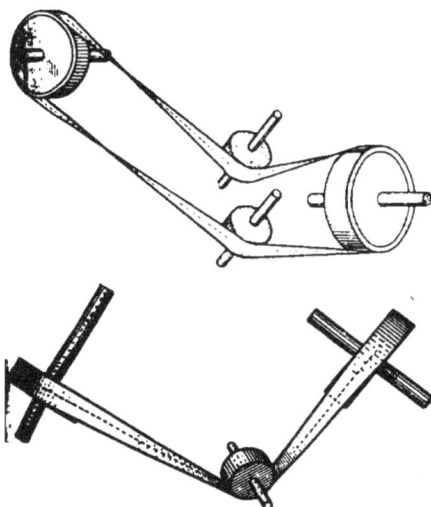

Fig. 15.   Quarter Turn Solution Using Two Guide Pulleys

oughly lubricated if expected to operate to best advantage.

FAST AND LOOSE PULLEYS

It is only in a few cases that several pieces of machinery operated from the same line shaft are all working at the same time, and in order to stop one machine while the others are still working it is necessary to provide some means by which the driving belt may be stopped without stopping the line shaft. One of the most common and satisfactory means is to provide a fast and a loose pulley upon the line shaft for each piece of apparatus, so that when a machine is to be operated the belt is shifted to the fast pulley, and when it is desired to stop the machine the belt is shifted back to the loose pulley, which is placed upon the shaft directly beside the tight or driving pulley.

The follower or pulley upon the machine is of double width, thus allowing the belt to run from one side to the other as the machine is working or idle. Aside from the fact that the driven pulley is of double width for the belt and has a flat face, its design is usually similar to standard pulleys. The driving pulley and the loose pulley placed upon the line shaft beside it are of standard design, except the width of the pulley face is not so great as that of an isolated pulley, thus making it easily shifted from one to the other. It is always best, where possible, to place the tight and loose pulleys upon the line shaft and the double width pulley upon the machine, by this arrangement, when the machine is not in operation the belt is not running, and its life will therefore be lengthened considerably. It frequently occurs, however, that the machine requires considerable power to start it, and it would be a difficult matter to shift the belt from one side of the pulley to the other without having it in motion. In such cases it is necessary to place the tight and loose pulleys upon the machine itself, and the belt must therefore be in operation all the time.

Another arrangement frequently resorted to is that of reversing a machine by means of shifting belts. In this case, however, it is necessary to employ two belts, one of which is open and the other crossed. The arrangement also requires the employment of one tight and two loose pulleys, thus having the loose pulleys on the outside with the tight pulley at the center, the follower pulley having a width a little more than three times the width of the belt. Where a machine is periodically reversed an automatic belt shifting arrangement is employed, thus relieving the operator of this responsibility.

Tight and loose pulleys are also frequently employed where a quick return movement of the machine, such as a planer, is desired, by providing two sets of tight and loose pulleys and belts, the ratios between driver and follower being

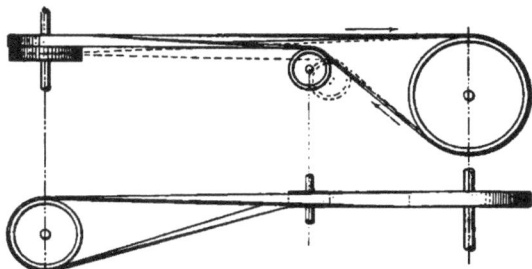

Fig. 16.   Guide Pulley Used as Belt Shifter

chosen to give the desired speed of the machine.

BELT SHIFTERS

The most common form of belt shifter consists of two parallel fingers placed one on either side of the belt with the space between a fraction of an inch greater than the width of the belt, these fingers being attached to a rod parallel to the shaft upon which the following pulley is placed. This rod is movable endwise the distance the belt has to be shifted and is always placed on the advancing side near the set of tight and loose pulleys. The shifter is operated by a lever or system of levers most convenient to the operator.

As stated above the shifters may be worked automatically, where the machine is to be run alternately in opposite directions or for a quick return motion which takes place periodically. This is usually done by means of a lever and weight upon the machine, which is operated by the machine until it gets into a vertical position when it drops over to the other side, thus shifting the belt while the machine is coming to rest and starting up in the opposite direction.

LOCATION OF PULLEYS

To secure the best service from the belt and pulleys and take advantage of the slack in the belt to give a greater arc of contact on the pulleys, they should be placed with their shafts parallel and

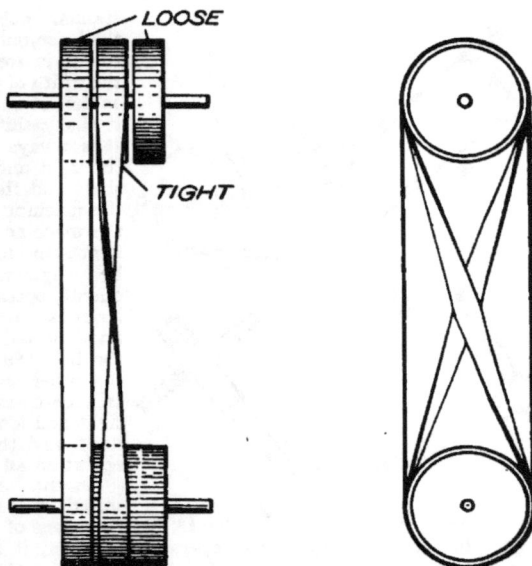

Fig. 17. Belt Arrangement for Reversing

in the same horizontal plane and the tight side of the belt at the bottom. This, however, is not always possible, and it is found necessary to run the belt at an angle from the horizontal, but vertical belts should never be employed where it is possible to run them otherwise. The difficulty here is that a belt must always have a belt tightener upon it, as any slack which shows up in the belt causes the lower pulley to slip in the belt. To the difficulty arising from high tension of vertical belts may be added the trouble with an over pressure on the upper shaft bearings and the lifting of the lower shaft in such a way as to wear on the upper part of the bearings. These conditions

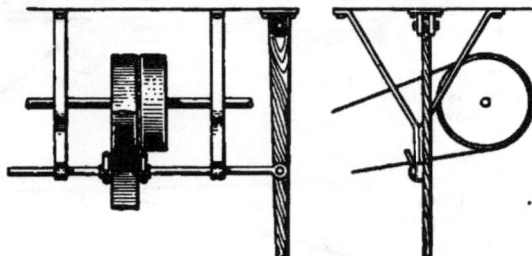

Fig. 18. Belt Shifter

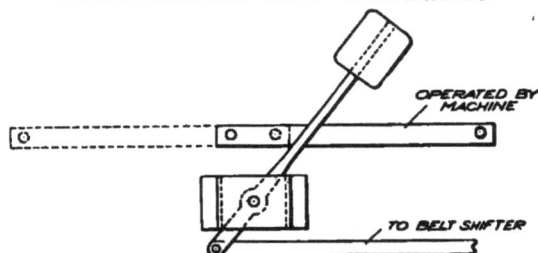

Fig. 19. Automatic Belt Shifter Device

are unnatural and should be avoided whenever possible.

A driven pulley located upon a long line shaft which has placed upon it several driving pulleys should be located as near to the center of the load as possible. In this way it will be seen that the diameter of the shaft can be reduced considerably, as only half the power of the whole shaft is transmitted in each direction, and also the vibration, which is likely to occur in long shafts.

IMPROVED LECLANCHE CELLS

Mr. J. G. Lucas, of the headquarters' staff of the British Post Office, read a paper the other day, dealing with the result of some investigations which had been carried out in practice and under laboratory conditions in connection with the ordinary types of signaling batteries used by the British Post Office and other telephone administrations. Dealing with the porous pot Leclanché type of cell, the author showed that the text-books were very much out of date on this question and that modern telephone requirements could be met much more economically by modifying the subdivision of the manganese dioxide and carbon in the porous pots. In the case of even such small current discharge rates as 40 milliamperes the result of the modifications was to decrease the total working costs not less than seven-fold, while at discharge rates higher than 40 milliamperes the decrease in costs was much greater. The agglomerate form of Leclanché battery was an expensive type, and the possible advantages which might be derived from its use were not commensurate with this expense. In connection with the use of dry cells as at present designed, the investigations went to show that the chief difficulties and expense lay in the fact that for discharges which must necessarily be spread over several months full use could not be made of the chemical energy contained in the materials used in these cells owing to the sudden rise, after a time, in the internal resistance. Details were given of experimental cells, the behavior of which indicated that the difficulties were not insuperable. The free ammonia gas evolved from Leclanché cells being found to affect seriously the metal work of telephones and to act deleteriously upon secondary cells, details were given of satisfactory experiments which had been conducted with an electrolyte of manganese chloride instead of the usual ammonium chloride solution. The author pointed out the practicability of sealing up the battery compartments in connection with cells charged with manganese chloride solutions and the consequent reduction in loss by evaporation as compared with ammonium chloride.

A first-aid fire engine which derives its propulsive energy from electric accumulators has been brought out in England. The cells are of sufficient capacity to enable the vehicle to travel for 24 miles. The current is supplied to two independent motors, each driving one of the rear wheels by means of a silent chain and a worm wheel reduction gear enclosed in an oil-tight case and running on ball-bearings. The vehicle will travel at 20 miles an hour, and will ascend a ten percent gradient. Accommodation is provided for six to eight men, for a 30 gal. water cylinder, which is kept under pressure by a cylinder of carbonic acid gas, and for 1,000 ft. of canvas hose, 30 ft. telescopic ladder, standpipes, etc.

THE LIGHTHOUSE SERVICE

Acts of heroism and instances of casualties in the line of duty among the faithful army of men—and women, too, for the late Ida Lewis, "America's Grace Darling," was one of them—who keep the government lights and beacons burning along the shores of the oceans, lakes, and larger rivers, are cited in the report for the fiscal year ended June 30 last, which Commissioner George R. Putnam, of the Bureau of Lighthouses, has submitted to Secretary Nagel, of the Department of Commerce and Labor.

These show the hazardous nature of the daily duties of fully three-fourths of the 5,500 present employees of the service, which is more extensive than any other lighthouse organization in the world. Notwithstanding this continual imperiling of limb and life, the average annual salary of a lighthouse keeper is but $600, although some of them earn more, while others make considerably less. These sturdy, brave and loyal guardians of navigation are drawn generally from among the hardy people who have been brought up by the sea and have taken up this dangerous work as it seems to appeal to them specially.

THE AGED AND THE DISABLED MUST BE DISCHARGED

Notwithstanding the small pay given, as the letter of the present law stands, those who are unable to perform their duties must be discharged. It does not matter whether the employee, advancing in years, is racked and twisted with rheumatism from exposure to the furious gales of summer and winter, or whether he, still in the prime of youth, is incurably crippled from a daring rescue of human lives from the waters of the sea.

THE REORGANIZATION OF THE SERVICE

The act of Congress of June 17, 1910, providing for a more direct administration of the service by placing it under a simple bureau form of organization, in the Department of Commerce and Labor, went into effect July 1, 1910, and the report, therefore, covers the first year of the new arrangement.

There has been, as a consequence of the reorganization, a saving in personnel, amounting to about 200 positions; in rents of offices and docks, and particularly in the use of lighthouse tenders; and yet the service has been steadily extended.

In carrying out the changes in the personnel an effort was made to avoid causing hardship to persons long in the service. When it was necessary to reduce the number of positions in any district the employees so discontinued were transferred to other places in the service, or, when practicable, to other branches of the government. To promote the efficiency of the service in general and to secure the best results in technical lines of work, a plan has been formed to have young men with suitable preliminary technical education enter the lower grades to be trained for the higher positions. In pursuance of this plan steps have been taken for the appointment of aids, cadet officers and cadet engineers.

Commissioner Putnam earnestly states, therefore, that there is great need for provision by law for the retirement of employees who, after long service, have lost their ability for active duty by reason of age or disability incident to their work. He considers it is not only a matter of humanity, but a business proposition, and in the best interests of the service. In England the lighthouse organization has a retirement system in the form of a life assurance policy. After three years of service each employee is given one, and in case of disability or superannuation the proceeds of it provide a means of maintenance for life. Germany, France, Denmark, Holland, Norway and Sweden have straight pension funds.

Provision should be made, the commissioner adds, for compensation to persons injured while engaged in hazardous employment in the Lighthouse Service, by extending to them the benefits of the act of May 30, 1908, "granting to certain employees of the United States the right to receive from it compensation for injuries sustained in the course of their employment." The only persons in the service now entitled to the benefits of this law are the artisans or laborers employed at the General Lighthouse Depot at Tompkinsville, N.Y., which has been construed to be a "manufacturing plant" under the terms of this act. Much of the work of the service is of a hazardous nature, such as the construction and repair of lighthouses and beacons, sometimes in very difficult locations, the handling of gas buoys and gas tanks, and of other heavy weights on vessels. It is believed that there is no class of government employees more justly entitled to consideration in case of injury while engaged on hazardous duty.

THE PAST YEAR'S WORK

The Lighthouse Service is charged with the lighting and maintenance of lighthouses, light vessels, buoys and other aids to navigation, along all the coasts and the principal interior rivers under the jurisdiction of the United States, with the exception of the Philippine Islands and Panama. There are 19 lighthouse districts, each in charge of an inspector. During the year three new districts were created, one including Alaska, one Porto Rico, and the third the Hawaiian Islands.

The service maintains at the present time over 12,000 aids to navigation, including lighthouses, light vessels, buoys, beacons and fog signals. During the year, 693 aids were established and 218 discontinued, leaving a net increase of 475. The service has 63 light vessels on 51 stations, the vessels in excess being used for relief. Some important improvements in the types of illuminating apparatus have been introduced on the light vessels.

There are at present 46 lighthouse tenders stationed along the coast, and used to supply light stations and vessels and for inspection and construction of new works. During the year five tenders were sold or transferred, being not worth repair for lighthouse purposes and not needed under the reorganization. One new tender and one new light vessel have been completed, and plans are under way for additional vessels to take the place of others as they become worn out.

The business methods of the service have been examined and various modifications made; a cost-keeping system introduced; and a general lighthouse inspector and an examiner, or traveling auditor, have been appointed to go systematically from district to district and examine both technical and business methods, the maintenance of vessels, etc.

Improvements in lights and other aids to navigation have been made as rapidly as means would permit. Two hundred and ninety-six lighted aids to navigation have been established during the year, including first-class stations at Cape Hinchinbrook, Alaska, and White Shoal, Rock of Ages and Split Rock, on Lakes Michigan and Superior. Twelve lights have been changed from fixed to flashing or occulting lights, giving a more distinctive characteristic. Incandescent oil-vapor lamps have been introduced in place of oil wick lamps at 29 stations, giving a much greater brilliancy of illumination for the amount of oil used. Acetylene lights have been installed in place of oil lights at 16 stations. These are mostly unattended lights, not requiring the services of light keepers. The marking and lighting of the new Livingstone Channel of the Detroit River is in progress, as this important channel will probably be open for navigation next year.

PROGRESS IN ALASKA

The need of additional aids to navigation in Alaska has been recognized. It was made a separate district and an inspector placed in charge with an office and depot at Ketchikan and Tonka. Two lighthouse tenders have been at work. There have been established 37 new lights, 2 unlighted beacons, 22 buoys, and 1 fog signal. A year ago there were 37 lights in Alaska, so that the number has been doubled since. Preliminary arrangements have been made for the installation of 27 additional lights during the current fiscal year, and an estimate is submitted for the establishment of a first-class light and fog signal station at Cape St. Elias. There are now 236 aids to navigation in Alaska, of which 74 are lights.

NEW TYPES OF LAMPS AND SIGNALS

Investigations and experiments have been continued throughout the year for the improvement of apparatus used in the service. A new type of oil-vapor lamp has been developed and put into use at various stations. The use of acetylene gas has been extended, especially for unwatched beacons and buoys, including large lighted buoys at sea, which will operate for long periods without attention. This gas has also been used for many of the unattended beacons in Alaska, where it is difficult to provide light keepers. Steps have been taken for the improvement of fog signals, particularly with a view to having them sounded instantly on the approach of fog. Investigation is being made of the availability of wireless telegraphy for fog signals.

The appropriations for the general maintenance of the service for the present fiscal year were $433,000 less than for the preceding year, this reduction being due largely to economies effected in the reorganization. The use of the appro-

priations of $290,000 for the construction of three lighthouse tenders was also deferred, as these vessels were not immediately needed.

The estimates submitted for the maintenance of the Lighthouse Service for the next fiscal year are practically the same as the appropriations for the present year. These estimates include new vessels required to replace light vessels and tenders as they become worn out in service, and important new lighthouses and depots for the conduct of the work of the service along all portions of the coasts of the United States and outlying territories.

A New Aeroplane Engine

The Yorkshire *Observer*, in an account of a lecture delivered before the Leeds University Engineering Society by R. J. Isaacson, gives his claims as the inventor of an improved aeroplane engine, as follows:

"He stated that it (his new engine) was based on the same general principles as the Gnome, but embodied many of his own devices, notably one which enabled the engine to be started slowly and run at almost any speed up to its maximum that the aviator wished. This was a vast improvement, because with all machines in use at present, it was only possible to work at one speed, and that the highest. Therefore, where an aviator, having attained a considerable height, wished to descend, he must shut his engine off altogether. But if the propeller once stopped revolving it was impossible to re-start the engine without help, and therein lay the reason for the awe-inspiring vol-planers, by which aviators descended from great heights. It was necessary to descend at great speed, so that the force of the air against the propeller might keep it in motion in order that when the aviator neared the ground he might re-start his engine, and thus control his movements. Mr. Isaacson claimed that the use of his engine would obviate all necessity for vol-planing."

The Man at the Door: "Madam, I'm the piano-tuner."

The Woman: "I didn't send for a piano-tuner."

The Man: "I know it, lady; the neighbors did."—*Chicago News.*

Calorized Electric Soldering Iron

The use of the ordinary soldering iron has two serious drawbacks: the impossibility of keeping it hot continuously, and the rapid wasting away of the copper. The development of the electric soldering iron obviated the former, furnishing the mechanic with an iron which not only stayed uniformly hot all the time, but one in which the heat intensity could be easily regulated by the mere turning on or off of the current. The second fault, that is, the rapid wasting away of the copper, still remained, to a large extent, necessitating frequent renewals, and consequently making no reduction in the cost of maintenance.

Therefore it is of much interest to metal workers to know that many experiments made in the research laboratories of the General Electric Company to mitigate this fault has resulted in the discovery of a process of treating the copper which renders the latter non-oxidizable under high heats and non-corrodable by the acids used in soldering. Furthermore, it reduces to a minimum the dissolving action of the molten tin, with which the working tip must always be kept coated.

This "calorizing" process or method of treatment does not merely coat the surface of the copper with a thin layer of non-oxidizable or non-corrodable substance, liable to scale off under the effects of heat and acids, but actually changes the characteristics of the copper to an appreciable depth. Thus the durability or practical working life of the copper is increased to such an extent as to provide a soldering iron of maximum economy and effectiveness.

SETTING DRILLS CENTRAL WITH SHAFTS

When holes have to be drilled transversely through shafts and spindles, some rapid and accurate means of setting the shaft truly in relation to the drill is

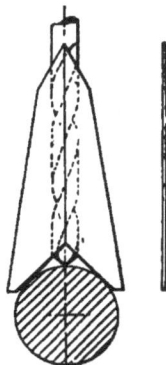

FIG. 1.

SETTING
DRILLS
CENTRAL
WITH
SHAFTS.

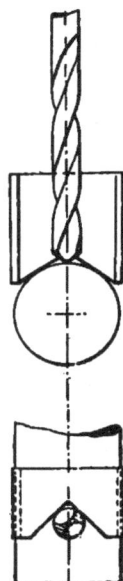

FIG. 2.

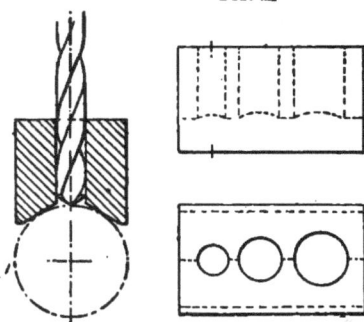

FIG. 3.

desirable. Sometimes the centering is automatically effected, because the vee block or blocks that receive the shaft are guided by a slot in the machine table, or in the poppet drilling-plate in the case of a lathe, or a solid vee block is used with a taper shank to fit the poppet barrel. But if none of these conditions are present, the drill must be adjusted with the help of some kind of device, of which we offer three as suggestions.

Fig. 1 is a center gauge, with a vee recess at one end and a point at the other. When set upon the shaft, as shown, the point should come into the center of the drill if the latter is standing correctly. This dodge is also useful for hand-drilling, with brace or hand drill.

Another method is to elaborate the gauge into a block, with two vee grooves at right angles (Fig. 2), one resting on the shaft, the other against the drill. The shaft is slid about until the contact is correct.

▸ Perhaps an easier block to use is that illustrated in Fig. 3, in which the drill fits into a hole, thus controlling the block rigidly while the shaft is being adjusted. Either three holes, as shown, or a greater number, may be made in the block.—
Model Engineer and Electrician.

A man was trying to call a party over the telephone. The two girl operators were discussing clothes and what they should wear, when the man interrupted. The girl was angry, and asked: "What line do you think you are on, anyway?" He said, "Well, it seems as if I am on a clothes-line."

STEEL PEN MAKING
Study of the Processes from the Rough Sheet to the Finished Pen, Packed Ready to Use

In excavating the ruins of Pompeii the earliest specimens of metallic pens were found. These were of bronze. Most of the early metallic pens were of this metal, although some were made of silver. The forms of the first steel pens were copies of the quill pen, being both pen and holder combined. They were slit like the quill pen.

EARLIEST FORM OF STEEL PEN

This style of pen was used for a number of years, but was very expensive, because as soon as the pen was worn out it was necessary to throw away practically both the holder and the pen; so the nib part was made separate, and the barrel part became the tip of the penholder of today. This was a great economy; and soon the pen took its present form, and the penholder was made to hold it.

One of the objections to the early steel pens was their stiffness. This was overcome by the introduction of the side slits; by varying the size, shape and position of these side slits a pen can be given any resiliency desired.

The steel pen industry did not make any rapid advances until the adoption of the foot, drop and screw presses, about the year 1825; then they were manufactured in fair quantities, but their introduction was by no means rapid, for even as late as 1860 to 1865 the Quartermasters' Department furnished the United States army with the quill pens. The first steel pens sold anywhere from 35 to 50 cents each, so that one pen cost as much as will now buy from one-third to one-half a gross of the better grades. In other words, they cost from 50 to 75 times as much as they do now. The consumption has increased very rapidly, and at the present time the world probably produces from ten to twelve million

This is far from being the case, as there are from 20 to 28 handlings, the number depending on the style of the pen.

The steel is imported from England and consists of selected sheets, 19 in. wide, about 5 ft. long and .023 of an inch thick. It is of the very highest grade—American manufacturers not having attempted to make this class of steel.

The first operation is to cut the sheets into strips 19 in. long and wide enough to cut two pens with their points interlapping. These strips, which are rolled hard and are too thick to cut a pen from, are annealed by packing them in iron boxes and heating them at a low red heat for a number of hours. They are then gradually cooled under a hood, to prevent drafts striking them. When cool the strips are soft and coated with a scale, which is removed by a pickle of dilute sulphuric acid. They are now ready to be put through the rolling mill and reduced to the required thickness, which averages about .009 of an inch.

ROLLING OUT THE STEEL

The rolling is known as "cold rolling," the strips not being heated after the first annealing. This gives an increased toughness to the steel. The number of times necessary to put it through the rolls depends on how thin the steel is to be rolled. Each strip is tested with a micrometer gauge, and should it be too thick it is again put through the mills, and if too thin it is laid aside for a pen for which a thinner steel can be used. The steel which started 19 in. long has been stretched to about 50 in., and is then ready for the pens to be cut from it.

Cheap pens are cut from steel that comes in large rolls ready for use. As it is impossible to roll this uniformly, the pens that are made from it are very ir-

determine the flexibility of the pen, and vary with the style of pen, some pens requiring two and three handlings in the piercing department.

SOFTENING THE BLANKS

The blanks having been cut from hard rolled steel, it is now necessary to soften them by annealing. This is done by putting them in large iron pots, heating them to redness for several hours, and then cooling gradually. They are then soft and pliable and ready to receive the name, which is the next operation, called marking. Some pens have a raised letter or design on them, called embossing. This is done in a marking press.

After marking the pen is raised; that is, brought to the form that it is to have when finished. There are on the market about 2,000 styles of pens. Raising is done in a peculiarly constructed screw press, and the pens are removed by compressed air.

Each pen is now carefully examined for imperfections in the previous operations, and, as they are soft, it is necessary to harden them by heating them red hot and dropping into cold oil. The oil is removed by centrifugal force and boiling lye, and the pens are then dried in sawdust. This makes the pen very brittle, so that it has no resiliency. In order to obtain the latter quality, the pen is tempered by gradually re-heating it until it has acquired the greatest toughness and elasticity possible.

The pen now has a coating of oxide which must be removed by scouring. This is done by placing the pens and a scouring material in tumbling barrels and revolving them until they are bright. Girls then grind the pens on emery bobs lengthwise and across the nibs. Some pens have only one operation in this department, while others have two and three. Pens are ground to enable them to hold ink better, and also give them

will always cut through the center of the point.

After the pens have been slit they can be used for writing, but they would be very scratchy and would stick in the paper. In order to overcome this, the points are rounded and made perfectly smooth.

The final examination is now given each pen; expert examiners sit before a slanting desk, on which is a slate of black glass; the pens lie on the desk and the examiners pick up one in each hand, pressing them on the glass and looking at the cutting, piercing, marking, raising, grinding, slitting, tempering, etc. Should the pens have any imperfections in any of these operations they are thrown into separate boxes, so that each room can be charged with the amount of its waste. This waste is then put in iron pots and heated, so as to prevent their being used when they are sold for scrap steel. There are 1,728 chances to make a bad pen in every gross, consequently its manufacture requires vigilant care and inspection.

THE FINISHED ARTICLE

The pens are now polished, and if they are to be left gray, are ready for the lacquering operation; if they are to be made bronze, blue, black or any of the various shades, they are sent to the tempering room and gradually reheated in a revolving cylinder until the required color appears upon them, when they are chilled quickly so as to prevent the color changing. The pen is now practically finished, but if put on the market in this form would rust very quickly. Each one is therefore given a thorough coat of lacquer, which preserves it. If the pens are to be plated with bronze, silver or gold, these operations are performed while the pen retains its bright polish.

The pens are counted by weight. It will be found impossible to put a gross of pens in the box intended for them unless they are laid parallel. They are put

FINDING THE DECIMAL POINT

Keeping Track of this Elusive Dot when Figuring with Pencil or Slide Rule

In making computations which involve the use of decimals, there is frequently some confusion in finding where the decimal point belongs. The danger of misplacing it is illustrated by the old story of the bridge designer whose bridge fell down when about three-quarters finished, and who, on seeking the cause, went carefully over his computations and finally exclaimed: "Confound that decimal point!" In the course of figuring he had misplaced it, and the bridge was only one-tenth as strong as he had intended, which brought a breaking strain on an important member of the structure.

FOR MULTIPLICATION

In many computations, says *Practical Engineer*, the reasonable value of the result will tell whether the decimal point is rightly placed or not; but in others where sizes are unfamiliar, we must rely on the accuracy of the figures for the correctness of the result. In the case of multiplication, while it needs some care to be sure that the decimal places are counted accurately, the rule is simple enough, namely to "point off as many places in the product as the sum of the places in multiplier and the number multiplied." For instance, in figuring the volume of 1.3148 lbs. of steam at

```
   0.02641
    1.3148
  --------
     21128
    10564
    2641
   7923
  2641
----------
.034723868
```

FIG 1

```
        0.0025 48
785 | 2.0000
      1570
      ----
       4300
       3925
       ----
        3750
        3140
        ----
         6100
         6280
```

FIG 2

```
          0.003097
177.6. | 55.00000
          5328
          ----
           17200
           15984
           -----
            12160
            12432
```

FIG 3

10 lbs. absolute pressure, we find that the volume of 1 lb., as given in the steam tables, is, from the November data sheet, 0.02641 cu. ft. The operation of multiplying is then as shown in Fig. 1, and counting the decimal places in the number and the multiplier, we find that there are 9 altogether, which means 9 decimal places in the result, giving us one cipher in front of the 3 and then the decimal point.

FOR DIVISION

In the case of division, the proper placing of the decimal point is not so easily expressed by rule, and a little kink in mechanical performances of the operation is helpful. Take the case of 2 to be divided by 785. If we set down the 2 and then the 785 to the left of it, divided by a line, as in the usual manner, and above the 2 draw a horizontal line, we may then add as many ciphers after the decimal point at the right of the 2 as needed for our purpose, and above the 2 we put a cipher, because our divisor is contained in 2, zero times. We then place a decimal point above the decimal point in the dividend and proceed with another cipher, because the 785 is contained zero times in 20, zero times in 200, and 2 times in 2,000. The method is shown in Fig. 2.

In the case where a number containing a decimal is to be divided by another one containing a decimal, there is an additional step. If it be required to find the number of pounds of steam at 26 in. vacuum, which will be contained in 5.5 cu. ft., we find from the data sheet, November tables, that 1 lb. occupies 177.6 cu. ft. To find the pounds in 5.5 cu. ft., we must, therefore, divide 5.5 by 177.6.

As before, we put down the 5.5, then draw a vertical line to the left and put down the divisor; we add zeros to the dividend as may be required, and then move decimal points in both divisor and dividend to the right until the divisor becomes a whole number, putting a cross in the former position of the decimal point to show where it was originally located. We then proceed as before, putting the decimal point in the quotient above the new decimal point in the dividend. The continuation of the calculation is shown in Fig. 3.

FOR COMBINED OPERATIONS

These methods cover all cases for single multiplication and division, and for operations that are carried on in series, that is, one after the other. There are, however, certain calculations which are made from formulas or equations in which we have indicated a series of multiplications and divisions, and where cancellation of factors is used. Take the case of the familiar horse-power formula represented by $\frac{PLAN}{3,3000}$, in which P is the mean effective pressure in a steam engine cylinder, L the length of the stroke in feet, A the area of the piston in square inches, and N the number of working strokes per minute; a considerable amount of arithmetical work can frequently be saved by indicating the entire series of multiplications and divisions, and then crossing out factors which are common to numerator and denominator.

Assume that the mean effective pressure is 25 lbs., the length of stroke 4 ft., diameter of piston 24 in., and revolutions per minute, 75. We then have all the factors of the horse-power formula, except the area of the piston which is found by squaring the diameter, multiplying by 3.1416, and dividing by 4. Substituting the values then in the formula, we get the expression as shown in Fig. 4, the number of working strokes being equal to the revolutions per minute when only one end of the cylinder is considered.

Now, canceling out like factors in numerator and denominator, we have both 4's out, the factor 25 will cancel into 33,000, leaving 1,320, the factor 4 will cancel into 24 leaving 6, and into 1,320 leaving 335, and the factor 5 into 75, leaving 15, and into 335 leaving 67. We can readily multiply 24 by 6 mentally, giving 124, and then to multiply by 15 we have 10 times 124 is 1,240, and half that will be 5 times 124 or 620, and the sum of the two gives 1,860. Then by long multiplication we take the product of 3.1416 and 1,860, giving, with 4 places pointed off, 5,843.3760. Performing the division and locating the decimal point by the method already given we get 87.21 as the horse-power.

The experienced engineer will know from the dimensions and data given, whether this is about right for the horse-power or not, but for the novice there might be a delightful state of uncertainty whether it ought to be 87 or 872 h.p., as the loss of the cipher anywhere in the cancellation would throw the position of the decimal point out of its proper place. It is useful, therefore, to make a rough

check to prove the decimal point's position, as shown at the bottom of Fig. 4.

Fill out the formula, raising or lowering values to bring them to the nearest figure having a zero or a 5 as the last digit, and taking care to raise about as many factors as are lowered. In this way an expression will be secured which can be quickly canceled to a few simple factors that can be mentally multiplied to indicate about the result that ought to be obtained, the process in the present instance being as follows:

In the numerator we have 25 and 4; then substitute 25 for 24, and 20 for the second 24, 3 for 3.1416 and 70 for 75.

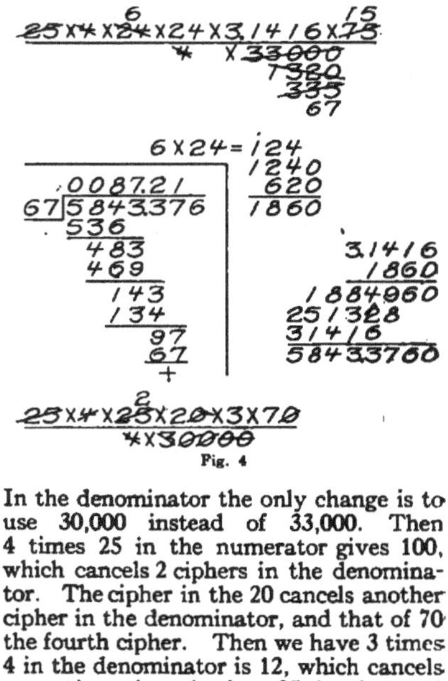

Fig. 4

In the denominator the only change is to use 30,000 instead of 33,000. Then 4 times 25 in the numerator gives 100, which cancels 2 ciphers in the denominator. The cipher in the 20 cancels another cipher in the denominator, and that of 70 the fourth cipher. Then we have 3 times 4 in the denominator is 12, which cancels approximately twice into 25, leaving us 4 simple factors, and we have 2 x 2 x 3 x 7, gives 84, showing that our result, 87.4, is correctly pointed off.

FOR THE SLIDE RULE

Coming now to the case of the slide rule, we must remember that the slide rule knows no decimal point, that is, in making slide rule computations, 125, 1,250, or 12,500, or 1.25 are all the same, so far as the slide rule manipulations and readings are concerned. It is necessary, therefore, to have some rule for locating the decimal point, and we may do this by the regular method for arithmetical multiplication or division, or by a special method which applies only to the rule.

Take the case of Fig. 1. It is necessary to remember that the slide rule does not read to more than 4 significant figures, and that it is not important that it should do so, for in practically all engineering computations we are working with values which depend on test observation, and these observations are not accurate to more than one-half of 1 percent. It follows, therefore, that the slide rule is not a good instrument on which to figure the interest on a million dollars; but we are assuming that the interest on a million dollars is of only academic interest to our readers.

Put 1 on the C scale at 2,641 on the D scale; remember that each division between the marks of the 2 to 4 section of the rule represents 2 points, and the last figure 1 means that the 1 on the C scale is just a hair to the right of the 264 division of the D scale. We then carry the runner to 131 on the C scale and note that the remaining figures, 48, is practically the same as 50, so that the runner will stand half way between 131 and 132. Under the runner on scale D we shall then find 347 and a small fraction which we would estimate at 3.

From Fig. 1 we find that the absolutely accurate answer is 34,723, but 3,473 is as close as we can expect to come to it with a slide rule reading. It is evident that with only these 4 figures known, finding the decimal point by the rule for arithmetical multiplication is out of the question.

Many rules have been devised for finding the decimal point in slide-rule computations, but perhaps the simplest is that given by Wm. Cox. The index of any number is the number of figures in the integral part of the number; that is, the part to the left of the decimal point. The index of 125 would, therefore, be 3, of 1,250, 4, of 12.5, 2, of 1.25, 1, of 0.125, 0. If we go still further we have to resort to the negative index; that is, 0.0125 has 1 place less than nothing to the left of the decimal point and the index is, therefore —1. 0.00125 would have the index —2, and so on.

The rule given by Cox is that for multiplication; if the final product is

read with the slide projecting to the left, the index of the product is the sum of the indexes of the factors, but if the slide is projecting to the right, the index of the product is the sum of the indexes of the 2 factors, less 1. In the example which we have just worked, Fig. 1, the index of 0.02641 is minus 1, and the index of 1.3148 is 1. The product is read with the slide projecting to the right, therefore the index for the product is minus 1 plus 1, which gives 0, less 1 equals —1, which indicates that there is one cipher to the right of the decimal point, giving 0.03473 as the product.

In case we want to know the foot pounds developed in a minute by an engine giving 8.7 h.p., we would multiply 87 by 33,000, which for the slide rule is 33. Setting 1 on the C scale to 87 on the D scale, we set the runner at 33 on the C scale and below it on the D scale we read 2,873, and the question arises, where is the decimal point. The index of 33,000 is 5; the index of 8.7 is 1; the answer is read with the slide projecting to the left, hence the index of the product will be the sum of the indexes of the factors, or 5 plus 1 equals 6, and there will be 6 places to the left of the decimal point, or the number will be 287,300.

For division, Cox gives the rule, if the quotient is read with the slide projecting to the left, the index will be the index of the dividend minus that of the divisor. If the slide projects to the right when reading the quotient, the index of the quotient is the index of the dividend plus 1, less the index of the divisor.

Take as an example, Fig. 2, to divide 2 by 785. We bring the runner to 2 on the D scale, now set 785 on the C scale to the mark on the runner. Under 1 on the C scale we then find on the D scale the reading 2,548, the slide projecting to the left. We then have the index of 2 is 1, and the index of 785 is 3; subtracting 3 from 1 gives us —2. The index —2 means that there are 2 ciphers to the right of the decimal point, and the quotient becomes then 0.002548, as shown in Fig. 2.

Again, suppose that we wish to divide 62.42 by 1,728. We set the runner to 6,242 on the D scale, and bring 1,728 on the C scale to the runner. Under 1 on the C scale, we find on the D scale the reading 361. The characteristic of

62.42 is 2, that of 1,728, 4, and the quotient is read with the slide projecting to the right. We have, therefore, that 2 the index of the dividend plus 1 equals 3, and less 4 the index of the divisor, gives —1, which is the index of the quotient. This means that there is one cipher to the right of the decimal giving 0.0361 as the quotient.

COMBINED SLIDE RULE OPERATIONS

These rules work well enough in the case of a simple multiplication or division, but where there is a series of operations, particularly where multiplication and division are combined, as is frequently the case in the slide rule computation, the carrying out of the rule becomes so involved that a determination of the decimal point position by inspection is quicker and more accurate. For the case of Fig. 1, it is evident that 1.3 times 0.02 would be still somewhere near 0:02, so that the product would evidently be 0.03473. In the case of multiplying 33,000 by 8.7, it is easy to see that 8 times 30,000 would be 240,000, which shows that the product must be 287,300. In the example of Fig. 2 it is easy enough to jot down the 2 and the 785, as if the division were to be performed by arithmetic, and the position of the decimal point can in this way be quickly located by inspection.

In the case of a computation by means of slide rule of the example shown in Fig. 4, it is evident that after performing the operations on the slide rule instead of by cancellation, the location of the decimal point could be quickly checked out by the method shown at the bottom of Fig. 4, for testing the accuracy of its location when doing the problem by arithmetic.

To take the case of another example on the horse-power formula, if we have a double-acting engine with mean effective pressure of 23 lbs., stroke of 3 ft., piston 18 in. in diameter, running at 80 revolu-

divided by 4. We set the runner to 23 on scale D, and bring 4 on scale C to the runner; then carry the runner to 3 on scale C, and bring 33 on scale C to the runner; then carry the runner to 3.1416 on C, which will be a trifle to the right of the 314 reading. All the divisions have now been accomplished, the remainder of the work is multiplication, and we bring 1 on C to the runner, then carry the runner to 18, bring 1 to the runner, carry the runner to 18, bring 1 at the right-hand end of the slide to the runner and carry the runner to 8, then bring 1 to the runner and carry the runner to 2. Then at the runner mark read on scale D the answer, which is 852. It remains to determine the decimal point, and this is most easily done by the same process of approximate cancellation as was used in the arithmetical

$$\frac{23 \times 3 \times 3.14 \mathrm{l} 6 \times 18 \times 18 \times 80 \times 2}{4 \times 33000} = 85.2$$

$$\frac{\cancel{23} \times \cancel{3} \times \cancel{3} \times \cancel{8} \times \cancel{2} \cancel{8} \times \cancel{8} \cancel{8} \times 2}{\cancel{4} \times \cancel{3} \cancel{0} \cancel{0} \cancel{0} \cancel{0}} = 96$$

Fig. 5

computation of Fig. 4. This is indicated in the lower part of Fig. 5. We first cancel the 4 ciphers in the numerator against the 4 ciphers of the 30,000 in the denominator. One of the 3's in the numerator cancels with the 3 in the denominator, and the 4 in the denominator cancels 2 of the 2's in the numerator, leaving as factors 3 x 2 x 8 x 2, which is easily figured mentally to equal 96, so that it is evident that the correct result is 85.2.

As a result of years of experience using the slide rule, the writer has adopted the method of inspection and approximate cancellation as the quickest and most accurate method of determining the position of the decimal point. While the method of index is easy in a single computation, it involves keeping track

FITTING HINGES

W. J. HORNER

The three types of hinges shown in Fig. 1 are those in common use. They vary in size and details, but not very much in their proportions. The shape of the butt-hinge adapts it for attachment to narrow edges of wood, as those of doors and box-covers. The length

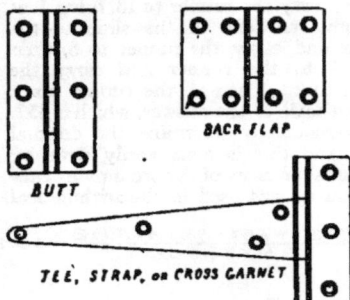

FIG. 1.

BACK FLAP

BUTT

TEE, STRAP, or CROSS GARNET

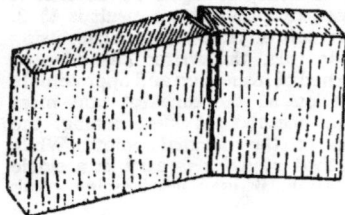

FIG. 3.

of its pivot or knuckle is greater than the extension of its flaps. In the back-flap hinge these proportions are reversed, the length of the knuckle being less than the measurement across the flaps. This type of hinge is used on broad surfaces, and chiefly for a rougher class of work than butts. The tee-hinge is used similarly, its extremely long extension serving as a brace and stiffener to a broad surface of comparatively thin wood.

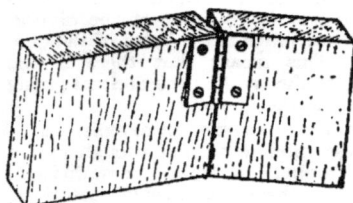

FIG. 2.

The butt-hinge is fitted as in Fig. 2, generally sunk into the wood, but not necessarily so. As far as the hinging of the parts is concerned, the effect is the same if it is screwed on the outer surface, as in Fig. 3, instead of in the joint, as in Fig. 2; but a back-flap hinge is more

suitably proportional for such a position. When a butt-hinge is used, as in Fig. 2, long screws can be inserted, and there is no risk of their tearing out or of the wood breaking away. If it is attached as in Fig. 3, the screws are rather close to the edge and strain on the hinge tends to split the wood away. For this reason a back-flap, or a tee-hinge, as in Fig. 4, is preferred for attachment to the outer surface. The former is occasionally

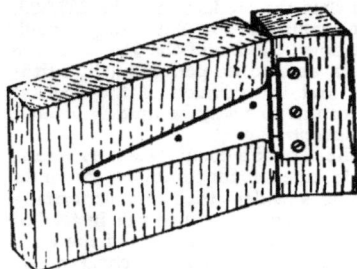

FIG. 4.

sunk flush with the surface, but the latter never is, being simply adjusted in the required position and screwed on. The butt-hinge, used as in Fig. 2, must be sunk in order to make a close joint between the hinged parts; but when the hinge is placed as in Figs. 3 and 4, its thickness does not interfere with the closeness of the edge joint, but only with

FIG. 5.

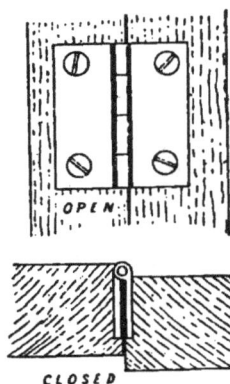

FIG. 6.

the closeness of the broad surfaces, if they are folded back against each other.

Generally it is easier and quicker to attach a hinge as in Fig. 3 than as in Fig. 2. The parts to be hinged simply have to be placed in position, with their edges in contact, and the hinges screwed on, care being taken to set their knuckles central over the joints. In many cases the parts can be laid flat on a bench while this is being done. If the hinges are to be sunk flush with the surface they are laid in position, lines marked round them, and the recesses chiseled out to correspond with the thickness of the flaps. Then the hinges are inserted and screwed.

In attaching butt-hinges, as in Fig. 2, the procedure is not quite so simple, for the parts can seldom be placed as conveniently. The hinges must be fitted

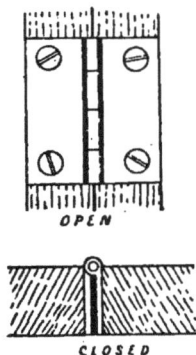

FIG. 7.

to one of the parts first, and this held in position for attachment to the other. They must be sunk to exactly the right depth or the joint will either be open or will bind and not close properly. The latter defect is shown in Fig. 5. The hinge there is sunk too deeply, and the edges of the wood at the knuckle bind and prevent the farther edges from coming into contact at all. The joint might be closed completely by the use of force, but it would spring open again as soon as released. The remedy is to take the hinge off and replace it with cardboard packing beneath, or plane the wood down to reduce the depth of the hinge recesses.

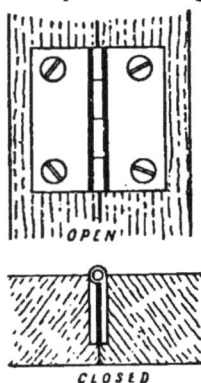

FIG. 8.

Another defect is shown in Fig. 6. The hinge there is sunk correctly, but is out of center. In the view showing the hinge open the center line of the knuckle is to one side of the joint. The consequence is that the surfaces of the parts are not flush with each other when the hinge is closed.

Figs. 7 and 8 show butt-hinges attached correctly. In Fig. 7 the hinge is the full width of the wood; in Fig. 8, it is less, and the recesses are, consequently, not cut right across. Fig. 8 is the neatest and most frequent method; but when the wood is thin the hinges sometimes have to correspond with it, as in Fig. 7. In other cases, where thin wood is hinged to a thicker piece, a combination of the methods is followed, one flap of the hinge extending the full thickness of the wood, and the other having wood extending beyond it.

The flaps of a hinge have to be sunk slightly more than their thickness, for the wood to make a close joint when closed. This is because the flaps are always made thinner than half the diameter of the knuckle. When a hinge is closed, with its flaps parallel, there is a space between them about equal to the thickness of another flap. This space must

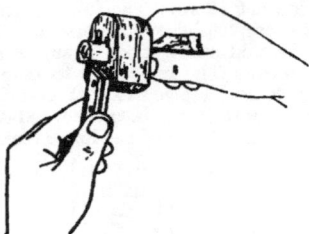

Fig. 9.

exist when the hinge is fitted; but the wood should form a close joint, as shown in Figs. 7 and 8. The correct depth for the recesses is marked on the wood with a gauge, which is set, as in Fig. 9, to the center of the knuckle. When this has been gauged on the wood the measurement in the other direction is taken, as in Fig. 10, from the outer edges of the flaps to the center of the knuckle. The length of the hinge is marked, as in Fig. 11, by laying it in position on the wood.

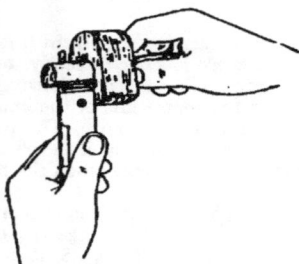

Fig. 10.

When hinges are simply screwed on the surface, without being let in, no gauging or marking is necessary, unless, perhaps, measurement with a rule, to get them at uniform distances. They are placed in position, screw-holes bored with a brad-awl, and the screws inserted.

Box covers are hinged, as in Fig. 12,

Fig. 11.

one flap of the hinge being on a narrow edge and the other on a broad surface. In this case also there is the alternative of putting the hinges on the outside, instead of in the joint; but the latter is the neatest and the usual way. The hinges are put on the lid first, as in Fig. 13, and this is held in position, first for

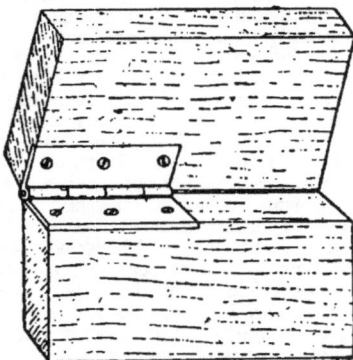

Fig. 12.

marking the lengths of the recesses on the box, and, finally, for screwing the hinges. It is better to complete the fitting of the hinges to the cover before marking their position on the body of the box, though the lines of thickness and width may be gauged on both simultaneously.

(Continued on page 196)

Fig. 13

*RECENT DEVELOPMENTS IN THE MANUFACTURE OF INCANDESCENT ELECTRIC LAMPS

Carbon Filament Lamps—Metal Filament Lamps—Osmium, Tantalum and Tungsten Types—Experiments with Tungsten Pastes—Ductility as a Factor

J. E. RANDALL

By common usage, the name, incandescent electric lamp, has been limited to a lamp whose light source is the glow of a wire heated in vacuo by electric current. This article will not use the name in any broader sense.

Incandescent electric lamps may be divided into two classes, depending upon whether their light-giving elements, that is, their filaments, are made of carbon or of metal. At present the best examples of each class stand rather far apart, both in appearance and in other features, although both are designed for the same service. One may be replaced by the other for nearly every use.

Lamps with carbon filaments have been supplied without any change in appearance for over eleven years. Within that period one notable improvement was introduced, namely, the metallized filament. Among the lamps with metal filaments, there has been, within the last five years, a procession of developments beginning with the osmium filament, the tantalum wire filament, the pressed tungsten filament, and ending with the drawn-wire tungsten filament. The author shall attempt to briefly review the advances that have been made in the quality of the most prominent members of the two classes.

CARBON FILAMENT LAMPS

The changes in quality of the regular carbon filament lamps of all standard wattages are shown in the subjoined table. Each year's quality is shown in comparison with the average of 1902.

Year	1902	1904	1906	1907	1908	1909	1910
Pct. of 1902	100	98.4	96.9	96.9	100	103.1	107.8

A sag in quality is indicated from 1904 to 1907. This is accounted for by the larger proportion of wattages below 50 and above 100, that were produced during those years. The large and small wattage lamps are known to be inferior to those between 50 and 100 watts. Within recent years the production of high wattage lamps has diminished, and doubtless will decrease still further. The proportion of low wattage lamps has been maintained and has held back the progress of average quality during the last three years. As a matter of fact, nearly every wattage shows a substantial improvement within the eight-year period.

No changes have been made in the processes of manufacture. The record exhibits the results of systematically following each detail, of rigid inspection, of thorough, exact and extensive tests, of the immediate use of the latest developments in equipment and the unhesitating discard of unsuitable equipment, of the services of trained operatives. The best lamps of ten years ago were as good as the best of the present year. The average has arisen due to the elimination of defectives.

The metallized carbon filament lamps, which are known as Gems, have made the advances shown by the following record: Calling the product of 1907 equal to 100; that of 1908 is 121; that of 1909 is 130; that of the year 1910 is 133.

All conditions favorable to advancement of the regular carbon filament lamps were of similar assistance to the Gems. A discovery in connection with the preparation of the carbons for these lamps resulted in a decided improvement in 1909. Heretofore, wattages lower than 50 have not been made successfully. Recent experiments show that wattages as low as 30 can be made.

The Gem lamp shows a sufficient superiority in quality over the regular carbon filament lamp to justify its more extensive use.

METAL FILAMENT LAMPS

As the developments in three metal filament lamps have been rapid, recent, and thoroughly published, no extended description will be given in this article.

The osmium lamp marks the beginning of development of metal filament lamps.

---

* Paper read at the fifth annual convention of the Illuminating Engineering Society, Chicago, September 25-28, 1911.

It reached a successful commercial stage in Europe. Its great fragility and the difficulties met in fashioning the filament would, no doubt, have been eliminated had its development not been arrested by the limited supply of osmium and by the advent of the tungsten filament.

The tantalum, nearly coeval with the osmium, was handicapped neither by fragility nor meager supply of metal. It is worthy of mention as an example of an article upon whose production years of research had been spent, upon whose design lavish experiments had been made. When first offered to the public, its design was finished and its qualities were thoroughly known. The inferior performance, due to offsetting, on alternating current was announced at the time the lamp was announced. The mechanical weakening of the tantalum wire, due to offsetting when kept on alternating current, has prevented the general introduction of the tantalum lamp in this country.

This lamp, however, was the first production of a real drawn wire lamp, and its development required a construction of the filament-supporting element different from any that had been used before. The design of support employed in the tantalum lamp has been followed, with slight modifications, in the drawn wire tungsten filament. The tantalum lamp cannot continue to compete with the drawn wire tungsten filament lamp in its present form.

The tungsten filament lamp was the immediate successor of the osmium lamp, and one of the most successful methods of producing the tungsten filament is based upon the process used in making the osmium filament. The completely developed process, however, has departed considerably from the method originally used, and it is doubtful whether the osmium filament could be produced by the methods now employed in the manufacture of the pressed tungsten filament. Various experimenters quickly discovered other methods of producing pastes from which tungsten filaments could be pressed. The most successful commercial methods are, however, really variations of the original Auer process.

The superiority of the metal tungsten for a lamp filament was immediately recognized, because of its extremely high melt-

ing point and because its boiling point is not greatly below its melting point. The brittleness of the pressed filament, especially when it is cool, has been a serious drawback to the general use of the lamp. The attachment of the filament rigidly to the circuit connections and to the intermediate connections between the filaments has probably been the chief cause for filament breakage in these lamps. The arced joint, while it was perfect electrically and mechanically, held the filament ends rigidly. Any jar to the lamp tended to make the filament vibrate and usually to break close to the joint. The schemes that were devised for avoiding this filament breakage were legion, but the author believes he is safe in saying that the loose contact at the bend of the filament with a support that was rigid made the hardiest lamp of the pressed filament type.

One of the most successful devices for preventing the breakage of filaments was that of introducing a short piece of piano wire between the center rod and the stem seal. This supported the filament structure with remarkable flexibility and prevented a breakage from blows on the lamp in almost any direction. A slight blow upon the base of the lamp was invariably fatal and this one weak feature served to prevent the general introduction of this method of support.

The pressed tungsten filament is not ductile when cold, no matter by what process it may have been produced. Although pressed filaments have been made that could be bent and that would take a permanent set if bent, these filaments were not truly ductile. It was natural, therefore, that immediate effort should be made to develop a quality of tungsten sufficiently ductile to be wrought into the form of wire. There was nothing to prevent success in this endeavor except lack of knowledge. It had been demonstrated that tantalum which had been known as an extremely brittle metal, could be so improved in purity that it would be ductile. This knowledge would naturally lead to the belief that many of the metals which had been considered as non-ductile could, if properly prepared, be made into ductile form. An epitome of the progress in developing ductile tungsten will read something like this:

In 1907 it was hoped that it would be

possible to produce ductile tungsten; in 1908 it was believed that it would be possible to produce ductile tungsten; in 1909 experimenters were sure that ductile tungsten could be produced; in 1910 it had been proven beyond doubt that ductile tungsten could be produced; in 1911 ductile tungsten was produced on an extensive commercial scale.

It is generally believed that the presence of carbon in tungsten is the cause of its brittleness. One well-known process for making pressed tungsten filaments does not involve the use of carbon, yet filaments produced by this process are as brittle as are filaments made by the use of carbon. As a matter of fact, the best pressed tungsten filaments have been those made by processes involving the use of carbon, yet they contain an amount of carbon so small that it can only be detected by the most delicate tests. For instance, filaments which are known to contain less than 0.005 percent carbon are no more ductile than those which are found to contain 0.1 percent. The elimination of carbon tended to reduce the length shrinkage of filaments when lamps were burned. You will doubtless recall that filaments produced in 1908 and 1909 sagged excessively and that the filaments often short-circuited due to the sag. The slack producing this sag was necessary because of the filament shrinkage. During the year 1909 decided improvements were made in this respect, and the basis of these improvements was the more complete elimination of carbon from the tungsten filament.

The progress during 1909 and 1910 did not indicate a material decrease in the fragility of the pressed filament.

It was evident, therefore, that to make the tungsten filament lamp a universal lamp, it would be necessary to have the filament in the form of wire which was sufficiently ductile to be wound, when cold, upon a spider structure. The drawn tungsten wire has met this need. While the wire before being placed in the lamp is amply ductile for the purpose of winding upon the spider and for all other manipulations needed in making the lamp, it loses much of this ductility when current is passed through it in a vacuum. The method of supporting the wire on the spider and of attaching it to the circuit terminals are, therefore, important factors in the hardiness of the lamp.

The wire may be considered to consist of pure tungsten. Chemical analysis does not find other elements. The ratio of resistance hot to resistance cold is as high as can be found in any form of the metal. The specific gravity is higher than that found for the pressed filament. The current and the candle-power peaks are low.

The structure of the metal appears to be fibrous. It changes to the crystalline form during the burning life of the lamp. This change may occur in some portions of a filament and not in others. Frequently, after the full burning life, small sections of filaments will be found that show ductility.

. Tests indicate that the wire is less brittle at every stage in the life of a lamp than are pressed filaments. There is no offsetting, either on direct or alternating current. The surface is the same in appearance, after the lamp has been burned, as that of a pressed filament. It looks as if the wire had been cracked into irregular pieces, and as if a cement of the same material had filled up the cracks. No fissures at the surface and no cavities in the body have been found.

While the wire, before being placed into the lamp, may be ranked with the toughest steel in tensile strength, ductility and elasticity, the decay of these properties after it is in the lamp makes it necessary to handle these lamps with reasonable care in order to prevent breakage. Breakage in transportation and handling compares with that for carbon filament lamps. Operatives in the lamp factories transfer lamps having wire filaments from operation to operation the same as if they had carbon filaments.

Another feature in which the drawn wire is superior is the wide range of sizes suitable for use. A piece of wire may be drawn to a size suitable for a 6.6 amperes series burning lamp or it may be drawn to a size suitable for a 20-watt, 110-volt multiple lamp. It will, when drawn to the proper diameter, be equally satisfactory for the largest or the smallest lamp. In addition, the wire may be shaped into helices, spirals or zigzags; thereby concentrating the light-giving element into a small volume. The latest automobile lamp is an example.

The number of contacts between the filament and supports, including terminals, as well as the size and material of these supports, will affect the performance and physical hardiness of a lamp.

The following results were secured from three series of tests in each of which more than 300 drawn wire tungsten filament lamps were used:

Number of contacts	11	13	15
Comparative strength, by pendulum test:			
Copper	91.5	100	96.5
Molybdenum	—	—	93.0
Comparative performance at normal efficiency:			
Copper	99.4	100	96.1
Comparative life at extreme temperature:			
Copper	107.0	100	87
Molybdenum	—	—	103

The lamps were standard in voltage and all were 40 watts. They were identical, except in the number of filament contacts. The results of the first and second tests confirm one another in indicating that 13 contacts are most satisfactory.

While no record is shown for molybdenum support lamps at normal efficiency, such lamps were tested, but their performance was much more poor than the corresponding copper support lamps.

The comparative lives at extreme temperature show that 11 contacts are better than 13 and that 13 are better than 15. Also that 15 molybdenum contacts are better than 13 copper, but inferior to 11 copper. These results are not in consonance with the results at normal efficiency. It is reasonable to believe that tests at, or near, normal efficiency indicate more accurately the behavior of lamps in service than do those at high temperatures. It has been observed that the wire in lamps burned out when at high temperature remains more ductile than the wire in lamps burned out at normal temperatures. The early failure of 15 contact lamps would not be explained by mechanical weakness. The wires usually "burned out," or melted, between the supports. The melting of the wire at the point of highest temperature has really controlled the life record of this test. The diameter of all copper supports was the same. The diameter of the molybdenum was 40 percent of that of the copper. Supports of copper having diameters 30 percent smaller and 30 percent larger than the size used in the above tests, both showed a lower strength by pendulum test. The author cannot explain why this should be so, but the tests were convincing.

Having traced recent developments up to the latest, it may not be amiss to consider the future. If the progress in lamp development may be gauged by the highest filament temperature at which each new lamp will show a given performance, one has a rational measure. For example, if 90 percent of the theoretical candle-power hours are developed in 1,000 hours burning, candle maintenance and mortality both considered, the advance from the raw carbon filament lamp to the tungsten lamp will show something as follows:

Raw carbon filament lamp (cellulose carbon)	100
Treated carbon filament lamp	119
Metallized carbon filament lamp	149
Tantalum filament lamp	206
Osmium filament lamp	270
Tungsten filament lamp	359

This comparison excludes many items, such as process difficulties, lack of wattage range, lack of voltage range, lack of suitability for both alternating and direct current, cost, etc., which affect commercial values. It is not a comparison of commercial values, although it is a comparison of the most important element in commercial values, namely, the energy wasted in doing equal work.

The change introduced by the metal filament lamp is noteworthy. Can carbon, with its many good qualities, reach or pass the record set by metals? The carbon deposited upon the treated carbon filament, when metallized, is dense, somewhat flexible, has a low vapor tension, has a fine quality of surface and has a cold specific resistance that is about 4 percent of carbon made from cellulose. All these qualities are favorable. Their further development may again place carbon in the race.—*The National Engineer.*

The total consumption of coffee in the United States in 1910 was 860,414,000 lbs. or an average of 9.33 lbs. per individual. There is only one other nation that consumes more coffee per capita than the United States and that is the Netherlands, where the consumption amounts to 15.12 lbs.

A man should never be ashamed to own he has been in the wrong.—POPE.

ELECTRO-CHEMISTRY

ERNEST C. CROCKER

Electro-chemistry is the branch of chemistry which concerns itself with the applications of electricity to chemical problems, and chemistry to electrical problems. In order to understand the inter-relations of chemistry and electricity, it is first of all necessary to have clear ideas concerning both, and these clear ideas are best formed when we attack the very heart of the problem, the ultimate nature of matter itself and of electricity.

To most people it must seem to be the height of presumption to speak of molecules, atoms and electrons as though we had seen them and knew they really existed. It is true that there is still much speculation concerning these; but a speculation that is founded upon experimental results and which always returns to experiment in case of doubt cannot be misleading. Although, in this department the different chemical and electrical concepts will be spoken about as though their existence and behavior were well known, this attitude is taken

only for the sake of clearness, and, furthermore, it will be an overwhelming discovery which will overturn the system upon which these assumptions are based.

We are in possession of many excellent and plausible theories, some the products of the world's greatest minds, which seek to explain the nature and origin of matter and electricity; but, for obvious reasons, we must turn them aside, unless they should be the expression of the results of the experimental scientist. If we hear the words of the experimenter, and doubt their truth, the only thing left to us is to experiment for ourselves and either disprove or verify his statements.

It is our aim in this article to set forth correct ideas concerning electricity and chemistry, which will stimulate the readers to attack investigations, and to attempt to satisfy that yearning to know the "why" of the problem which is so strong, particularly in the younger readers.

All pertinent questions will be cheerfully answered.

THE ULTIMATE NATURE OF MATTER

Experiment shows us that it is possible to cause two substances to combine and form a third substance which has few properties in common with the original substances. It furthermore shows that substances combine in a definite proportion, and that if we have an excess of either original substance it remains uncombined. In some cases two substances can combine under different conditions, in two or more proportions, but these are always in simple relation to each other as 2 to 1, 3 to 2, etc. After observing hundreds of cases of definite combination, by weight, it occurred to chemists that when two substances united, they formed a definite "compound," rather than a simple mixture.

To illustrate: Let us add 56 parts by weight of iron filings to 32 parts by weight of sulphur and mix them well. If, after mixing, we hold a magnet near the mixture, we can separate all the iron from the sulphur. Let us mix the substances together again, and put the mixture in a

glass tube and heat the tube over a gas flame; all at once a glow begins in one spot and quickly spreads throughout the entire mass. When the glow is over and we cool and examine the substance, we find that a magnet can no longer pick out the iron, nor can we find any evidence of the sulphur. A chemical change has taken place and we have a definite compound, iron mono-sulphide, which is expressed in chemical shorthand by the symbol FeS. This substance can be used in place of the very similar compounds Chalcopyrite and Bornite, in wireless telegraphy, in a "Perikon" detector. In nature there occurs a mineral "iron pyrites," which analysis shows to be a compound of iron and sulphur, but here the amount of sulphur which combines with 56 parts of iron is 64 and not 32 parts, and we have iron di-sulphide FeS_2. This is the mineral "Pyron" or "Ferron" used in wireless telegraphy. Iron mono-sulphide can be formed artificially by heating iron and sulphur together, but iron di-sulphide is found in nature, in rocks, as a deposit from certain mineral waters.

Water, a compound of hydrogen and oxygen, two gases, has the formula H_2O. If we add a drop of acid, or a little salt, to a glass of water (to make the water an electrical conductor) and pass an electric current through the water, using two platinum wires as leads, we shall find that gases are evolved at each wire. If we collect the gases evolved, in the manner shown in the illustration, it will be found that there is evolved just twice as much hydrogen gas at the negative terminal as there is oxygen gas at the positive terminal. On testing tubes full of each gas with a lighted match, it will be noticed that one gas, the hydrogen, burns, while the other, the oxygen, makes the match burn more vigorously. This is an application of electricity to chemistry, as a method of analysis; but at the same time can be used as a chemical method of finding the direction of an electric current, for if ever two terminals of an electric circuit are plunged into water, the most gas is always given off at the negative terminal.

The fact that compounds contain definite proportions of different "elements" or simple substances, led to the belief that all elements are composed of definite small particles, "atoms," which unite in twos, threes, etc., to form the compounds. This belief has been so completely justified that, although nobody has seen the atoms, it is no longer reasonable to doubt their existence. The little groups of two or more atoms of which substances are composed are called "molecules," or little masses.

By careful observation of the combining weights of the different elements, it has been possible to obtain the relative weights of the atoms of the elements, and a table of such values is here given. Originally, hydrogen was taken as unity, but it was later found more convenient to take oxygen as exactly 16, which makes hydrogen 1.008. By carefully looking over the table it will be seen that, with the single unaccountable exception of iodine (126.97) and tellurium (127.6), there is a gradual increase of values from 1.008 for hydrogen to 238.5 for uranium. The "valence," an electro-chemical property of the atom which we shall consider

later, determines the vertical column (group), and the atomic weight determines the series, in which an element is placed.

This remarkable table of progressive weights suggests that perhaps all atoms of "elements" are made of an original mother-substance, the atomic weights indicating the relative amounts of this substance present. Some of the recent work with the heavier elements, uranium, thorium and, most particularly, with radium, has shown that the atoms of these elements (?) are slowly disintegrating into lighter atoms, of which helium is one. Although radium is the most prominent example of a disintegrating atom, this disintegration is found in the case of a number of other atoms, including even the light-weight element potassium, and there is reason to believe that disintegration is a common property of all elements.

There have been many elaborate theories advanced as to what the original mother-substance may be, some having it "ether," some "electrons," some a single simple substance, and some a combination of two or more elementary substances. All we can say about this, just at present, is that we must wait for a few years until there is advanced some generally acceptable theory which agrees with the facts. All that we know is that what we call atoms are really not elementary particles, but are complicated structures.

Many attempts have been made to disintegrate atoms, but up to the present time, nothing but failure has resulted— we only know of disintegration where it occurs spontaneously, and even there we cannot yet hasten or retard it. It is difficult to say what may be done by, for instance, a more intense heat than any which we can now produce, or by some peculiar application of electricity.

Although we cannot clearly understand all about the atom, it possesses some peculiar properties which are of interest both to the chemist and to the electrician. All atoms do not combine with each other in equal numbers, although always in a simple ratio, nor do they have equal tendencies to combine— in fact, there are elements which will combine with no other elements, or their valence is zero (group 0 in the table).

TABLE OF SYMBOLS AND ATOMIC WEIGHTS

NOTE: The names in parentheses are the Latin names of the elements

Group	0	I	II	III	IV	V	VI	VII	VIII
Valence	0	1	2	3	4	3 or 5	2 or 6	1 or 7	2 or 8
1		Hydrogen H 1.008							
2	Helium He 4	Lithium Li 7.03	Beryllium Be 9.1	Boron B 11	Carbon C 12.00	Nitrogen N 14.04	Oxygen O 16.000	Fluorine F 19	
3	Neon Ne 20	Sodium (Natrium) Na 23.05	Magnesium Mg 24.36	Aluminium Al 27.1	Silicon Si 28.4	Phosphorus P 31.0	Sulphur S 32.06	Chlorine Cl 35.45	Iron (Ferrum) Fe 55.9 / Nickel Ni 58.7 / Cobalt Co 59
4	Argon Ar 39.9	Potassium (Kalium) K 39.15	Calcium Ca 40.1	Scandium Sc 44.1	Titanium Ti 48.1	Vanadium V 51.2	Chromium Cr 52.1	Manganese Mn 55	
5		Copper (Cuprum) Cu 63.6	Zinc Zn 65.4	Gallium Ga 70	Germanium Ge 72.5	Arsenic As 75	Selenium Se 79.2	Bromine Br 79.96	Ruthenium Ru 101.7 / Rhodium Rh 103.0 / Palladium Pd 106.5
6	Krypton Kr 81.8	Rubidium Rb 85.5	Strontium Sr 87.6	Yttrium Yt 89.0	Zirconium Zr 90.6	Columbium Cb 94	Molybdenum Mo 96	—	
7		Silver (Argentum) Ag 107.93	Cadmium Cd 112.4	Indium In 115	Tin (Stannum) Sn 119.0	Antimony (Stibium) Sb 120.2	Tellurium Te 127.6	Iodine I 126.97	Europium Eu 151.79
8	Xenon Xe 128	Caesium Cs 132.9	Barium Ba 137.4	Lanthanum La 138.7	Cerium Ce 140.25	Didymium Di 140.5	Neodymium Nd 143.6	Samarium Sm 150.3	—
9		Gadolinium Gd 156	Terbium Tb 160	Erbium Er 166	—	—	—	Thulium Tm 171	Osmium Os 191
10		—	—	Ytterbium Yb 173.0		Tantalum Ta 183	Tungsten (Wolfram) W 184	—	Iridium Ir 193 / Platinum Pt 194.8
11		Gold (Aurum) Au 197.2	Mercury (Hydrargyrum) Hg 200.0	Thallium Tl 204.1	Lead (Plumbum) Pb 206.7	Bismuth Bi 208.5	—	—	—
12		—	Radium Rd 226.5	—	Thorium Th 232.5	—	Uranium U 238.5	—	—
Higher Oxides	None	R_2O	RO	R_2O_3	RO_2	R_2O_5	RO_3	R_2O_7	RO_4
Hydrides	None				RH_4	RH_3	RH_2	RH	

Save this table, as it will be referred to in succeeding articles.

Oxygen and fluorine are very active elements and not in this group, yet it has never been possible to make them unite, and even though both are of such activity that either unites with hydrogen with explosive violence.

The simplest explanation of the attraction (sometimes explained by simply calling it "chemical affinity") is that of electrical attraction between the atoms of the different elements. Some are negatively and some positively charged,—and the greater the available potential, the more active the element. Oxygen and fluorine are practically equally charged, but both negatively, hence their mutual repulsion. The electrical potential is supposed to be due to the presence of little particles of negative electricity, "electrons," which, while they belong to the atom, are free to move about within the atom, which itself bears the positive electricity. Evidently, if there are just enough electrons present to form an amount of negative electricity just equal to the positive electricity of the

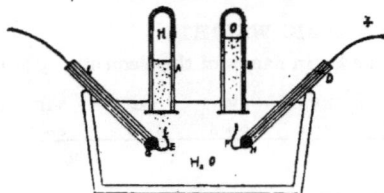

atom, the atom as a whole will be neutral, while, if not enough are present it will be positive, or if too many are present, negative. Very few atoms are exactly neutral, there being usually an excess of one or other kind of electricity. Actual justification for this electrical explanation of chemical attraction will be found when we consider the inner workings of electric batteries.

"Valence" is easily explained, for it also appears to be electrical in its nature. Valence is the combining power of an element; for instance, an atom of valence 4 needs either four atoms of valence 1, two atoms of valence 2, or one atom of valence 4 to form a compound. Four atoms of valence 3 also unite with three atoms of valence 4. It seems that a valence 1 atom has one electron in excess or lacks one, an atom of valence 2, two electrons, etc.

Although chemistry and . electricity are ordinarily separate studies, they may advantageously be combined to explain each other. We have considered the granular structure of matter and the existence of the electron, and from time to time we shall find things which demand a knowledge of the electron as well as the atom or molecule and shall find many places where, as in X-rays, electrolytic rectifiers and batteries, the electrical action can only be understood when we know the chemical action.

BELL WORK IN NEW HOUSES

W. F. PERRY

Ordinarily an electrician who is wiring in a new frame house lets the bell-work remain until last, then this is pulled in with wonderful rapidity, simply because it does not have to be insulated. Owing to this fact, the wires are run over, around and through anything and everything, joints being made between walls and ceilings. In one year, or possibly less, the bell is out of order, the electrician sent for, the test made and a line found to be broken inside somewhere, so that another wire has to be run. Many times this line has to be run in plain sight, the owner knowing that pulling his house to pieces for so small a job is a needless and a foolish undertaking. In a few years' time, the closets, back-stairs, and cellar-way is a maze of dusty and dirty bell wire, some of which are useless and one or two of which perform the duty which each in its turn had done.

To overcome this difficulty, the writer installs all bell wire, no matter how long or how short, in a manner such that, if any trouble is found on a line, that one line may be pulled out without disturbing any other wire. The plan is to take each and every wire into the cellar in a perfectly straight line, there being no fastenings whatever for them.

A piece of $\frac{7}{8}$ in. board should be fastened flush with the studding and a hole bored through it with a No. 11 bit. Holes should be bored right through to the basement, care being taken to get them somewhere near in line. A small piece of lath will suffice to hold the wire from falling back through the hole by twisting the wire around it. In the cellar, a nail may be driven in and the wire tied to it. The button wires at the door should be done as nearly as possible in the same manner. Of course they have to pass through one or more studs at right angles; but if care is taken in cutting the holes, the wires will be free to slide back and forth.

After the house is completed and the bells and buttons in place, the loose ends in the cellar should be connected. This requires a little work in the basement, but all the joints are in plain sight.

Now, if, after a year or so, trouble is found on any concealed wire, that wire is disconnected from its bell or button and from the nearest joint in the cellar, and a new piece of wire of the proper length tied on—by loosening the plaster away from it, where it passes through the wall, the old wire may be pulled out, and the new one pulled in behind it.

With this method, the closets and stairways are always free from loose wiring, and the electrician always knows it is but a simple matter to repair it.

MOVING PICTURE MARVELS

Every day fifteen million people attend the moving picture shows throughout the country, but despite the evident popularity of the amusement, there are few who understand how the many mystifying effects are produced.

Men walk on ceilings like flies, or run up the sides of houses in apparently entire disregard of the law of gravity. Horses, dogs, cats and other animals walk backwards or sideways in a manner quite contrary to their usual nature; coffee pots proceed to pour themselves without any visible interference; automobiles dash through crowded streets in defiance of all speed laws; men and women fall down steep inclines and over the edges of precipices and get up again, apparently none the worse for their experience; real trains collide, and real ships are wrecked without regard to the expense involved—and all for the modest price of a nickel or a dime, says *Boston American*.

How is it done? The answer may be told in two words: It isn't. It's all make-believe. Real persons and animals pose for the pictures, and the scenery is sometimes genuine, but that is all. The rest, for the most part, is just make-believe, as a visit to the great studios where moving pictures are made would readily demonstrate.

Some of the most amazing effects are easily accomplished by a tactful manipulation of the film; others require more elaborate preparations. To make an animal walk backwards, a moving-picture is taken of it walking in the usual manner, and to produce the desired effect the film is simply reversed. In the same way the really astonishing pictures of brick apparently flying into a bricklayer's hod are obtained, the simple operation of a man dumping a hodload of bricks being sufficient to give the fantastic effect when the film is reversed.

Everyone who has visited a moving picture exhibition is familiar with the ease with which moving picture heroes and heroines run nimbly up the sides of houses when pursued, and, no doubt, everyone has wondered at one time or another how the feat is accomplished.

There are two ways of doing this: One is to have the person posing for the picture drawn up the side of a real house by means of a rope, moving his feet all the time, as though he were walking, the rope being afterwards painted out on the film; the other and more common way is to make the film in a specially prepared studio. On the floor a canvas picture of the house in question is spread and the man pursued just scrambles along it on his hands and knees.

The effect of inanimate objects moving themselves, such as coffee pots pouring themselves, chairs and similar objects jumping up in the air, chimney pots falling off and flying back into position, and typewriters working of their own accord, is produced by means of wires which are either too fine to appear in the picture or, if they show, are readily painted out.

Railroad collisions are frequent enough, one would imagine, to enable the moving picture concerns to obtain genuine pictures of them, but the thrilling pictures seen on the moving picture screen are obtained in a far less realistic manner. Miniature trains constructed and staged with great fidelity to actual conditions, and which run automatically, are used for the purpose, and serve very well. Sometimes an auto is made to collide at a crossing with a locomotive in a similar manner.

The familiar film showing a painter stencilling a ceiling, to which he appears to be clinging in a most unnatural manner, while an assistant is holding a pot of paint up to him, never fails to create wonderment among the uninitiated, but is easily made. The pictures are taken in a make-believe room, the walls of which are painted upside down on a four-sided screen, and the floor of which is painted white to resemble a ceiling. To a rafter across the top a man is suspended by his feet and holds an empty paint pot towards the floor, upon which the man posing as the painter kneels. The latter holds a stencil to the floor with one hand and with the other dips a brush in the paint pot which the suspended man holds towards him. After the pictures are taken the film is run off upside down and gives the topsy-turvy effect desired.

In a similar way the film which shows a man holding himself to the ceiling by the top of his head and the palms of his hands is made, the man simply standing on his head. To make the picture realistic, tables and chairs are attached to the make-believe room and an elaborate chandelier is attached to the floor, so that when the film is reversed the room will appear to be fully furnished and equipped.

In two out of three moving pictures there is a pursuit in which men, women and children are made to scamper over hill and dale at phenomenal speed, horses and wagons and automobiles tear pellmell through the streets, knocking over fat policemen in their path, and everything moves with a hustle and bustle that is little short of amazing. It is needless to say that neither the animate nor the inanimate subjects of the pictures ever actually covered space at the rate indicated. When the pictures are taken, the persons posing for them may move as leisurely as they please, the effect of speed being produced by cutting out numerous sections of the film.

The super-imposed negative, a process familiar to every photographer, is often resorted to to produce weird effects. In this manner are made the films showing normal-sized men and women watching a contest between what appear to be men and women no bigger than a thumb. Two sets of pictures are taken. First, the full-sized spectators are photographed while making the gesticulations and motions to be expected of interested spectators, and then another set of pictures of the contestants is made, the persons posing for them being stationed at such a distance from the moving picture machine that they come out very small in the pictures. It is then a comparatively simple matter to combine the two films.

To construct a sky-scraper in the short space of a minute or two seems easy enough after observing the operation on the moving picture screen. The foundation is dug, the steel skeleton construction is completed, the masonry and woodwork are added, the scaffolding is removed and, lo and behold, the tenants are filing in and out, all within the space of time it takes to run off the film!

But to make the film is a much more tedious operation than might be supposed. Every day during the progress of the actual building of the structure the moving picture man must photograph the work, and when the building is completed pictures of the tenants going in and out must be taken. Then the various films are united, and the effect when they are run off on the screen is little short of marvellous.

The adventures of "Alice in Wonderland" have suggested some most fantastic ideas to the moving picture man. In one of the films Alice grows so large that she literally bursts through her house, her legs and arms bulging through the windows and walls.

This effect was obtained by means of the multiple process: separate, normal-sized pictures of the house being taken first, and then pictures of Alice with her arms and legs enlarged to such an extent that they fairly filled the window and doors. When the two films are combined Alice is prodigious enough to suit the most fastidious.

Most of the moving picture concerns employ regular stock companies, the members of which receive high salaries. There are few moving pictures displayed in which some deception of this kind is not practised, but nobody minds it. When we go to the land of make-believe we are willing to be fooled.

No effort is spared to produce weird effects of this kind. Skilled actors are employed and elaborate properties constructed to make the pictures successful, and the moving picture man is always looking for something novel and unique.

FITTING HINGES
(Continued from page 186)

Doors are treated in the same way, the hinges being put on the door first, and then it is supported in position against the post for marking the height of the hinges, and, finally, for screwing them. The weight of the door is supported during these processes by packing it up with thin bits of wood or wedges beneath. The door should be turned back as far as possible, so that the hinges are wide open, as in the upper views in Figs. 7 and 8. There is then no difficulty in attaching it properly. In house doors an open joint is often allowed, so that the door will open reasonably wide, without binding against the moulding of the doorway. The open joint is not made by screwing the hinges on the surface, but by letting their knuckles stand out a little farther when the door is closed.

THE SIMPLIFIED TURBINE

At the meeting of the National Electric Light Association, last spring, Nikola Tesla announced for the first time tests on an improved and simplified form of turbine prime mover which, from the success so far achieved and from the probabilities, seems to embody a long step forward in the utilization of the energy of steam. Two difficulties have handicapped the turbine, the complication in reversing, and the great number of small parts which must be assembled to make the complete machine. Effort has been expended on all classes of prime movers to secure simplicity, gradual abstraction of the energy of heat and high speed in order to reduce waste. If the development of the future bears out the present promise, the new form devised by Tesla will go far to remove all these difficulties, and secure desirabilities.

To give an idea of what he has accomplished, a small turbine having a rotating member 9¾ in. in diameter and 2 in. wide, developed 110 h.p. with free exhaust, and it had no blades, vanes, valves or sliding contacts of any kind, except the journal bearings. Furthermore, to reverse the direction of rotation, all that is necessary is to close one valve and open another. This power was developed when exhausting to the atmosphere and taking steam at 125 lbs., and another unit on which rather extended tests have been made, and which had a rotor 18 in. in diameter by 3½ in. wide, running under like conditions of pressure and exhaust has developed 200 h.p. at a steam consumption of 38 lbs. per horse-power hour. While this does not seem a very remarkable steam economy, it must be remembered that this is a first experimental engine, that the turbine form of prime mover does not operate with the best economy unless the exhaust is at low vacuum, and that the weight of engine per horse-power developed was only 2 lbs. The space occupied by this 200 h.p. unit is only 2 ft. x 3 ft. and 2 ft. high.

At first thought it might seem that the speed used, 9,000 revolutions per minute, would be a drawback; but here again it must be remembered that the rotating member consists only of thin steel discs fastened firmly to the shaft,

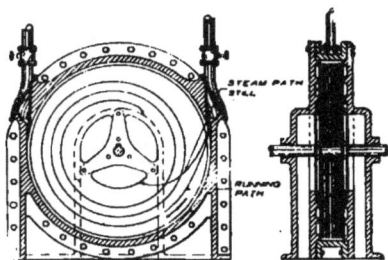

Sectional View of the Simplified Turbine

and having no blades, vanes, or other loose or attached parts, and that even at the highest speed the tensile strain is not over 50,000 lbs. per square inch on any part of the machine.

Tesla predicts that with a low-pressure stage added, exhausting into a 28½ or 29 in. vacuum, as is the case in steam turbine practice, he will secure a steam consumption of less than 12 lbs. per horse-power hour on a 200 h.p. unit, and that he will be able to get 10 h.p. per lb. weight of prime mover.

In accomplishing this result, Tesla has made use of a new principle in steam engineering, namely, the viscosity and adhesion of the fluid.

Heretofore, in abstracting the energy from steam, the reaction of steam to push the moving part forward has always been utilized. In the Tesla turbine the steam drags the rotating member around with it by means of the adhesion of the steam to surfaces of the thin discs.

A somewhat similar process in the reverse direction has been utilized in the water brake for testing turbines in which discs are revolved rapidly in a chamber filled with water and the drag of the water on the discs produces a braking effect which loads the turbine. In that application, however, the discs are kept at some considerable distance apart and usually blades are introduced between them to retard the water so that the friction of the disc on the water is depended on to produce the drag.

It is easy to conceive that by placing the blades close together the water which will be caught between them would be thrown out as in the case of a centrifugal pump, and if there were an inlet at the center, a centrifugal pumping action would be produced. This is exactly what Tesla has done in the pump application of his principle; and the motor application is simply a reversal forcing the steam in at the edge of the discs in a tangential direction and permitting it to work its way inward in a spiral path until it exhausts at the center of the disc. In the 200 h.p. unit there are 25 discs, each 1/32 in. thick, and so placed on the shaft that the entire distance occupied along the shaft is 3½ in. As there are no guide plates or vanes, the steam takes its natural path from the circumference to the center, and when the rotor is at rest, it flows by a short-curved path, as indicated by the line in the end view, across the face of the disc. When the rotor is up to speed, the steam, in making its passage across the face of the disc, makes several revolutions, so that its path is from 12 to 16 ft. in the form of a spiral as indicated.

It is evident that as the direction of rotation of the shaft is determined entirely by the direction of the entering jet, all that is necessary to reverse the flow and the rotation is to shut off the nozzle at one side of the disc and turn on the nozzle at the other side. The drawing, which is an illustration of the patent principle, shows how this is accomplished, but does not, of course, indicate the actual detail of the working machine.

To show the reversibility of the operation as prime mover or pump, two 200 h.p. machines are coupled together through a torsion spring, and one of these is driven as a turbine while the other one has steam admitted in the direction opposite to that of rotation, and is used as a brake. Evidently the amount of twist of the torsion spring is an indication of the load carried and this has been used to construct a direct reading horse-power indicator which shows at a glance just what power is being developed.

Mr. Tesla, in an interview given to the *Electrical Review*, has stated that the 110 h.p. turbine has attained a performance of more than 2 h.p. per pound of material, and by careful construction for lightness might have been much improved, also that the steam consumption for this little unit was only 36 lbs. per horse-power hour with free exhaust and back pressure of 1 to 2 lbs. He states further that this economy could further

be greatly increased by using a 2-stage machine, so that it would not take over 12 lbs. per horse-power hour, which will be recognized as a remarkable performance for a 110 h.p. prime mover.

To demonstrate the efficiency of the principle as a pump, Tesla has in his offices a 1/12 h.p. electric motor driving a Tesla pump, which pumps 40 gal. a minute against a 9 ft. head. Evidently the motor is carrying some over load, as the water horse-power would be 0.09, while the rated capacity of the motor is 0.083 h.p., but the demonstration serves to show the high efficiency of the principle when applied to pumping.

It is evident that so simple a device, and one which has in it the elements of high efficiency, has wide application in the industries, and, the simplicity of construction and the strength that may be secured, make it seem probable that a solution of the gas turbine proposition is to be worked out along this line. It is quite conceivable that by applying this method the simplicity of the steam turbine and the high efficiency of the gas engine will be combined, giving us a more efficient heat motor than has yet been known. Much remains to be determined as to distance between discs, diameters and speed of revolution for different pressures, but the workability of the principle seems to be proved.— *Practical Engineer*.

***

X-Ray Apparatus of Today

(*Continued from page* 161)

volts are required to jump them, and this is, of course, wasted energy.

The results attained with this extraordinary arrangement, as regards the resultant wave form, is seen in Fig. 8, as contrasted with the wave-form of the secondary current supplied by an induction coil, Fig. 9. As will be evident, there are dead intervals between each pulsation of direct current as supplied by the "Interrupterless" outfit, but the current does not fall as rapidly in value as the induction coil current, and there is no inverse current, whatever, to injure and disintegrate electrodes of the X-ray tube.

This type of machine gives extremely good results in connection with a high-frequency therapeutic or treatment set, the regular connections being given in Fig. 10 and 11 for Tesla coils and Oudin coils, respectively.

CONCRETE AND CEMENT

P. LEROY PLANSBURG AND L. BONVOULOIR

One 'of ¦the 'most important materials used in the building construction of today is concrete. This material and its method of use are by no means new and date back to the Romans more than 2,000 years ago. In 500 B.C. the Romans made use of concrete and secured most excellent results from a mixture of slaked lime, volcanic dust, sand and broken stone.

The Romans used concrete for many purposes, such as pavements, floors, walls, arches and domes. Almost invariably the Roman walls were built of concrete, faced with brick, stone or marble, the facing being very thin. Really, the concrete was employed as a core between the inner and outer facings. An example of early concrete construction is the Pantheon in Rome which was built by Agrippa about the time of Christ. The dome of this building is 142 ft. in diameter, and is built of concrete in a framework of brick arches. To construct a similar dome would be regarded as an engineering feat even at the present time. The development of domed structures among the Romans was probably in a large measure due to the use of concrete. Other examples of early concrete construction are the aqueduct of Vejus and an upper floor in the House of Vestals. This floor has a span of 20 ft. and is 14 in. thick. In the Baths of Caracalla there are remains of extensive concrete vaults.

The Normans well understood the use and value of concrete and it was extensively used by them for walls and foundations. It is an interesting fact that the castle of Badajos in Spain still shows the marks of the boarded frame into which the concrete was poured. In many of the Norman and English cathedrals, concrete was successfully employed in the foundations, some of the best examples being Ely Cathedral, Westminster Abbey, and Salisbury Cathedral. During the latter half of the twelfth century, Guilford castle was erected with concrete walls, 12 to 14 ft. thick at the base.

The concrete of today is a mixture of lime or cement, water, sand and small irregular pieces of stone, brick, cinders, etc., and differs from that of the Romans only in that cement is usually employed in place of the lime. The ingredients of concrete may be divided into two classes: the first being the active or cementing materials, namely, lime or cement and water; the second, the inactive materials such as sand, small pieces of stone, brick and cinders. The active agents which cause the concrete to solidify are termed the matrix, while the inactive ingredients are called the aggregate. Since the aggregates are of irregular shape and varying sizes, there would ordinarily exist air spaces between the separate pieces, and it is the duty of the matrix to fill all such spaces (which are known as voids) and form the whole into a solid mass.

While we have seen that concrete has been used for over 2,000 years for building purposes, it was not until the invention of "Portland cement" by Joseph Aspdin of Leeds, England, that concrete assumed a foremost place in the list of available building materials. The cement was discovered by Aspdin in 1824, and is without doubt the strongest cement that the world has ever known. For the next twenty-five years progress was slow, but about 1850, due to both improved methods of manufacture and improved quality of the finished product, there was a more general recognition of its merits and its commercial success was assured.

Until quite recently, lime was practically the only material employed as a matrix. The Romans made use of lime for all of their construction work, both for concrete and mortar, and good evidence is shown of how well they understood its nature by such of their structures as still exist. When lime is used as a matrix, it should be ground, not slaked and sieved, as is done in some places. There are conflicting opinions as to whether it should be used fresh or allowed to season.

Probably the inventor of Portland cement never for a moment imagined what a revolution it would make in all building and engineering works. The Thames River tunnel was the first engineering work of importance in which Portland cement was used. Originally, chemistry had little or nothing to do with the manufacture of Portland cement, and, as a result, the original cement was a weak product compared with that of today.

An average analysis of some of the best modern cements is as follows:

Lime, 60 to 65 percent; silica, 20 to 25 percent; alumina, 6 to 9 percent; oxide of iron, 2 to 5 percent; sulphuric acid, 1 to 4 percent; magnesia, 1 to 3 percent.

The magnesia is simply an impurity in the chalk and it may cause expansion and disintegration if it exists to a greater extent than 1 to 3 percent. Nearly all governments limit the presence of this impurity to 4 percent in the finished product.

Repeated experiments have shown the importance of uniform fineness in cement, and that it cannot be ground too

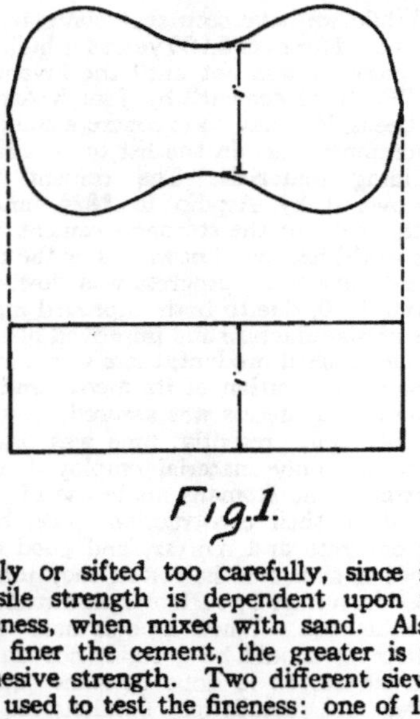

Fig.1.

finely or sifted too carefully, since its tensile strength is dependent upon its fineness, when mixed with sand. Also, the finer the cement, the greater is its adhesive strength. Two different sieves are used to test the fineness: one of 100 meshes to the inch (10,000 to the square inch), and the other of 200 meshes to the inch (40,000 to the square inch). The cement is first passed through the 100-mesh sieve until not more than 1 percent passes through after sifting one minute. The sieve is moved back and forth at the rate of about 200 strokes per minute, tapping at the same time with the other hand. The part remaining in the sieve is then passed through the 200-mesh sieve in the same manner.

In order that the tests made shall be the same everywhere and that the manufacturers may have a definite standard to work from, the American Society for Testing Materials has issued a pamphlet containing standard specifications for cement. These specifications are used when any concrete construction work is being carried on. The following extracts are taken from the pamphlet: "Portland Cement is the term applied to the finely pulverized product resulting from the calcination to incipient fusion of an intimate mixture of properly proportioned argillaceous and calcareous materials, and to which no addition greater than 3 percent has been made subsequent to calcination. The specific gravity shall not be less than 3.10. Its fineness shall be such that it shall leave by weight a residue of not more than 8 percent on the No. 100, and not more than 25 percent on the No. 200 sieve. It shall not develop initial set in less than thirty minutes; and must develop hard set in not less than one hour, nor more than ten hours. The minimum requirements for tensile strength for briquettes 1 sq. in. in section shall be as follows:

Age	Neat Cement	Strength
24 hours in moist air		175 lbs.
7 days(1 day in moist air, 6 days in water)		500 lbs.
28 days(" " " " " 27 " " ")		600 lbs.

The cement shall contain not more than 1.75 percent of anhydrous sulphuric acid (SO_3), nor more than 4 percent of magnesia (MgO)."

It is, of course, desirable to test every new batch of cement to see that it comes up to the specifications, as different batches are apt to vary somewhat in quality, or may have become damp in transit and have had its quality impaired thereby. In order to obtain a representative value of the batch, a small quantity of cement is taken from a certain number of barrels, say one in every ten. A hole is drilled in the side of the barrel, and the cement so obtained is mixed thoroughly. The results of the tests are taken as the average value of the batch.

The most important test is the tension test. Briquettes of the shape shown in the sketch, Fig. 1, and having their smallest cross-section of 1 sq. in., are made either neat or in the ratio 1 to 1 (one part of cement to one part of sand). After being carefully tamped in the molds

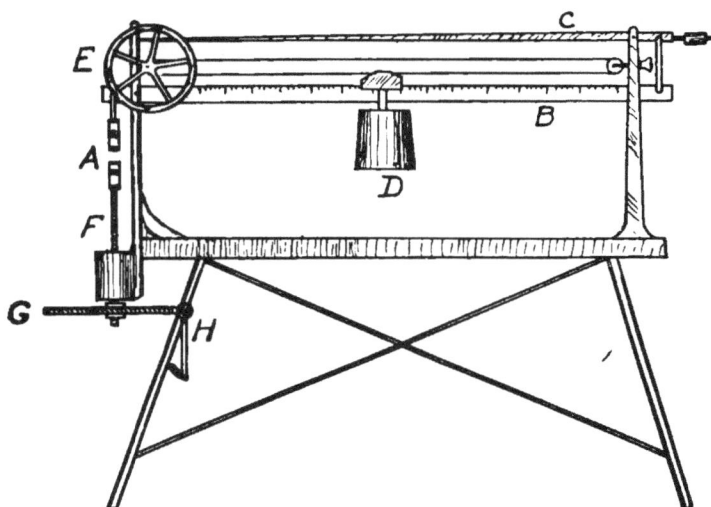

Fig. 2. Machine for Testing Briquettes

they are allowed to set for one day in moist air and then placed in water for periods varying from one day to 28 days. Several different types of machines are used to test the briquettes, but in principle they are all alike. The machine shown in Fig. 2 is one of the common types. The briquettes are placed in the jaws marked A. The upper jaw is connected to a system of levers B and C. D is a weight mounted upon a roller and travels on B. It is propelled by means of a cord running over pulleys and to one of the pulleys is attached the handwheel E. The end of the lever C comes opposite a small fixed pointer and is used merely to magnify the deflection and thus make possible a more perfect balance. When a briquette in A is under tension, C drops and by turning the handwheel E, the weight is moved until balance is secured and the end of the lever C comes directly opposite the pointer. The tension may then be read on B, which is graduated in pounds. The lower jaw is attached to a long screw F, which passes through a nut in the center of the gear G. H is a worm engaging in G and turned by a handle. When a briquette is being tested, H and E are turned at the same time, thus applying the load with one hand and keeping it balanced with the other until the briquette breaks.

The breaking strength per square inch may then be read directly from the scale.

Compression tests are rarely made outside of the laboratory, since the crushing strength of cement is much greater than its tensile strength, and therefore such tests would require heavier and more expensive machines. It is true that concrete or cement is rarely used in tension, but since the results are merely comparative and have no relation to their actual strength when in use, the results obtained from the tensile tests are the only ones needed.

Next, in importance, comes the time of setting. Two sets are recognized: the initial and the final sets. A Vicat needle is usually employed to determine the time of set, but home-made needles may be used with as good results. In Fig. 3 a wire or rod 1 mm. in diameter is fastened to a weight so that their combined weights shall be 300 grams. A screw permits raising or lowering the weight. A pat of cement is placed and the needle is carefully brought down to the level of the pat. When the needle no longer penetrates to a point .2 of an inch above the glass plate, initial set is said to have been obtained, and when it no longer sinks into the surface, final set is reached.

The other tests described in the speci-

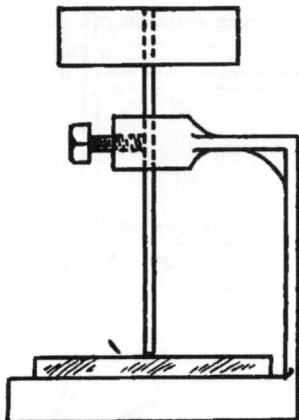

Fig. 3

fications of the American Society of Civil Engineers are not usually made in the field, since they require somewhat elaborate apparatus.

A matter of great importance is the testing of the voids in the sand and aggregates to see that with the given proportions, all of the voids will be filled. The following is the method usually followed. A water-tight vessel of known volume is filled up level with the aggregates, and water is poured in until it comes level with the top, taking care to notice the amount of water poured in. This volume of water divided by the volume of the vessel will give the proportion of voids in the aggregates. The voids in the sand are measured in the same way. Suppose we have a 1:2:4 mixture. The vessel contains 1,000 cu. in., and 400 cu. in. of water have been poured in after it was filled with the aggregates. Then there are 40 percent of voids in the aggregates. From the proportion of 1:2:4 it is seen that 1,000 cu. in. of aggregates would be used with 500 cu. in. of sand and 250 cu. in. of cement. 40 percent of voids would require only 400 cu. in. of sand, so that we have an excess of 100 cu. in., and we are certain that all of the voids are filled. If the voids in the sand are found to be 35 percent and 250 cu. in. of cement are to be used with 500 cu. in. of sand the proportions would be correct. True, this would give us an excess of cement,

but this is desirable, as we must allow for shrinkage of the cement and sand when mixed with water.

In measuring the sand care should be taken thoroughly to dry it as there is always some moisture present in the sand, and this would cause the voids to appear to be smaller than they really are. In case the voids in the aggregates exceed the amount of sand to be used, the aggregates should be broken up into smaller sizes or else a different kind used.

Aeroplanes in the Postal Service

With the aeroplane's practical military value being so clearly demonstrated before our very eyes, we only need to turn our attention in another direction to see an even greater practical triumph accomplished by the aeroplane for another branch of the Government which means no less for the immediate progress of our rapidly moving civilization in time of peace than the wonderful accomplishments of the aeroplane toward the practical solution of some of the great problems of war. I refer to the recent establishment at Allahabad in India of an aerial postal service with special aerial postmarks to be put upon all letters, and a complete postal installation embracing all the red tape attached to the transportation of His Majesty's "Royal Mail." This has lately been done under the personal supervision of Captain W. Windham, who organized this excellent undertaking to demonstrate the absolute practicability of maintaining postal communication with a city, even though it may be undergoing a state of siege and is completely surrounded by the forces of the enemy. Over six thousand letters were carried in specially-constructed mail pouches which were carefully loaded upon an aeroplane and transported to a neighboring postal station.—*Augustus Post on " Practical Uses of the Aeroplane" in June Columbian.*

Prisoner (stuttering painfully): "Tz-tz-tz-st-st-st!"

Magistrate: "Dear, dear, what's he charged with, Constable?"

Constable: "Sounds like soda water, your worship.—*Everybody's Weekly.*"

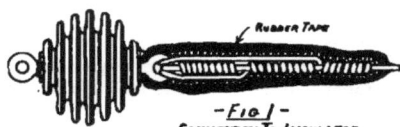

In this department will be published original, practical articles pertaining to
Wireless Telegraphy and Wireless Telephony

THE USE OF ALUMINUM WIRE FOR AERIALS

CHAS. HORTON

Wireless men as a rule discourage the use of aluminum wire for aerials, on the ground that it is too apt to break under strain, and that good connections cannot be had, because it cannot be easily soldered. The use of copper wire eliminates these troubles, but, as copper is much more expensive than aluminum, length for length, it would be very desirable, if possible, to use aluminum, especially for amateurs.

The author's experiences in the use of aluminum for aerial construction shows that this wire may be so used and with very good results, if proper precautions are taken.

The author has such an aerial of four No. 12 aluminum wires, which span a distance of almost 200 ft. The wires are almost 4 ft. apart and about 80 ft. from the ground where the station is located. This aerial has been in use about a year, and has weathered some very severe storms, including the hurricane that we had early in the spring. One thing in favor of aluminum wire is the fact that, owing to its somewhat oily surface when oxidized, it does not allow the formation of ice in sleety weather, thus eliminating one of the causes which break down many telegraph and telephone lines every winter.

Now, as to the methods to be used, and the precautions to be observed:

1. In order to prevent galvanic action, which is very apt to occur with aluminum, see that all contacts of the aluminum with any other metal be prevented as far as possible. For instance, where the wire loops through the galvanized ring of the insulators, the ring should be

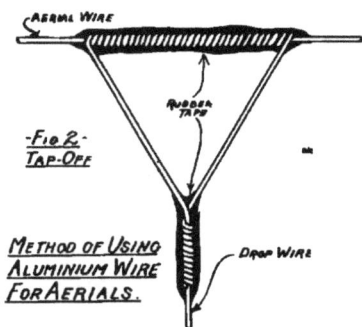

- Fig 1 -
Connection To Insulator

- Fig 2 -
Tap-Off

Method of Using Aluminium Wire For Aerials.

first wound with tape and then the wire looped in and fastened so that the tape prevents any electrical connection with the zinc.

2. All joints should be soldered with special aluminum solder on the market, or if this is too difficult, the wires may be scraped clean and wound together tightly for a distance of 3 or 4 in. The finished joint is then wound over tightly with rubber (Okonite) tape which soon forms a water-tight one-piece coating and effectually protects the wire from oxidation. The author, on examining such a joint after a hard season, saw that the surface of the wire was perfectly bright. The rubber may be protected by a layer of ordinary tape.

3. Where the wire is sharply bent or

subject to strain, it should be reinforced as follows:

The wire is first looped through the taped insulator ring and twisted about itself as usual. Referring to Fig. 1, the aerial wire is shown in the middle passing through the ring and twisted about itself close up to the insulator ring. Next, a piece of aerial wire is looped through the ring and each end twisted about the aerial wire beyond the first twist. This is readily seen in the drawing. Now a long piece of aerial wire is very tightly wound over the whole and fastened in the insulator ring on one end and twisted about the aerial wire at the other. This winding is shown in section in the drawing by the two rows of circles. To finish, the whole joint is wound tightly with rubber tape, and, if desired, the rubber wound with ordinary tape. The rubber tape is here shown in section by the heavy black part. Soon after the rubber tape is wound on, it becomes a solid piece of rubber, and it

is impossible to unwind it again. When it is cut off for inspection the joint will be found perfectly bright.

4. The aerial here referred to is of the shifted T-type, and where the drop wires are taken off the joint is made as follows. Referring to Fig. 2 it will be seen that the drop wire is brought to the aerial wire, wound about for a distance of 3 or 4 in., and finally brought back and twisted about itself, thus forming an open triangle. This form of joint prevents a right-angled connection, which would sway back and forth until the wire breaks off.

5. When using pincers to twist the wire, be very careful that the cutting edge does not nick the wire, for even a slight nick is apt to cause a break.

· If all the points given above are carefully observed, no trouble will be experienced. No. 10 or 12 should be used unless the span is very short, when No. 14 may be safely used.

IMPROVED MINERAL DETECTOR STAND

P. MERTZ

A mineral detector stand that has been designed to eliminate several disadvantages of the ordinary type, and yet be as simple, compact and efficient as possible, is shown in the illustration.

To make this detector get a base A, preferably of hard rubber, $5\frac{1}{2}$ in. long, 2 in. wide and $\frac{1}{2}$ in. thick. Then drill two $^{11}\!/_{64}$ in. holes in the positions shown

at E and F in the plan view, and counterbore them on the under side with a $\frac{3}{8}$ in. drill to a depth of $\frac{1}{4}$ in.

Get two pieces of brass tubing (C and D) $\frac{5}{8}$ in. outside diameter, $^{11}\!/_{64}$ in. inside diameter, and $1\frac{1}{8}$ in. and $\frac{1}{2}$ in. long, respectively. As this tubing is hard to get and is expensive, thin tubing can be used if filled with melted solder or lead,

IMPROVED MINERAL DETECTOR STAND.

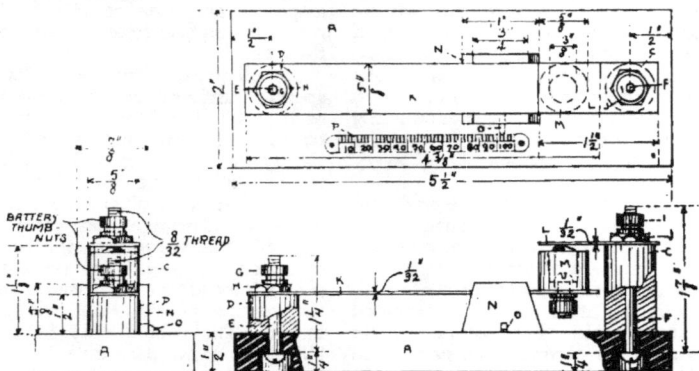

while an $^{11}\!/_{64}$ in. bar is held in the center of it, forming a core, which is removed when the solder is hard.

Procure two 8-32 round head brass machine screws E and F, respectively $1\frac{1}{4}$ and $1\frac{7}{8}$ in. long, with two thumb and hexagon nuts G and H, I and J, to fit each.

Get two pieces of spring brass or phosphor-bronze strips K and L, $\frac{5}{8}$ in. wide, $^{1}\!/_{32}$ in. thick, and $4\frac{3}{8}$ and $1\frac{1}{2}$ in. long, respectively, with a $^{9}\!/_{16}$ in. hole drilled $\frac{3}{8}$ in. from one end of L, and each end of K.

Also obtain a piece of brass rod M, $\frac{5}{8}$ in. in diameter and $\frac{1}{2}$ in. long, with an $^{11}\!/_{64}$ in. hole drilled right through and counterbored with a $\frac{3}{8}$ in. drill, to a depth of $\frac{3}{8}$ in. Pass an 8-32 round head brass machine screw $\frac{1}{2}$ in. long through the small hole, as shown in the illustration, and fasten the sensitive crystal in the cup in the usual manner. You can now pass the screw through one of the holes in K and fasten the cup to the spring with a battery thumb-nut.

When this has been done the screw E may be put through its hole in the base A, and the brass tube D and brass strip K are slipped over it. Then the nut H is screwed over this and G put on to act as binding-post. The same is done with screw F, tube C, strip L and nuts J and I, on the opposite side of the base.

Now the slider N can be made. This is cut out of a piece of hard rubber or black fiber 1 in. long, $\frac{7}{8}$ in. wide, and $\frac{5}{8}$ in. thick to the shape shown, not forgetting the saw kerf to admit the indicator O. This indicator consists of a piece of $^{1}\!/_{16}$ in. sheet brass cut out as shown and forced into the kerf made in N.

A scale P is now to be made of thin sheet brass and divided into 100 parts, as shown.

Then the slider N is placed in the position shown in the illustration and the springs K and L bent so that when N is at the extreme left (O on the scale) the crystal in cup M will not touch L. When, however, N is brought to the right, different degrees of pressure can be brought upon the mineral according to the number shown on the scale by the indicator.

When the contact between L and the crystal in M is to be broken for sending, or for any other reason, the number shown by the indicator must be noted, and the slider pushed to the extreme left O.

When contact is to be restored, N is simply pushed to the right until the desired number is shown by the indicator.

NEWS OF THE WIRELESS SOCIETIES

The Massachusetts Institute of Technology Wireless Society

The Mass. Institute of Technology Wireless Society.—In response to the request for names to aid in the compilation of a directory of wireless amateurs near Boston, as published in the January, 1912, *Electrician and Mechanic*, a considerable number of letters have been received by the Secretary.

It will be of great advantage to Boston amateurs if a complete and accurate list of their stations can be compiled and published; and it is sincerely to be desired that every amateur who possesses a station powerful enough to communicate readily with Boston should send in to the Secretary his name, address, call letters and power. It will also be appreciated by the Society if he will send in the same information about other amateur stations which he knows to be in constant operation.

It has already become impossible to answer all the letters which have been received; but notice will be given when the list is ready for publication, and operators whose names appear in it will be given a chance to secure copies at cost price.

J. H. ELLIS, *Secretary*.

Pittsburg, Pa.—The Signal Corps of the National Guard of Pennsylvania, with headquarters in Pittsburg, will be equipped with two complete United States Army type portable wireless outfits. The issue of the equipment will be made about February 1, and wireless practice stations will be installed immediately.

The armory of the Signal Corps is located on Mt. Washington, south of the city, more than 400 ft. above the surrounding valleys and overlooking the business district. The other stations

The Foot of the Tower

will be placed at other points in the city, and the signalmen will be taught thoroughly in the uses and technical operation of wireless telegraphy. The system will be the same in use on government vessels and in the United States army.

Practice hikes, the transmission of messages and field maneuvers with the cavalry and infantry organizations of the Pennsylvania Guard have been planned, and in consequence there has been a rush for enlistment in the Signal Corps. Among the applicants are many men among the workmen and electrical engineers connected with the big local Westinghouse plants. Capt. Frederick C. Miller of the Signal Corps says he has a big waiting list of applicants and so great has become the enthusiasm for the service that the organization of another company has been suggested.

The first meeting of the fourth year of the Chicago Wireless Association's career was held on Friday, January 12, 1912, at the club room in the Athenaeum Building, Chicago, Ill.

The following officers began their duties at this meeting: John Walters, Jr., *President;* E. J. Stion, *Vice-president;* C. Stone, *Treasurer;* F. Northland, *(Continued on page* 208)

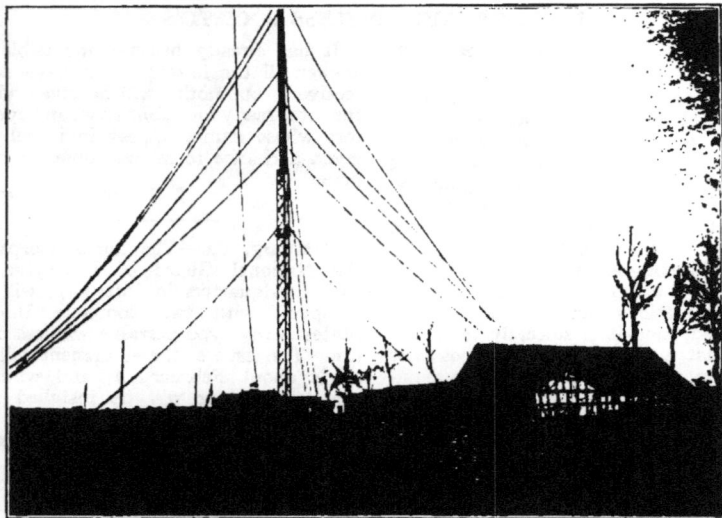

General View of the Highest Station in the World

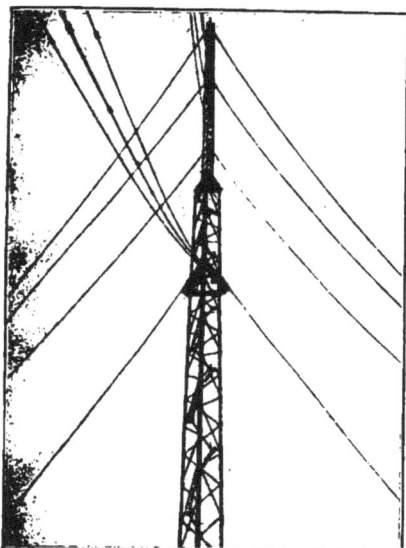

Addition Made Above the Triangle

The Highest Wireless Station on Earth
FELIX J. KOCH

One of the most wonderful achievements in mechanical engineering has been completed lately at Reinickendorf in Berlin. The most striking feature of this wireless station is the 200 meter high mast (about 660 ft.). It was only 330 ft. recently, when, just lately, another 330 ft. has been added. Except for the Eiffel tower in Paris, this station is the highest in the world. Owing to the fact that the power of the machine has been increased four times, this station can communicate with all stations within a range of 6,000 (nearly 4,000 miles) k.m., so that Germany can get easily in touch with all its possessions.

Play Full Game of Checkers by Wireless

A checker game by wireless, the first on record, was played at Minneapolis, Minn., recently, between C. L. Holton and James A. Coles. The distance between the homes of the contestants is two miles. From the time the game was declared on until Mr. Holton took Mr. Coles' last man, the players were in almost continuous communication.

A Steel Cable Anchoring

News of the Wireless Societies

(Continued from page 206)

Recording Secretary; R. P. Bradley, 4418 S. Wabash Ave., Chicago, Ill., *Corresponding Secretary.*

The Association will be pleased to correspond with any wireless club in the country. All communications should be addressed to the Corresponding Secretary.

The Tufts Wireless Society gratefully acknowledges the receipt of copies of the January number of *Electrician and Mechanic*, and sends its appreciation of the notice concerning the society. The station is now working, and we should be very glad to entertain anyone you care to send out.

J. A. PRENTISS, *Secretary.*
3 East Hall, Tufts College.

The Wireless Association of Canada, 189 Harvard Ave., N.D.G., Montreal, Que., Canada, has been organized for the sole object of furthering the interests of amateur wireless telegraphy. Officers: Wm. C. Schnur, *President* and *Corresponding Secretary;* Thomas Hodgeson, *Financial Secretary.* For further information inquire by letter to above address.

WM. SCHNUR, *Corresponding Secretary.*

SPITZBERGEN'S WIRELESS STATION

The Norwegians are making the experiment of maintaining a wireless station at Green Harbor, Spitzbergen. Six men are stationed there, and for half of the year they will be cut off from the rest of the world. Dr. Sigel Roush of Troy gives the following account of the station:

"Strangely incongruous was the upstretching iron arm of this wireless telegraph station, as it gradually, foot by foot, arose, pointing to the sky. It is a Norwegian government project, and when completed will constitute one of the most powerful wireless stations in the world.

"The promoters hope by means of this 300 ft. tower and powerful batteries to be able to span the Arctic and touch by the waves of the atmosphere the ear of the listening instrument on the mainland.

"The great steel tower was braced by scores of metal guys, which were fastened to anchor pins deep-driven into the earth, for the winter winds even in the sheltered cove of Green Harbor are supposed to be terrific. Alongside the tower, and nearly completed, stood the combination office, living and battery house.

"This building was a marvel of construction, exhibiting walls more than 3 ft. thick, composed in order of timber, tar paper, felt and asbestos. Over the roof and around each of the four corners heavy iron cables were passed and made fast, after which they extended to the ground and were anchored in buried foundations of rock and cement.

"Thus every precaution was taken in constructing this desolate habitation to defy the cold and the winds, while in the cellar underneath sufficient food was stored to maintain those six polar prisoners through the weary length of the dismal night.

The government's excuse for such an expensive experiment is found in the explanation of the overseer. Said he:

"'First and most important of all is the value the station is hoped to prove in predicting those severe northern storms that, especially in winter, sweep unheralded down the Norwegian coast and work tremendous damage to Norway's shipping. Our fishing, is done mainly in winter, and thousands of fishing craft frequent the western coast to follow the one means of support to vast numbers of our people. If by means of this Arctic station we can give due warning of a destructive storm, the government, if for no other reason, would be justified in the investment.

"'Then again, the Norwegians have always been among foremost polar explorers. By means of this wireless station polar parties equipped with their own outfit can be traced and watched, and in case of danger can be located and reached. Greater risks can therefore be taken, and greater results be obtained.

"'Then, there is the commercial phase of the subject, for coal of a fine quality has been found in Spitzbergen, and even now an American-Norwegian company is mining coal here and shipping it to the Continent at a handsome profit.'"

EDITORIAL

It is with a feeling of deep regret and personal loss that we find ourselves obliged to chronicle the unexpected death of Mr. William C. Getz, well known as a wireless expert. Mr. Getz had been ordered to the Philippines to take part in the erection of government wireless stations there, but was taken ill just after the transport left the Hawaiian Islands. Upon arrival at Manila he was so ill that the hospital authorities ordered his return to the United States. He grew steadily worse on the return trip and died in San Francisco on the 13th of January, 1912, a few days after landing. Mr. Getz leaves a widow and an infant daughter, to whom we are sure the sympathies of our readers will go out.

Although Mr. Getz would not have attained his twenty-fifth year until March 10th, he was an expert electrician and an authority on wireless installations. His early life was spent in Baltimore, and he began experimenting with wireless telegraphy among the earliest of the American amateurs. Becoming proficient in the days of needle detectors, he began manufacturing apparatus for sale to amateur friends and evolved a complete series of improved, tuned circuit instruments. These instruments he described for *Electrician and Mechanic's* readers in a series of articles which were the first published for amateurs on the subject of tuned circuit instruments. From the publication of this series dates the wide-spread interest of amateurs in wireless in the United States, and subsequent writers, as well as many builders of apparatus, have found a mine of information in these articles. Through their publication the attention of the United States Signal Service was called to his achievements and he was asked to enter the government's employ. This rendered it necessary for him to dispose of his rapidly-growing electrical instrument business, and as a signal service operator he traversed the entire United States, erecting and testing stations for the government, and it was in pursuance of his duty in this direction that he started on the trip which ended fatally.

Personally Mr. Getz was modest and unassuming, and in spite of his manifold duties his knowledge and time were always freely at the command of every earnest seeker after information.

Electricity's Part in Photography

The primary element of photography is light, either the bright, penetrating light of day or the intense artificial light of the electric arc. But aside from the boon of artificial light, electricity is a wonderful help to photography. During the rush seasons it is often necessary for a good photographer to work night and day to keep up with his orders. In this case, of course, all the night photos are made under the rays of powerful electric lights. But the simple taking of negatives is only the beginning of the artist's hard work. The plates have to be developed, dried, retouched and reprinted. Where there is time enough and to spare, the negatives are merely racked up and allowed to stand until they are dry. Often it is desirable to print up the orders quickly, and in such cases electricity is brought in to dry the plates. A simple way to effect this is to let the breeze from an electric fan play over the negatives, and other photographers have used to good advantage the electric hair dryers. This is a device combining an electric air heater and a small fan. This steady current of dry, hot air, will speedily dry the wet plates and prepare them for quick printing. An electric light is also used by the retoucher to bring out the detail of the negative. The printing itself can be done just as well and with a great deal more certainty beneath the rays of large arc lamps. Again the electric fan or the hair dryer is used, when rush orders are demanded, to dry the developed prints, and a small electric flat-iron is used to smooth out the prints in place of the regular press.

Photo Wireless Station by Gets

Transmitting side of the United Wireless Co.'s station at Hill Crest,
San Francisco, which recently worked with a station in Japan.

We print herewith a bill of materials which, through an accident on the part of our printer, was omitted from the February issue, it being a necessary part of the article entitled "A 100-Watt Step-Down Transformer." In the meantime we have discovered that the major part of this article had been published in an earlier issue, by *Modern Electrics*. This we find was due to the fact that the author had submitted nearly identical articles to both publications, and in our case the type matter has been standing for some months. The bill of material follows:

BILL OF MATERIAL

	Name	Material	No. Req.	Remarks, Sizes, etc.
No. 1	Core leg	Soft sheet iron		Enough to make a pile $2\frac{1}{2}$ in. high
No. 2	Core yoke	Soft sheet iron		Enough to make a pile $2\frac{1}{2}$ in. high
No. 3	Spool tube	Hard fiber $\frac{1}{16}$ in.	2	Bend into a square tube, $1\frac{1}{4}$ in. inside
No. 4	Spool end	Hard fiber, $\frac{1}{8}$ in.	4	Put 1 on each end of spool tubes
No. 5	Insulator	Hard fiber, $\frac{1}{64}$ in.	2	Bend into a square tube, $1\frac{3}{8}$ in. inside
No. 6	Clamp-bar	Hard wood	4	
No. 7	Bolt	M. steel	4	Standard $\frac{5}{16}$ x 4 in. square-headed bolt
No. 8	Base	Wood	1	Make to suit requirements
No. 9	Washer	Copper	8	For bolts No. 7
No. 10	Wire	Copper		570 ft. No. 20 B.&S. d.c.c. for primary
No. 11	Wire	Copper		77 ft. No. 10 B.&S. d.c.c. for secondary

100-WATT STEP-DOWN TRANSFORMER

Primary voltage, 100–110   Primary turns, 980
Secondary Voltage, 10–11   Secondary turns, 100
Frequency, 60 cycles   Weight, about 12 lbs.

QUESTIONS AND ANSWERS

Questions on electrical and mechanical subjects of general interest will be answered, as far as possible, in this department, free of charge. The writer must give his name and address, and the answer will be published under his initials and town; but, if he so requests, anything which may identify him will be withheld. Questions must be written only on one side of the sheet, on a sheet of paper separate from all other contents of the letter, and only three questions may be sent at one time. No attention will be given to questions which do not follow these rules.

Owing to the large number of questions received, it is rarely that a reply can be given in the first issue after receipt. Questions for which a speedy reply is desired will be answered by mail if fifty cents is enclosed. This amount is not to be considered as payment for reply, but is simply to cover clerical expenses, postage and cost of letter writing. As the time required to get a question satisfactorily answered varies, we cannot guarantee to answer within a definite time.

If a question entails an inordinate amount of research or calculation, a special charge of one dollar or more will be made, depending on the amount of labor required. Readers will, in every case, be notified if such a charge must be made, and the work will not be done unless desired and paid for.

1730. Detector Connections. C. A., New Rochelle, N.Y., asks: How can I connect two mineral detectors so that I can receive a message on one and then switch off on the other detector and receive the same message? Ans.—By the use of a two-point switch with its arm connected to the instruments in place of one of the detector

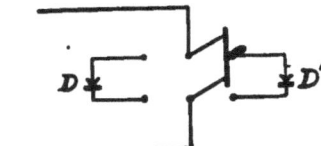

D.D'-detectors

terminals. One terminal of each detector would be connected to the respective switch points, while the remaining detector terminals are bridged with a piece of wire which is in turn connected to the instruments in the usual manner as per diagram herewith.

1731. Loose-coupled Transformer. X. Y. Z., Springfield, Ill., asks: (1) Is a coil with a vibrator ever called a make-and-break coil; if not, what is? (2) Would like to have data for making a protection device to use in the primary of a wireless transformer. (3) I have a tuning coil 3½ in. in diameter, 14 in. long (double slide), 100 ft. aerial, 150 ft. high, composed of four aluminum wires spaced 3 ft. apart; 2,000 ohms Holtzer-Cabot receivers, Clapp-Eastham ferron detector and condensers. I can hear amateurs all over this city, but cannot hear any station such as a commercial out of town. Do I need a loose-coupler, or what? All joints soldered, etc. Ans.—(1) Yes, but not usually. (2) If you mean to protect the primary of a wireless sending transformer, a fuse is the proper thing; but if you mean a receiving transformer, use a lightning arrester. (3) Your apparatus is all right for local work, but your longer-distance receiving would be much improved if you would use a good loose-coupled transformer.

1732. Induction Coil. D. W. D., Montrose, Col., asks: (1) What is the rating of this coil in watts: Core 1 x 9 in., primary two layers of 12 d.c.c., secondary 6 lbs. of No. 30 enameled, to be run on 110 volts, if it was made a closed-core transformer? (2) How big a primary condenser should be used? A secondary shunted around spark gap? (3) How big and thick a spark should this coil give? (4) Where can I purchase a good set of wireless instruments for receiving? Ans.—(1) It is very difficult to give the voltage of your coil, either as open-core induction coil or closed-core transformer. Perhaps the output in the first case would be 1-12 k.w., and in the second about double this. (2) As an induction coil your primary condenser should have a capacity of about 1 or 2 m.f., the exact amount not being very important. A secondary condenser is always needed if you wish to use tuning instruments. (3) As an induction coil, using batteries, but not the 110 volts, you could probably get a thin 2 in. spark, but, as a closed-core transformer on alternating current, not over ¼ in., but very thick. (4) There are many good, reliable concerns now making wireless apparatus. Send to some of the advertisers in this magazine for their catalogs.

1733. Pump. H. D. K., Erie, Pa., is proposing to make a table fountain, to be operated by a small pump and electric motor, and asks several questions as to the practicability of the scheme. Ans.—Of course a double-acting pump should be used, but the one to which you referred, described in the March, 1909, magazine, is rather an extravagant construction, except for a person who has access to a plumber's scrap heap. It would be well for you to get a catalog of a manufacturer of small hand pumps, and the address of such can easily be obtained by consulting such periodicals as you may find in a plumber's or steam fitter's office, and observe some of the simple cuts displayed in the lists. If you have tools, you can easily construct patterns and make a pump of your own design. You might be able to purchase a second-hand one at a garage, perhaps of the sort used for getting up pressure in the tank of a Stanley steam carriage. Your proposition to drive such a pump by use of a fan motor is not practicable, for such a motor running without its fan gets very hot. You would need a small "power" motor.

1734. Solenoid. S. O. H., Mosgrove, Pa., writes that he is constructing an apparatus which makes use of a large number of small solenoids in a very restricted space, and would like to have some information on the subject. What will be the dimensions of a solenoid which will give a "pull" equal to the lifting of a weight of 5 or 6 oz., not including the weight of the core? What size wire should be used, and how much? Will want to use about 10 or 12 volts furnished by dry cells. The solenoids will have to be as small as it is possible to make them, and at the same time give the above requirements. Ans.—Such a design would involve considerable calculation and perhaps experiment. We would advise you to procure a copy of Underwood's book on the "Electromagnet," which we would be pleased to furnish at $1.00. Mr. Underwood has made a special study of such topics and has been the chief designer for a large manufacturing company.

1735. Armature Construction. W. H. B., Chicago, Ill., asks: For a book in regard to armature and field calculations. I want a book that will tell me just how to calculate the size and amount of wire necessary for armature and fields of motors and dynamos, also how to calculate how much current a dynamo is generating. Ans.—This question involves the whole field of dynamo design, and no one book or short course of reading will give the information desired. Underwood's book on the "Electromagnet," Hobart's "Armature Construction," and "The Electric Motor," are first-class, but perhaps beyond the scope of the enquirer. I would recommend that he take a course in direct-current dynamo design in one of the Correspondence Schools.

1736. Wireless. H. A. W., Chicago, Ill., asks: Would consider it a great favor if you would give me the facts of possibilities of danger in a receiving station in wireless telegraphy. I expect to use your statement as a standback in obtaining permission in erecting a receiving station. Ans.—We believe there would be no danger in receiving stations in wireless telegraphy except from lightning, and in this, as in other similar stations, it is well to show prudence, and we would advise grounding aerials outdoors during a thunder storm.

1737. Spark Coil. S. R. W., Stoneham, Mass., asks: I have purchased a home-made coil, the design of which I do not like, especially as regards the core which is 6 in. long and 1¾ in. in diameter. Independent interrupter is used. Do not know the dimensions of primary, but it is insulated from secondary by ordinary mailing tube. The secondary contains 3 lbs. of No. 32 and ½ lb. of No. 36 B.&S. wire, and is wound in seven sections. Distance between heads 5½ in. At present gives ⅝ in. spark between needle points. Will you kindly advise me as to the best way to rebuild this coil so as to give greatest efficiency for wireless purposes? Ans.—Your coil is not appropriately designed for good work—the coil is too large in diameter for its length. You will get 3 to 4 in. spark with the amount of secondary you use. An excellent series of articles on a 6 in. coil were written by Mr. Thomas Stanleigh in magazine numbers for February, March, April and May, 1911, and an article particularly on the insulation of coils in the May number, 1910.

1738. Wireless Diagram. F. M. L., Portland, Ore., asks: (1) Please tell me where I can find a diagram for the following set: aerial, six aluminum wires No. 14 B.&S., 2 ft. apart, 60 ft. high, 75 ft. long; three slide tuning coil, No. 24 B.&S. enameled wire on core 3 in. in diameter, 9 in. long; silicon detector; variable and fixed condensers. (2) The wave length of the set. (3) How can I improve the set? Ans.—(1) See Fig. 48, page 97, of the "Manual of Wireless Telegraphy for Use of Naval Electricians," 1909 edition. This will give you a choice of several

diagrams employing the instruments mentioned. (2) Impossible to state, owing to variation caused by local conditions. (3) By using an inductively coupled tuner.

1739. Slide Wire Bridge. R. F. A., Carmine, Tex., asks: I wish to make a resistance set for a slide wire bridge with steps as follows, viz., 5, 10, 25, 50, 100, 300, 500, 1,000 and 5,000 ohms, and wish to know what size German silver wire to use. (2) What is the largest number of telephones that can be used on one line (bridging); phones are equipped with 5 bar generators and 2,500 ohm ringers? Ans.—The resistance of German silver wire varies greatly with the percentage of nickel contained in it. Since this is very uncertain in ordinary German silver wire, you cannot make coils which will be close enough to the values you desire to make reliable standards. You had better purchase the coils.

1740. Storage Battery. H. McC., Omaha, Neb., asks: (1) How to remove sulphate from the plates. (2) Why is it allowable to speak of positive and negative terminals of an induction coil, when the current is confessedly alternating? Ans.—(1) Several means are common. One consists in charging the cells for a considerable time, but with a small current, in the reversed direction. This is also an effective remedy for restoring the capacity of old negative plates. Another method is to charge the cells in the usual direction, using a caustic soda solution in place of the acid. After restoring the proper color of the plates, the soda must be washed and soaked out, and the ordinary solution replaced. Still another method that is about as cheap, and eminently successful, is to throw the sulphated plates away, or sell them to a junk man, and buy new ones. (2) It is true that the currents are alternating, but the impulse following the break of the primary circuit is so much greater than that following the make, that the above expressions are understood.

1741. Switchboard. A. H. M. G., Toronto, Can., has two direct current generators that operate in parallel. One of these is of twice the ampere capacity of the other. It is proposed to install a third machine, of twice the capacity of the larger now in use. From limitations of space, the switchboard for the new one will be 10 ft. distant from the others. What will be the best way to regulate the voltage when "cutting in" or "cutting out"? Ans.—You did not state whether the machines are compound or plain shunt. If the latter is the case, you will of course need no equalizer. If compound, the new panel may well have one double-pole switch for the positive side of line and equalizer, and a single-pole switch for the negative line, ammeter being included in this side. Only one voltmeter for the three machines should be used, and this mounted on a bracket in such a manner as to be visible from the three panels. Bus bars will, of course, extend between the two boards. A dummy hand wheel can be placed on the present board and connected by means of a sprocket chain to the rheostat on the new panel. All the operations, except closing and opening the main switches, can, therefore, be controlled from the present position. If this suggestion is not feasible, we will be pleased to explain some other means, though not so simple or cheap.

1742. Train Lighting. E. W. M., New Springfield, N.Y., asks how trains are electrically lighted. Ans.—This is one of the most troublesome problems in electrical engineering. The three methods you ask about are all used, but the expense is considerable, and it is difficult to see how the railroad companies can afford them for all trains. Each has its advantages and disadvantages. A small generator in the baggage car or on the locomotive may be used, but when shifting locomotives the lights go out. Some trains use storage batteries alone, but without special attention the voltage does not remain sufficiently constant, and for long runs, the batteries are heavy and expensive. The scheme of letting each car have a small dynamo driven from the axle, in connection with a storage battery is largely used, and with increasing success. Only during stops at stations are the batteries called upon, therefore they need not be of large size. Automatic devices connect or disconnect the cells for proper charging.

1743. Loss of Voltage. G. C., Upper Sandusky, O., asks several questions about the flow of current in a conductor, and as to whether all the current comes back to a generator. Ans. —Your difficulties come from a serious confusion of the meaning of volts and amperes. It is certain that all the current a machine generates comes back to it. This is easily proved by putting two ammeters in circuit,—one in each lead from generator. If the instruments are right, they will both indicate the same. A single ammeter may be used, first in one main, then in the other, but you might not wish to interrupt the service during the change, or there might be some doubt if the actual load was the same at the two different times. Your trouble comes from considering how the whole 240 volts are used. It is a simple idea that the lamps may get 220, for they purposely have a high resistance, while the line must require some volts, for there is no perfect conductor. The 20 volts constitute the "line drop" in pressure, just as there is a drop in pressure in a water system when a large draught is made. All the water that is pumped into the pipes, after they are once filled, has to come out. That constitutes the current (of water). In the electrical case there is an analogy that matches the requirement of first filling the pipes, for a charging current is initially supplied to the line,—of no consequence for low voltages, but of measurable and serious effect in high voltage systems. Your question is merely how the expenditure of 240 volts is distributed in the system. The dynamo may really be generating 260, for there is an appreciable waste in the armature due to its resistance. There is a loss of about 1 volt at each set of brushes. There may be equal or greater losses in the connections and instruments at the switchboard.

1744. Choking Coil. W. J. K., Lombard, Ill., asks for directions for making such a device for controlling the lamp for a moving-picture apparatus, to pass a current of from 30 to 50 amperes, when taken from 60 or 133 cycle circuits at 110 volts pressure. Also would this coil be more appropriate than a transformer? Ans.—The two-coil transformer would not answer, for at the starting of the arc, a practical short circuit would exist on the secondary, and the primary always imitates what the secondary is doing. You would have to have a resistance or choking coil in this secondary circuit, so the best thing is to comprise the double function of voltage reduction and control in one device— *i.e.*, the choke coil. Sheet iron 28 x 84 in. may be obtained and cut into 7 x 12 in. sheets, a sufficient number being used to give a stack 2¾ in. high. Remove a piece 2½ x 9¾ in. from each, as cut from the narrow end, so as to leave square cornered U-shaped pieces, 2¼ in. wide in every part, with an opening 2½ in. wide. Shears will take two of the cuts, while the third may be managed with a sharp cold chisel against the edge of a vise. From half the sheets clip off one of the ends 2¼ in. shorter than the other. The purpose of this will appear later. Paint the sheets with thin asphaltum, and clamp them together, the short ones alternating with the long, but with the latter all on the same prong, and bind them with a complete layer of strong twine, excepting on the end where the sheets are incomplete. Wind tough Manila paper on the other prong, to a thickness of ⅟₁₆ in. Wind, either directly in place over this insulation, or else on a suitable form, nine layers of No. 10 d.c.c. wire, five wires in parallel, eleven turns (55 wires) per layer. This sub-division of the conductor will be easier to wind than a single conductor of equivalent section, and will greatly reduce the heat resulting from eddy currents. Now slip the pieces of sheet iron that were cut from the center in between the openings in the exposed prong until they come within ¼ in. or ½ in. of the other end of the U. This provision gives the adjustment you need for meeting the varying conditions, for this separation, or air gap, in the magnetic circuit will prevent the iron from saturating with too small currents. If the hum resulting from the vibration of the sheets is troublesome, you can fill in the empty spaces in the adjusting strips with cardboard, and clamp the mass together. Do not use any volts, however, through the sheets, nor indeed outside the sheets in the vicinity of the air gap.

CORRESPONDENCE

Dear Sirs: In the February number on page 139, a question was asked by R. B., Oak Park, Ill., for the dimensions of a 2 k.w. transformer, and, having built one of the said capacity, I wish therefore to state its dimensions which I hope I shall see in print: length of iron, 17½ in.; width of iron, 8¾ in.; thickness and width of core, 2¼ in.; primary wire, No. 8 d.c.c.; length of primary coil winding, 10¼ in.; primary wire d.c.c. 13½ lbs., three layers; 18½ amperes taken at full load; No. 28 secondary wire d.c.c., 21 lbs.; 30 pies ¼ in. thick, 920 turns per pie; size of square opening in pie 2¾ in.; about 15 amperes lowest voltage; range, 300 miles, doubled by night or over water.   Yours very truly,
WM. STENGLE.

P.S.—Finally, I wish to state that any person without much knowledge of making transformers would better not attempt it, as it will most surely result in failure.   W.S.

Editor of *Electrician and Mechanic.*

Sir: In reference to your editorial in regard to the wishes of subscribers as to the subjects they would like to have you publish in your very valuable magazine for the ensuing year. I, for one, would like to see some of the subjects you mention in your article, such as mechanical drawing, with the mathematics to go with the different designs; telephone engineering, with the troubles met with in every-day practice, with the addition of an article on the every-day troubles of the dynamos and motors; one on storage batteries; arc rectifiers; rotary converters; also an article on electrical calculations of wiring for lighting and motors for single-phase, two-phase and three-phase distribution.

As I find that a very large number of men working at the electric business today do not know how to figure wire at all and those that know how to figure common wiring do not know anything about three-wire service or the figuring of two and three-phase calculations.

There does not seem to be at the present time a practical magazine on the market other than yours that seems to cover the subjects that I mentioned in a practical way for the every-day workman is installing all classes of this apparatus all the time. If some magazine would treat these subjects in a good practical way that one could understand and so that it would be of some practical use to them, I think that the magazine would find a very large increase in its sales.

I have been buying your very valuable book now for about six months at the newstands, and find it as near to my idea of a magazine for the practical every-day worker as I have found so far, and I have been a subscriber to several other technical magazines for years, and they either do not go deep enough into the subject, or it is made too much of a laboratory and not a practical treatise of the subject.

I am a practical electrician, working at the business every day and belong to the Electrical Worker's Union, and I know that there are hundreds of men working at the business that are very anxious for a good practical magazine on these subjects.

Respectfully yours,

FRED'K. A. COKER.

Editor of *Electrician and Mechanic.*

Sir: In the January, 1912, issue of the *Electrician and Mechanic* I noted under the heading "Correspondence" the item by Mr. Cecil A. Wanner, regarding the present form of the *Electrician and Mechanic* as compared with the "old style," as he terms it; also his suggestions to you that the magazine should specialize in certain articles.

I heartily congratulate the management and publishing department of the *Electrician and Mechanic* on the many excellent articles that are published monthly in the magazine. The articles are written in simple semi-technical language, and a feature which I think makes them extremely useful and practical is that mathematics is employed in sufficient quantity to make them a real tool in every-day work.

This is a marked contrast to the many so-called popular articles in which one is usually at a loss to know just what the author means to convey. In my experience in the mechanical field, and in my work as civil engineer I subscribe to a number of magazines and journals, each treating a special line of work, but I enjoy to peruse the pages of the *Electrician and Mechanic* for the many practical articles of a *general* interest, in which I always find some suited to my particular needs.

In answer to the inquiry of the publisher under the caption "Editorial," I beg to say, that a series of articles in mechanical drawing and simple machine design, including elementary practical mathematics, ought to prove of general interest and value to a great number of readers, also the suggested articles as further named. Personally, I always appreciate an article on some subject of applied science. Articles on marine engines, steam, gas and water turbines and oil-consuming engines will interest me particularly.

It seems to me that your correspondent ought to remember that not all of the subscribers are interested in wireless telegraphy, etc., and that the *Electrician and Mechanic* in its present form during the period of the year provides a good many articles of special interest to the varied tastes of its subscribers and fills a niche of its own in the Temple of Modern Science and Invention.

Yours very truly,

C. O. THON.

Editor of *Electrician and Mechanic.*

Sir: May I, as a foreign reader of your excellent magazine, be permitted to give my opinion on the paper in general—the articles contained therein?

I am inclined to agree with your correspondent, Mr. Wanner, whose letter you published in the January number, that perhaps the style of your former issues was preferable to that now followed. Such a number of different subjects are now treated that not enough of one is given at one time. No doubt the diversity of subjects tends to increase the subscription lists, but it is at the expense of readers such as myself, who are desirous of knowing all about one subject. However, barring this one growl, I have nothing but praise for the lucidness, conciseness and general excellence of all your articles. Of all the papers I read, yours caps the lot for general interest and usefulness, and I must congratulate you heartily. Personally, I should like to see more space devoted to Wireless Telegraphy for Amateurs and to the construction of electrical apparatus for amateurs; but of course that may not be the opinion of the majority.

With regard to your editorial, a department for Telephone Engineering would be exceedingly interesting; also the articles on Mechanical Drawing. The other departments suggested have little interest for me.

Wishing you all the success your paper deserves, I am,

Yours truly,

LAURENCE D. HILL.

TRADE NOTES

This year there will be held in Boston the largest Electric Show that America or the world has known.

The 1912 Boston Electric Show—international in scope—will occupy the entire Mechanics Building, which is one of the largest permanent show buildings in the world. The whole plan and scope of this great Show will make it the most complete and most comprehensive electrical exposition ever seen. The area occupied—105,000 sq. ft. of exhibit floor space—is twice as great as that of any previous show.

Duration—Five Saturdays and four full weeks—twice as long as any previous show.

The exhibition has already been advertised in every civilized country in the world where electric machinery is manufactured.

Every single electric light company in New England will actively lend its co-operation to the Boston Show, for the benefit that will come to its own business.

During the entire month of the Show there will be excursions continually operated from all New England points into Boston, both by train and trolley.

The Edison Electric Illuminating Company of Boston stands its sponsor, and acts as financial manager of the Show, not intending, however, to make a profit either for itself or for any individual, or group of individuals, but is undertaking this huge Exhibition to illustrate the practical adaptation of electric current for every walk in life.

An idea of the scale on which this Show has been projected can be gained from the fact that the exhibition itself has been planned on a basis of securing at least 250,000 paid admissions, in order to recompense the management for the actual expenses incurred before opening the Show.

The tremendous way in which this entire enterprise has been laid out is based on the fact that big things succeed best.

The opportunity of making an exhibition of your apparatus upon a wholesale scale at The 1912 Boston Electric Show is not just unusual—it is unprecedented.

The Department of the Interior, Bureau of Mines, announces the following new publications. (List 7—December, 1911.)

Bulletin 6.—Coals available for the manufacture of illuminating gas, by A. H. White and Perry Barker. 1911. 77 pp. 4 pls.

Bulletin 16.—The uses of peat for fuel and other purposes, by Charles A. Davis. 1911. 214 pp. 1 pl.

Bulletin 19.—Physical and chemical properties of the petroleums of the San Joaquin Valley, California, by I. C. Allen and W. A. Jacobs, with a chapter on analyses of natural gas from the southern California oil fields, by G. A. Burrell. 1911. 60 pp. 2 pls.

Reprints

Bulletin 21.—The significance of drafts in steam-boiler practice, by W. T. Ray and Henry Kreisinger. 62 pp. Reprint of United States Geological Survey Bulletin 367. Copies will not be sent to persons who received Bulletin 367.

Bulletin 26. Notes on explosive mine gases and dusts, by R. T. Chamberlin. 67 pp. Reprint of United States Geological Survey Bulletin 383. Copies will not be sent to persons who received Bulletin 383.

Bulletin 29.—The effect of oxygen in coal, by David White. 80 pp. 3 pls. Reprint of United States Geological Survey Bulletin 382. Copies will not be sent to persons who received Bulletin 382.

Bulletin 30.—Briquetting tests at the fuel-testing plant, Norfolk, Va., 1907–8, by C. L. Wright. 41 pp. 9 pls. Reprint of United States Geological Survey Bulletin 385. Copies will not be sent to persons who received Bulletin 385.

The Bureau of Mines has copies of these publications for free distribution, but can not give more than one copy of the same bulletin to one person. Requests for all papers can not be granted without satisfactory reason. In asking for publications please order them by number and title. Applications should be addressed to the Director of the Bureau of Mines, Washington, D.C.

New Cunarder for Boston Service

The *Laconia*, the new Cunarder, built primarily for the Boston-Liverpool trade, left Liverpool January 20th, on her maiden trip for New York, from which port she makes a cruise to the Mediterranean before entering the Boston service, on March 26.

The *Laconia* was launched July 27, 1911, and next to the *Mauretania*, she and her sister ship the *Franconia* are the two largest ships that have ever been constructed on the river Tyne. The leading dimensions of the *Laconia* are 625 ft. in length, 72 ft. breadth; her gross tonnage is about 18,100, and her displacement 25,000 tons. To add to the comfort of passengers by increasing the steadiness of the ship, Frahm's anti-rolling tanks have been installed. The *Laconia* is the first British ship and the first North Atlantic liner to be fitted with these tanks.

In internal arrangements the *Laconia* is as comfortable as anyone can desire. The cabins, as well as the public rooms, are lofty, spacious, well ventilated and heated. The general style of decoration in the ship is Georgian, known in America as "Colonial." The second-class accommodation has been carefully planned, and is in every way equal to what was provided for first-class passengers only a few years ago. The third-class passengers have also been well catered for. Throughout the ship every care has been taken in minutest details to ensure the safety and comfort of all on board, and it is confidently expected that the *Laconia* will enhance the reputation of both her owners and builders for producing ocean steamships which take their place in the very front rank of modern liners, and will prove equally as popular as the *Franconia*, which vessel during the first eight months she has been in commission has carried over 21,000 passengers, a sufficient proof of the public's favorable verdict.

These steamers will appeal to all New Englanders, since they are the largest steamers to enter Boston Harbor, and mark the first step towards re-establishing Boston as one of the great Atlantic ports.

The Globe Ear-Phone, an advertisement of which appears elsewhere in the pages of this magazine, is really the same as a telephone of commerce. A microphone transmitter and an electromagnetic receiver, connected by small flexible cords with a source of electric current. The transmitter is worn on a chatelaine hook, or a loop of any sort, on the clothing. In particularly favorable cases, it will be even satisfactory for hearing in a church or at the theatre, but this is rather an extreme test for the portable instrument and calls for a special church earphone equipment (not portable). The receiver is usually worn with a head band. Worn in this way, it is not very conspicuous, especially since the transmitter may be worn underneath any ordinary clothing, and the battery is carried in a pocket. It is a very effective, and, at the same time, a very convenient form of hearing instrument. The effectiveness is caused partly by the simple increase in the intensity of the sound, and partly by the direct application of the sound impulses. Furthermore, it is effective on account of the slight sharpening of every part of the wave without otherwise altering its form or phase.

It has been stated in Boston, very recently, that a woman in one of the nearby suburbs who has attended church for twenty years, heard her first sermon in fifteen years by using one of the church instruments, through which she heard every word. A small transmitter at the pulpit caught the words of the sermon and invisible wires carried the sound to the receivers which were in the church pews.

The above-mentioned Company are looking for agents to further the sale of this excellent working apparatus for deaf people, and full information can be obtained by addressing the Company at Boston, Mass.

BOOK REVIEWS

"The Story of the Slide Rule" is the title of a little booklet issued by Geo. W. Richardson of Chicago, and in which is given a description of his new Direct-Reading Slide Rule. The slide rule has been in use for more than 300 years, and is a most valuable instrument to any individual who has to do with figures, regardless of whether he is a technical or a practical man. The Richardson Slide Rule has embodied in it all of the latest slide rule improvements, and combines in a single simple instrument a great deal of the information that the ordinary practical engineer would need to look up in his handbooks. Unlike most slide rules, the Richardson is quite inexpensive, costing but $2.50. The stock and runner of the rule are made from aluminum, while the scale is made of celluloid, highly polished and washable. Since there is no wood in its construction, the rule is free from any changes which wooden rules suffer caused by the condition of the atmosphere. The distinctive feature of this rule is that more than forty every-day problems may be solved on it with but a single setting for each. This is accomplished by means of keys and key-holes. All that is necessary is to place the key of the problem wanted (reference to which is found on the back of the rule) under the key-hole, and the answer to the problem is found directly upon the face of the rule without any further manipu-

lation of the rule. The rule is sold in a durable cloth-bound case, is about 11 x 1¾ x ¼ in. in size, and a book of instructions (written in clear language) is furnished with each rule.

We are in receipt of ten most instructive and interesting treatises upon the use and care of the various parts of self-propelled machines, such as motor cars, balloons and aeroplanes. These various works are all reprints of articles which have appeared in the *Automobile Journal*, and are written by Victor W. Page, M.E. The text is simple and non-technical while the illustrations are especially good. All of the different articles are bound in paper covers and range in price from 25 cents to 50 cents. A list of them is as follows:

The ABC of Motor Car Operation	$0.50
Overhauling, Rebuilding and Equipment of the Motor Car	.50
The ABC of Aerial Navigation	.50
The ABC of Internal Combustion Engine, Maintenance and Repair	.35
The ABC of Carburetor Construction, Maintenance and Repair	.35
The ABC of Magneto Systems	.35
The ABC of Battery Ignition Systems	.30
Lighting the Motor Car by Electricity	.25
Maintenance and Repair of Motor Car Tires	.25
The ABC of Motor Car Chassis, Maintenance and Repair	.25

The publishers are "The Automobile Journal," Pawtucket, R.I.

The Second Boy's Book of Model Aeroplanes. By Francis A. Collins. New York, The Century Co., 1911. Price, $1.20 net.

This interesting book is a most worthy successor to the author's earlier book, "The Boy's Book of Model Aeroplanes," and brings up to date both the science and sport of constructing and flying model aeroplanes. Drawings and photographs of over a hundred new model aeroplanes are shown while detailed instructions are given for building fifteen of the latest types of these little machines. Mr. Collins' entire treatment of the subject is simple, straight-forward and thorough, and the book will undoubtedly appeal to everyone who is interested in experimental aeronautics.

Hand-Forging and Wrought-Iron Ornamental Work. By Thos. F. Googerty. Chicago, Popular Mechanics Co., 1911. Price, $1.00.

While many books have been written on the subjects of hand-forging and wrought-iron work, it is doubtful if any of them treat these subjects in a more helpful or practical manner than is done by Mr. Googerty. The illustrations used throughout the book are particularly clear and well chosen; the correct positions of holding the tools when working being clearly shown. Art-Craft Ironwork is steadily increasing in its popularity, but it has been difficult for the amateur to obtain any text-book which clearly pointed out the principles and methods which underlie all of this type of work. This difficulty, however, at last seems to have been overcome.

ELECTRICIAN & MECHANIC

VOLUME XXIV APRIL, 1912 NUMBER 4

LATHE WORK: MILLING ATTACHMENTS—Part IV

"SIGMA"

The writer feels that an apology is due those readers who may have been interested in my former articles, for the unreasonably long wait for this installment, but personal affairs have not the sooner allowed me the time necessary to their continuance.

It will be remembered that the former installments on milling attachments which appeared in the March and April issues of *Electrician and Mechanic*, Vol. XXII, described a simple milling spindle that was intended to be bolted to the tool rest and driven from an "overhead," such as described in the April issue referred to above; also reference was made to two other types of milling attachments, viz.: the vertical slide or elevating spindle and the elevating milling slide rest. This installment describes the construction of the elevating spindle type.

The design submitted is original, so far as the author is concerned or as far as such things can be original, and embodies all the features and ideas gained by a couple years' experiment with several different designs.

The great fault of many appliances of this kind is their lack of material where strength is essential; that is, they are perhaps strong enough structurally, but too often no provision is made in the way of excess material to keep the deflections within reasonable limits; as, for instance, consider a boring bar that is required to stand a tool pressure of 50 lbs. at 10 in. from the point of support. The bar would be sufficiently strong, considered merely as a cantilever beam, if made 1⁵⁄₁₆ in. in diameter, but if an accuracy in its boring of ±.001 in. was required it would have to be made 2³⁄₁₆ in. diameter, because the smaller bar would have so

great a deflection at the assumed tool pressure as to be unable to turn out work with an accuracy greater than ±.006 in.

I merely make this digression to answer the criticisms of any who may object to the apparently liberal use of material in the design, and before leaving this discussion I wish to point out that the weak point in many small lathes is their tool rest or saddle design, which are too often poorly designed, thus limiting the use of the lathe and any appliances that may be attached thereto. Though this attachment was originally designed for my own 10 in. lathe, it is as strong relatively as the slide rest of many 13 to 15 in. lathes; and while an attachment of smaller dimensions would be advisable, nevertheless it can be used in a limited way on 9 in. lathes.

SPECIFICATIONS

Spindle No. 1 Morse taper socket, ⁵⁄₁₆ in. hole clear through, is ¾ in. by 5½ in. in ball bearings and will take every tool and cutter made with this taper, while No. 1 taper shank arbors with blank ends (¹¹⁄₁₆ in. diameter by 1⅛ in. long) can be obtained and turned down to fit the various small milling cutters having holes to ⅝ in. diameter. Vertical movement of spindle 4 in. The attachment is back-geared and by the use of eight gears a total speed ratio of 16 to 1 can be obtained relative to any one speed of driving belt, which, together with the various speeds that may be applied to the back-gear shaft through the driving belt, gives a wide range of available speeds for the spindle. The gears required are one 1 in., one 1½ in., one 2 in., two 2½ in., one 3 in., one 3½ in., and one 4 in., costing 30, 38, 60, 65, 70, 80 and 90 cents each, total $4.98. These gears are all 16 pitch with $\frac{1}{2}$ in. hole by $\frac{1}{2}$ in. face and give speed ratios of 1:4, 1:2$\frac{1}{2}$, 1:1$\frac{1}{2}$, 1:1 and in inverse ratio to 4:1.

CONSTRUCTION

The Knee.—Have the slide of the knee casting planed or milled to the dimensions given and also have the under side of the holding-down foot planed exactly at right angles to the face of the slide, leaving only the hole for the holding-down bolt to be drilled to suit the tool rest of your lathe and the drilling of the hole for the feed screw, which should be located exactly half way between the two edges of the slide and $\frac{9}{16}$ in. from the face of the slide. To drill this hole correctly, place blocking on the lathe bed until its top surface is exactly $\frac{9}{16}$ in. below lathe centers, lay the slide on this, entering a $\frac{3}{8}$ in. drill in the center hole of the boss, line up the casting and feed to the drill with the tail center, thus completing this part.

Spindle Slide.—This is shown in Figs. 27 a, b, c. Secure this casting with the slide planed to the given dimensions and I might add that the hole for the bearings and the spindle should be cored out. Lay out and center drill the tap-hole for the feed screw and block up on the lathe bed just as was done for the knee, but have the blocking $\frac{9}{16}$ in. *above* the lathe centers; and after lining up carefully, drill out the boss with a $1\frac{9}{64}$ in. drill, tapping the hole $\frac{3}{8}$ x 16 in. Now make a holding-down bolt to fit your lathe and a capstan nut, like Fig. 26, and bolt the knee to the tool rest with the spindle slide assembled and fit a piece of $\frac{1}{8}$ in. steel for a gib, as shown in section in the assembled drawing. This done remove and drill the holes for the gib screws with a No. 26 drill tapping $\frac{3}{16}$ x 32 in., and it might be explained here that the angular boss for the bottom screw is provided in order to clear the hub of the back-gear bracket. Replace the slide on the knee put the gib in place and spot through the holes in the slide so the points of the gib screws, Fig. 28, will enter and hold it in place.

Feed Screw.—Put a centered piece of $\frac{5}{8}$ in. cold rolled or machinery steel in the lathe and turn down all over with the exception of the shoulder to $\frac{3}{8}$ in. diameter, after which the feed screw can be threaded $\frac{3}{8}$ in. x 16, and lastly the

$\frac{1}{4}$ in. end can be turned down and threaded $\frac{1}{4}$ x 20 in., and a standard hexagon nut fitted. The ball handle can either be made to the dimensions shown in Fig. 32, or purchased from any machinery supply house. It is prevented from turning on the feed screw by a round key, Fig. 25a, let half way into both the feed screw and the handle, the key-way being made with a $\frac{1}{8}$ in. drill, the handle being held in place.

You are now ready to bore out the bosses to receive the roller bearings. Provide a $\frac{3}{4}$ in. boring bar for this job, and one round nose tool and a square nosed tool cranked to the right and one cranked to the left.

Bolt the knee to the lathe, assemble the feed screw run on a $\frac{3}{8}$ in. nut and assemble the slide. Now get the knee parallel to the axis of the lathe, and with the feed screws of both the lathe and milling slides, center the bearing hubs; and with the boring bar in place, bore out both hubs with the round nose tool, after which bore out the fillet left by the round nose tool and face off the bearing seats with the right and left tools. For fear the bearings might vary somewhat from the dimensions given, it would be best to check them up before taking the finishing cuts. These bearings are called the "Knipe" pat. ball bearing, and are 1$\frac{5}{8}$ in. diameter by $\frac{9}{16}$ in. thick, and are listed at 50 cents each. In boring the hubs, see that the lathe gibs are tight and also tigthen up the milling slide gibs securely and run down the nut on the feed screw referred to in order to make the job as stiff as possible and prevent chattering. Assemble the bearings, and a couple of $\frac{1}{8}$ in. pointed grub screws should be put through each boss and let into the soft steel outer casing of the bearings to prevent them from turning. Now drive a hard wood plug in each bearing hole and carefully center, place in the lathe and driving with a stud from the face-plate turn down the bearing hub over which is slipped the back-gear bracket, as shown in the assembled drawings.

Spindle.—The spindle is made from a No. 1 Morse taper socket, catalog No. 100. These are regularly made with the socket and knock-out slot finished, and the socket end turned down to about $\frac{3}{4}$ in. diameter, for a distance of 3$\frac{1}{8}$ in., the balance of

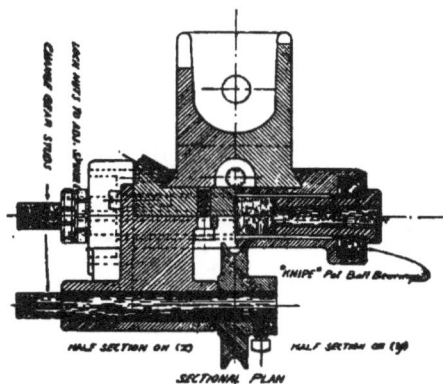

HALF SECTION ON (X)   HALF SECTION ON (Y)

SECTIONAL PLAN

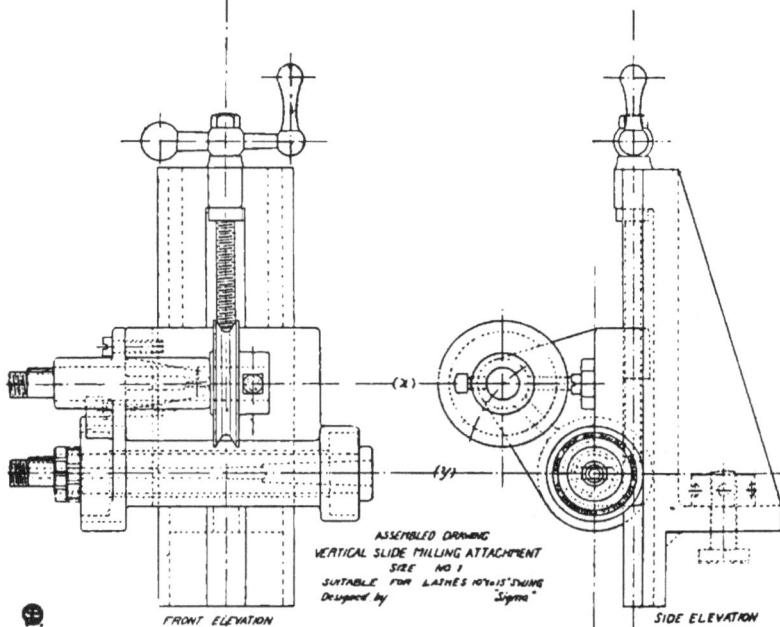

ASSEMBLED DRAWING
VERTICAL SLIDE MILLING ATTACHMENT
SIZE NO 1
SUITABLE FOR LATHES 10"x15" SWING
Designed by   "Sigma"

FRONT ELEVATION

SIDE ELEVATION

the length (3⅞ in.) being left blank and 1⅛ in. diameter. The shoulder will have to be brazed or shrunk on—I turned down the end slightly and turned up a ring about the right size for the shoulder and bored it out for a shrink fit and when it was shrunk on, it was turned up to finished dimensions, making a good job. I might add that a plug for the socket end is furnished for turning up the socket, both the plug and the blank end being centered. Drill out this ring with a 11/16 in. drill and very slightly bevel off the edge of one end of the hole to guide the spindle in entering. Turn down the end of the spindle for a distance of ¼ in., leaving a square shoulder, and to such a diameter that it won't quite drive into the ring, then heat the ring to a dull red and try your fit. If the spindle won't drive in put it back in the lathe and rub down slightly with a file, and try again; and when a fit is secured quench in water at once. This cut and try method presumes the absence of a micrometer caliper, in which case a fit might be tried with the spindle .01 in. larger than the hole in the ring. The ring can also be brazed on if care is used in doing it, but in either case the ring must be trued up after attaching.

The knock-out slot is hardly necessary, but it is regularly furnished and might be useful sometime in case a shank should become stuck in the socket and could not be removed by way of the regular 5/16 in. hole, provided for this purpose. The other end of the spindle is threaded ¾ x 20 in., as shown in Fig. 30, and two lock-nuts ¼ in. thick fitted as shown in the assembled drawings. These are for the purpose of taking up any slackness and maintaining a light pressure on the bearings. The end of the stud for the change gears is threaded ½ x 13 in. threads, and a slot cut and a ⅛ in. diameter x ½ in. long key let in. The 5/16 in. through hole will have to be drilled in from each end, thus completing the spindle.

Back Gear Shaft Bracket Bearing.—This is shown in detail in Figs. 24 a and b. Hold the large ring in the chuck with the short hub of the bearing next the faceplate and bore out to 1⅞ in. Next lay out and drill the three ¼ in. holes for the cap screws, Fig. 29, and fit to the slide. Spot the slide through these holes, run a 3/16 in. drill through and tap ¼ in. x 20 thread and put the screws home. Now assemble on the lathe again and with both feed screws center (at a radius of 2½ in. *exactly* from the center of the spindle), the bearing and drill with a ⅝ in. drill, following the work up with the tail center. A ⅛ in. oil hole completes the part.

Back Gear Shaft.—Fig. 31 is made from a piece of cold rolled shafting turned to the dimensions shown. The stud for the change gears is exactly similar to that of the spindle.

Drive Wheel.—Fig. 33 is turned up from a simple casting, and though easily made one might be bought. This completes the device with the exception of splining the gear wheels for the ⅛ in. round keys. This is best accomplished by drilling a ⅛ in. hole in the end of a centered piece of round steel, say ¾ in. diameter, and at a distance of exactly ¼ in. from the center, then by turning this down so that one-half the hole is exposed, or in other words to ½ in. diameter, a counterpart of the groove in the studs will be obtained. Make this end dead hard if of steel tool, or case harden if made of soft steel, by bringing to a red heat and sprinkling well with yellow prussiate of potash (potassium ferrocyanide), when this has fused, re-heat and sprinkle again, doing this several times (only the groove need be hardened), and after the last fusing, heat and quench in water, when it ought to resist a file. Now chip a nick in the edge of the hole in the gears, slip them on this stud so the nick is opposite the groove and drill out the groove with a ⅛ in. drill, letting the drill follow the groove in the stud. You will get nice straight holes in this way, and they will register with those in the studs. Your attachment is now complete and it is hoped that it will prove most serviceable.

A small drill chuck fitted with a taper shank makes a useful adjunct, and for some operations a necessary one, and when a full set of tools and cutters are acquired you will be pleasurably surprised at the variety of operations that are possible with it. I might add that by replacing the spindle slide with a milling vise slide you can also have an elevating milling slide, thus making two tools of one and thereby further increasing its utility. I will describe the construction of such a vise in the near future.

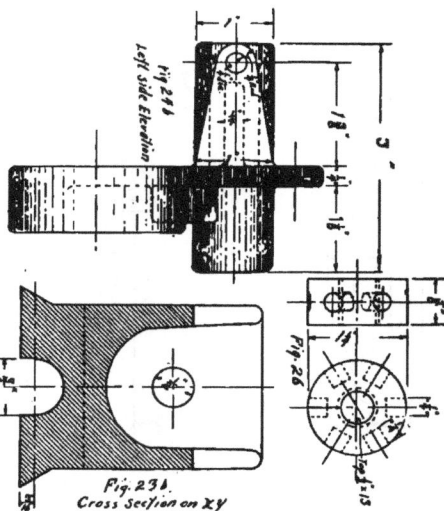

Fig 24b.
Left Side Elevation

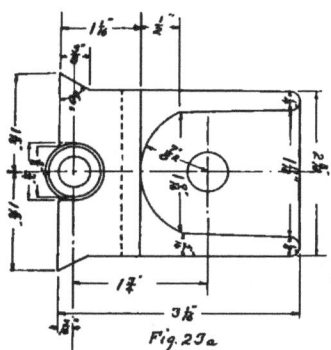

Fig. 23a

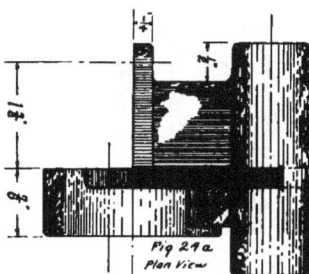

Fig. 23b.
Cross Section on XY

Fig. 26

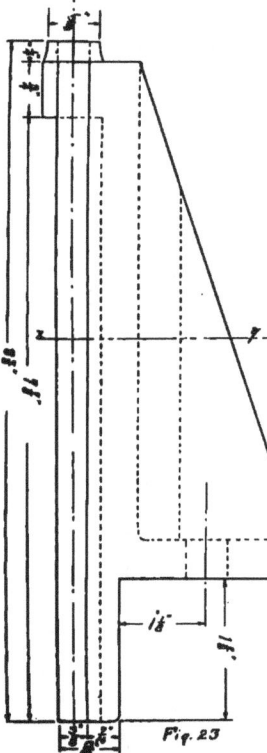

Fig 24a.
Plan View

Fig 25a

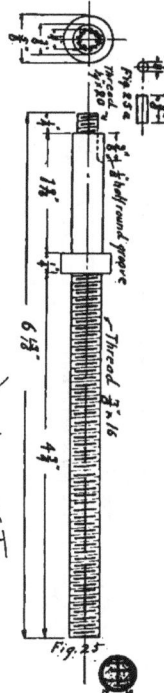

Fig. 25

Fig. 23

Front Elevation

Fig. 24

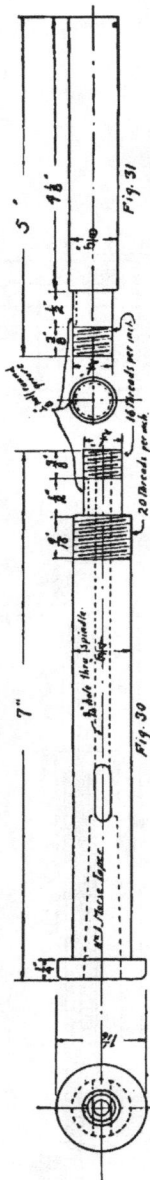

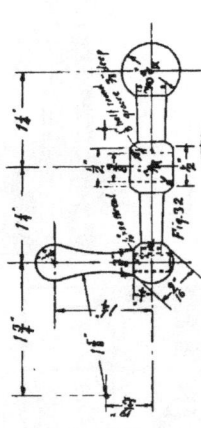

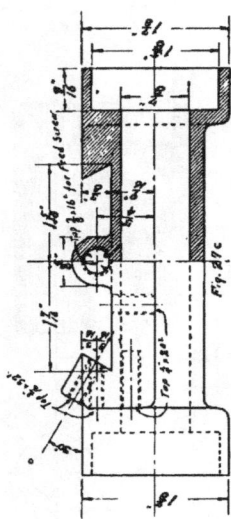

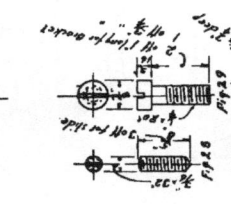

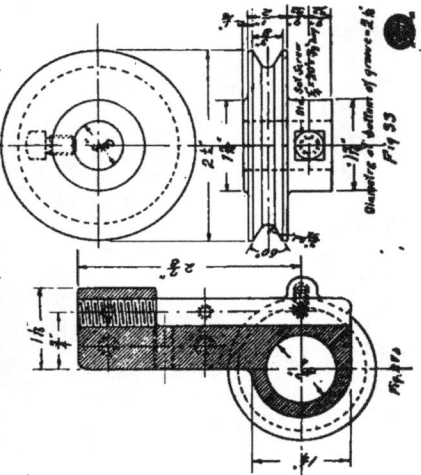

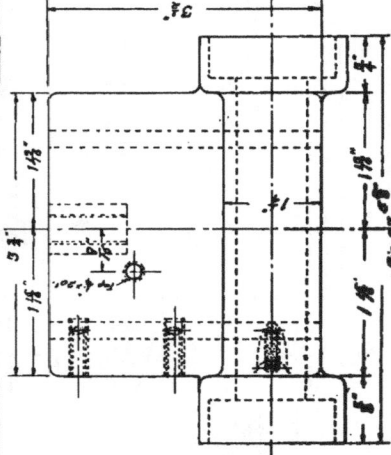

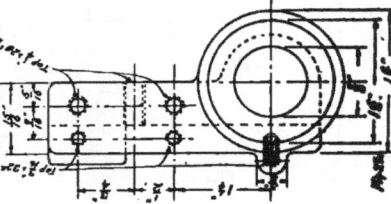

THE BURGESS-WRIGHT BIPLANE AND HYDROPLANE

AUSTIN C. LESCARBOURA
Member of the Aeronautical Society

The pioneer experiments of the Wright brothers, combined with the years of technical experience possessed by the famous yacht builder, Mr. W. Starling Burgess, have produced America's most successful aeroplane.

The Burgess-Wright biplane is built at Marblehead, Mass., and differs from the standard Wright biplane only in the minor construction details It is licensed under the Wright patents. For instance: the spruce wood used for many of the parts in the Wright biplane has been replaced with ash; heavier control wires replace those of smaller size and strength; turn-buckles are employed throughout in place of wires of the exact length; and, in short, many other details have been improved, which render this type the realization of a reliable and efficient biplane.

The biplane is of the headless Wright type, which made its debut at the Asbury Park, N.J., aviation meet during the middle part of 1910. Two seats are arranged on the front edge of the lower plane, while the motor is placed next to the seats. The motor drives two wooden propellers in opposite directions through chain drives. Above the motor are placed the radiator and gasoline tank.

From the two main planes, four main beams are trussed so as to form a boxed structure which extends to the elevating plane and rudders, placed at several feet to the rear of the main planes. The elevating plane is connected with one of the controlling levers by wires. Two vertical planes are connected so as to move in unison, and are controlled by wires leading to another lever. These rudders enable the steering of the machine to the right or to the left. Heavy ash skids extend from the rear edge of the main planes to several feet in front of them, and are joined by two wooden struts extending to the leading edge of the upper plane. A small surface has been placed between the skid and the strut, in order to eliminate the tendency of the biplane to skid sideways when negotiating a sharp turn. On the skids are mounted pairs of heavy rubber-tired wheels with shock absorbers. All the surfaces are double coated, so that the ribs of the planes are entirely covered. The rear edge of the

Howard W. Gill Driving a Burgess-Wright Hydro-aeroplane

Phillips W. Page Driving a Burgess-Wright Hydro-aeroplane and Taking Motion Pictures

two main planes can be pulled down, or warped, by means of control wires leading to one of the control levers. This enables the biplane to retain its stability when turning curves or when assailed by brisk winds.

The motor is of the standard Wright type, 30–35 h.p., 4 cylinder, and weighs 180 lbs. The body of the motor is cast aluminum, while the cylinders are individually cast of iron. The crank shaft is nickel steel, cut and machined from a block, as is likewise the cam shaft operating the exhaust valves. The cylinders are lubricated by a pump, and the intake valves are automatic in action. Gasoline is taken directly into a mixing chamber without passing through a carburetor, by means of a gear pump and injector. A centrifugal pump circulates the water through the aluminum water jackets of the cylinders, and the ignition is furnished by a high tension magneto. Many of the Burgess-Wright biplanes are being equipped with the Gnome rotary motor and prove speedier and more reliable.

The control is effected by two levers; but, by having an additional lever, which is a duplicate of one of the two necessary

levers, it is possible for either passenger or aviator to drive the biplane while in flight. One of these levers controls the rear elevating plane, so that the biplane may be guided upwards or earthwards. The other lever is arranged with a movable handle, so that the rudder planes can be moved; either side of the main planes warped, or both operations performed together by one movement. The independent movement of the handle and lever enables these separate operations. The combined warping of the planes and the turning of the vertical rudders in one operation, is the basis of an important Wright patent.

The principal dimensions of the Burgess-Wright biplane are:

Spread over wings.............39 ft. 6 in.
Length over all................30 ft. 0 in.
Height on wheels..............8 ft. 0 in.
Depth of wings................6 ft. 3 in.
Elevator ....15 x 3 ft.
Dimensions for transporting, 39 x 8 x 8 ft.

Of the many important flights made with the Burgess-Wright type of biplane, those made by Harry N. Atwood are the most remarkable and noteworthy. His

Standard Burgess-Wright Biplane

flights from Boston to Washington; from St. Louis to New York; and from Boston to the mountains of New Hampshire with a passenger, have placed American aviation progress on par with the accomplishments of the European aviators.

By replacing the skids and wheels with two metal boat-shaped, air-tight pontoons, the Burgess-Wright biplane has been converted into a hydroplane of no less a success. The hydroplane parts of the machine have been designed by Mr. Burgess, and may be seen in the accompanying illustrations. By having the wheels and pontoons on the biplane, it is possible for the aviator to start and alight on either the water or land.

Harry N. Atwood recently made a 126-mile flight along the New England coast, while Phillips W. Page, and Howard W. Gill are also making interesting hydroplane flights in similar machines. Page has made flights with a motion-picture camera mounted on his hydroplane and arranged so as to operate by the motor, and thus require no attention for turning the crank. Meanwhile New York city has had another novelty in the witnessing of the remarkable flights of Coffin, the Wright pilot using a standard Wright hydro-aeroplane. A motion-picture camera has also been mounted on his machine for taking pictures of New York harbor.

From the recent activity in hydroplaning by several of the American aviators, little doubt remains but that it will become the sport of yachtsmen. Many sportsmen who would not risk aeroplaning are turning their attention to the hydroplane, which offers the same excitement and pleasure, but with the danger practically eliminated.

TELEPHONE vs. TELEGRAPH

Some people entertain the erroneous idea that the telephone will supplant the telegraph. It is, however, evident that the transmission of articulate speech between two distant points by any means whatsoever is a problem involving difficulties far greater than those of telegraphy, where transmission of single waves or impulses following each other in the proper succession is all that is required. A telegraph circuit only requires one wire, and in quadruplex telegraphy four circuits may be obtained from the one wire, two operators sending and two receiving. In order to obtain absolute quiet on a telephone circuit, a metallic circuit of two wires is required. In telephonic transmission, not only must the constantly varying rate and amplitude of vibration be faithfully reproduced at the receiving end of the line, but the fundamental tone and all overtones must be reproduced, giving each its proper value and without altering the phase relations between them.

PICTURES OF LIGHTNING

Three years ago the Smithsonian Institution received a letter inquiring for a publication. Framed in a foreign hand upon a scrap of paper and expressed in quaint English, it incidentally mentioned some curious experiments which the author had made in odd moments. With a small camera which he held in his hand and revolved from side to side, he had taken some photographs of lightning. He enclosed a print in his letter and wanted to know if his results had any scientific value. The idea of photographing lightning with a moving camera was a novel one to the Smithsonian experts, and, after an investigation, these officials decided that such experiments were worthy of assistance. A grant was therefore voluntarily made to enable the continuance of this photographing with more accurate apparatus.

In letters which followed it was learned that Mr. Alex Larsen, for such was the author's name, was a Danish immigrant, educated in physics, chemistry and electrical engineering at a night school, and that all his experiments were performed for the pure enjoyment of doing something new.

With the aid of the Smithsonian grant Mr. Larsen constructed special apparatus for his work. Upon a revolving table turned by a timed motor were placed cameras in such a position that they would not miss a flash when one occurred. To secure the photographs desired, the tabletop was then revolved at a certain speed. The results are interesting. Where the flash appeared perpendicular, the negatives show naturally a broad sheet for a mere streak of lightning. By calculating from the speed of the camera's motion and measuring the width of the sheet, the time of the flash is easily determinable. But the photographs showed at the very start that a single flash is not one big vibration. It is made up of very many minor flashes, or rushes, following usually in the same channel as the first, and herein lies the special value of the work. In the best of the negatives there are easily counted as many as forty separate rushes which, as the whole flash lasted little over half a second, followed each other in marvelously rapid succession. By measurements and by subsequent calculations, Mr. Larsen determined the actual time between each rush; the figures, as may well be imagined, are almost inconceivably small, varying from the largest, three one-hundredths of a second, to the smallest, as low as two one-thousandths of a second.

There appear many peculiarities of these separate rushes which might bear scientific investigation, but the most salient feature over which meteorologists and electrical engineers may puzzle is recorded on some of the plates among all the bright oscillations, as a marked black rush of lightning. The idea of lightning producing the extreme of darkness is repugnant to the actual name of "lightning." Yet the black rush is plainly visible. Mr. Larsen, after refuting a number of suggestions that might be made to account for it, ventures his own theory to solve the puzzle. In discussing the record of a particular flash in which the mark of the black rush is very distinct, he concludes: "The flash must have given out light of a wave length much shorter than the wave lengths of visible light and with a power sufficient to render the portion of the plate struck by it non-sensitive to ordinary light." "Such a flash," he says, "would appear black on a partially illuminated background, or be invisible."

Invisible lightning, therefore, seems to be a term perfectly proper in view of the results recorded in some of these photographs. At the suggestion of the Institution officials, and with their help, Mr. Larsen carried his researches still further into the actual makeup of lightning. Photographs and studies of the light spectrum of electric flashes in the air were compared with sparks produced by a static machine. The conclusions, in line with the century-old observations of Benjamin Franklin, show that there is little perceptible difference.

The latest ideas outlined by Mr. Larsen to the Smithsonian Institution are no less remarkable than the first. He proposes to photograph electric sparks reflected in a rapidly revolving mirror, and thereby secure records, a study of which may add materially to our knowledge of electric action.—*Boston Transcript.*

NEW METALLIC FILAMENT LAMPS*

G. S. MERRILL

The author deals with the lamps themselves rather than with their applications in commercial service, and discusses the following physical properties of the materials: Electrical resistance, vapor tension and melting point, emissivity, selectivity, mechanical strength.

The following figures are given for the effect of 1 percent increase in voltage of lamps operating at normal efficiency:

	Increases Wattage Percent	Increases c.p. Percent
Tungsten	1.59	3.73
Tantalum	1.72	4.27
Metallized carbon	1.77	4.90
Treated carbon	2.07	5.69
Untreated carbon	2.32	7.10

The filament of an incandescent lamp must not only meet certain requirements of an electric conductor, but it must also possess certain characteristics as a luminous radiator. The latter condition greatly limits the number of materials which can be used as filaments of incandescent lamps. The light of incandescent lamps is produced by temperature radiation as distinguished from luminescence or fluorescence. It is therefore desirable to operate the incandescent filament at as high a temperature as possible in order that the proportion of the energy radiated within the visible spectrum, and hence the efficiency of light production, may be high. The maximum temperature at which it is possible commercially to operate a certain material in the high vacuum of an incandescent lamp bulb is determined by its rate of disintegration if the vapor tension is high, and by its melting point if the vapor tension is low. The high vapor tension of the carbon filament, which is shown by its tendency to evaporate at temperatures far below its melting point, makes it impossible commercially to operate carbon filaments at the normal working temperature of the tungsten filament.

Not only do the various filament materials differ in their ability to withstand high operating temperatures, but they differ also in their radiating properties. The emissivity or mere ability to radiate energy does not affect the luminous efficiency, but it plays an important part in determining the lengths and diameters of filaments in commercial lamps.

The metal osmium has been shown to possess the property of selective radiation in the visible spectrum to a high degree, but the difficulty of working it and its relatively low melting point have made its commercial use impracticable. Tungsten, while not quite as selective as osmium, is more plentiful, and, moreover, it can be operated commercially at a higher temperature and efficiency, so that it is a more desirable filament material than osmium, in spite of the greater selectivity of the latter substance.

The greatest problem in the production of metal filament incandescent lamps has been to produce the filament material in a high state of purity and in the proper physical form, and the ingenuity of the manufacturer has been taxed to the utmost in devising means by which this can be accomplished. It is difficult to secure a high degree of purity in tungsten metal, principally because its atomic weight is high and, in consequence thereof, a very little carbon (which is the most common impurity) is sufficient to form a considerable amount of tungsten carbide. The Paper contains a number of micro-photographs which illustrate the changes during the various processes of manufacture and during the subsequent burning. Microscopic inspection of the longitudinal surface of tungsten filaments after various periods of operation shows that a gradual wrinkling or breaking up of the surface takes place. Apparently the effect is confined almost entirely to the surface material, the filament itself suffers no serious distortion and no marked difference in appearance is produced by burning on alternating current. The appearance of the tantalum filament after burning on direct current is not very different from that of a tungsten filament, but on alternating current a gradual breaking up of the tantalum filament takes place. This can hardly be ascribed to vibration due to rapid heating and cooling, because tungsten filaments, operated under similar conditions, do not show this tendency.

* Abstract of a Paper read before the Franklin Institute.

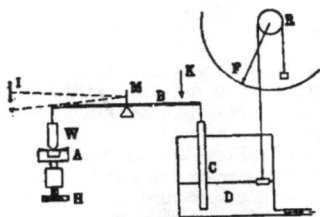

FIG. 1 — FILAMENT TESTING MACHINE.

The study of the change in appearance of filaments throughout their life naturally leads to a consideration of the effect of burning on the strength of the filament. In order to investigate the relative strength of lamps of different types, various methods of testing them have been devised. Strength tests are made with the filaments cold, because when they are heated it is practically impossible to break them, the most serious damage that can then result from shocks and blows being a twisting together of one or more loops.

One method of testing strength consists of mounting the lamp rigidly on a block which can be bumped by a second block allowed to swing as a pendulum. The impact is regulated by the arc through which the pendulum is allowed to swing, and is increased by small increments until the filament is ruptured. The method has been modified by one investigator so that the impact is obtained through a ball rolling down an inclined plane. Another method requires the lamp to be mounted on a block which can be dropped from gradually increasing heights, the apparatus resembling somewhat a miniature pile-driver. Other tests have been made by packing lamps in boxes as for shipment, and dropping the boxes from increasing heights, the lamps being frequently inspected. These methods have the advantage of breaking the lamps under conditions at least approximated in service, and, moreover, give some indication as to the value of the method used in supporting the filament. The results obtained are at best somewhat erratic, and a large number of lamps must be averaged in order to obtain a proper appreciation of relative strengths.

In order to obtain some idea as to the strength of the filament itself, considered apart from its supporting structure,

several methods of testing have been commonly used. One method consists in laying the filament on a pad of paper, smooth but not rigid, and allowing a small metal cylinder to roll lengthwise along it. Rollers of increasing weights are used until the filament breaks. In another device the filament is drawn by a hook at the center between two converging guides until it breaks; the distance which it was drawn into the "V" before breaking gives an arbitrary measure of its elasticity. This method has been varied in another device so as to draw the filament through a spiral guide, thus giving the sample a greater and greater curvature until it breaks.

The number of ways in which filament strengths can be measured with precision is somewhat limited by the physical nature of the filaments and by their extremely small diameter. The filaments readily lend themselves to cross-bending tests, and in order to carry out certain investigations a device for making such tests with some degree of accuracy was constructed. Short pieces of filament were supported at points 0.5 cm. apart and a load slowly applied at the middle until the filament broke, the load and deflection both being measured. By taking short pieces of filament it is possible to test from 10 to 20 samples from a given lamp, and the average load and deflection shown by these 10 or 20 samples gives a very good measure of the filament strength in that particular lamp. In addition to determining the breaking strength, the deflection under various loads could be measured and deflection load diagrams could thus be obtained.

The machine, which it is believed involves a new principle, consists essentially of a horizontal beam balanced on a knife edge (see Fig. 1), supporting at one end a cylindrical plunger C and at the other end a cylindrical weight W, of the shape shown. The weight W at one end of the beam hangs with its lower edge midway between the points of support of the filament which is laid horizontally across an opening 0.5 cm. long. The block A carrying the filament can be raised through measured distances by means of a micrometer screw.

A small concave mirror M on the beam reflects light from the source to form an image at I. By keeping this image on a certain point, the beam may be kept horizontal and the deflection of the filament may be read from the micrometer screw.

The plunger at the other end of the beam hangs within a vessel which may be filled slowly with water. The plunger is buoyed up by the water as the latter rises and a load proportional to the weight of water displaced by the plunger is thrown upon the filament. As the plunger is uniform in section, the volume of water displaced can be most easily measured by its height upon the plunger. In testing, the beam has been kept balanced with a small part of the plunger submerged, and consequently the load upon the filament is determined by a difference in water levels. To enable the water level to be read with accuracy, a carefully balanced float rides upon the surface of the water in vessel D, and by means of a silk thread passing over the small pulley R, its motion is magnified by the long pointer F traveling over an arbitrary scale. The water level readings can be reduced to the equivalent load in grammes on the filament by means of a calibration curve.

After putting the filament in place, and taking the initial reading of the pointer F and the micrometer head H, water is allowed to flow into the vessel D. As this throws a load on the filament and produces a deflection, the operator raises the filament with the micrometer screw, in order to keep the image I at a fixed point. The loading continues until the filament breaks, at which instant the water supply is automatically cut off by electrical contact of the unbalanced beam with the point K. The pointer F indicates the water level at breaking. The deflection of breaking is indicated by the difference between the initial and final readings of the micrometer head.

In order to compare strengths and deflections of filaments of varying size, it was necessary to express the results in terms of some unit length and diameter. Consequently all measurements of strength and deflection have been reduced to the equivalent load in grammes and deflection in centimeters of a filament 0.5 cm. long and 0.005 cm. in diameter. In some cases filaments were found to be slightly elliptical in cross section rather

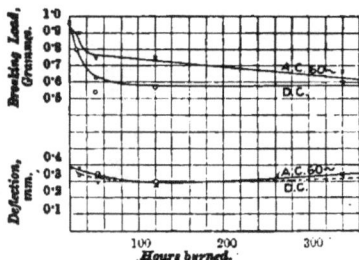

Fig. 2

than circular. (The difference between the maximum and minimum diameter, however, rarely exceeded 10 percent). In reducing the results of the cross-bending tests to a common basis, the diameter of the filament was therefore taken as the mean of two diameters at right angles in order to reduce the error due to possible deviation from a circular section.

Some of the most interesting tests made with the device have reference to the change in strength of the filament during its life. Three lots of 40-watt 110-volt tungsten filament lamps representing modifications in manufacturing processes were run through such comparative test. Each lot was divided into two parts, one of which was burned on 60-cycle alternating current and the other on direct current. All lamps were burned at 1.23 watts per candle-power. At certain intervals lamps of each lot were removed from the burning racks and the strength of the filaments tested. Fig. 2 shows the results of the tests for one lot, the general tendency of the curves being the same in each case.

Each point on the curves is the average of 10 determinations of strength of the filament from a single lamp. The deflection at breaking indicates that the flexibility of the filaments does not change very much with burning, especially after the first few hours.

The decrease in strength with burning could be most easily shown by either some change in the structure of the filament or by some change in the effective cross section. Although the results have all been reduced to terms of a uniform gross diameter, a microscopic inspection of the filament shows that there is a tendency for the surface to become wrinkled with burning, which would, in all probability tend to reduce the effective diameter as far as strength is concerned. In general, the surface, which was originally smooth, gradually becomes rough during the first 100 hours' burning. Beyond this point the changes are not so marked. The strength curves indicate in a general way that the greatest reduction in strength occurs during the first 100 hours of inspection, and thereafter the change is less marked. This seems to point to the possibility that the strength is dependent at least to some extent upon the surface roughening. It might be noted that during the period covered by the test the gross diameter of the filaments of the several lots remains practically constant, i.e., no marked reduction in diameter with burning could be noted.

In regard to changes in structure, there is a tendency for the crystalline structure to become somewhat coarser with burn-ing. This change would naturally tend to weaken the filament, due to the greater ease with which the fracture could follow the larger crystalline surfaces. A compact small-grained crystalline structure is apparently the more desirable from the standpoint of strength. The investigation, however, should be carried further before any definite conclusions are reached.

In order to give the relative idea of the strength and stiffness of a tungsten filament and a glass thread, a glass rod was drawn down to a diameter approximately equal to that of the filaments and a number of samples were tested. The ultimate strength of the threads of glass was almost three times that of the filament, but the flexibility of the glass was considerably less, giving about 60 percent as great a deflection as a tungsten filament under the same load.

NEW USE FOR ELECTRICITY

Electricity as an agency to destroy the codling moth is the latest innovation of modern apple-orcharding in the Spokane valley, where W. M. Frost, inventor of the device, and J. C. Lawrence, a practical grower of Spokane, made what is declared to be the first demonstration of its kind in the world the evening of August 18. The test was made in a six years' old orchard at Opportunity, Wash., where a score of second-brood moths and hundreds of green aphis were killed in a few minutes. The apparatus consists of a storage battery to charge incandescent light globes of 6 c.p., which are netted with fine steel wires, coated with copper and tin, alternately. Attracted by the bright light in the tree, to which the globe is strung by a covered wire, the moth flies against the net work, completes the electric circuit and is instantly killed, the body dropping into a receptacle beneath the globe. Mr. Frost thinks that one battery to an acre of trees will keep the moths under control, thus eliminating spraying and saving many dollars for equipment and fluid. If electric light wires are extended to the orchard tracts, as they are in the Spokane valley, the expense of batteries may be saved by making direct connection and using the commercial current. The cost of covering the globes with wire nets is a small item and any electrician can do the work. Growers in various parts of eastern Washington are preparing to equip their orchards with the new pest destroyers the coming season.

Repairing A Stripped Thread
A. G. D. C.

The thread on one of the studs holding down the cylinder of a petrol motor had partially stripped, and as to replace the stud meant taking the engine down it was necessary to make a temporary repair. It was not possible, owing to the position,

to cut another thread on the stud, but the job was done as follows: The nut was taken and a saw-cut made in one face as shown, the nut was then pinched up in the vise till the saw-cut was closed. On threading the nut on the bolt it was found to grip the remaining scraps of thread quite tightly and made a satisfactory temporary repair.

ENGINEERING LABORATORY PRACTICE—Part V

The Determination of the Percentage of Moisture in Steam

P. LE ROY FLANSBURG

When steam is delivered to an engine or pump it may be in any one of three conditions, and it is highly important that the engineer should know in which of these conditions the steam is when admitted. The three possible conditions are: 1st—wet saturated steam; 2d—dry saturated steam; and 3d—superheated steam. If the steam is perfectly free from water it is called dry saturated steam. Provided the boiler from which the steam is taken has a sufficiently large steam space and is making steam slowly, it is possible that the steam may be entirely free from moisture; but under ordinary conditions it is much more common to find a certain percentage of moisture in the steam. Steam containing moisture is known as wet steam, and it is then important to know the exact percentage of moisture which is present. The third condition is obtained by taking the steam as it comes from the boiler and then heating it to a higher temperature than it was at in the boiler. Steam which is so treated contains no moisture and is known as superheated steam.

In engineering work it has been found convenient to speak of the amount of steam which is present in each pound of wet steam, as the quality of the steam or the dryness factor. This factor is ordinarily represented by the symbol x.

There are several methods of determining the quality of steam, but the best modern practice favors the use of one of four types of calorimeters. These four types are called "throttling" calorimeters, "superheating" calorimeters, "separating" calorimeters, and the "Thomas Electric" calorimeter.

In its simplest form the throttling calorimeter can be readily made up from pipe fittings and a throttling valve. The steam is led from the steam mains through a throttling valve into a small cylinder which is well-covered with a material such as hair felt or some other non-conductor of heat. The steam then exhausts from this cylinder, and where there

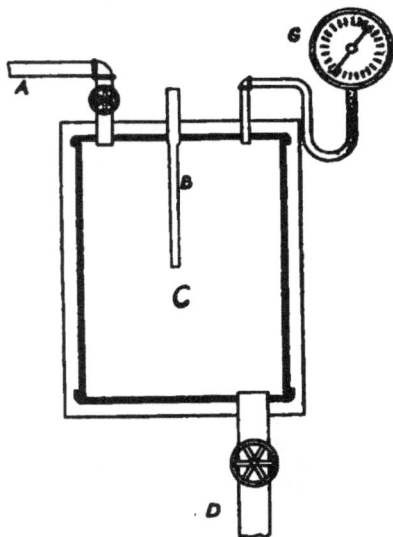

Fig. 1

is no valve on the exhaust the back pressure is practically atmospheric. A thermometer is placed inside of the cylinder and by means of it the temperature of the steam inside of the cylinder may be read. Where accurate work is desired, an improved form of the throttling calorimeter is used. One of the improved forms of instrument is the Carpenter Throttling Calorimeter, and another the Peabody Throttling Calorimeter. The Carpenter type of instrument is very similar to the instrument which has just been described, the only real difference being that a "manometer" (a device for measuring pressures) is attached to the body of the calorimeter. By means of the "manometer" it is possible to determine the pressure of the steam which is inside the calorimeter.

The Peabody Throttling Calorimeter is shown in Fig. 1. It consists of a chamber or reservoir C into which the

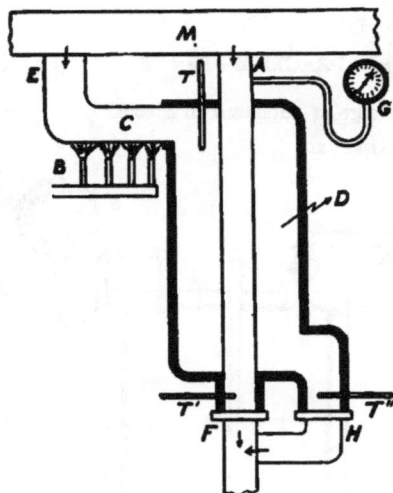

Fig. 2

steam is admitted through a throttling valve, and from which it is exhausted through a pipe D. The pipe A is connected directly to the mains where it is desired to know the quality of the steam. Pipe A is a ½ in. pipe, while pipe D is a 1 in. pipe. The pressure in the mains is read with a guage. Due to the large diameter of the exhaust pipe D, the pressure in the reservoir C (given by the gauge G) is far lower than the pressure in the mains. The total heat of saturated steam increases with increase of pressure, so when the saturated steam is expanded through the valve and has its pressure decreased, the excess heat is liberated and will evaporate any moisture present in the steam. Provided the amount of moisture is not excessive, the dry steam will then be superheated.

As the calorimeter is carefully protected by a covering of hair felt (which is a good non-conductor of heat), there will be no loss of heat during the test. Therefore, the total heat of the steam in the mains is equal to the total heat of the calorimeter steam, pound for pound. Stating this same thing in another way: during an expansion where there is no heat lost, the specific heat of moist steam remains constant. At the higher pressure the specific heat is not sufficient to

vaporize all of the water, but at the lower pressure it is sufficient not only to do this but perhaps to even superheat the steam. This type of calorimeter cannot be used unless there is sufficient excess heat liberated to superheat the steam.

Let the boiler pressure or pressure in the mains equal p. Look up r, the latent heat, and q, the heat of the liquid at this pressure. Let p' equal the pressure inside of the calorimeter; r', the heat of vaporization; q', the heat of the liquid; and t, the temperature of saturated steam at that pressure. By means of the thermometer B, read the temperature t_s of the superheated steam within the calorimeter. Now call x the quality of the steam. Then $xr + q =$ total heat at entrance.

$r' + q' + C_p (t_s - t) =$ total heat at exit (where $C_p =$ the specific heat of steam). Now equating the total heat at entrance to the total heat at exit, you obtain

$$x = \frac{r' + q' + C_p (t_s - t) - q}{r}$$

The throttling calorimeter is by far the simplest calorimeter to install and operate.

The Superheating Calorimeter is shown in Fig. 2. In this type of calorimeter the steam to be tested is allowed to enter the calorimeter through A, and after flowing down the tube leaves it through an orifice F of cross-section a. Just before the steam passes through F, its temperature is measured at T'. Surrounding the tube A is a jacket D which is filled with superheated steam. The superheated steam is obtained in the following manner. At E, steam from the mains enters a pipe C and is superheated in this pipe, by means of Bunsen burners, to an amount which is determined by the thermometer T. The superheated steam then flows through the jacket D and leaves the jacket through an orifice H of cross-section a. Just before passing through H the temperature of the superheated steam is taken by a thermometer T''. The pressure in the calorimeter is measured by the gauge G.

Since the area of the two exit orifices is the same and the pressures are the same, if we neglect the differences of volume due to exit temperatures, then equal weights of steam pass out in a given interval of time. In passing through the jacket, the superheated steam loses some of its heat by radiation, and if we run the apparatus admitting steam only to the jacket, this radiation factor can be obtained. The difference in temperature at entrance and at exit of the jacket shows the amount of heat which is lost in passing through the jacket. If from this is subtracted the loss in heat due to radiation, you can at once find the amount of heat given up to the sample during any interval of time. From this data the quality of the sample of steam may be calculated.

When there is more than 3 percent moisture in the steam, a separator is used which will remove all of the moisture from the sample of steam by some mechanical process of separation. To find the percentage of moisture in the steam use the formula:

$$1 - x = \frac{W - R}{W + w}$$

where:—

W = water drawn from separator.

R = water thrown down during run, by radiation.

w = weight of dry steam discharged at exit orifice.

One of the best forms of Separating Calorimeters is the type designed by Prof. Carpenter.

The Thomas Electric type of calorimeter is shown in Fig. 3. Although somewhat similar to the superheating type, the Thomas Electric type would probably be preferred, owing to its ease of operation and the fact that the determination of the quality of the steam is easily computed.

The steam is allowed to enter the cylindrical vessel or chamber from the mains through the pipe G. It then passes up and over a heating coil C, which is supplied with electrical energy either from batteries or from electric mains. It is possible to govern the amount of energy put into the coil by means of a rheostat or other electrical resistance connected in the circuit as shown in the drawing. The actual watts input is measured by means of an ammeter and a voltmeter. After passing over the heating coils the steam is superheated and the number of degrees of superheat is measured by a thermometer T. The steam then passes through D, and, by inserting a glass tube E, you have a means of observing the

Fig. 3

condition of the exit steam. For instance, if the steam is wet the glass will become moist, while if the steam is dry, no moisture will be present to fog the glass. Now, knowing the heat added to make the steam dry saturated and the amount of steam flowing through, you can compute the quality of the steam.

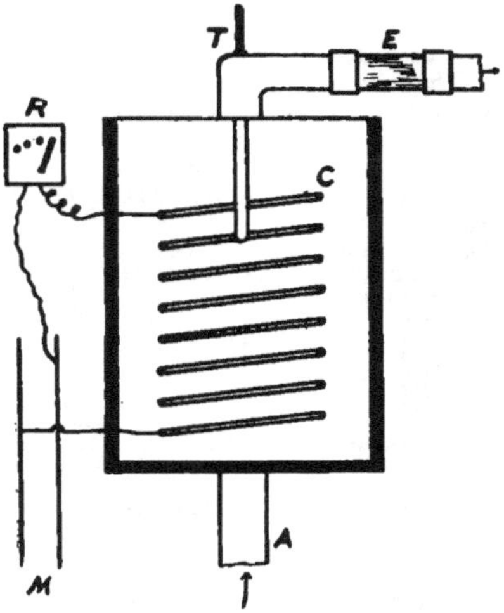

Let E_1 = the number of watts needed to dry the steam.

E_2 = the number of watts increase which are needed to superheat the steam to some such temperature as 30 degrees Centigrade.

W = weight of steam flowing per hour under first conditions.

S = amount of electric energy needed to superheat 1 lb. of steam from saturation at various pressures to 30 degrees Centigrade.

Heat required to dry one pound of steam = H,

$$H = K\frac{E_1}{E_2} B.t.u.$$

In finding the quality of the steam, use the formula

$$x = \frac{r - H}{r}$$

r = the heat of vaporization at the pressure used.

K = a constant, and is obtained for all pressures from a plot supplied with the calorimeter.

The following test is one made by the author with a Peabody Throttling Calorimeter.

TEST MADE WITH A PEABODY THROTTLING CALORIMETER

Barometer—30.41 in.

Boiler pressure in lbs. gauge	Calorimeter press. in inches gauge	Temperature of Calorimeter
41.8 lbs.	2.55 in.	120.0°
41.8	2.60	120.0
41.7	2.55	120.5
41.7	2.45	120.0
41.7	2.55	120.5
41.7	2.55	120.0
41.73 lbs.	2.54 in.	120.16°C = 248.4°F

Calorimeter pressure = $(30.41 + 2.54) \times .491 = 16.2$ lbs. abs.

Boiler pressure = $41.73 + (30.41 \times .491) = 56.6$ lbs. abs.

$T = 217°F$ = temp. caused by pressure in cal.

$C_p(T_s - T) = (248.4 - 217) .463 = 14.8$ B.t.u.

185.6 $1151.8 + 14.8 = 1166.6$ B.t.u.
966.2
1151.8

$q + xr = H = 1166.6$
$258.6 + 917x = 1166.6$
1166.6 $917x = 908$
258.6 $x = .99 = $Ans.
908.0

Readings taken at 3-minute intervals.

Alcohol as Fuel

A process of burning alcohol as a fuel under conditions similar to those obtaining with the use of gasoline (petrol) has recently been developed, in which the alcohol is used in combination with acetylene gas. In a recent demonstration in New York, use was made of a single cylinder de Dion Bouton motor of 3½ h.p., coupled direct to a dynamo having a separately excited field. A bank of incandescent lamps was used as resistance.

An ordinary carburetor was employed, the mixture being formed of alcohol, acetylene gas and air in about equal quantities. Acetylene gas alone burns too quickly and alcohol too slowly for direct use in gasoline motors, but the combination is said to have given practically the same results as would have been obtained by the use of gasoline. The process is a method of producing the gas for power purposes, in which a spray of atomized diluted alcohol is brought into contact with calcium carbide, resulting in the formation of an explosive vapor.

The apparatus is constructed to carry the atomized alcohol, on its way to the motor, through a chamber having a bed of calcium carbide. The resulting product, consisting of the three gaseous substances, has been called "Alkoethine." This is passed by suction into the engine cylinder to be exploded. The power developed is said to be about equal to that from gasoline, and to be produced at a small cost. In addition is the ability to start a motor while cold, which it is impossible to do with alcohol under ordinary conditions. Denatured alcohol contains about 10 percent of water, which would be a deterrent where alcohol only was used. In the new process, however, it is a decided advantage, because not only is this water utilized, but still more is required in order to produce the chemical effect of forming acetylene gas from the carbide. In this connection it is said that up to a certain point the more water added the greater is the power obtained. The new process when fully developed will doubtless have a very wide field of operation.—*The Engineer*.

Testing Oils for Household Use

Petroleum for use in lamps, stoves, etc., should be white or light yellow in color with a blue reflection; clear yellow indicates imperfect purification or adulteration with inferior oil. The odor should be faint and not disagreeable. The specific gravity at 60 degrees Fahrenheit ought not to be below 0.795 nor above 0.84. When mixed with an equal volume of sulphuric acid, of the density of 1.53, the color should not become darker, but, if anything, lighter. A grade of oil that will stand these tests and possesses the proper flashing point may be safely used. It is of great importance to know that this oil is pure and safe for home use, as loss of life and property has been caused many times by the use of inferior grades.

Earnest effort increases your employer's business; you should not talk too much during business hours nor close up shop too soon when the day's work is done.

ELECTRO-CHEMISTRY
Electricity and Electrical Conduction
ERNEST C. CROCKER

STATIC ELECTRICITY

Through experiments with static electricity which were carried out by Benjamin Franklin and others over a century ago, it was seen that there were two distinct kinds of electrical charges. These two kinds were named "positive" and "negative" charges, for convenience. From the way in which charges of electricity could be conveyed from one body to another, and from the presence of visible sparks accompanying such transferences, it was considered that electricity must be material and a fluid or fluids. By "fluid" was meant "fluid" in its larger sense, which means anything which flows, as water or air.

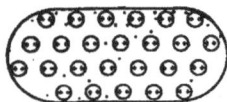

Fig. 1
Relative positions of atoms and electrons in an uncharged (neutral) conductor.
Circles containing + signs are atoms; small dots electrons. An atom is represented as having a charge equivalent to that of two electrons. This representation will be the same in all the figures.
Note.—Atoms are stationary, hence cannot form current. Electrons are free to move, and can form current.

The two different kinds of electrical charges suggested two things: electricity is a one fluid substance, or electricity is composed of two different fluids. In the first case, a body is neutral if just the normal amount of fluid is present, positive if the fluid is in excess, and negative if deficient. In the second case, neutral if the same amount of each fluid is present, positive if positive is in excess, and negative if negative is in excess.

Recent work points out that each theory is right in part: there are two kinds of electricity, but one is a "solid"

Fig. 2
Relative position of atoms and electrons in an uncharged (neutral) non-conductor or insulator.
Note.—Atoms are stationary, hence cannot form current. Electrons are so bound that they cannot flow to form current. The result is that no current can flow through this body.

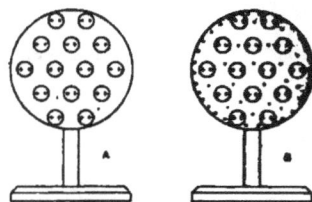

Fig. 3
A—Positively charged body.
B—Negatively charged body.
Note.—There is a deficiency in electrons in the positively charged body (less than two per atom), and an excess in the negatively charged body (more than two per atom).

and the other a "fluid." The so-called "negative" electricity is the fluid. The "positive" electricity, as we shall see later, is ordinary matter; possessing rigidity, as it does, it cannot move around. A negative charge is then an excess of negative fluid; a positive charge, a deficiency; and neutrality, just enough negative fluid to neutralize the effect of the positive. The positive electricity, being immovable, need not be considered in static electricity (see Fig. 1 to 5).

Reliable experiments have shown us that the negative fluid is composed of small particles, just as is matter. These small particles bear the name of "electrons." Positive electricity seems to be matter itself,—in fact, no particles of positive electricity have yet been detected which are smaller than the atom of hydrogen. Some have been found to be much larger. An "electron" has only about ᴛ¹₇₀₀ of the mass of a hydrogen

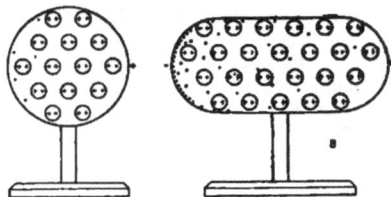

Fig. 4
A—Positively charged body.
B—Neutral conductor brought near the charged body.
Note.—Electrons in the conductor B have been attracted by the + charge on A, and, accordingly, have piled up on the end nearer A. If the body be cut in two and the charged body A be then removed, each half would be found to retain its charge (excess or deficiency in electrons).

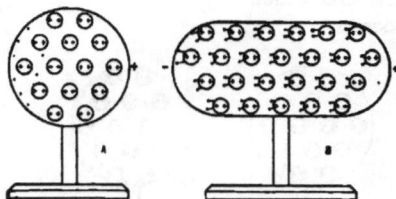

Fig. 5
A—Positively charged body.
B—Neutral insulator brought near the charged body.
Note.—In this case the electrons are held to the atoms by restraining bonds. If the body be cut in two and the charged body A be then removed, no trace of charge can be found on either half (showing that electrons did not leave their places to pile upon the end).

atom, yet it may be measured with considerable accuracy. The hydrogen atom itself is exceedingly small, as must be evident from the fact that there are about 4×10^{21} or 4,000 trillion atoms of hydrogen besides half as many more of oxygen in a single drop of water (H_2O.)

According to this new view, we should reasonably expect that a charged body would weigh more or less than it did before it was charged. Theory says that it does, but up to the present it has not been possible to detect any difference in weight, and for two reasons: the difference is so small, even with great charges, that present balances are not delicate enough to measure it, and, from the fact that a charged body exerts attractive force on all nearby bodies, we probably could not weigh it if a balance were found which is sensitive enough. A charged Leyden jar or other condenser certainly cannot have its weight altered because of the charge, for it loses as many electrons from one plate as it gains on the other, thus keeping the weight constant (see Fig. 6).

CURRENT ELECTRICITY

Metallic Conduction. — An electric charge in motion is an electrical current. In a metal we have the rigid body of the metal formed from atoms, and this constitutes the positive electricity. Evidently this framework of metal cannot flow when we have a current in the metal. Between the spaces of the framework of atoms, which is the structure of the metal, there are, however, the electrons, the particles of negative electricity, and it is upon these we must rely for the transmission of the current.

Since electricity (here only the negative

electricity is considered) is made up of "grains" it is somewhat comparable with sand: a current of electricity through a metal wire is like a flow of sand through a pipe. A remarkable completion of the analogy is the fact that if we have a small enough opening, we can have an electric current come through in individual "grains," one at a time in slow succession, just as would be the case with a stream of sand through a small hole. "Electrometers" (special kinds of "electroscopes"), can be constructed which give a distinct indication for each individual electron.

ELECTROLYTIC CONDUCTION

The kind of conduction which occurs in the solution of a salt or acid in water is called "electrolytic" conduction. In metals we have to deal with the positive electricity "frozen" into a rigid body and only the negative free to move; but here we have both positive and negative electricity in a mobile condition.

For an example of electrolytic conduction, let us consider the case of a solution of copper sulphate containing copper electrodes. A molecule of copper sulphate ($CuSO_4$) is composed of an atom of copper ($Cu + +$) lacking two electrons, thus positively charged by two units, and a "sulphuric radicle" ($SO_4 - -$) bearing two free electrons. When in the crystal condition, the copper atom and the sulphuric radicle are closely combined but the moment the crystal is dissolved in water, they "dissociate" into the two ions ($Cu + +$) and ($SO_4 - -$). We have now a liquid in which are movable charges of electricity riding on special carriers called "ions" (Fig. 7).

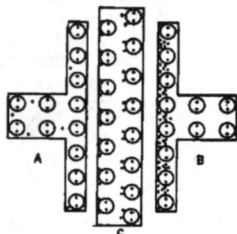

Fig. 6
Condition in a charged condenser.
A—Positive plate.
B—Negative plate.
C—Dielectric.
Note.—There is excess of electrons at negative plate, and deficiency at positive; also, dielectric is under electrical strain.

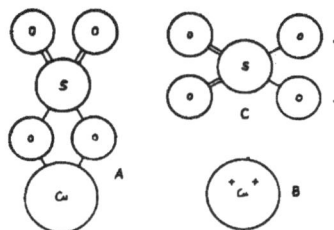

Fig. 7
A—Neutral molecule of copper sulphate (CuSO₄).
B—Copper ion (Cu + +).
C—Sulphuric ion (SO₄ – –).
Note.—The approximate size of the different atoms and of the electrons is correct. The actual arrangement of the atoms and electrons in the molecule is probably far different from that in the drawing.

In a metal we have only one kind of movable charge, but in a liquid we have two. In order that a current pass through the liquid, both positive and negative particles must become active, one going one way, the other the opposite. When a copper ion $(Cu + +)$ arrives at the negative electrode, it draws two electrons from it and becomes a neutral atom of copper, which then attaches itself to the electrode. When a sulphuric ion $(SO_4 – –)$ arrives at the positive electrode, it gives up its two free electrons, and seizing an atom of copper from the electrode, becomes a molecule of copper sulphate which dissolves in the liquid to take the place of the one just decomposed (Fig. 8).

In our solution of copper sulphate there is a whole cycle of changes during which every sulphuric radicle changes its partner many times. The positive electrode wears away, its substance being used in furnishing new partners to grasping sulphuric radicles which pay the price of two electrons; the negative electrode is the resting-place of all the tired-out and jilted copper ions which are each paid two electrons to stay and become neutral atoms. The result is that the negative electrode is built up by copper which goes through the solution, with the assistance of the sulphuric ions, and a stream of electrons is carried to the positive electrode. Each time a neutral sulphuric radicle captures a neutral copper ion to form a molecule of neutral copper sulphate $(CuSO_4)$, the sulphate dissolves in water, the sulphuric radicle takes two electrons from the copper and both become ions, $(Cu + +)$ and $(SO_4 – –)$.

VACUUM CONDUCTION

There is good reason to believe that a perfect vacuum is a perfect insulator or non-conductor, but since no vacuum even approximately perfect has ever been attained, we need not consider a perfect vacuum. At the so-called vacuums which we find in X-ray tubes and Geissler tubes there is considerable conductivity. A very excellent vacuum may contain only about one millionth as much gas as the same space before exhaustion and yet contain an astounding number of atoms of the gas. A cubic inch of any gas contains about 65×10^{19} atoms at ordinary pressure and even if only one millionth of the gas is left there will still

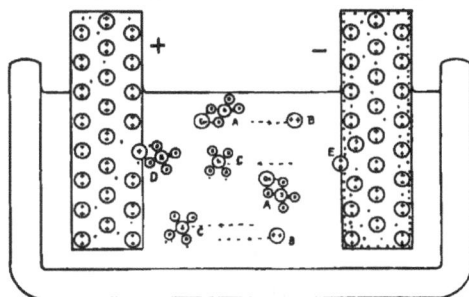

Fig. 8
Electric current flowing through CuSO₄ solution.
A—Undissociated molecules of CuSO₄.
B—Copper ions (Cu₄ + +).
C—Sulphuric ions (SO₄ – –).
D—Sulphuric ion just deposited on positive electrode, combining with an atom of copper to make a molecule of CuSO₄.
E—Copper ion which is attaching itself to negative electrode.

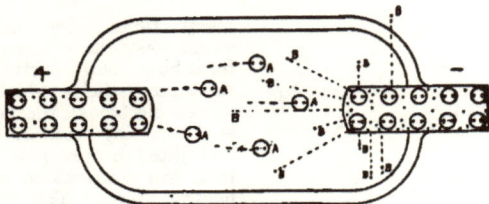

Fig. 9
Electric current flowing through vacuum tube (with metal electrodes.).
A—Copper atoms (acting as ions) crossing vacuum to deposit themselves on negative electrode.
B—Electrons, shot off at right angles to surface of electrode.

be 65 x 10¹⁸ or 650 trillion atoms left, and surely this is far from a perfect vacuum.

Experiment shows that when a current is flowing through a vacuum tube, the positive electrode (anode) wears away, while the negative electrode (cathode) increases in weight. This passage of matter through the tube shows that positive electricity acts here as in the case of solutions. Here, the outside atoms of the metal of the anode detach themselves and move through the "vacuum" and deposit themselves on the cathode (Fig. 9). Meanwhile the negative particles, the electrons, are not idle, although they do not act as they do in solutions where they have convenient "ions" on which to be ferried across. In this case, they have the singular property of shooting off at exactly right angles to the surface of the cathode and traveling away at a rate not much less than that of electricity in a wire (186,000 miles per second). In a short space they are slowed down by friction, but are usually able to actually get beyond the walls of the glass tube. A stream of these negative electrons is called a "cathode ray," more about which we shall consider under "X-rays."

CONDUCTIVITY OF AIR AT ORDINARY PRESSURE

Ordinarily, air is very nearly a nonconductor, but there are influencing factors such as flames, ultra-violet light, X-Rays, cathode-rays, and radium rays which "ionize" the air and make it conductive. This kind of conductivity will be taken up in detail under "Radium," and it will not be further dealt with at this time.

ELECTRICAL RESISTANCE

The electrical resistance of a conductor is the friction which the particles of electricity must encounter as they move through the conductor. Resistance varies with the length and cross-section of the conductor, and also with another quantity called the specific resistance, or resistance of a piece of given size. As a rule, the longer a conductor the greater its resistance—there is, however, an apparent exception in the case of air at ordinary pressures.

Under metallic conduction, we considered a piece of metal as a framework of the atoms of the metal which could not move to convey the current. As an illustration of the virtual condition of the atoms and electrons in a piece of metal, let us consider the particles of positive electricty, the atoms of the metal, as a honeycomb structure and the electrons, little insects which can fly around through the interstices of the honeycomb. An electric current is then a swarm of the insects flying in one direction, and electrical resistance is the opposition which they meet as they fly.

A curious fact in regard to the resistance of metals is that an alloy is always of higher resistance than would be expected from its components, and often higher than any of them. This is easily explained when we consider that electrical resistance is the opposition which the little insects of our illustration encounter as they fly through the honeycomb—the different metals form crystals of different shape, some crystallizing first and then others crystallizing into the holes which the first metals leave, thus tending to block up the passage. It will be noticed that, as a rule, the more different metals there are in the composition of an alloy, the higher its resistance. To state a few instances: alloys like German silver and "nichrome" have greater resistance than the two-metal alloys like brass, bronze, etc. This should follow directly from the above illustration.

The electrical resistance of "electrolytic" conductors, like solutions, decreases as we have more and more free "ions" present; in other words—depends upon the degree of dissociation of the dissolved salt. The degree of dissociation, as well as the friction which the ions encounter as they move through the solution, depends much on the temperature; liquids are better conductors when hot than when cold.

Temperature influences metallic as well as electrolytic conduction, but as a rule the resistance increases with the temperature. A notable exception is carbon. Recent experiments on the resistance of metals at very low temperatures have shown that all pure metals are practically perfect conductors at and near the absolute zero of cold ($-273°C$); in other words, their resistance becomes too small to measure. This is not true in the case of alloys. Insulators, as a rule, become conductors of the "electrolytic" type when they are strongly heated; for instance, the filament of a "Nernst" lamp is almost a perfect insulator when cold, and ordinary glass, at a red heat, is a fair conductor.

There is one element, selenium, which has its resistance altered not only by heat but by light. Another element, bismuth, has its resistance increased as much as 50 percent by a very strong magnetic field, although not proportionally as much by weaker fields. The last two cases, those of selenium and bismuth, are unique, and show properties which are much sought after.

SUMMARY

We have seen from the foregoing, that electricity is not a vague "something" which is so incomprehensible that we dare not express an opinion concerning it, but is something real and tangible. We saw that, as far as has been ascertained up to the present, there are two kinds of electricity; one kind, the positive, having atoms of ordinary matter for its ultimate particles; while the other, the negative, is made from perfectly real, but smaller particles called electrons. We considered the rather homely analogy of a honeycomb of matter through which the insect electrons fly, in the case of metals. We saw how those chemical ferry-boats, the ions, convey the little electrical passengers in solutions. We saw how the pieces of the electrode themselves go through the vacuum tube to carry the current, and, particularly, how the little electrons go on a headlong dash, away from their electrode, not seeming to care where they go.

CONCLUSION

Substances possessing electrical properties like those of selenium and bismuth are always in great demand. Just at present, if a substance could be found which was more sensitive to light and more reliable in behavior than is selenium, there would be a revolution in the development of apparatus which enables one to see the person with whom he is talking over the telephone, and apparatus for telegraphing photographs, etc. A substance possessing the power of greatly varying its resistance in a weak magnetic field, as bismuth does in a strong, would be in great demand for the construction of telephone relays, wireless detectors, and many other similar instruments.

In our study of electro-chemistry, we shall consider many of the peculiar chemical and electrical properties of substances which may have great bearing on the development of new kinds of electrical apparatus in the future. Surely the recent views of electricity do much, at present, to clear up the doubtful views which we may now hold as to the "why" of many electrical phenomena.

Quality of Leather used in Belts

So much inferior leather is sold for belting that a test of some sort is of great importance to the user of belting. Cut a small piece from the belting to be tested and place in vinegar. If the leather has been well tanned and is of good quality, it will remain in the vinegar without any change other than a slightly darker color. If, on the other hand, it is of inferior grade, the fibers will promptly swell, and after a short time be converted into a gelatinous mass. This variety of leather is of no use as belting, and should be avoided, as it will not wear well and will prove an expensive proposition to the purchaser.

ADVANTAGES OF A HIGH-SPARK FREQUENCY

ROBERT E. BRADLEY

If two alternating currents of the same intensity but of different frequencies be sent through a telephone, it is found that the sound in the telephone produced by the current of higher frequency is much louder than that produced by the lower. This fact is due in part to the peculiarities of the human ear, which is more sensitive to high-pitched sounds than to low, and also due in part to the diaphragm of the telephone, which is usually of such a weight that it will vibrate most readily to a sound of rather high pitch. This fact has an important bearing on wireless telegraphy, for the pitch of the sound produced in the telephone connected to the detector at the receiving station depends simply on the number of sparks per second at the sending station. In order to determine exactly what is the relation between the strength of current required to produce an audible sound in the telephone and the frequency, a series of experiments were recently carried out on a pair of head telephones of the type usually used in wireless telegraphy.

It was found that it required about a thousand times as much voltage at a frequency of 60 to produce a sound as it required at a frequency of 900. We may assume, therefore, that if the number of sparks per second at the sending station be increased from 60 to 900, and the spark length kept the same, the effect at the receiving station would be increased one thousand times. If the number of sparks per second be increased in this way without reducing the spark length, it is evident that the energy made use of at the sending station must be greatly increased. If we assume that the energy is proportional to the number of sparks, and divide the relative increase in loudness of sound in the telephone for any frequency by the relative increase in the number of sparks per second, we will have a fair comparison of the efficiencies at the two frequencies. The results show that there would be a very slight advantage in replacing a 60-cycle alternator giving 120 sparks per second with one giving 240 sparks or a 120-cycle machine, but that the advantage increases rapidly as the frequency is increased. The maximum sensitiveness of the telephone appears to be in the neighborhood of 900.

In addition to the increase in sensitiveness of the telephone at high frequencies, there are other quite independent advantages in the use of a high-pitched spark. First, it is found in practice that a high-pitched musical signal is much more readily distinguished at the receiving station in the midst of ordinary interference and atmospheric disturbances; and second, at the sending station a shorter spark gap, which would generally be used with a high-frequency spark, puts less strain on the insulation of the condensers and other parts of the circuit, and reduces the losses due to brush discharges, which, in many stations, amount to a considerable share of the total power used.

A third advantage is that larger amounts of energy can be radiated from a moderate sized antenna without subjecting it to excessively high potentials.

Experiments have recently been carried out in which it has been shown that in moderate spark frequencies with stationary gaps there are nearly always secondary discharges, irregular, but giving very high tones, so that the advantage of high spark frequency, from the standpoint of telephone sensitiveness, is usually less than that indicated above. The advantages of ease of reading, the lessening of the strain on the instruments, and the increase in effective energy capacity of the antenna, especially when the latter is small, are very marked, so that it has been found possible to use small sets of 2 k.w. where formerly 5 and 10 k.w. were used.

Handy Cement for the Laboratory

Small pieces of gutta percha, which are sometimes discarded as useless, can be used to good advantage by dissolving them in benzole, and adding a little carmine or other color. This solution when brushed upon corks or other caps forms a tight-fitting and very efficient cement, impenetrable to air, dampness, alcohol and acids. When desirable to remove any article coated with this solution, a simple turn is all that is necessary and no difficulty is experienced from sticking, as is often the case with other cements.

SOME TYPES OF SLIDE VALVE MOTORS
"CHASSIS"

The term "slide-valve" includes the rotary valve (which simply slides in a circumferential direction) and the piston valve. It is the purpose of this article to describe some of many varied designs which fall under this broad definition of the slide valve.

The Knight motor should, perhaps, be described first, because it was the first type of slide valve motor to score a pronounced success in the motor-vehicle field and because it was, in great measure, the success achieved by this motor that set the designers of many a big concern to work developing the slide valve we have today.

The distinctive feature of the Knight motor is the pair of sleeves which reciprocate between the piston and the walls of the cylinder. The sleeves (A and B, Fig. 2) each have two ports on opposite sides near the top. The ports I and I_1 register with each other, and with a port I_2 in the cylinder wall during the suction stroke. Similarly, during the exhaust stroke, ports D and D_1 register with each other and with the port D_2 in the cylinder wall. During the compression and the working strokes none of these ports coincide in such a way as to permit communication between the combustion chamber and the outer air.

The sleeves are actuated by short connecting rods E, which join the sleeves to eccentrics carried on a shaft S. The latter is positively driven from the crankshaft (either by chain or gearing) in such a manner as to make one revolution to two revolutions of the crank. The eccentric driving the outer sleeve is displaced from 60 to 90 degrees from the eccentric driving the inner sleeve. This arrangement results in a motion of the two sleeves which is difficult to follow

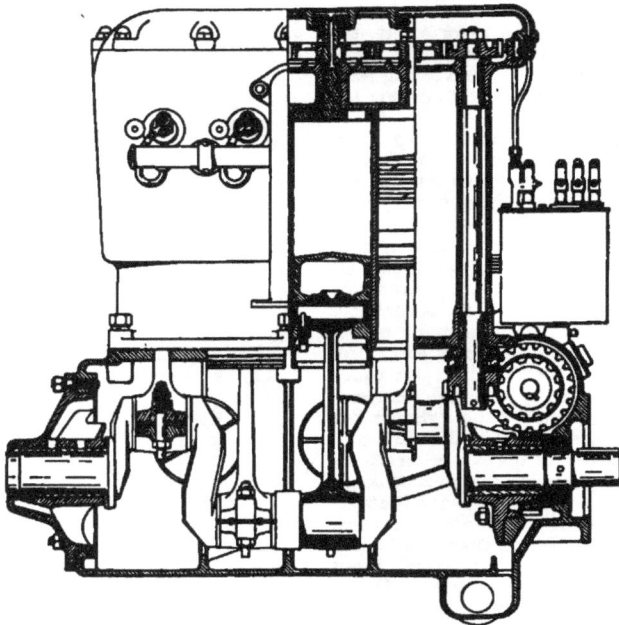

'Fig. 1
Reynolds' Rotary Plate Valve Motor

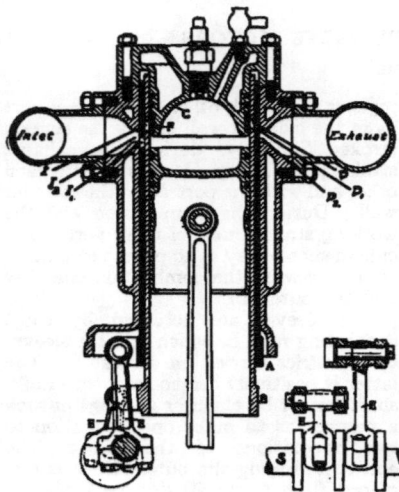

Fig. 2
Section of a Panhard-Knight Motor

in the mind, but which is a most advantageous one in respect to proper functioning of the motor.

The junk ring F is, in principle, the same as a wide piston ring. It bears against the inner sleeve, preventing leakage, and protecting the ports during the firing stroke.

It is worthy of note in this design, first, that the port opening may be made very large without sacrifice to the shape of the combustion chamber, thus yielding great power output for a given bore and stroke. Second, that the pressure occurring in the cylinder is balanced, i.e., causes no pressure on the valve seat against which the valve must move. Wear is therefore but slight, and the problem of proper lubrication is correspondingly simplified. Third, that the ports of the valve which might be injured by the high temperature occurring within the cylinder are protected at the time of ignition and during a greater part of the expansion stroke. This fact also tends to minimize leakage.

The rotary valve motor has advantages in the way of simplicity which no other type possesses. Many engineers believe that some valve of this type will prove to be "the valve of the future." One of the simplest forms of the valve consists of two cylindrical members; which are really nothing more than pieces of solid cast iron "shafting" with slots drilled through one diameter. These valves are driven at one-quarter crank-shaft speed and run in a seat cast in the cylinder and so water-jacketed that excessive heating is eliminated.

One valve functions the exhaust, while the second controls the inlet. In a multi-cylinder engine using this construction the cylinders are cast en-bloc, and the two valves extend the length of the casting (parallel to the crank-shaft), one on each side of the cylinder head. Opposite each cylinder is placed the slot which uncovers, at the proper moment, the port through which the gases enter or leave the cylinder. These two pieces perform, in the case of a four-cylinder motor, the same functions performed by eight poppet valves, each with its spring, push rod, cam follower and cam—certainly a big saving in parts and a great gain in simplicity.

The operation of the valve is so simple that it hardly need be explained. The slots are so placed that they come opposite the ports in the cylinder wall at the proper time to permit the entrance of the charge and the exit of the burnt gases. During the compression and working strokes the valves seal the ports, preventing leakage. The valves are driven by worm gearing or some other form of positive mechanism.

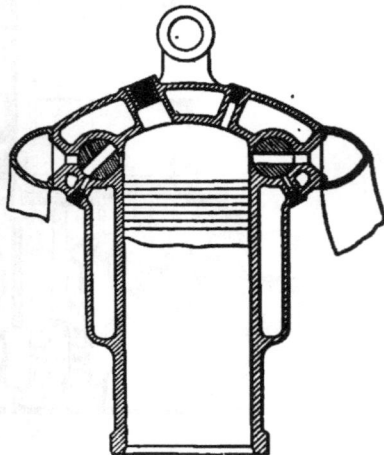

Fig. 3
Mead Rotary Valve Motor

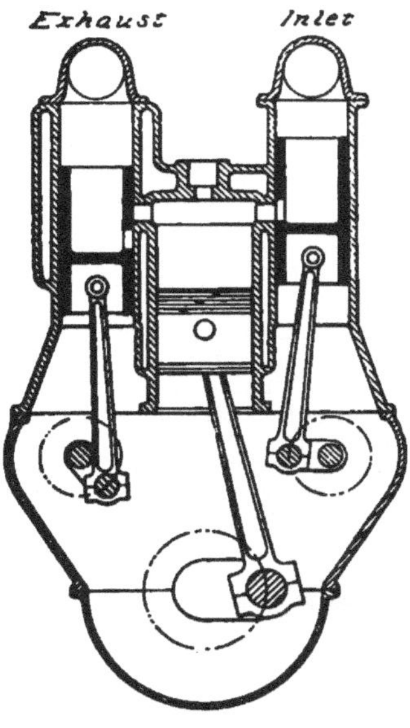

Exhaust   Inlet

Fig. 4
Piston Valve Motor

At each side of the opening through the valve is a solid portion which rests in a bearing. One manufacturer states that he makes the clearance between valve and bearing about one and one-half thousandths, while the clearance in the zone opposite the ports is two thousandths of an inch. This clearance is apparently sufficient to allow for expansion by heat.

The valve is, of course, open to the pressure occurring in the combustion chamber and must, therefore, have adequate bearing surfaces. The latter are easily provided for, however.

Among the advantages of the construction are the following:

1. Simplicity in design (both of valve and cylinder casting) and small number of parts.

2. Silence due to uniform rotary motion and absence of striking parts, none of which have a reversal of strain due to a reciprocating motion.

3. Ample port area without sacrifice to improper shape of combustion space combined with positive functioning at all speeds. Also low speed of those surfaces of the valve which rub against the housing.

This valve mechanism has the disadvantage of tending to wear out-of-round and of a tendency to leak around the seat, short-comings which are not difficult to minimize (almost to the point where they become negligible) by good design. Some trouble is likely to be encountered also in boring and reaming the long valve seat so as to be perfectly true. This difficulty is not unsurmountable, however, requiring simply the proper tools and a reasonable amount of care.

A modification of this type,—which is an improvement in some particulars, although it is open to the criticism that it increases complication and cost—has been suggested. It contemplates placing a single sleeve between the piston and the cylinder wall, the sleeve to be operated in a manner similar to the sleeve of the Knight engine. In this case, the rotary valves are relieved from pressure occurring during the compression and working strokes, the sleeve ports registering with the ports in the cylinder wall only when the gases are entering and leaving the cylinder. This arrangement practically eliminates leakage and should reduce the wear on the valves very materially.

Another form of rotary valve is the disc-type, shown in Fig. 3. It consists of a circular plate rotating on a stem which is co-axial with the valve. Valves of this type are usually flat but may be "dished," i.e., conial in form, the seat having, of course, the same shape as the face of the valve. Disc valves are placed against the cylinder head which forms their bearing surface. In the latter are cut the ports with which the openings in the valve register when admitting the charge and letting out the exhaust.

Disc valves are actuated by keying to the stem a gear which is positively driven from the crank-shaft. In a multiple cylinder design, such as is shown above, a chain of gears makes a very neat construction. The advantage of this type of valve over other types of rotary valves is largely one of construction. It can be made without special machine equipment. It possesses most of the advantages of the rotary type in general, however, although its disadvantages are perhaps somewhat greater. Chief among these is the fact that it operates at high temperatures and under high pressures, although the latter are, fortunately, periodically reversed, reducing to a value below atmospheric as often as they reach the high maximum. This condition tends to prevent the oil film from being forced out. Again, the rubbing speed of the valve near its periphery is high, unless its diameter is kept small, and lastly, the friction is likely to be rather high because anti-friction bearing surfaces cannot be used where the temperature is high, and because abundant lubrication (such as the connecting-rod bearing gets in a splash system) would create a smoky exhaust and deposits of carbon.

Still another type of sliding valve is the piston type—a form of valve which, like some of the other types mentioned above, had its prototype in certain forms of the steam engine valve. The piston valve has been applied in many forms, as have also the rotary valves just mentioned. One form is shown, in section, in Fig. 4. The pistons, driven from half-time crank-shafts, cover and uncover the ports in much the same manner as the sleeves of the Knight motor. There is no apparent reason why the piston used as valves should give any more trouble than the working piston. To this extent, therefore, the design is favorable, and should work well. It is, however, a rather cumbersome and bulky construction, being costly, also, in its manufacture. Nevertheless, it has many points in common with other slide valves which render it superior to the poppet valve. As a result, it has met with some favor abroad, and may prove its worth after more general use.—*Gas Review*.

SILENT LANGUAGE OF THE SAW MILLS

The accompanying set of illustrations, showing some of the silent signs used in the mills in the United States which are generally understood by mill men, is reproduced from the *West Coast Lumberman*.

The signs, up to and including twelve, are given simply by raising the hand, as indicated. From 13 to 19, inclusive, they are given by placing the hand in position as indicated, and then drawing the same across the body from left to right.

The illustrations showing the fractions are given as examples of how the signs are are combined. In some cases it is not possible to give these signs, where there are combinations, in one movement. For instance, $3\frac{1}{4}$ cannot be given at one time, as the three first fingers represent three, and the little finger a quarter, so, given at the same time, it would be four; it is given, therefore, by first giving the sign of three, then closing the three fingers and raising the little finger for the quarter. Three-quarters following any unit is given by first giving the sign of three, then following with the little finger. The same thing pertains to $\frac{1}{2}$, the thumb representing the half. For example, $4\frac{1}{2}$ cannot be given with one motion, as a combination of the four fingers and thumb make five; it is given, therefore, by first raising the four fingers, with thumb closed, then closing the four fingers and raising thumb.

In giving the sign for an eighth, the sign for eight, index finger down, is used. Take $7\frac{3}{8}$ as an example; hand closed with thumb up for 7, followed by three fingers up, then index finger down for $\frac{3}{8}$.

Instructions to turn the log are given by raising open hand with palm out, then dropping same to side.

The order to set log for cutting off slab is given by raising closed fist and holding same up until the log has been set at proper place, then dropping fist to side.

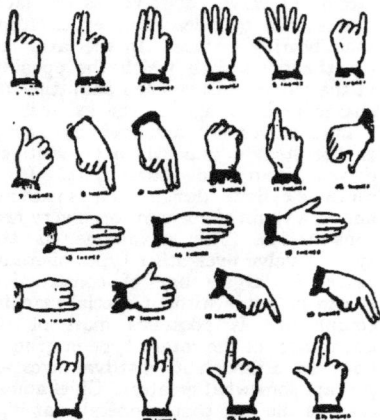

Signal Code Between Sawyers and Cutters

EASILY MADE CHUCKS FOR SMALL BENCH LATHE

The purpose of this article is to describe some simple chucks, that are easily and cheaply made, and that are suitable for one of the many small bench-lathes now on the market. Also they may be made from stock material, without the bother and expense of obtaining castings, which is a simplification the amateur usually appreciates. For the purpose of giving exact measurements, as a guide, it has been assumed that the screw-thread on the mandrel nose is ½ in. Whitworth, and ½ in. in length.

Assuming that the size of the mandrel-nose is as given above, it should be pointed out that the over-all dimension given for these chucks described are not rigid, in one way, at any rate. They may be made larger, if desired, but not smaller, as the dimensions given are practically the minimum to gain the requisite strength. Of course, if the stock material at hand is

larger, it can be used as it is, without wasting time by turning down. The metal to be used, in most cases, may be either brass or mild steel (except where the metal to be used is specifically stated).

Of course, if the mandrel-nose on the lathe is larger than ½ in., the outside dimensions of the chucks given on the drawings will necessarily have to be larger in some cases; but the maker should easily be able to determine that point for himself.

FORK CHUCK

The first chuck required for wood-turning is the fork chuck, shown in Figs. 1 and 2. For this a piece of metal 1 in. in diameter by 1 in. in length is used, being drilled and tapped right through ½ in. to fit the mandrel-nose, as shown in the section Fig. 1. A piece of mild steel rod, ½ in. in diameter by 1 in. long,

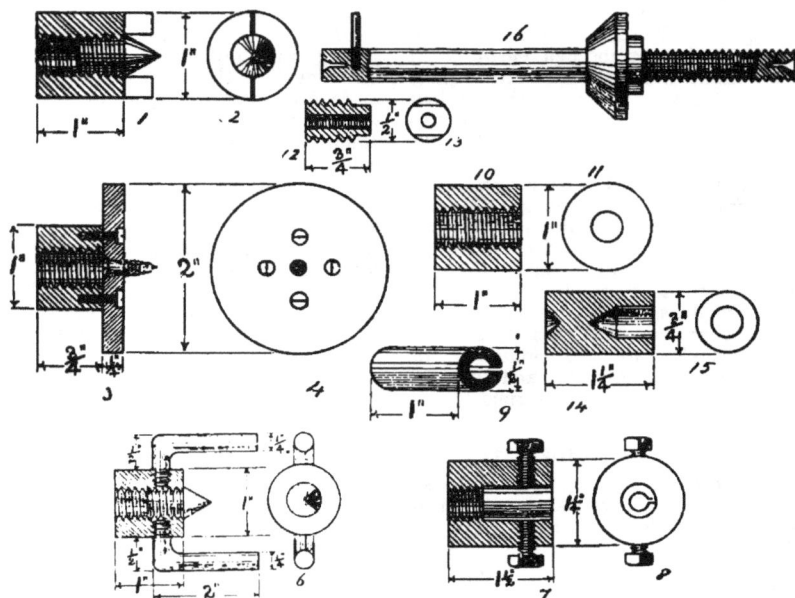

Fig. 1—Section of Fork Chuck, used for Wood-turning. Fig. 2—Front View of Fork Chuck. Fig. 3—Section of Taper Screw Chuck, used for Wood-turning. Fig. 4—Front View of Taper Screw Chuck. Fig. 5—Section of Carrier Chuck, used for Metal-turning. Fig. 6—Front View of Carrier Chuck. Fig. 7—Section of Drill Chuck. Fig. 8—Front View of Drill Chuck. Fig. 9—Metal Sleeve, for holding Drills in Drill Chuck. Fig. 10—Section of Chuck for Holding Screws. Fig. 11—Front View of Chuck for Holding Screws. Fig. 12—Section of Screwed Sleeve, for Chuck for holding Screws. Fig. 13—Front View of Screwed Sleeve. Fig. 14—Section of Female Center. Fig. 15.—End View of Female Center. Fig. 16—Spindle for Circular Saws and Emery Wheels.

is screwed next for ½ in. up, and screwed into the body of the chuck, which is then placed on the mandrel, and the projecting piece of mild steel rod is turned to form a center as shown in Fig. 1; also the face of the chuck should be trued up. Next the center is unscrewed from the body of the chuck, and a saw-cut made across the face of the chuck (or a groove made by a thin file instead) for the accommodation of two rather thin pieces of flat steel, as shown in Fig. 2. These pieces of steel should be embedded firmly in the chuck, one way of doing this being to file a small groove along the sides of the pieces of steel exactly on a line with the face of the chuck, which may be then riveted over into the grooves.

TAPER SCREW CHUCK

Another necessary chuck for wood-turning is the taper screw chuck, shown in Figs. 3 and 4. For this the best material to use is brass. One piece 1 in. in diameter by ¾ in. in length, and another 2 in. in diameter and ¼ in. thick, is required. The smaller piece is drilled and tapped right through ½ in. to fit the mandrel-nose, and the larger piece is then attached by means of three or four small screws, as shown in the section Fig. 3. The chuck is next placed on the mandrel, and a small hole made through the exact center of the large plate, while revolving in the lathe, which is then unscrewed from the smaller portion, and the small hole is reamered out from the back (using a taper reamer) to fit a wood-screw, as shown in Fig. 3. This hole should be countersunk at the back, and the screw is then securely soldered in place. Replace the circular plate, screwing tightly in place, and then the face of the chuck may be turned true in the lathe.

CARRIER CHUCK FOR METAL

For turning metal work between centers, a carrier chuck, as shown in Figs. 5 and 6, is used. For the body of this a piece of metal 1 in. in diameter by 1 in. in length is required, being drilled and tapped right through ½ in. to fit the mandrel nose, as shown in the section Fig. 5. A piece of mild steel rod 1 in. long by ½ in. diameter is screwed for ½ in. up, and screwed into the body of the chuck, which is then placed on the

mandrel, and the projecting piece of steel rod turned to form a center, as shown in Fig. 5. Next, two pieces of mild steel rod, 2¾ in. long by ¼ in. diameter, are bent as shown in Fig. 5, and have a thread put on the bottom end for a short distance, being then screwed very tightly into the body of the chuck, as shown.

DRILL CHUCK

A drill chuck is a very handy addition to a lathe, and a simple one is shown in Figs. 7, 8 and 9. The body of this chuck consists of a piece of metal 1¼ in. in diameter by 1½ in. in length, and is drilled and tapped right through ½ in. to fit the mandrel nose. Next, the body is placed on the mandrel, and the projecting portion of the screwed hole is bored out to the full ½ in., which will leave a smooth, central hole in the body 1 in. long by ½ in. diameter. Next, two holes for ¼ in. set-screws are drilled and tapped, opposite to each other, as shown in the section, Fig. 7, these screws serving the purpose of gripping the drills securely. There are two ways of completing the drill-chuck. The first is, assuming that the drills to be used are made from steel rod ¼ in. in diameter, to make a metal sleeve 1 in. long by ½ in. diameter, as shown in Fig. 9, with a ¼ in. hole running through the middle (this hole being bored in the lathe, with the sleeve in position in the chuck), and with a saw-cut down one side. The use of this is obvious, the metal sleeve being placed inside the hole in the body of the chuck, and the drill inside the hole in the sleeve, when, owing to the slit, the drill will be securely gripped on tightening the two set-screws. The set-screws also offer a means of very fine adjustment for centering the drill accurately. The second method is, assuming twist-drills are to be used, to make a separate metal sleeve (1 in. long by ½ in. diameter, as described above) with a central hole to fit each different size of drill used.

CHUCK FOR HOLDING FINISHED SCREWS

For facing and turning heads of finished screws a chuck that will not spoil the thread on same is required, and such a one is shown in Figs. 10, 11, 12 and 13. The body of this chuck consists of a piece of metal, 1 in. in diameter by 1 in. in length, drilled and tapped right through

½ in. to fit the mandrel-nose, as shown in Figs. 10 and 11. For holding the screws, screwed metal sleeves are used, ¾ in. long by ½ in. diameter, as shown in Figs. 12 and 13. These sleeves screw into the body of the chuck, and each one has a hole drilled in the exact center (while revolving in the lathe, to insure accuracy), and tapped to fit the screw being turned. Of course, a number of sleeves with different sizes of hole will be required to fit varying screws. For removing these sleeves, a flat is filed on opposite sides of same, as shown in Figs. 12 and 13, to enable a spanner to be used for the purpose.

FEMALE CENTER

In drilling work mounted on the face-place a female center will be required to fit over the center on the loose head-stock. This may be made from a piece of metal 1¼ in. in length by ¾ in. in diameter. A hole is drilled at one end as shown in the section, Fig. 14, to fit over the loose headstock center, on which it is placed next, and a center drilled, as shown in Fig. 14, by means of a drill (with a suitable angle at the point) placed in the drill chuck on the mandrel.

SPINDLE FOR CIRCULAR SAWS, ETC.

A simple method of making a mandrel for holding a circular saw or an emery wheel, is as follows: A piece of mild steel rod, of suitable diameter, has the ends marked, and center holes drilled, as shown in Fig. 16, as accurately in the middle as possible. Next, a brass or iron collar is either shrunk on or made a driving fit, and securely fixed with a screw right through into the steel rod. After screwing a small steel rod into the end, as shown, to act as a carrier, the mandrel may be placed in the lathe, and the portion to be screwed together with the collar turned true and to size. Finally, screwing the portion required completes the mandrel.

All of the chucks described will be found to answer their purpose very well in practice, although being of easy and simple construction, the assortment given should be found quite sufficient, with the addition of a face-plate (which should be purchased, as being more satisfactory) for all ordinary work required to be done in a small lathe.—*English Mechanic and World of Science.*

The World's Best Lighted Tower Clock

FELIX J. KOCH

Those qualified to know claim for the great clock in the tower of the Metropolitan Life Insurance Building, on Twenty-third Street, New York City, the distinction of being the best-lighted tower-clock in the world. The clock is connected with an electric cont roller

Metropolitan Building

which announces the arrival of the hour by 88 white lamps bursting into flame, and of the quarter and half hours by the illumining of 56 red lights.

Self-sufficiency is soldered down by useful knowledge, and men's minds become less arrogant in proportion as they become better informed.—BISHOP OF LITCHFIELD.

HOW TO MAKE AND INSTALL A STATIONARY VACUUM CLEANING SYSTEM

H. M. NICHOLS

The writer was lead to take up this subject by an inquiry from the owner of a light automobile for a vacuum cleaning system that he could operate from his automobile engine. It was his idea to buy a pump and rig up the rest of the system himself. This idea appeals to the writer as being the simplest solution of the problem, as outside of the pump the apparatus is quite simple and easy to make. With this idea in view, this article will be confined to the description of a suitable dust collector and other parts necessary to form a complete vacuum cleaning system, outside of the pump, which, it will be assumed, is already on the job.

The complete vacuum cleaning system can be divided into several heads, as follows: the pump or vacuum producer, the dust tank, the piping system, and the hose and cleaning tools.

The vacuum pump may be either of the rotary, the reciprocating, or centrifugal fan types. If one of the first two types of pumps mentioned (high vacuum types) is installed, it should have a displacement of at least 35 cu.ft. of air per minute, and should give a maximum vacuum of not less than 5 in. of mercury. If a centrifugal or fan exhauster (low-pressure type) is employed, it should have a displacement of at least 75 cu.ft. of air per minute, and be capable of producing a vacuum of at least 25 in. of water.

With the pump set in place, build a solid framework of timber around it. Use long lag screws to hold the pump firmly in place. Locate the pump as near the dust tank as possible, in order to save frictional losses in the piping. Make sure that the pump has the proper sized pulley to run it at its rated speed, when belted to the source of power. Use a piece of vacuum hose to connect the air inlet on the pump to the discharge pipe *J* of the dust tank, or if the run is a long one, use iron pipe and make the end connections with short lengths of hose. If it is desired to remove the foul air from the house, along with the dirt, pipe the exhaust side of the pump to the chimney or carry it through the wall, and allow the air to exhaust out-of-doors.

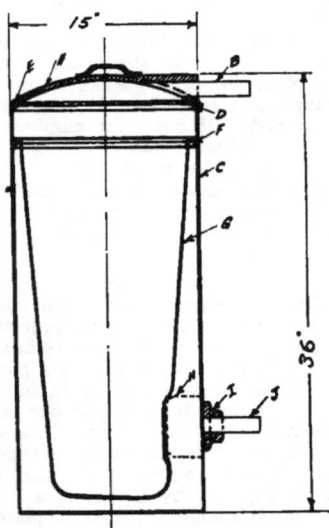

Dust Tank For Vacuum Cleaner

The dust tank is made from heavy sheet iron, with the seams either welded or soldered together. All the joints around the seams and rivets should be filled with iron cement or putty. Give the tank two coats of black shellac, soaking it in well around the joints.

The particular tank shown in the drawing is 36 in. high and 15 in. wide. However, the builder will probably find it more convenient to make the size of dust tank and parts fit the materials that he may have at hand, rather than attempt to build the tank to a given set of dimensions. With this idea in view the detail dimensions have been purposely left off from the tank drawing.

The dust is removed from the tank by lifting off the cover *A*, and then taking out the cloth bag *G* and shaking the dirt out.

The tank has an iron ring *D* riveted ¼ in. below the top. This ring forms a channel or groove which should have a strip of rubber cemented in it. This rubber ring serves as a packing, to insure a tight joint between the cover and the top of the tank. About 6 in. below this ring a second ring F is riveted to the inside of the dust tank. This second ring serves as a support for the dust bag. The tank can be held securely to the floor by riveting several angle straps to the side of the tank and fastening them to the floor with screws.

The dust tank cover A is made from either a brass or aluminum casting. The surface that comes in contact with the rubber packing should be machined smooth. In addition, the cover has a projecting ring on the inner side that holds it from slipping off the tank. A boss is provided on the side of the cover to carry the inlet pipe B. This boss is drilled to such a size that the inlet pipe can be driven in place, and make a snug fit. Of course the size of this pipe will depend on the size and kind of vacuum hose used. For a handle for the cover, use a malleable iron door handle, held on with small machine screws.

The dust bag G is made from extra heavy muslin or cotton cloth. It is gathered around a brass ring at the top and is tapered slightly to the bottom. A piece of packing is put between the dust-bag ring and the flange F, to keep any dust from leaking by the bag and thus getting into the vacuum pump. Make certain that all seams are tight and that there are no imperfect places in the cloth where the dust can leak through.

The exhaust pipe for the dust tank is shown at J. It is fitted into a flange fitting F, which is riveted to the tank. On the inside of the tank and over the inlet to this pipe there is a wire mesh screen H. This screen is intended to keep the dust bag from being drawn into the exhaust pipe by the rush of air. This screen should be about 10 in. in length and should hold the bag about 3 in. away from the mouth of the pipe.

Unless it is possible to reach all parts of the house from the dust tank, with the length of hose available, it will be necessary to run a pipe to the various floors. For this purpose, use standard weight wrought iron pipe and long turn recessed drainage fittings. These fittings are known as "Durham" fittings, and are the ones used by plumbers. They are especially adapted to vacuum cleaning piping, as when they are screwed up they

make a bore that is smooth and continuous with the pipe, there being no sharp edges or ridges left for dirt to catch on and stop up the system. Wherever there is room to do so, always use the long turn fittings in preference to short turn fittings, as the long turn fittings are much less liable to stop up, due to their greater radius. Take care to have the pipe free from burs and pins, and screw the joints up tightly, using red lead on them. Provide inlets having air-tight covers for the hose on each floor. Use a short length of flexible vacuum hose to connect the bottom of the piping to the dust tank. The hose should be fastened to the inlet pipe on the dust tank in such a manner that it can be easily removed when it is desired to take the cover off for cleaning out the dust bag. If the hose makes a tight fit, just a slip joint will be sufficient to keep it from leaking.

For high vacuum pumps use $1\frac{1}{4}$ or $1\frac{1}{2}$ in. pipe, and for low vacuum pumps use 2 or $2\frac{1}{2}$ in. pipe, for the air line.

The size of hose and tools will vary with the type of pump used. The high vacuum pumps (rotary and reciprocating pumps) use $\frac{3}{4}$ in. and 1 in. hose, with tools to correspond. The 1 in. size will do far better work than the $\frac{3}{4}$ in. size, and in places where extra long lengths of hose are required, it would be desirable to use $1\frac{1}{4}$ in. hose. The low pressure type of exhauster (centrifugal fan) will require $1\frac{1}{4}$ in. or $1\frac{1}{2}$ in. hose. The latter size will do the best work, but it is somewhat awkward to handle.

For sweeping carpets and rugs some type of carpet renovator and connecting rod will be required. These can be obtained from some local dealer in vacuum machines and tools. For cleaning and polishing hard wood floors, make a wooden shoe to fit on the cleaning face of the carpet renovator. Cover the rubbing surface of the shoe with heavy felt, leaving a slot to correspond with the slot in the renovator. For cleaning corners and places hard to get at with the carpet renovator, use the end of the vacuum hose. A good scheme would be to look at a set of cleaning tools and then duplicate any that might be desired, using heavy sheet copper or brass bent to shape and held together by soldering the joints.

When the vacuum system is all connected up and before starting the pump for the first time, prime the dust bag. This can be accomplished by putting about a quart of ordinary flour in the inlet to the dust tank and starting the pump slowly. The inrush of air will carry the flour into the pores of the cloth bag and render it dust proof. Then bring the pump up to full speed, stop up all the inlet valves on the system, and go over the piping and tank, listening carefully for any hissing noises, indicating leaks. If any leaks are found, stop them with putty. There will probably be a slight leakage that it will be impossible to locate and stop. To determine about what percentage of the total capacity of the pump is represented by this leakage, close all the inlets to the system, reduce the exhaust opening on the pump to about ½ in. and feel the strength of the breeze created by the discharged air. Then open the inlets to the vacuum piping and feel the rush of outflowing air again. These two comparisons will give a rough idea of how much air is being lost through leakage.

AERIAL NAVIES READY

France and Germany Well Matched as Regards Dirigibles

There is wide interest in the revelation just made of the real aerial military forces of France and Germany. While it has generally been believed that in case of war aeroplanes and dirigibles would be advantageously used by both sides, only well-informed persons have realized that two powerful rival aerial navies are already equipped.

It is understood in both countries that immediately after a declaration of war all private aeroplanes and airships will be requisitioned by the Government and they are already incorporated into the military aeronautic corps. Of the dirigible class France today possesses twenty-four airships and Germany, twenty-five; and it was to obtain appropriations to perfect the aerial corps that the French military authorities made known the exact strength of the French and German forces. Public opinion in France demands supremacy in the air, just as English public opinion demands supremacy on the sea.

The following table gives the clearest idea of the two rival airship forces:

FRENCH DIRIGIBLES—MILITARY

FRENCH DIRIGIBLES—PRIVATE			
Clement-Bayard VI	Clement-Bayard	6,200	76
Astra I	Astra-Clement	4,475	18.60
Astra-Torres	Astra-Chenu	16,000	47.72
Croiseur-Transaerien	Astra-Chenu	9,000	76.25
Zodiac I	Zodiac-Ballot	910	35
Zodiac II	Zodiac-Ballot	1,600	42.50
GERMAN FORCES—MILITARY			
Z. I.	Zeppelin-Durr	12,000	134
Z. II, 1911	Zeppelin-Durr		132
M. I, 1908	Gross-Basenach	5,200	74
M. II, 1909	Gross-Basenach	5,200	74
P. I, 1908	Parseval	4,000	60
P. II, 1909	Parseval	6,000	70
P. III	Parseval	10,000	
M. IV, 1910	Gross-Basenach	9,200	
M. V	Gross-Basenach		
M. VI	Gross-Basenach		
P.L. XI	Parseval	10,000	96
Z. XI	Zeppelin	13,000	135
GERMAN DIRIGIBLES—PRIVATE			
P.L. VI, 1910	Parseval	6,800	70
P.L. VIII, 1910	Parseval	5,600	68
P.L. IX, 1910	Parseval	1,500	40
P.L. X, 1911	Parseval	1,510	40
Z. IX, Schwaben 1911	Zeppelin-Durr		
Siemens-Schuchert, 1910	S. Schuchert	13,000	120
Schutte-Lanz	Schutte	20,000	128
Veeh, 1911	Veeh	6,800	70
Bellonsansbelice, 1911		3,800	
Clouth, 1909	Clouth	19,000	42
Gaus Fabrice			
Suchard		12,000	76
Frans Steffen			

Three French airships do not figure in the chart, although they complete the number 24, while all the available German airships are mentioned. The Germans claim to have 29 units, but of these four are not military possibilities.

As to the relative value of the opposing air navies, this has yet to be proved by a conflict, although it may be said that

THE SLIDE RULE

C. W. WEBBER

The purpose of this article is to give the amateur an idea as to what a slide rule is, what its uses are, and why it performs mathematical computations mechanically.

The slide rule is very little known among non-professional men and seems to surprise them exceedingly when they are shown what it will do. The use of the rule is confined almost entirely to engineers and students in scientific schools.

The slide rule was made possible by the invention of logarithms about 1614, by John Napier, Baron of Merchiston, and their improvement by Professor Briggs. In 1630 the invention of the slide rule, by Edmund Wingate, followed.

The slide rule is an instrument by means of which mathematical computations can be performed mechanically. It consists of scales so arranged that one can be moved past the other.

In order to explain the instrument it is necessary that the reader understand the principle of logarithms. Any number can be expressed as the power of any other number; for example, 4 equals 2 raised to the second power; that is 2^2; and 64 equals 4 raised to the third power; that is 4^3. Now suppose we express all numbers as powers of 10; that is 10 equals 10^1, 100 equals 10^2, 1,000 equals 10^3, etc. It is seen that the numbers between 10 and 100 can be expressed as powers of 10, but that the powers will be greater than 1 and less than 2. For example it has been worked out that 45 equals $10^{1.6532}$. The exponent is called the logarithm of the number and the number which is raised to the power is called the base. Thus 1.6532 is the logarithm (commonly called log) of 45 to the base 10. The system of logarithms used in ordinary practice has 10 for a base, although the base used in higher mathematics is 2.718. The first system is called the Briggs or Common System, while the other is the Napierian or Natural System. We will, however, only concern ourselves with the Common System. Logarithms of numbers have been worked out and tabulated, so that to find the logarithm of any number it is only necessary to refer to a table.

A few logarithms are given below:

Number	Logarithm
1	0.0000
2	0.3010
3	0.4771
4	0.6021
5	0.6990
6	0.7782
7	0.8451
8	0.9031
9	0.9542
10	1.0000
20	1.3010
30	1.4771
100	2.0000
200	2.3010
300	2.4771

There are two parts to every logarithm: the part to the right and the part to the left of the decimal point. The part to the right is called the Mantissa, and the part to the left is called the Characteristic. Thus, in the above table log 300 equals 2.4771; 2 is the characteristic and 4,771 is the mantissa. The characteristic simply shows the location of the decimal point, while the mantissa shows what the number is. In the above table it is seen that 3, 30 and 300 all have the mantissae equal to 4,771, so that if we did not have

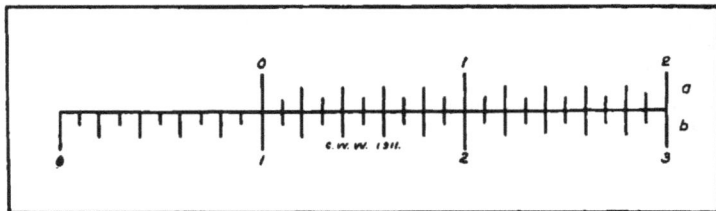

Fig. 1.

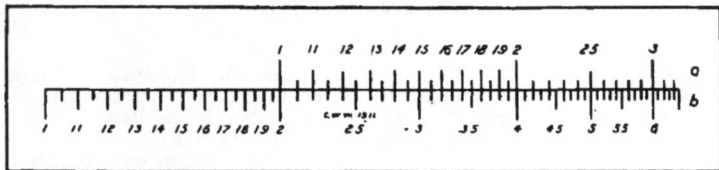

Fig. 2.

the characteristic, we could not tell what the number was.

In algebra, to multiply two numbers together we add their exponents thus, $X^3 \times X^2 = X^5$. Now as a logarithm is an exponent, if we wish to multiply two numbers together we add their logarithms. Suppose, for example, we wish to multiply 5 by 2. From the table we get

$$\log 5 = 0.6990$$
$$\log 2 = 0.3010$$
$$\overline{\hspace{2cm} 1.0000}$$

In the table we find that the number having a mantissa equal to .0000, and a characteristic equal to 1 is 10.

In dividing numbers we subtract exponents, so that in dividing 10 by 5 we get

$$\log 10 = 1.0000$$
$$\log 5 = 0.6990$$
$$\overline{\hspace{2cm} 0.3010}$$

which gives us 2, for 2 is the number whose logarithm is 0.3010.

Let us now apply this to the slide rule.

Suppose we have two scales equally divided and arranged so that one can be made to slide past the other. In Fig. 1 let the zero of scale a be the index. It would be well to follow this through, using two rulers and sliding one past the other. Now, if we place the index of scale a at the 1 on scale b, we have measured off one unit on scale b. Look along scale a until we come to the division marked 2, and under it on scale b we find 3; that is, we have laid off one unit on scale b and added to this two units by means of scale a, giving us three units as a result on scale b. In the same way any two numbers can be added.

Now, suppose that instead of the scale Fig. 1 we have a scale in which the divisions represent the logarithms of the numbers instead of the numbers themselves. Such a scale is shown in Fig. 2.

The divisions are, however, marked with the numbers corresponding to the logarithms. In this case 1 will be the index. Suppose we wish to multiply 2 by 3, we lay off 2 on scale b, add to it 3 by means of scale a in the same manner as above, and the result is 6 on the scale b. In order to follow this through, the writer would advise that the reader obtain from a dealer in mathematical instruments a sheet of logarithmic plotting-paper and cut out two strips about $\frac{1}{2}$ in. wide, and slide one past the other. The cost of a sheet of this paper is five cents.

To divide one number by another the operation is the reverse of the above; that is, to divide 6 by 3 (Fig. 2), we lay off log 6 on scale b and subtract from it log 3 by means of scale a, giving us 2 on scale b.

Any two numbers can be multiplied together in the same manner as above, so it is seen that the process of multiplication has been reduced to moving a slide. This is a great saving of time and labor.

There are many other things that can be done with the slide rule; for example, finding the squares and square roots of numbers, working out problems in proportion, finding the logarithms of numbers, and several other things. These will not be taken up, however, as they are explained at length in instruction books which come with each instrument.

In closing, the writer would advise the student to buy a cheap slide rule for practice, so that he may become familiar with this wonderful but simple instrument.

"Kind lady, I'm just merely trying to keep
  Soul and body together!"—he did look thin;
But the lady did neither smile nor weep,
  As she handed the tramp a safety-pin!

PRACTICAL OPERATING FAULTS OF THE ALTERNATING AND DIRECT CURRENT MOTOR

WM. G. MEROWITZ

It is always interesting to the practical mechanic to understand the construction and principles of operation of all machinery, especially those mechanisms which are associated with his daily work or to which most of his study is given. But a mere understanding of operation is not the only essential to a thorough knowledge of the machine performance. With this point in view, a few words on every-day faults in the operation of electric motors will not be amiss.

For alternating current circuits we have the synchronous motor, commutator motor and the induction motor, while for direct-current circuits we are using only a commutator motor. Of the A.C. motor, the type most commonly used at present is the induction motor, either of the single- or poly-phase winding, and of the squirrel-cage or wound-rotor type.

ALTERNATING CURRENT MACHINES

Before taking up the common faults in A.C. motor operation, it may be well to review briefly just what is going on in the performance of the squirrel-cage induction motor. If by closing a 3-pole switch, we connect the stator winding of a 3-phase motor to the line, then the armature, while at rest, corresponds to the secondary winding of a transformer, although the construction is somewhat different from that of an ordinary transformer. And the field does not pulsate like that of a transformer, but rotates. This rotating field produces electromotive forces both in the stator and rotor windings. The counter e.m.f. now produced in the stationary winding is like that of the primary coil of a transformer, nearly equal to the applied voltage, so that only the magnetizing current flows in the primary winding, when the secondary windings are not short-circuited. But the magnetizing current of a motor must be much larger than that of a transformer, for the lines of force do not only flow through iron, but have to pass through two air gaps. Although the space between rotor and stator is kept as small as possible, a much greater magnetizing

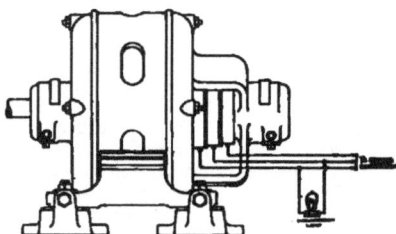

force is required than in the case with a magnetic flux having a path entirely of iron. On starting the motor, the field rotates with full speed around the still stationary armature; hence, an excessive current will be produced in the short-circuited armature or rotor winding, which reacts on the stator field with the effect of so weakening it that a large current is drawn from the line. This lasts only for a short time, for the current flowing in the rotor winding causes the rotor to start with considerable torque. The quicker the rotor runs, the nearer it approaches the synchronous speed of the rotating field, and the fewer lines of force it will therefore cut. Consequently the e.m.f. and current induced in the rotor decrease, the reaction on the field becomes smaller, and the stator absorbs less current from the mains. To avoid these rushes of current, we must not short-circuit the rotor windings, but connect them with slip rings, so that a resistance may be placed in the rotor circuit, which is used for starting only, for when actually working under a load, there is no difference between a "short-circuit" and "wound rotor" secondary, as we finally short-circuit the starting resistance when the rotor has reached its maximum speed. With the foregoing explanation of operation principles, a discussion of the faults of A.C. motors will be better understood.

Sometime we may want to know whether the motor has a correct rotating field connection or not. With a properly connected 2- or 3-phase motor of the wound rotor type, this can be readily determined by observing the voltage induced in the windings of the rotor. A lamp connected with the slip rings, as shown in Fig. 1, will burn regularly as long as the rotor circuit, through the starting resistances, is not closed. The position of the rotor does not make any difference, since the field rotates with a uniform speed about the stationary rotor.

With a single-phase motor, however, we have no rotating, but a pulsating field; and the lamp would burn either brightly, or with little light, or no light at all, according to the position of the rotor coil in the pulsating field. Hence, if we connect a lamp or voltmeter with two slip rings of a wound rotor, and with a slow rotation of the armature, we observe that the voltage between the two slip rings varies considerably, then we can infer that there is something wrong with the revolving field. In the case of a squirrel-cage motor, one of the phases of a 3-phase motor may be connected in, in a reversed order, as in Fig. 2. This would be a mistake made in the shop where the machine was built, and would be recognized by the motor failing to attain its full speed. Then the three currents entering the motor are not alike, as they should be, and often at starting a considerable humming will be noticed. Another fault with induction motors is a sudden shut-down and resulting blowing of fuses. Upon investigating, all circuits may be O.K., and, in fact, the motor was probably operating satisfactorily for some time. But the common cause is the rotor rubbing on the stator teeth, on account of no air gap. Air gaps of induction motors are generally made very small, to insure good power factor for the magnetizing current, which lags behind the applied voltage and thus lowers the power factor. The magnetizing current is used principally to force the lines of force through the air gap, the iron parts of the magnetic circuit not taking much. Therefore, the smaller the air gap, the better the design, and in the case of a shut-down or blowing of fuses, the air gap should be investigated. This rubbing is also injurious to the winding, and, since the energy represented by it is shown as load, it may be of sufficient magnitude to destroy the insulation, introducing short-circuits and grounds. Low voltage on the line is another cause of induction

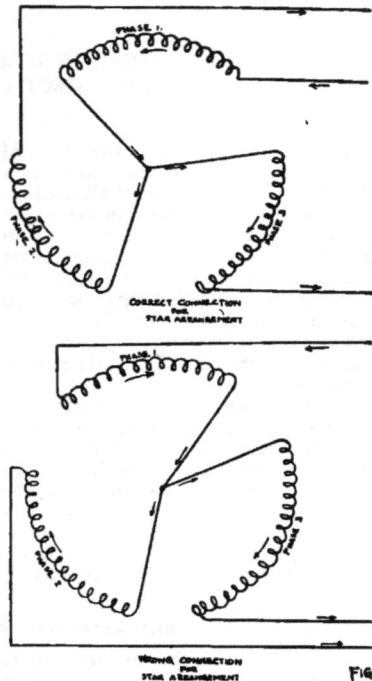

CORRECT CONNECTION FOR STAR ARRANGEMENT

WRONG CONNECTION FOR STAR ARRANGEMENT   FIG. 2

motor trouble, since the output of an induction motor is proportional to the square of the applied voltage. Hence, if a motor has swings of load, carrying it up for a moment to someting near its maximum output, it may break down under such load conditions if the voltage is low. In starting a motor, too, a low starting torque will result from a low voltage. Low voltage on a motor may be caused by an unbalanced voltage on the line.

Short-circuits in the stator winding are a source of considerable bother with poor insulation. Such a short-circuited coil does not burn out at once, since the current induced in it by the pulsating flux, opposes the flux. About three times the normal voltage flows and creates a local heating, which may affect other coils, until finally the motor becomes inoperative. In a plant using motors it is well to measure the insulation resistance once a month, at least, to locate such faults soon after they appear.

If a single-phase motor will start free but will not start under load, it shows that either the line voltage is too low, the load too great or the frequency too high. The speed of the motor is directly proportional to the frequency of supply, therefore with high frequency we have a high speed. But the output is also proportional to the product of torque and speed; hence for a constant horse-power output the torque will decrease if the speed increases, and, as the torque is the real turning effort of the motor, the machine will not start if the torque is too low, due to a high frequency. The voltage and frequency with motor running should be within about 5 percent of the nameplate rating, and the voltage within 10 percent to 15 percent while starting. Vibration in an induction motor may be due to a shaft that is sprung or by a steady unbalanced magnetic pull caused by uneven air gap clearance, or the eccentricity of the rotor. If one phase of the stator is open-circuited in a Y-connected 3-phase motor, the motor will not start alone, and if started mechanically, it will operate as a single-phase motor, with a material reduction of power. An ammeter connected in each phase will give current readings in only two of the phases. If the motor is delta-connected current will flow in all three leads, but they will be unbalanced. If, however, the motor is not started mechanically, the rotor will remain stationary, acting as the secondary of a static transformer. In such events the whole machine will rise to an exceedingly high temperature, due to the heavy current drawn from the line; and if the line switch is closed for a long period, the conductors on the rotor will be badly burned or even melted in as many places as there are poles on the stator.

DIRECT CURRENT MACHINES

In direct current motors there is a liability of more operating faults occurring than in alternating current motors. This is owing to the simplicity and ruggedness of A.C. motor design and the more complicated D.C. construction, with its commutator and sparking troubles. All faults likely to occur in D.C. motors will produce one or more bad effects, which may be classified as follows: (1) Sparking at commutator; (2) Heating of armature; (3) Heating of commutator; (4) Heating of bearings; (5) Noisy operation; (6) Voltage. not right; (7) Speed not right, or (8) Motor stops or fails to start. Any of these effects is evident to a person making a careful examination, and the next step is to find out if two effects are not being produced by one cause.

We will take up the causes and remedies of the common faults enumerated above. (1) *Sparking at Commutator.*—This is a common trouble that is not usually objectionable, if moderate in duration and amount. If allowed to continue beyond these limits it will burn and roughen the commutator, thus encouraging the difficulty. Sparking may be caused in several ways, among which are: too much current on armature, brushes not set at neutral point, commutator has high bars or poor brush contact. To decrease the effect of sparking on account of armature carrying too much current, the driven load should be reduced, the strength of the magnetic field should be increased or decrease the size of driving pulley. If the motor starting-box has too little resistance, it will cause the motor to spark badly at first, owing to a very sudden start. In this case the only remedy is an increase of resistance to cut down the voltage applied to the armature terminals.

To find · the correct neutral position for the brushes, carefully shift the brushes backward or forward until sparking is minimized. If the brushes are not exactly opposite in a two-pole machine or 90 degrees apart on a four-pole machine, they should be made so by counting the commutator bars and dividing by either two or four, as the case may be. To remedy high bars on the commutator or flat bars or projecting mica, smooth the commutator with a fine file or fine sandpaper, the latter being applied by a block of wood which exactly fits the commutator. If, however, the commutator is very rough or eccentric, it should be turned down in a lathe. On large machines, a slide-rest attachment is usually provided for either turning off or grinding the commutator without removing the armature from the bearings. To improve the brush contact, draw a strip of sandpaper back and forth beneath the brush with the rough side scraping the carbon. See that the brush holders work freely; this may be the cause for poor brush contact.

(2) *Heating of Armature.*—This may result from either moisture in armature coils, eddy currents in armature core and conductors, unequal strength of magnetic poles, or excessive current in the armature coils.

Moisture in armature coils is not a very common occurrence, but a motor that stands in a damp place, or one that has been inoperative for a long time, is likely to absorb some moisture in the armature windings. A sympton of this fault is that the armature takes considerable power to run free. This is really a case of short-circuit, as moisture has the effect of short-circuiting the coils through the insulation. To remedy this, the armature should be placed in an oven or other sufficiently warm place to drive out the moisture, but not hot enough to injure the insulation. A convenient way is to pass a current through it, about three-quarters of full-load current, and occasionally turn the armature over by hand.

Excessive amount of eddy currents in the armature iron will make the core hotter than the coils after a short run, and considerable power will be required to run the armature when the field is established and there is no load on the motor. Sparking is absent with temperature rise, due to eddy currents. To eliminate these stray currents, the armature core must be more perfectly laminated, which is a matter of first construction. If there are excessive eddy currents in the conductors and not in the iron, the conductors will become hotter than the core on no load. This trouble is due to the difference of voltage induced on the two sides of each armature conductor. In this case the conductors should be reduced in thickness or split up into a number of strips or strands, which should be twisted to equalize the effects. Beveling off the edges of the pole pieces may also reduce the trouble.

Unequal strength of magnetic poles will cause excessive currents to flow in the armature, thereby heating it abnormally, in the case of multipolar machines with parallel winding. This unequal strength is usually due to the fact that the armature is closer to one or more poles. This condition may be corrected by slightly shifting the bearings, but in some cases, especially when direct-connected, it is preferable to shift the field magnet. When the armature gets out of center from too much bearing wear, however, the proper procedure is to replace the bearings with new ones.

(3) *Heating of Commutator.*—This fault, like sparking, may occur in D.C. machines, and in the commutator types of A.C. machines. There are various causes for commutator heating, among which are: Heat spread from another part of machine, sparking, carbon brushes heated by current, friction of brushes on commutator or bad connections in the brush holder. If the heat in the commutator comes from another part of the machine, start up the motor with the parts cool and run for a short time. The seat of the trouble is in the part that heats first. Any of the causes of sparking will cause heating, which may be slight or serious.

An overheated commutator will decompose carbon brushes and cover the commutator with a black film which offers resistance to the efficient collection of current. Carbon brushes require less attention than copper, because they do not cut the commutator and their high specific resistance usually reduces sparking, but it may also cause them to heat more than copper brushes. The friction produced by the brushes will generate heat, which can be detected even when the brushes carry little or no current. Reduce the spring tension and decrease voltage, keeping up speed by weakening field strength. A little lubrication of a high-grade mineral oil, applied sparingly, will help to reduce the friction.

(4) *Heating of Bearings.*—The cause of bearings heating should be found and removed promptly, but may be reduced temporarily by applying cold water or ice to them. This should be resorted to only when it is absolutely necessary to keep running, and great care should be taken not to allow any water to get upon the commutator, armature or field coils, as it might short-circuit or ground them. If the bearing is very hot, the shaft should be kept revolving slowly, as it might stick to bearing, if stopped. Heating of bearings is due to a lack of oil, shaft rough, grit or other foreign matter in bearings, shaft and journal too tight, shaft sprung,

or bearings our of line, too great a load on belt, or bearings heated by hot armature, commutator or pulley. A rough shaft should be turned to smoothness in a lathe and the bearing fitted to the new diameter. It is very difficult to straighten a bent shaft; it might be bent back or turned true, however, but usually a new shaft will be necessary. In lining up bearings by either raising or lowering the bearing on its seat, or by moving it sideways, it is well to note that an even air gap must be maintained at all times, to avoid any trouble due to an unequalized magnetic pull on the armature. If there is too great a load or strain on belt, which would cause heating of bearings, reduce the belt tension or use larger pulleys and lighter belt, so as to relieve the side strain on the shaft. The slipping of the belt on the pulley may heat one or both bearings.

(5) *Noisy Operation.*—This may arise from various causes among which are: vibration due to pulley or armature out of balance, shoulder on shaft or shaft collar, strikes against bearings—this results from the armature being out of magnetic center—squeaking of brushes, flapping of belt or noise of belt joint going over pulley and slipping of belt on pulley.

To detect vibration due to unbalancing of any of the revolving parts, place the hand on the machine while it is running and change the speed. The vibration almost disappears at some speeds. The proper balance should be effected in such a case.

If the noise comes from a rattling against the bearings, it is evident that either the armature core is not properly located on the shaft or the bearings are too long. Squeaking brushes are usually the result of a rough or sticky commutator. The brushes on a new machine may be noisy, but this will be reduced after the machine has been running for a day or so.

For a noisy set of brushes, which may be detected by a sound of high pitch, apply a very little oil to the commutator with a cloth free from lint or threads. Carbon brushes are apt to squeak in starting up or at low speeds, but as the speed increases the noise diminishes. Always clean the commutator thoroughly after sandpapering it or filing it to smooth up the surface.

The flapping of a belt or the pounding of the belt joint can be distinguished from any other sound about the motor by its periodic occurrence: once for each complete revolution of the belt. To remedy such unnecessary noise, use either an endless belt, that is, one with ends glued together, or smooth up the joint by one of the many ways of lacing a belt.

(6) *Voltage not Right.*—This is a common difficulty which may arise with any machine. The main trouble is some fault of the generator supplying the energy to the line on which the motor is running. This generator trouble may result from the speed of the prime mover being too high or too low, field strength not right, brushes not in proper position or short-circuited armature or field coils. It is seldom that any trouble is experienced with the motor starting-box. However, if the line voltage is too low, the inexperienced operator would naturally think that something was wrong with the starter, when the motor would not come up to speed. It is advisable when the motor speed is low to test out for the line voltage the first thing, and a great deal of time may be saved, which would be important in a shop depending upon the motor for its driving power.

(7) *Speed not Right.*—This is generally a serious matter in an establishment where reliable speeds are essential. The speed may be either too low or too high. A low speed may come from either a low voltage, overload, short-circuit or ground in armature or shaft tight in bearings. A high speed may result from either a weak magnetic field, voltage too high, or motor too lightly loaded. Any of these faults can be easily remedied by adjusting the voltage, field or load to the name-plate rating. A short-circuited armature coil is often caused by a piece of solder or other metal getting between the commutator bars or their connections with the armature. If the short-circuit is within the coil itself, the only effective remedy is to rewind the coil.

(8) *Motor Stops or Fails to Start.*—This fault may arise from a variety of conditions, and usually causes some worry at the outset, because one knows that there are 101 things that can occur in a direct-current motor to make it stop or fail to

start. Oftentimes, after closing the switch and pushing over the starting-box lever, the operator will begin his hunt for trouble on the interior of the machine, when he finds it failed to start. As a matter of fact the real trouble is generally on the outside and is usually an open circuit of some kind. Such an open circuit may either result from a fuse blown out, a circuit breaker open, wire broken or slipped out of connection, circuit supplying motor open or brushes not in good contact with commutator. However, a motor may not start with an extreme overload or from very excessive friction due to armature touching pole pieces or shaft, bearings and other moving parts being jammed.

If the load is too great when the switch is closed, the motor will draw an excessive current, which will either melt the fuses or open the circuit breaker. Without these safeties in the circuit, the armature would likely burn out. The field circuit of a shunt-wound motor may be open, in which case the poles would not be magnetized and the armature would be drawing a large current. In such a case, or even when the field is only weak, the motor armature is apt to burn out, unless there are fuses or a circuit breaker in the circuit.

From the foregoing discussion of the practical faults in electric motor operation, it is easily seen that a thorough knowledge of the principles of operation of electric motors is of utmost importance to the efficient, economical operator. He need not be concerned with the design of these machines, but he must know what makes the machine go and how to keep it going.

SOME THINGS WORTH KNOWING

H. W. H. STILLWELL

To Make Corks Air Tight and Water Tight

In experimenting, much trouble is often experienced in the chemical laboratory in getting corks to be perfectly air or water tight, or to drill holes in them for the insertion of glass tubing, etc. Melt a quantity of paraffin in any suitable vessel, and allow the corks to be treated to remain beneath the surface of the melted paraffin. The corks may be held down by a perforated lid, wires, or any other convenient arrangement. Corks which have been treated in this manner can be cut or bored with ease and the exterior is perfectly smooth. When introduced in the neck of a flask or other piece of chemical apparatus, they form a perfect seal.

A Few Practical Hints

Clean and oil leather belts without breaking them off their pulleys. If taken off they will shrink, then a piece must be inserted in them and removed again after the belt has been run a few days.

For leading steam joints, mix the red lead or litharge with common commercial glycerine instead of linseed oil.

Too little attention is sometimes given to the bearings of engines, shafting, and other machinery. Often from 25 to 50 percent of the power is consumed through the lack of good quality of oil, or the lack of oil altogether. Machinery requires common-sense care, and if it is received, then you will be repaid many times by the increased efficiency of your equipment, and the life of same will be prolonged indefinitely.

The decay of stone, either in buildings or monuments, may be arrested by heating and treating with paraffin mixed with creosote. A common paint burner may be used to heat the stone.

In tubular boilers the hand-holes should often be opened and all the collections removed from over the fire. When boilers are fed in front, and are blown off through the same pipe, the collection of mud, sediment or scale in the rear end should be often removed.

Nearly all smoke may be consumed without the aid of any special apparatus by attending to a few simple common-sense rules. Suppose we have a battery of boilers and "soft coal" is the fuel. Go to the first boiler, shut the damper nearly up and fire up nearly one-half of the furnace, close the door, open the damper and go to the next boiler and repeat the same thing. By this method nearly all the smoke will be consumed.

RESTORING PICTURE FRAMES

G. F. RHEAD

The better class of ornamental picture frame, decorated by means of composition ornaments in gilt, is well worthy of restoration, especially the old-fashioned ones which may often be picked up, in a damaged condition, quite inexpensively from second-hand stores. They are quite within the scope of the amateur worker to repair, providing reasonable pains are taken and sufficient of the modeled ornament is left from which to make up missing parts. A fair sample of this class of frame was recently purchased by the writer for a mere song, same having been rather badly knocked about. Fig. 1 will give some idea of its condition, when it will be noticed that in many parts the ornament has been entirely broken away, although one corner is fairly complete, which fact will make it a comparatively easy matter to repair the rest. As the four corners of these frames are in almost every case alike, the method of repairing

is to make a plaster mould of the undamaged portions, and from this mould to take "squeezes" in composition to replace those that are missing.

The first matter to see to is whether the miters of the old frame are firmly secured. Very often the wire nails which held them together have lost their hold, and in such a case new holes should be opened with the brad-awl and fresh nails entered. After this the frame should be given a thorough washing with warm soap and water, working in all the interstices with a small hog's-hair brush, the most convenient shape being shown in Fig. 2. Then set the frame aside to dry.

Usually with a little touching up one corner can be made pretty complete. In the case of the frame illustrated, the corner marked x had been only slightly damaged, so the missing parts, which were hardly noticeable, were built up by applying small pieces of composition

Fig. 1. Frame for Repairing

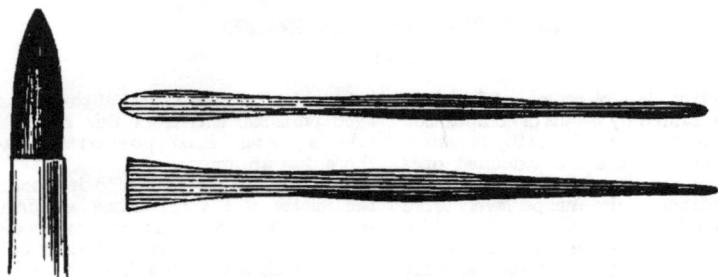

Fig. 2. Hog-hair Brush and Repairing Tools

formed into the shape of the missing parts as near as possible, with the aid of a modeling tool shaped from a piece of hard wood, two useful forms of which are shown in Fig. 2. If the best corner is rather badly damaged, molds are taken from the others of the required parts and impressions made, and so it is built up complete. The whole corner is then oiled and a little plaster of Paris mixed up to a fairly thick consistency and applied, care being taken to expel air bubbles from the hollows by blowing the wet plaster well into the crevices. This is left for a couple of hours to harden, when it should leave the frame quite easily if the portions were carefully oiled with linseed oil in the first instance. Fig. 3 shows two molds taken from the frame illustrated in Fig. 1.

In the case of frames that have been so badly damaged that the ornament is unrecognizable, the only method of restoring the frame is to entirely model a new corner and from this make a mold, as described, and take squeezes for the others. If the worker has a little knowledge of drawing and form, the modeling of such designs as shown in Fig. 4 will be a matter of no great difficulty to him. With the aid of a few tools and a suitable plastic material the production of frame corners will be a positive delight.

Composition or "compo," as it is known in the trade, may be obtained from dealers in picture frame makers' materials, fibrous plaster workers, etc. It may be made at home quite easily and then costs very little. The components are whiting, oil, glue and a little resin. Set in a jar to melt over the fire: four parts boiled linseed oil, six parts glue, and one part resin, allowing the mixture to boil together until all is thoroughly liquified, then add sufficient whiting to work up into a stiff dough. This makes a good "compo" that sets as hard as stone, but needs to be warmed and worked up each time, before using, to bring it to a plastic condition. In making it, the working up should be continued for some time, for the more it is worked the better

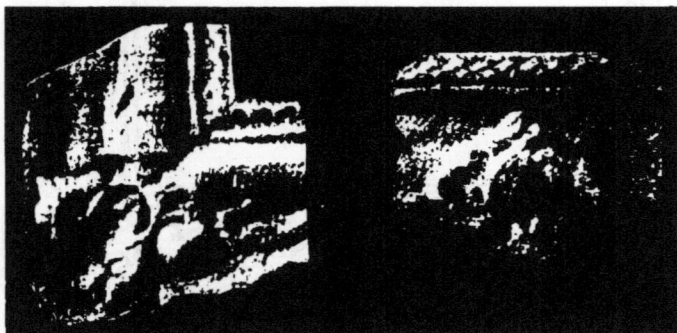

Fig. 3. Plaster Molds of Details

Fig. 4. Composition Frame Ornaments

it will become, and the less likely are lumps to interfere with a good impression.

The plaster molds having been made and dried (the drying being necessary to prevent sticking), an examination of the frame should be made to note the missing portions. If a great deal of a corner has been demolished, perhaps the best way will be to take an impression from the mold of the whole corner in "compo" and apply it to the frame, having previously removed the existing portions. In such a case, take a small piece of the compo, press it out flat, and then work it well into the depressions of the mold. Back it up with more if necessary, and cut the back smooth with fine wire. It will be found to leave the mold quite easily, if the latter is dry, by gently drawing away from one corner. The ornament is then applied to the frame, the latter having been given a thin coating of glue to afford an attachment. In the case of a small break, an impression is taken of this part only, and the piece of composition is cut in shape as near as possible to the piece broken away. Extreme care is necessary to avoid damaging the ornaments while in a plastic state, by pressure or otherwise, and should such damage occur it will be found more expeditious to take another copy from the mold than to attempt to restore the modeling to any extent, by hand. The wooden tools previously referred to, however, will be found most useful for pressing the sections in their places, and supplying small pieces of ground if they do not just fit.

The whole of the missing enrichments are replaced in this manner and cracks and fissures filled in with the composition, also holes where the brads have been entered (if any), when the frame should be left for a day or so, to thoroughly harden, and is then ready for gilding.

A SIMPLE RULE FOR FINDING THE DIRECTION OF A CURRENT

ROBERT E. BRADLEY

The novice often asks for some simple method for telling the direction of a current in a circuit. The ordinary rules for the deflection of a magnetic needle as the "swimming rule," the "hand and thumb rule," and the "cork-screw rule," are confusing to anyone, and especially so to the novice. The following rule is, perhaps, the simplest to be found:

"To tell the direction of the current in an ordinary galvanometer, look at the coil from either the east or the west side, and if the south pole of the needle be the nearer of the two poles to the observer, the current flows clockwise in the coil. If the north pole be the nearer, the current flows counter-clockwise."

The rule appears confusing at first, but it is easily remembered in the tabular form:

"South pole nearer, current flows clockwise.

"North pole nearer, current flows counter-clockwise."

By the "north pole of the needle" is meant the pole normally pointing toward the north pole of the earth, the "south pole" meaning, of course, the other pole. By "clockwise" is meant the direction in which the hands of a clock turn; "counterclockwise" meaning the opposite direction.

The same rule may be applied to a current flowing in a straight wire or other conductor. The wire having been placed in a north and south position, with a small pocket compass either above or below it, by regarding the wire as an arc of a circle having the compass needle at its center, the rule is easily applied.

A SQUARE-CUT CHAIR

A chair suitable for the living room can be made from any one of the furniture woods. The materials can be secured from the planing mill, and the list can be made up from the dimensions given in the detail drawing. The front legs, as well as the back ones, are made from $1\frac{3}{4}$ in. square stock, the back ones having a slope of 2 in. from the seat to the top.

All the slats are made from $\frac{7}{8}$ in. material and of such widths as are shown in the detail. The three upright slats in the back are $\frac{3}{4}$ in. material. The detail drawing shows the side and back, the front being the same as the back from the seat down. All joints are mortised in the posts, as shown. If making dowel joints they must be clamped very tight when glued and put together.

The seat can be made from one piece of $\frac{7}{8}$ in. material, fitted with notches around the posts. This is then upholstered with leather without using springs. Leather must be selected, as to color, to suit the kind of wood used in making the chair. The seat can also be made with an open center for a cane bottom by making a square of four pieces of $\frac{7}{8}$ in. material about 4 in. wide. These pieces are fitted neatly to the proper size and doweled firmly together. After the cane is put in the opening it is covered over and upholstered with leather.

Before the leather is put on, however, the chair should be stained and French polished, if a plain wood has been used; or polished only in the case of a better variety.—*Hobbies.*

THE CONSTRUCTION OF A SMALL WATER MOTOR

B. F. DASHIELL

Wherever the water pressure is sufficiently large, a water motor is one of the cheapest and easiest methods of obtaining light power.

It is especially desirable for running small dynamos for any length of time, such as in charging storage batteries, operating window displays, etc.

It is the purpose of this article to show in a simple manner, the construction of a water motor to operate any small dynamo

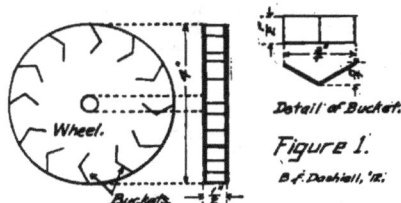

Detail of Bucket.

Figure 1.

B. F. Dashiell, '12.

Wheel. Buckets.

having an output of 40 watts and under. For the construction of a water motor to operate larger machines, see the table of dimensions.

I have used a 25 watt dynamo in connection with this water motor and have had excellent results with same.

The water motor is very easy to construct, as it has no complicated parts.

Fig. 1 shows the wheel and buckets, with all necessary dimensions. It is made of heavy tin and all parts should

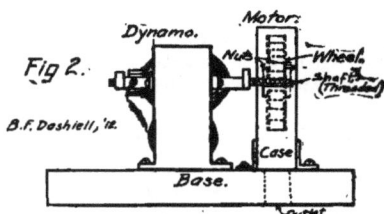

Fig 2.

B. F. Dashiell, '12.

be well soldered. As shown, the wheel has two sides, which gives stronger construction, but if the extending end of the dynamo shaft is not long enough, then the wheel may be made with only one side. This makes a somewhat weaker wheel.

The dynamo shaft should be threaded

to within ¼ in. of the bearing. This is done so that the wheel may be clamped between two nuts. See Fig. 2. The wheel should be painted with red lead so as to prevent rusting.

The containing case is made out of a tin box or can. It should be about 6 x 6 x 2 in. in dimensions. The place where the shaft goes through is found by experiment. A ¼ in. hole is punched through at this place.

The nozzle is shown in Fig. 3. A cross sectional view is given. One end is soldered up and has a ⅛ in. hole drilled in it. This is to reduce the diameter of the issuing jet of water and to increase its kinetic energy.

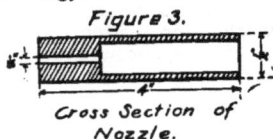

Figure 3.

Cross Section of Nozzle.

B. F. Dashiell, '12.

A round hole is cut in the case and the nozzle soldered in place. Fig. 4 shows this and the correct direction of the nozzle. Care must be taken to direct the nozzle correctly as an error in its direction would seriously interfere with the successful operation of the motor.

The dynamo and water motor should be mounted on a heavy wooden base and connections made to a drain pipe so as to provide for the waste water. The

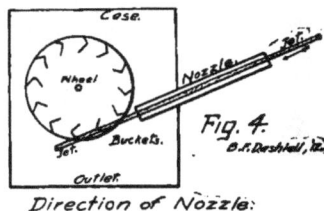

Fig. 4.

B. F. Dashiell, '12.

Direction of Nozzle.

Table of Water Motor Dimensions.					
Greatest Horse-Power.	1/8	1/6	1/4	1/2	3/4 to 1½
Dia. of Wheel.	4"	6"	9"	12"	16"
Greatest Speed.	3500	3000	2400	1800	1200
No. Buckets on Wheel.	12	12	16	20	24
Length of Bucket.	3/4"	3/4"	1"	1"	1⅛"
Dia. Nozzle Inlet. (Jet).	1/8"	1/8"	3/16"	7/8"	1/4"
Diameter of Water Mains.	1/2"	1/2"	1/2"	3/4"	1"
Material in Wheel.	Heavy Tin	Heavy Tin	Sheet Brass 1/16"	Brass 1/4"	Brass 1/4"

nozzle is connected to a faucet by means of a hose. All connections should be very tight, as there is a high pressure in the hose and nozzle. The water motor ought to be provided with a front cover so that the interior may be readily inspected. The whole should have a coat of red lead on the outside and inside.

This motor attains a very high speed, owing to its small diameter. Its power will depend on the pressure. The faucet may have to be partly closed to prevent the dynamo from running too fast.

THE SMALLEST PAPER-MAKING MACHINE IN THE WORLD

THOS. H. GROZIER

The smallest paper-making machine in the world, actually producing a finished and continuous web of paper, is shown in the accompanying illustration. It is 8 ft. long by 1 ft. wide, and manufactures paper having a width of 4½ in.

The drying cylinders, which are eight in number, are heated by coal gas, an entirely new method, giving excellent results in this instance. When coal gas is not available, as sometimes happens, a special gasoline burner is attached to the cylinders, and the paper is then dried in exactly the same manner as with the coal gas. The machine uses all sorts of pulp and can turn out any kind of paper that is ordinarily produced on a large sized paper machine. The machine possesses a great advantage on account of its portability, and it can be sent from one place to another—no matter what distance—without being disconnected in any way. Owing to its self-contained prop-

erties it is possible to demonstrate the machine without creating any uncleanliness whatever, and it requires the attention of but one person. The motive power for driving is supplied by a ¼ h.p. electro motor, but when electric power is not at hand, a small gasoline engine substitutes.

The Miniature Paper-Making Machine was designed as a means of demonstrating in a practical way the entire process involved in converting fibrous pulp into the finished paper. It has fulfilled this task so well that the original model has been several times duplicated, and these are now being used by practical paper-makers, technical institutions, and even as a means of advertising the stationery departments of large stores, also by publishers who have used it in connection with public exhibits showing the scientific side of modern newspaper production.

MEASURING AN ELECTRICAL CURRENT MECHANICALLY

JAMES A. SEAGER

Every mains engineer and installation contractor knows how important it is to be able not only to detect the presence of an electrical current in a cable, but also if possible to find out how much current there is actually flowing. Furthermore, it is sometimes inconvenient or impossible to disconnect the cable in order to use the necessary instruments of the ordinary type, and where such measuring instruments are not installed, or at any rate the proper calibrated resistance strip which can be used as a shunt to a millivoltmeter is not already in circuit, a considerable amount of ingenuity is necessary in order to find out what exactly is happening in the circuit under observation. It is, therefore, a very good thing for electrical engineers that the London firm of Messrs. Drake & Gorham, Ltd., have introduced an electrical current gauge of a strong and simple character, which is able to give a mechanical means of detecting the passing of a continuous current. This gauge, which is illustrated in front and side view in Fig. 1,

is made under the Frisby patents, and consists of a device somewhat similar to a pair of pliers. One leg of this combination is a U-shaped bend of soft iron, which, under the circumstances becomes an electromagnet. The other leg at B is straight and forms the armature of this magnet, being pivoted at the center J. The continuation of the armature forms the lever, having attached to it a flat spring D, which, when B is closed on A, presses against the screw E. At the other end of this screw is a knob or handle F, to which is attached a pointer P. K is the main body of the instrument terminating in the handle H. Mounted on it is a dial over which the pointer plays, and the dial is calibrated so that the readings can be taken directly in amperes. When it is desired to bring the instrument into use, the pointer is first brought to the zero position and then the conductor which it is desired to test is encircled by the U-shaped bend A. The armature B is then pressed home against the magnet poles, care being taken to see that the

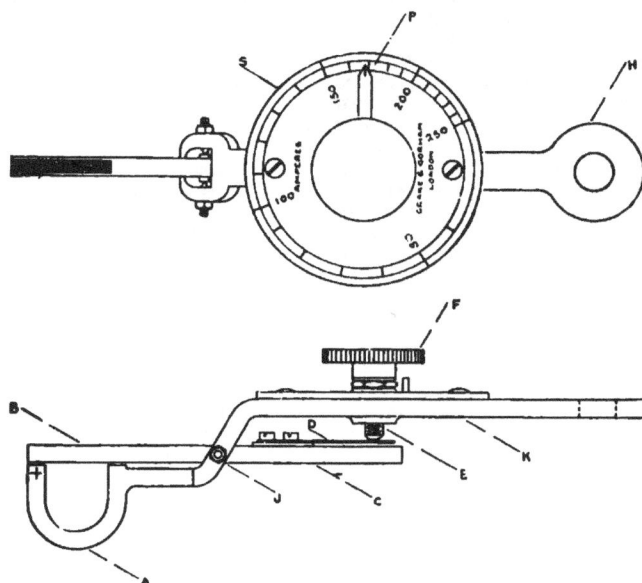

Fig. 1

conductor or its covering is free in the opening thus formed, so that the magnetic circuit does not come into mechanical contact with anything which would impede the movement of the parts. If the cable or conductor presses against the armature, the accuracy of the reading will be destroyed. When this is done the handle is held in one hand, and with the other the knob F is turned as slowly and evenly as possible without any jerking until the armature springs away from the poles of the U-shaped piece of iron. At this instant the pointer will indicate on the scale S the strength of the current which is passing through the conductor encircled by the device.

The readings can be obtained equally well in whatever way the cables are insulated, and are only slightly affected by the presence of steel armoring around the insulation. It is easy to make an experiment or so on a piece of armored cable in order to determine the correction to be applied to such an instrument for the presence of the steel. Vertical and horizontal conductors can be tested with equal ease, and it will be seen that the instrument is a simple application of first principles, namely, the utilization of the magnetic field which always surrounds a conductor carrying the electric current. Hence, the apparatus which depends on magnetic and mechanical effects and is independent of electrical contacts can be used for determining the current passing in a conductor without cutting or making any fresh connections, and this is an advantage which is peculiar to this kind of device alone. Moreover, there are many situations in which cables are placed which are difficult of access and in which the use of an ammeter would be impossible, and here again the advantages

of this appliance are evident. Yet another point to be remembered is that as the action depends upon the field enclosed by the iron loop and its armature, the readings are unaffected by current passing in neighboring conductors.

It is, however, necessary to keep the poles and the armature perfectly clean, and when the direction of current in the detector is unknown, it is necessary to employ a pole finder which may now be described. This is shown in Fig. 2, and is designed on the same lines as the gauge above mentioned, but has a permanent magnet instead of a soft iron U. It is not a current measurer, but is employed to detect the presence and direction of the current in conductors, and two obvious applications of such a gauge may be mentioned. The first is when there is a fault in the cable network; the location of the fault can be easily traced by utilizing the gauge on the various junction boxes and finding the direction in which the current is flowing. It can also be employed for picking out one cable from a bunch, as when a number are run all together in one trench, by passing the current through the cable to be picked out, and testing for the presence and direction of current by means of this detector. The general description of the gauge is the same as the detector with one or two modifications. The detector is held by the handle H, and is first set by turning the knob F until the spring D just overcomes the magnetic attraction on the armature B. This determination of the "critical point" is, of course, done away from the cable to be tested. The conductor is encircled by the U loop, first one way round, pressing the armature B home against R and R_1, and then removing the hand from the armature B.

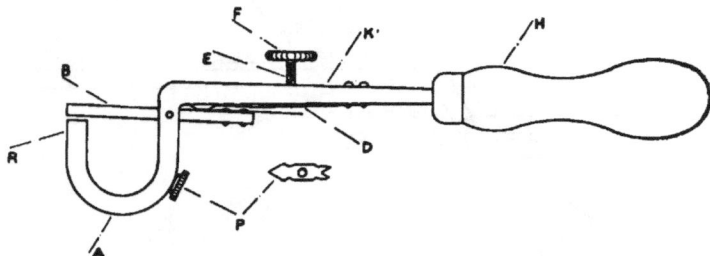

Fig. 2

If the armature now remains against the pole R, the direction of the current as shown by the pointer P mounted on the detector can be read. Supposing, however, that the armature does not remain against the pole R, this means that the magnetism surrounding the cable is to some extent neutralizing the permanent magnetic field of the U loop, and the position of the magnet should then be reversed so as to encircle the conductor again, but in the opposite direction. If it is now found that the armature is holding the magnet against the pole piece, the direction of the current can be read from the pointer. There is yet a third case. Supposing that after trying both ways it is found that the armature does not cling closely to the pole piece

in either position it may be taken, on ordinary power circuits, that there is no current flowing in either direction in the cable. It should not be assumed, however that there is no pressure on the cable, as pressure may be applied at either end of the conductor, and if there is an open circuit there will be a static potential on the conductor resulting in no current, and therefore it would be dangerous to cut into the cable without first ascertaining that pressure was off. For determining actual presence of current, its value and also its direction, the two portable instruments which can be carried in any mains man's pocket or kit bag are invaluable to the saving of time, and will be greatly appreciated by practical men.

A QUICK SHAFT REPAIR
W. F. PERRY

While employed in a shop where motor repairing was a specialty, motors were frequently brought in, the armatures of which were frozen in the bearings, owing to the lack of proper oil.

On taking down the motor and breaking the tight-box away from the shaft, it was found that the shaft at this point was so badly scored and worn down, that it was a good deal smaller than at any place on the shaft. Owing to this a new box of the proper fit could not be made.

After swinging the armature in the lathe, the shaft (on the worn end), was turned down until it was of uniform size the whole length. A piece of steel tubing was sawed off the same length as the shaft, and (being of correct diameter) was forced onto the turned end. Two small holes were drilled through the tubing and shaft and pins driven through to make the sleeve secure. These pins were of a good driving fit. The armature was again placed in the lathe and the sleeve turned off to the regular size of the shaft, which is generally to be gotten from the opposite end of the shaft. When the job is finished up with emery and oil, and all rough places removed, the keyway should be cut through, if there is one, and the new bearing made. Upon re-assembling the motor will run as good as ever.

Shafts usually catch in the bearings, on the back or pulley end, owing to improper care in oiling or having too tight

a belt. A bearing should always be renewed when the greatness of wear is first noticed. Failure to do this usually results in broken binding wires and torn armature coils.

How to Cut Glass Tubing
HOWARD TUCKER

Often, as in the making of oil cups, one wants to cut glass tubing that is over an inch in diameter. This cannot be broken by filing a nick in one side and snapping it by pressing the thumbs on the other side, as may be done in the case of smaller sized tubing. Neither will the often-told way of winding a cord saturated in alcohol around the tube and lighting it do. But, I have used the following way to 'good advantage, and have known it to work nine times out of ten. Bend a $\frac{1}{8}$ or $\frac{3}{16}$ in. round iron rod in the form of a quarter-circle and corresponding to the circumference of the tube to be cut. Then wind a wire around the tube at the point to be cut and heat the iron rod to a cherry red. Then bring the red-hot rod into contact with the tube and slowly revolve the tubing so as to heat all parts evenly, letting the rod press against the wire on the tube so as to make the portion cut off even.

When the rod shows signs of cooling plunge the tube down into cold water and it will break evenly all around at the point meant for it to break.

THE HOME CRAFTSMAN

Ralph F. Windoes

FERNERY

Few articles combining simplicity of construction with beauty of appearance can be offered in a series such as ours, but we feel free to say that the fernery illustrated is foremost in the few. The main body is made of copper and the legs and binding strips around the top are composed of brass, fastened with large copper rivets. Inside is placed a zinc box into which the fern is set.

The material needed is the following, some of which the craftsman undoubtedly has on hand from previous projects:

1 pc. No. 18 gauge copper $14\frac{1}{2}$ x $14\frac{1}{2}$ in.
4 pcs. No. 18 gauge brass 1 x $6\frac{1}{2}$ in.
4 pcs. No. 18 gauge brass, $2\frac{1}{2}$ x $4\frac{1}{2}$ in.
Some large copper rivets.
Steel Wool.
Lacquer (banana oil).
Zinc box, 3 x $6\frac{1}{4}$ x $6\frac{1}{4}$ in.

First, the pattern of the box should be laid out very accurately on a piece of paper, full size, and working from center lines so that both sides will be symmetrical. Transfer this to the copper and cut it out, bending it on the dotted lines shown on the drawing. Sharp bends may be secured by bending it over a hardwood block $7\frac{1}{2}$ in. square.

Next rivet the top strips into place. These may be bent between two hardwood blocks placed in the vise, and afterward hammered nearly together. It may be easier for the craftsman to rivet them before bending up the sides of the copper piece, as he would then have nothing to interfere with his rivet set or hammer.

The corner strips come next. They should be cut and shaped as shown in the detail. A suggestion as to the forming of the correct angle on the bottom and on the projecting pieces would be to cut them straight, bend in the middle and place them on the box where they belong. Then) the correct angle can easily be secured by laying them out parallel with the top of the box. This could be done with one piece and transferred to the other three.

The last operation consists in polishing and lacquering. This may be done with steel wool and banana oil. It is not a good plan to etch or otherwise decorate a piece of metal work in which both copper and brass are combined, as the contrast between the two when highly polished gives a most pleasing effect.

The zinc box should be made by the craftsman if he has mastered the art of soft soldering, but it is suggested that he have it made by a tinsmith if he has not. If the dimensions given on the drawing have been accurately followed, the zinc box will be 3 in. high and $6\frac{1}{4}$ in. square, but it is advised that the maker measure his fernery before ordering the box that size. The box should have five holes in the bottom of it, about $\frac{1}{2}$ in. in diameter; also two rings soldered on the inside for handles.

PLATE RACK

The plate rack illustrated is a combination of small, sharp angles, which give a most pleasing appearance to the whole. The material list is as follows, the lumber, quarter-sawed oak, to be planed and sanded at the mill:

2 pcs. $\frac{7}{8}$ x 3 x 20 in.
2 pcs. $\frac{7}{8}$ x $2\frac{1}{4}$ x $7\frac{3}{4}$ in.
2 pcs. $\frac{7}{8}$ x $3\frac{1}{2}$ x 8 in.
8 pcs. $\frac{3}{8}$ x 4 x $6\frac{1}{2}$ in.
1 pc. $\frac{7}{8}$ x $4\frac{1}{2}$ x 34 in.
1 pc. $\frac{7}{8}$ x 2 x 34 in.
1 pc. $\frac{7}{8}$ x 3 x 34 in.
1 pc. $\frac{7}{8}$ x $4\frac{1}{2}$ x 36 in.
1 pc. $\frac{7}{8}$ x $3\frac{1}{2}$ x $35\frac{1}{2}$ in.
1 pc. $\frac{1}{2}$ x $\frac{7}{8}$ x $33\frac{1}{2}$ in.
$1\frac{3}{4}$ in. 10s flat head screws.
Glue, stain, filler and wax.

The working drawing in the center section of the sheet gives the overall dimensions and shows mortise and tenon joints as most of the fastenings, the sides and shelves being screwed into place.

Cut each piece out as given in the details, and to exact dimensions. The upper shelf is not shown, but it is exactly like the lower, excepting that it is $3\frac{1}{2}$ in.

FERNERY.-

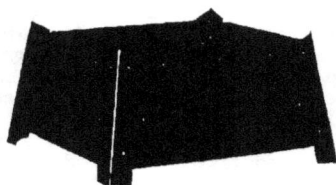

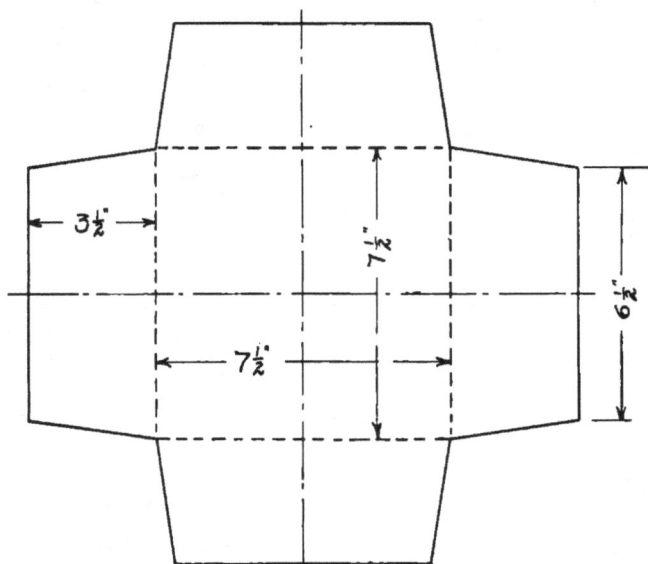

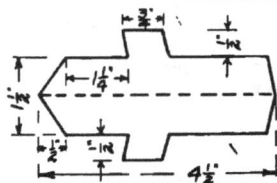

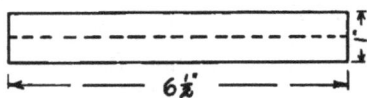

wide, 35½ in. long, and has but one groove down the middle of it. The eight back pieces of the lower section have their front edges beveled with a ⅜₀ in. chamfer, giving them a matched wood appearance.

Where but one dimension is given on an angle, it is taken to be an angle of 45 degrees. The tee-bevel should be set at this angle and each laid off from it.

The back pieces must be placed before the mortises are glued up, or else it will be impossible to get them in. A suggestion might be given that would save the craftsman some time. He could have a groove ¼ in. deep and ⅜ in. wide sawed out of the bottom of piece C and the top of piece D when he orders the lumber, as by so doing he will save much chiseling, unless he happens to possess a grooving plow. The same suggestion might be given for the two shelves, but in case this is done, do not let the grooves be cut entirely across the pieces, as the ends should be covered by the side pieces.

As was mentioned before, the shelves are screwed to the sides and these latter to the back, which conveniently covers all screw heads, except in the bottom shelf, and these will not show when the rack is put to use.

Any style of finish may be applied to the piece, but a dark mission, well filled and waxed, is to be recommended, as it brings out the quartered flake to a better advantage than some of the other finishes.

Cypress as a Wood for Craftsman Furniture

In these articles, and others, quarter-sawed white oak has been recommended to craftsmen as the proper wood to use for furniture construction, and we still believe it to be the best. But in recent years another wood has come into popularity, which, on account of its cheaper price, lasting qualities, ease in working, and ability to take a good finish, is a close contender for the honors oak has gained. This wood is cypress, ordinary American bald cypress, as cut from the cypress swamps of our southern states.

Cypress varies in color from a near-white to a near-black, but the average is a yellowish brown, with the heart wood darker than the sap wood. It is com-

paratively cheap, can be secured from any dealer in the country, and if the craftsman does not care to invest much in a piece of furniture, cypress is to be highly recommended to him; and for outdoor furniture, such as garden seats, porch swings, etc., it stands in a class by itself.

It is a very easy wood to work, as it is rather soft, straight grained, and yields readily to sharp tools.

One of its greatest features lies in its beauty of finish. It may be left natural or given a deep brown color, a really beautiful tone. But before any finish whatever is applied, it is essential that the wood be perfectly dry. Of course the different pieces in the article should present the same natural color and character of grain.

For the natural finish, we recommend three or more coats of good white shellac, each coat be smoothed down with fine sandpaper, while the final coat may be rubbed with pumice and oil, giving the popular dull "egg shell" finish, as it is called. For the artificial color, asphaltum varnish, thinned out with turpentine, applied to the wood and soon rubbed off, gives it a wonderfully striking appearance. This should be left to dry thoroughly, then finished as above, or waxed.

(To be continued)

———

Lead in Enamels

Unscrupulous manufacturers sometimes add lead to their preparations for the coating of various household utensils, which is injurious, and vessels containing it should never be used. A very simple test and one which may be applied very rapidly is as follows: The vessel to be tested is cleaned to remove all grease, etc. A drop of strong nitric acid is then placed upon the enamel or tinning, and evaporated to dryness by gentle heat. The spot where the acid has acted is now wet by a drop of solution of potassium iodide (5 parts iodide to 100 water), when the presence of lead is at once shown by the formation of yellow lead iodide. Tin present in the enamel, etc., does not give a yellow spot when the potassium iodide is added, the stannic oxide formed by the nitric acid not being acted upon.

PLATE RACK.

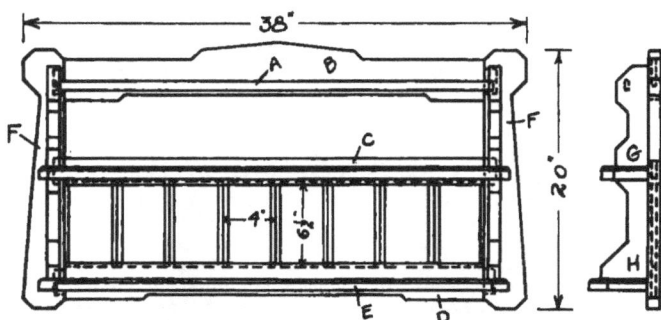

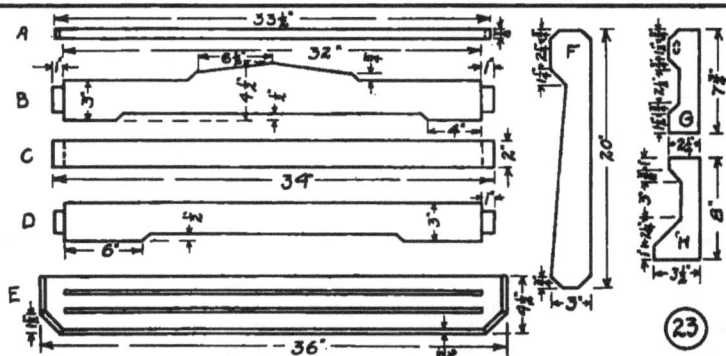

WIRELESS TELEGRAPHY

In this department will be published original, practical articles pertaining to
Wireless Telegraphy and Wireless Telephony

NAVY TO WAR ON WIRELESS NOVICES

Investigation to be made as Result of Interference with Message of Distress— To Seek Federal Law—Officials, by Requiring Licenses, Hope to Check Amateur Operators who Hamper Seaboard Business

Serious and flagrant interference by amateur wireless operators in the transmission of legitimate messages along the Atlantic seaboard has aroused the Navy Department to such an extent that an official investigation will be begun today. Immediate action was prompted when a message of distress from the torpedo boat destroyer *Terry*, recently was interrupted by novice operators here, causing a delay of more than an hour.

Officials in the Brooklyn Navy Yard say they have the names of several young men who were responsible for the interference in the transmission of the message from the disabled *Terry*. While the pernicious interference cannot be stopped by law now, the naval officials hope to check it by a personal canvass of the amateur operators and by continued agitation cause the enactment of a federal law requiring all operators to obtain a license.

"The incident of the *Terry* is argument enough for a federal license law," said one of the navy investigators. "For more than an hour amateur operators interfered with the receipt of the message of distress. They were asked repeatedly to cease their activity in sending messages to each other. Instead of complying with the request, several of them retorted with impudent replies.

"During the delay the fierce gale and high seas that battered the distressed destroyer put her wireless outfit out of commission, and we were unable to learn her exact position to rush aid to her.

"Our country is the only one in the world in which all wireless operators are not required to have a license. There are approximately more than five hundred amateur operators in and around New York. Their interference is a serious menace when vessels are in distress, and exasperating to professional operators who have difficulty in receiving and sending legitimate messages.

"On Saturdays and Sundays the amateurs keep the air charged with messages, and it is next to impossible to carry on the regular business. It is from this fact that we deduct that most of the amateurs are schoolboys, who then have time to carry on their experiments. We have estimated that among 500 young men or amateurs who have outfits, at least half that number own and operate a sending equipment. It is the sending that causes the 'break' in messages being received by professional operators. Sometimes it is possible to check the amateurs by a process known as 'tuning them out.' But for the most part, the operators are powerless to 'call them off.'

"We do not wish to be represented as discouraging young men who are ambitious in carrying on experiments in wireless operation. For the most part, they are young geniuses who have built their own stations. But when it is realized how serious their interference is at times, and what it might cost if some vessel was in distress, it can be appreciated that some action must be taken. The final solution lies in the federal license, but in the meantime we will do all in our power to discourage interference with legitimate messages."—*Aerogram*.

A THREE-CIRCUIT INDUCTIVE TUNING APPARATUS FOR WIRELESS TELEGRAPHY

With Description of Experimental Wireless Station at Richmond

G. G. BLAKE

Fig. 1 shows the wooden framework of the three-coil tuning inductance. The two frames, *A* and *B*, are made out of American whitewood, $1\frac{1}{4}$ x $1\frac{1}{4}$ in., in which is cut a groove $\frac{3}{4}$ in. wide; frame *A* is 14 in. square. *X*, Fig. 1A, is a

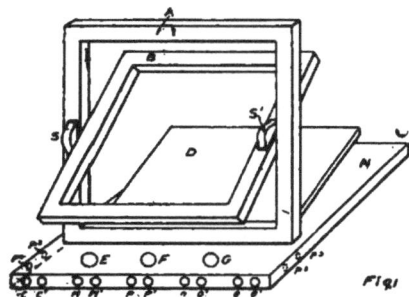

Wooden Framework of Three-Coil Tuning Inductance

section of one of the frames, and *Y* is the capping, which is screwed in place after the inductances have been wound in the grooves. The inductances *A* and *B* each consist of 12 turns of No. 14 insulated copper wire. *S* and *S'* are two switches (not shown in diagram, Fig. 4), by means of which either 3, 5 or all 12 turns can be brought into use as may be desired. The third inductance is wound on a board *D*, 12 in. square, in the form of a

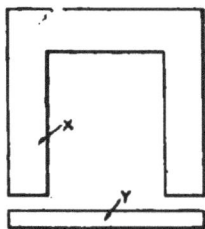

Fig. 1A
Section of Framework in Fig 1.

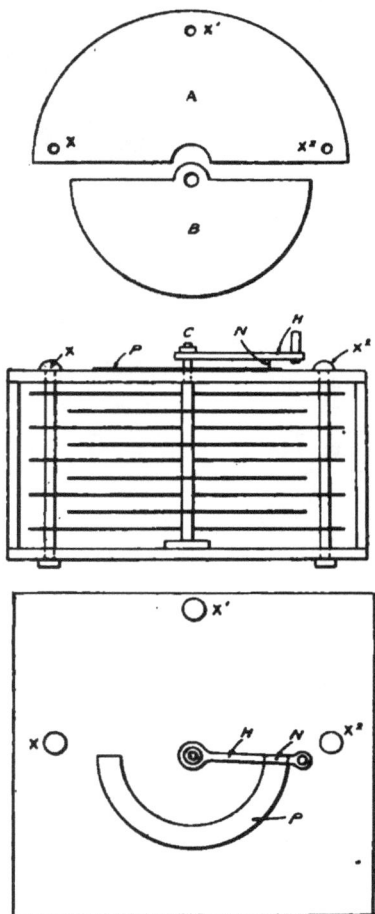

Fig. 2
Showing Construction of Condenser

spiral. No. 16 electric light wire is used for this, and is fastened in position as it is wound, with small pins driven through its insulation into the board.

Inductance frame *A* is screwed in an upright position near to one end of base-

Fig. 3
A Group of Some of the Wireless Telegraphic
Apparatus

board N, and inductance D is hinged to its base; inductance B is pivoted at its center. $P^1 P^2$, $C C^1$, $M M^1$, $P P^1$, $O O^1$, $Q Q^1$, $P^2 P^2$, are terminals for connection to condensers, aerial, etc.; E, F and G are respectively two-, three- and four-way double pole switches.

Two variable condensers, V and V^1, Fig. 4, are used in conjunction with this instrument. I have made my own of sheet zinc, separated by cycle washers, as in Fig. 2. Each condenser is made up of 25 large sheets, A, 8½ in. in diameter, and 24 smaller ditto, B, 6½ in. in diameter. Fig. 2 gives an idea of the construc-

tion. The large sheets, A, have a semi-circular hole cut at each of their centers to allow the rod C (upon which the smaller sheets are fastened) to turn, and three bolt holes, X, X^1 and X^2, near their outer edges; the smaller sheets are cut with a projection and a hole in its center, through which the rod C passes.

Both the condensers are mounted, as can be seen by reference to the photograph, Fig. 3, in glass cases. The smaller sheets can be revolved round between (but without touching) the larger ones by means of small handles H, Fig. 2, an ordinary draughtsman's protractor P, which is calibrated to 180 degrees, serves as a scale; one is fixed on the top of each of the cases under a needle point N, soldered to the underside of each of the handles. Fig. 4 is a diagram of connections exactly as used at my station; 4 is a direct-coupled tuner, all the coils of which are connected in series, and which can be used in place of the 3-coil inductive tuner for stations of very great wave length. To use this, the aerial is disconnected from C^1 and connected to W^1, as shown, and the four-way detector switch is turned over to its fourth position, thus connecting P to L, and P^2 to L^1. The rest of the diagram, though somewhat complicated, needs very little explanation.

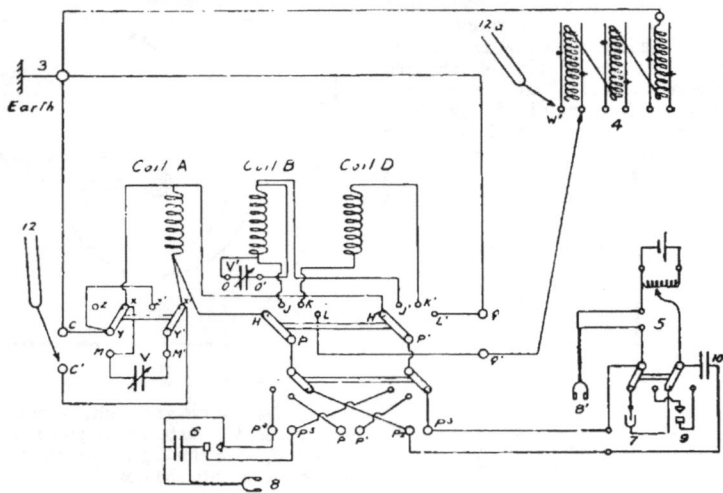

Fig. 4
Diagram of Connections

Fig. 5
General View of Mr. Blake's Coils

Fig. 7
Diagram of Transmitting Connections

lytic detector 7, or to a crystal and carborundum detector, 9, by means of a two-way switch. 10 is a small condenser of fixed capacity, 3 an earth connection, and 12 and 12A represent the aerial connections. P and P¹ are two extra connection terminals for testing any new form of detector. The two-way switch, marked E in Fig. 1, puts the variable condenser V, Fig. 4, either into parallel or series with the inductance A, as can be seen by following the connections shown in Fig. 4. The three-way switch F allows the tuner to be used with any

Coils A, B, and D are shown well separated from each other for the sake of clearness. V and V¹ are the two variable condensers; 5 is a potentiometer, with its battery and pair of phones 8, 8¹, which can be connected to either an electro-

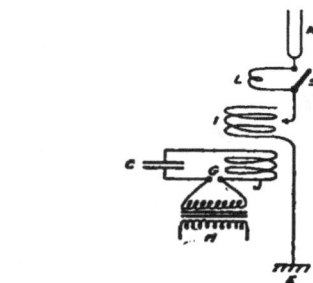

Fig. 8

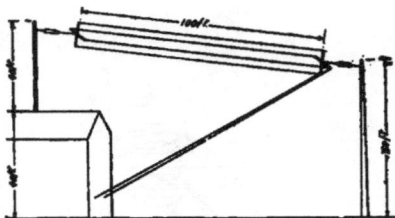

Fig. 9
Diagram of Author's Aerial

detectors connected to terminals P^4, P^5, P, P^1, or P^2, P^3.

The four-way switch G serves to connect up the detector across coil A as "stand by," or across coil B, in which case the signals are received inductively from coil A, and are, in consequence of their being more sharply tuned, much louder. The third position of the switch connects the detector across coil D, the third inductive coil, which is used to cut out interference when very sharp tuning, is necessary.

When switched over to its last position, the switch disconnects the three-coil tuner and connects the detector to the direct-coupled tuner 4. I think it is almost unnecessary to add that the various inductive circuits are brought into tune with each other by suitable selection of the number of turns of inductance by use of switches S and S^1 (Fig. 1), and by altering the capacities of the variable condensers V and V^1 by moving their handles.

The photograph, Fig. 3, also shows a magnetic detector on the shelf just above the right-hand variable condenser, and the motor starting switch (for mercury break used to interrupt current in the primary of coil when signaling), also the transmitting key.

Fig. 5 shows my coils, and Fig. 6 my transmitting inductances and oil condenser, which is in the box under the spark gap. Sometimes I use a rotary spark interrupter (not shown in the photograph) in place of the spark gap.

Fig. 7 shows the transmitting connections: A is the aerial; L a 16 c.p. 110 volt electric lamp, with short-circuiting switch S; I aerial inductance; and E earth connection.

The closed oscillatory circuit is composed of a condenser C, inductance J and spark gap G; M represents the spark

coil which supplies the energy. The coupling between the two inductances can be altered at will to ensure sharp tuning. The usual wave length which I use is 300 meters, and my code call is "B O X."

Fig. 8 shows apparatus for producing undamped waves for wireless telephony and telegraphy, with one or two modifications; it is similar to the Lepel generator. It is worked from an alternator shown on the ground. The microphone used in telephone experiments is shown in the center of the photograph, and the transmitting inductances can be seen at the top of the room above the alternator. The switchboard and motor starting switches are also shown.

Fig. 9 is a diagram of my aerial, which is constructed with 7-22 copper wire, which I had made up especially for the purpose. Each strand is insulated separately; they are then twisted together, and the whole coated with insulating material.

The separation of the wires from each other allows the use of all the surface of the wire, and gives the aerial as large a capacity as possible. The guy wires holding up the masts are not shown in the diagram.—*Model Engineer and Electrician.*

SUMATRA, North Coast, Pulo Weh. The Netherlands Government has given notice that a wireless telegraph station, call letters SAB, has been established on Pulo Weh, north coast of Sumatra, near Tapa Gaja Point Light, in (approximately) latitude 5° 54′ 30″ N., longitude 95° 20′ 00″ E. The wireless signal is made from an iron mast 212 ft. high. (*Bericht aan Zeevarenden No.* 280 (2366), '*sGravenhage*, 1911). (N.M. 3, 1912.)

H. O. Charts Nos. 854b and 1595.

B. A. Chart No. 2760.

China Seat Directory, Vol. I, 1906, page 63.

H. O. Publication No. 87, International Code of Signals, 1911, page 120.

If a thought does not suggest other thoughts to the mind that receives it, either the thought or the mind is not worth much.

He who knows least, doubts most; at the same time he doubts not but that he knows best.

EXPERIENCES IN WIRELESS TELEPHONING

AUSTIN C. LESCARBOURA

During the latter part of 1908 and the beginning of the succeeding year, a number of wireless telephone experiments were conducted by several wireless companies, in view of proving the practicability of their respective sets and receive an order for equipment from the United States Signal Corps. The tests were between Fort Hancock, Sandy Hook, N.J., and Fort Wood located on Bedloes Island in New York Bay, the latter being on the same island as the Statue of Liberty. The total distance between the stations was 18 miles, with the high hills of Staten Island separating both. The writer at the time was in the employ of one of the competing concerns, and aided in the operating of the transmitting wireless telephone set at Fort Hancock. The experience, both from a technical and humorous point of view is interesting, and in the following paragraphs, a few incidents and descriptions are faithfully given.

Fort Hancock, as stated before, is located at Sandy Hook, a long and narrow stretch of sandy waste extending into the Atlantic Ocean and forming the lower portion of New York Bay and entrance. In summer the heat is extreme, for the sun heats the sand to a tropical heat, while in winter, the cold wind blows in from the open sea with Arctic vigor. The temperature is usually in one extreme or another, but during several months in the spring and autumn, the weather may be fair at intervals.

On a morning in the latter part of October, 1908, another man and the writer started for Sandy Hook from New York in the U.S.S. *Ordnance*. This "steamer" in reality is an overgrown and comfortable tug boat, equipped for carrying freight and passengers to the many forts in New York Harbor. After a rough trip, we arrived at 9 o'clock and immediately walked to the wireless station. This station consisted of a two-story concrete structure, with a wooden mast in the rear supporting an umbrella aerial. The wireless apparatus consisted of a 1 k.w. transformer mounted in a cabinet with the spark gap and condensers. A large desk contained the receiv-

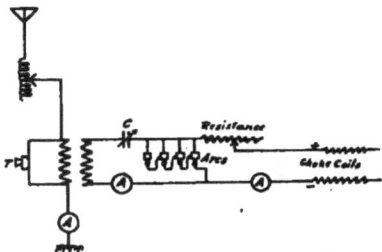

Wiring Diagram

ing apparatus consisting of a large tuning coil with silicon and electrolytic detectors. The key was mounted on a rubber base, with a long lever passing through a slot and into a tank of oil in the desk, where the contacts were located.

Our telephone set was mounted on a large table with a back board. The transmitting set consisted of ten arc units in series, each unit comprising of a copper cylinder which was filled with water and mounted on a wooden frame; and a large carbon rod held on a long spring. The carbon rods could be adjusted by a thumb screw located at one end, which also contained a large handle, so that all the springs could be pressed down and thus start the arcs if desired. Two sets of five arc lights each were placed at both sides of the table, while the oil condenser was located in the center, in back of a switch for connecting the receiver or transmitter to the aerial and ground. On the backboard, two hot-wire ammeters were mounted, one of these indicating the high-frequency in the aerial circuit, while the other indicated the energy in the oscillation circuit. The third ammeter was of the standard magnetic type, and indicated the amperage consumed by the arcs. The current was furnished by a 500 to 600 volt C.&W. generator, directly connected to a 110-volt, 7 h.p. motor. Suitable field control, enabled us to obtain voltages from 400 to 600 volts. The microphone, which is one of the "missing links" in all wireless telephone sets, consisted of a round enclosed case with a diaphragm in front and an insulated contact in the rear. Having a number of these microphone units, we were able to slip a new one into place whenever necessary, by merely giving the mouthpiece a slight turn, and replacing the old microphone with a new one.

The first morning we arrived, the apparatus had already been delivered and was unpacked in the operating room, only the temporary wiring being necessary. We connected the motor-generator and the regular operator of the station called the power house on the telephone with orders to start up another generator for the peak load to follow. One wire from the generator was then attached to the tin side of a can, and the other wire attached to a voltmeter and then allowed to dip in the can of water. The meter then indicated whether the connections were correct, so that the positive pole could be identified. The meter did not read higher than 125 volts, and for this reason the water resistance had to be inserted. The positive and negative wires were then connected to their proper terminals on the transmitting apparatus. The water having been placed in all the copper cylinders, the arcs were started, and the condenser adjusted. This condenser consisted of 24 stationary and 23 rotary plates, the glass containing jar being filled with paraffin oil. After an hour or more of adjustment and changes, the ammeter in the oscillation circuit indicated that the current was steady. The aerial circuit was then connected to the aerial, and immediately the needles on both ammeters began to flicker again, finally coming to rest after another period of adjustment. The microphone then being slipped into place, the words were shouted into the mouth-piece. At every sound the needles on both ammeters fluctuated, the variation being more pronounced the higher the pitch, and phonograph conversation or music pro-

noticed when the microphone was continually being turned; which would suggest that a carbon microphone being slowly revolved by a mechanical device would be more suitable to withstand the high amperage, since it is continually moving the carbon grains. At one time a large dog was brought into the operating room and placed on a large box with his head near the mouth-piece. He was finally coaxed into barking, which, judging from the deflection of the ammeters, must have been heard by the stations within our range. This is the first record of a dog "speaking" over a wireless telephone!

On a cold November morning we again set out for Sandy Hook on the same boat as before. On nearing the fort, the writer became worried in failing to see the wires of the aerial. A gale had swept the coast the night before, and it was not impossible that the wires had been blown down. The pole was plainly visible, but no wires could be noticed. Both of us became excited, for we knew that without the aerial, we would have a whole day wasted with nothing to do but to walk around the reservation. However, on reaching a few hundred feet from the fort, we noticed that the wires were still there and that these happened to be of a very small gauge. In fact, we believed at the time that our failure to cover a greater range was due to the inefficient aerial, which was composed of small wire and had but a single wire lead to the aerial from the station. On making the necessary connections, the starting-box lever was moved, but the motor did not start. An instant later the 60-ampere fuse in the cut-out went off, and indicated something wrong. On trying to turn the armature by hand, it was found to be firmly held by the bearings so that it could not turn. This was probably due to the

thoroughly cleaned and placed in another jar. We substituted another condenser of the same type and started the arcs. After the customary adjusting and dickering, the meters finally came to a reasonable rest, and the phonograph started. After a short while, in which the phonograph was continually playing and only the occasional knocking or replacing of the microphone was found necessary, there suddenly came a slight noise similar to that caused by escaping steam, but just for an instant, and immediately the arcs' and ammeters went wrong. It proved to be the short-circuiting of the variable condenser, a spark having jumped between two plates, and a little black dirt appeared between the plates. This dirt is a compound of carbon from the paraffin oil, and conducts the current from one plate to another, thus rendering the condenser useless for high voltage currents. Happily, we still had a large condenser belonging to the Dönitz wavemeter, which was inserted in place. It might be stated here, that this condenser, though it had the plates separated only ¼₆ in. apart, withstood the potential without breaking down, while condensers with plates spaced ⅛ in. and built in this country were continually breaking down. This illustrates the accuracy of German mechanics, for the plates of the German condensers were perfectly true, which cannot be said of the others.

During the course of the afternoon, a battleship, which was passing Sandy Hook on its journey to the Hudson River where it was to anchor, called the operator and asked him the name of the set being used. On being told, he telegraphed back: "The music is fine, give us some more." We heard later that all the officers on the battleship had been called by the operator to hear the music in the telephone receivers.

In all our tests, between times when the phonograph was not working, the conversation usually ran: "Hello, Hello, Hello Fort Wood, how do you get me now? One, Two, Three, Four, Five," and so on, most of the words being shouted very slowly and drawn out. It is rather a peculiar feeling to be talking into the mouth-piece of a wireless telephone, and not knowing whether the speech is being heard or whether it is not being heard. The phonograph is

used the greater portion of the time, for it carries better than the human voice. That afternoon our concert was heard at 20 miles, the Brooklyn Navy Yard operator having listened the greater part of the morning and afternoon. At four o'clock we returned to the boat for New York.

The third and last test of this series, if the writer correctly recollects, occurred in January of 1909. After all the preliminaries, such as the wiring, adjusting, substituting, and swearing, the phonograph was started, and for upwards of an hour we did not think of changing the record. The one playing happened to be, "The Anvil Chorus," from the opera, "Il Trovatore." This selection was played by a band with a number of persons whistling, and proved to be a very effective record for fluctuating the ammeters, which was a desired feature. After an hour had passed with the continual playing of the same record, we shut down the generator and arcs, while the operator listened in to hear whether Fort Wood would call us. Upon calling Fort Wood, he received no reply from that station, but Manhattan Beach (DF) immediately called, and upon being told to go ahead, telegraphed: "For —— sake change the tune." When asked whether he had received all we had spoken and played, he said that he could get all of it without trouble, but was disgusted with the same record continuously. We did change, and for the rest of the morning played different records. At 12 o'clock, a telegram arrived via the Postal Telegraph station telling us to abandon the tests as the interference was too powerful in the upper bay and while we had been heard clearly at times, the extreme proximity of the other stations completely overcame our signals. We had to wait until 4 o'clock for the boat, so devoted the time to visiting the buildings and looking over the various interesting features of the reservation. A humorous incident, the writer recalls, is the reporting of steamers sighted at Sandy Hook. The two telegraph companies, Postal and Western Union, are located at the end of the Hook, and both have tall buildings resembling lighthouses. The one we visited was maintained by an old time operator who had six wires to handle beside the reporting of the steamships sighted. There is great rivalry between the two companies as to which one reports a steamer first. The old-time operator had erected a few wires from a pole to his telegraph station, and with the aid of the wireless station operator and other local talent had succeeded in constructing a simple receiving set. He would then listen with the telephones placed on his head, and hear the different steamers report to Fire Island, about 40 miles away. Upon the first sign of smoke over the horizon about an hour and a half later, he would immediately telegraph to New York that the steamer was sighted. Meanwhile, the other operator in the other building was straining his eyes through a 5 ft. telescope to get a glimpse at the funnels of the boat. For sometime the competing operator was at a loss to understand how the veteran operator could report the ships before they could even be seen, and on asking him was informed that the operator recognized them by the smoke only! The news finally leaked out, and the wireless was abandoned for the purpose, only the long telescopes being used.

Through these tests the practicability of the wireless telephone was found to be uncertain. Though these tests were performed over three years ago, no definite advancement has been made in the art. The greatest difficulties are in the arc, condenser, and microphone. The arcs will never become practical as they exist at present, for there are periods when the oscillations are perfectly steady, but in the middle of an important conversation, the arcs will suddenly sputter and the words are lost. The condensers are a continual source of worry, and, unless accurately constructed, will break down rapidly. The microphone, likewise, is unreliable, and continually requiring attention. These weak points cause the wireless telephone to be uncertain.

The points to be learned from these tests are: to employ an aerial having a large capacity and many leads to and from the aerial; that a microphone of the carbon grain type with a continuous rotating device will overcome the "baking" to a great extent; that the condenser should be made with rotary plates in oil, and that these plates may be larger and separated by a larger gap to overcome the breaking down as experienced with smaller gaps; that many arcs give greater results than a single arc; and, finally, that hot-wire ammeters are necessary in both the aerial and oscillation circuit to determine whether the set is actually transmitting high-frequency waves and whether these are smooth so that the conversation and music will be heard at the receiving end.

Timing Fast Trains

Perhaps you have often been curious to know just how fast you were traveling on a railroad train. Many roads have little white posts beside the track, marking the miles and usually the quarter and half miles also, but these may not be on your side of the train.

There is another way to tell the miles. The telegraph poles are almost invariably placed fifty yards apart, except when they carry a very large number of wires, and if you count thirty-five of them it will be a mile. If you have a watch with second hands on it you can tell just how many miles the train is traveling in an hour.

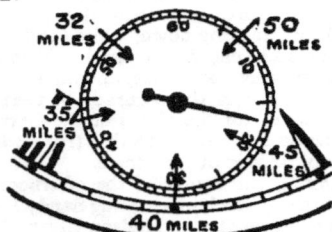

Showing Method of Timing the Train

Note the time from one mile post to the next. Anything more than a minute is slower than sixty miles an hour. If the second hand gets past the minute and down to thirty seconds you are going forty miles an hour. If it gets only twelve seconds past the minute you are going fifty miles an hour and so on.

You may cut this out and take it with you on the train next time you make a railway journey, and see if you can determine your speed.—*New York Sun.*

Be thankful every day; don't pile your gratitude all onto one day. The man who is thankful only when the Governor says he must, never is very thankful any day.

WIRELESS NEWS

The Amateur Wireless Telegraph Association of New Bedford has at present twenty-four members, and is desirous of extending to other Amateurs in New Bedford and vicinity the benefits of same through their becoming members. The purposes of the Association are to promote wireless telegraphy among individuals for both knowledge and pleasure; through becoming members of the Association amateurs aid in preventing interference and also extend their knowledge of wireless telegraphy and other affiliated arts; Practical demonstrations and lectures being arranged for every meeting of the Association; The Association has measuring and testing instruments of various kinds which are loaned to members free of any charge, also a technical library to which members can refer, including all magazines pertaining to wireless.

The rating of members' transmitting sets are from a 1 in. coil to a 3 kw. open core transformer, the Association having the use of a 1 kw. set with which they conduct experiments of research.

The officers of the Association are: Chas. Praihanzels, president; Chester Dable, vice-president; Edw. De Mello, Secretary; Wm. Isherwood, treasurer; Herbert Charnley, collector; Lester Jenkins, operator and librarian; Chester Dahl and Geo. W. Pope, auditors.

Vancouver, B.C., Sept. 18, 1911.

On August 17th, at a meeting of the Wireless Amateurs of Vancouver, B.C., a Wireless Association was formed.

The object of this club is to assist its members in the study of Wireless Teleg-

same purpose. Communications should be addressed to Corresponding Secretary, 1934 William St., Vancouver, B.C.

North Carolina.—Frying Pan Shoals Lightvessel—Wireless telegraph station established.—A wireless telegraph station, call letters "NLC," has been established on board Frying Pan Shoals Lightvessel, No. 94, seacoast of North Carolina.

Approx. position: Lat. 33° 33′ 30″ N. Long. 77° 48′ 20″ W.

(See Notice to Mariners No. 47(3401) of 1911). (N.M. 6, 1912.)

(*Notice to Mariners* No. 5 (146), *Bureau of Lighthouses, Washington*, 1912).

H. O. Charts Nos. 21a, 1411 and 943.

U.S. Coast Survey Charts Nos. 11, 149 and 150.

List of Lights on the Atlantic and Gulf Coasts, 1911, Nos. 977 and 978.

U.S. Coast Pilot, Part VII, 1906, page 16.

List of Wireless Telegraph Stations of the World

A complete list of the naval wireless shore stations and ships of the Navy equipped with wireless apparatus is published in the "List of Wireless Telegraph Stations of the World," copies of which may be obtained from the Superintendent of Documents, Government Printing Office, Washington, D.C., at a cost of 15 cents each (money order).

The above takes the place of Special Notice to Mariners No. 47a, of November 22, 1904.

QUESTIONS AND ANSWERS

Questions on electrical and mechanical subjects of general interest will be answered, as far as possible, in this department, free of charge. The writer must give his name and address, and the answer will be published under his initials and town; but, if he so requests, anything which may identify him will be withheld. Questions must be written only on one side of the sheet, on a sheet of paper separate from all other contents of the letter, and only three questions may be sent at one time. No attention will be given to questions which do not follow these rules.

Owing to the large number of questions received, it is rarely that a reply can be given in the first issue after receipt. Questions for which a speedy reply is desired will be answered by mail if fifty cents is enclosed. This amount is not to be considered as payment for reply, but is simply to cover clerical expenses, postage and cost of letter writing. As the time required to get a question satisfactorily answered varies, we cannot guarantee to answer within a definite time.

If a question entails an inordinate amount of research or calculation, a special charge of one dollar or more will be made, depending on the amount of labor required. Readers will, in every case, be notified if such a charge must be made, and the work will not be done unless desired and paid for.

1745. Electroplating. H. S. M., Syracuse, N.Y., has been trying to plate iron with copper, using an ordinary solution of sulphate of copper. The results are not satisfactory, and he asks for advice. Ans.—While the sulphate solution may be used for the body of the deposit, a preliminary coating should be made in a bath consisting of the double cyanide of potassium and copper. To make this, proceed as follows: In 500 parts of warm water, dissolve 30 parts of neutral copper acetate, 30 parts of crystallized sodium sulphite, and 5 parts of ammonium carbonate. In another 500 parts of water now dissolve 35 parts of 98 percent or 99 percent potassium cyanide, and pour this into the former solution, while heating. All parts are by weight. Be sure you recognize the poisonous character of the cyanide.

1746. Transformer Control. H. B. P., Franklin, N.H., (1) has a 1½ kw. open-core transformer for wireless use that seems to draw too much current from the 110-volt 60-cycle lighting mains, and subjects him to severe criticism for interference with the regular electric lights. He asks if a choke coil can be inserted that will reduce this trouble without cutting down the power. (2) In regard to examinations for securing a position in the U.S. Navy as a wireless operator, are there examinations to take other than those specified in the January magazine? Ans.—(1) Evidently you are drawing a pretty large current, and the electric lighting company ought to require you to remedy the trouble. This you cannot do and retain present location and wiring without reducing the power of the apparatus. We would advise you to try a water rheostat, using merely a wooden pail with an iron grate or pulley at top and bottom, and a solution of carbonate of soda. By varying the strength of solution and distance between plates, you can get a wide range of control, and perhaps find that less current than you have been using will suffice. (2) Write to the U.S. Civil Service Commission, Washington, D.C. They will furnish all the information desired, even to sending sample examination papers.

1747. Fly Wheel Enigma. F. R. W., Independence, Mo., asks where he can buy a device to conserve the power in a fly wheel. Ans.— We do not know whether you are in earnest in this question or merely joking, for it is only another way of asking for perpetual motion. In this search you will meet the inevitable fate that the laws of nature have prescribed.

1748. Induction Motor. F. M., Corvallis, Ore., wishes to rewind a 7½ h.p. 3-phase, 60-cycle, 220-volt motor that has been through a fire. He asks for information. Ans.—The simplest and altogether sufficient method will be to take off the burned coils, make a suitable wooden form on which to wind new coils of the same shape as original ones. From the name-plate on motor, copy all the data and send it to the manufacturer with the request that they send a blue-print showing the method of connection. You can purchase coils from them all ready to put in place, and perhaps cheaper than you can make them yourself.

1749. Reciprocating Motor. C. B. D. asks what there is impractical about such a device. Electromagnets would seem to be enormously strong and capable of producing vigorous rotative power. Ans.—This was the first form of electric motors, and by all odds they were the poorest that have ever been made. The magnetic pull can be enormous, and even give the builder difficulty in providing sufficient strength of structure to withstand it. Time, however, is required for a current to establish itself in the electromagnets, and this means a low speed of rotation. Wasteful eddy currents will flow in magnet cores and armatures, and this means low efficiency. The worst feature is the requirement for make-and-break contacts. Of all things in an electric motor there must be a continuity of the circuit. Even the feeble sparking at the brushes of an ordinary motor is objectionable. In your motor, you would require a continuous performance of what you see when a trolley wheel leaves the wire. It is impossible to conceive any motive power acting under more favorable conditions than in the ordinary electric motor, where the action of the field flux on the currents in the armature conductors produces a continuous and highly efficient torque. See our article on electric motors in the January. 1907, magazine.

1750. Patents. G. T. C., Saco, Me., asks if there are now any patents on the "squirrel-cage" rotor for induction motors, or on the "split-phase" scheme of making single-phase induction motors self-starting, or indeed on any other part or scheme of single-phase motors? Ans.—All the fundamental patents on three-phase motors have now expired, as also those on the split-phase construction. It is of interest, however, that patents were issued on January 30, 1911, covering the "compensated repulsion-induction" single-phase motor. While the patentee, Latour, of France, made his application in 1904, it has been the subject of an interference of a similar application by Winter and Eichberg, of Germany. In this country the motor appears as one of the products of the General Electric Company, and is designated as the "RI" type.

1751. **Induction Coil.** G. M., Jr., Philadelphia, Pa., asks: (1) I am building the coil described in the February, 1911, issue. Will you tell me how much wire I will need for primary wire? Please tell me in pounds so I can go to the store and buy the right amount. (2) Will black fiber do instead of red? (3) What do you mean by "oiled linen" for the core? (4) What kind of oil will I use for the linen? (5) The primary has three layers of wire; what will I do with the ends? Ans.—(1) The primary will require approximately 2 lbs. of wire. (2) Black fiber will answer the purpose. (3) By "oiled linen," the author refers to the high-grade insulating medium known to the trade as "empire cloth," which may be obtained from any large electrical supply house. (4) The object of bringing out the loops in the primary winding is to enable the user to take taps from any one of the three layers, or, in other words, to vary the number of turns in primary winding at will. This will permit of the use of several types of interrupters. For instance, with an electrolytic interrupter, the best results will be obtained by using the full three layers of wire, making connection at the ends of the winding. The ordinary vibrator interrupter will show better results on either one or two layers, depending upon the speed of vibrator and the impressed voltage. (5) There is no reason for cutting the loops as you suggest, and, indeed, such a proceeding is very likely to result in confusion, as it is evident that you do not understand the underlying principles of the work you are doing. We suggest that you bring leads to four binding posts, lettering them A, B, C and D for convenience as per the diagram. By connecting to A and B you will then include one layer in your circuit; from A to C will give two layers and from A to D, three, or the total number of layers.

1752. **Wireless.** L. V. A., Homer, Mich., asks: (1) Are the Thos. M. St. John wireless instruments up to the standard? (2) Will I have to get a permit to put up a wireless station with a sending range of 17 miles and receiving range either 700 or from 1,500 to 2,000 miles? (3) I live about half way between the east and west borders of Michigan (lower peninsula) and in the second row of counties from the bottom. Do ocean ships send far enough to reach me from 200 or 300 miles out in the Atlantic? Ans.—(1) We have heard nothing against them and believe they are worth the prices asked. We would advise you to buy the highest-class instruments you can afford. (2) Not at present. Legislation to this effect is sought by the navy department, but is warmly opposed by several interests. (3) Not under ordinary circumstances.

1753. **Indoor Aerial.** L. D. H., Manchester, Eng., asks: (1) Is it possible to telegraph over the distance of one mile, using an indoor aerial and suitable apparatus? If so, please give details of aerial. (2) Any American handbooks or

books dealing with construction of wireless apparatus from an amateur's point of view with prices of same. Ans.—(1) It is possible, but hadly practical to use an indoor aerial to send one mile should you happen to live in a house with a high attic and have the apparatus on the first floor. If you can attain an antenna height of 30 ft. above the instruments, a 2 in. spark coil used in conjunction with loose-coupled sending set and electrolytic or crystal detector should work very nicely. (2) We can supply a list of good books on the subject of wireless, and one can also be found in the current number of the magazine.

1754. **Diagram for Wireless Set.** P. M. L., Portland, Ore., asks: (1) Where can I find a diagram for the following set: aerial, six aluminum wires No. 14 B.&S., 2 ft. apart, 60 ft. high, 75 ft. long, three-slide tuning coil, No. 24 B.&S. enameled wire on core 3 in. in diameter, 9 in. long, silicon detector, variable and fixed condensers? (2) The wave length of the set? (3) How can I improve the set? Ans.—(1) See Fig. 48, page 97, of the "Manual of Wireless Telegraphy" for Use of Naval Electricians, 1909 edition. This will give you a choice of several diagrams employing the instruments mentioned. (2) Impossible to state, owing to variation caused by local conditions. (3) By using an inductively coupled tuner.

1755. **Wireless.** H. A. M., Silver Creek, N.Y., asks: (1) If the following dimensions are reasonably correct for the construction of a 1 kw. open-core transformer for wireless use, when used on 60-cycle, 110-volt current, please advise approximate output when used on 120-cycle 110-volt current. Core 1½ in. diameter, 14 in. long, primary winding two layers of No. 12 d.c.c. wire, secondary winding 12 lbs. of No. 32 s.s.c. wire. (2) What would the corresponding dimensions and data be, for construction of a transformer of 1 kw. output, same to be used on 110-volt, 120-cycle alternating current, transformer to be of the open-core type? Approximate secondary voltage desired 15,000 to 20,000. (3) If answers to above questions require an inordinate amount of research or calculation, will you kindly advise me approximately the proportion of difference between the dimensions and windings of transformers, open-core, when same are made to be used on 60-cycle current, or when made to be used on 120-cycle current? That is, if the core for a given size of 60-cycle transformer is 1½ in. x 14 in., as above, how would the core for a 120-cycle instrument compare in size, if output is same as the 60-cycle one? Ans.—(1) While you could undoubtedly use the above transformer up to 1 kw., so far as input is concerned, we are of the opinion that the output would be very disappointing. The core is too small and there are not enough turns in the primary. By using the inevitable impedance coil, you can use the transformer on either 60- or 120-cycle circuits. (2 and 3) The design and calculation of open-core transformers are by no means as simple and reliable as in the case of the closed-core type, and for your purpose the calculation becomes even more complex. We would suggest that you look up an article in our August, 1911, issue, on Experimental High-Frequency Apparatus, for data in regard to a ½ kw. open-core transformer; and in the following number, an article on the design and calculation of closed-core instruments. The advantages offered by the open-core transformer are very nearly offset by its low efficiency and difficulty of securing correct design for the purpose. The closed-core type is easily built, and is much less expensive, owing partly to its smaller size for a given output, and when properly designed is superior to the open-core for charging condensers. A transformer designed for use on 60-cycle circuits may be used at somewhat lower output on 120-cycle circuits. It is common practice, however, to reduce the volume of iron in the core when the higher frequency is used.

1756. **Transformer.** W. H. H., Pittsburgh, Pa., says: I am constructing the transformer and rectifier, as described in the January, 1911, issue, and have run into a little trouble. Instead of getting 100 turns per layer of the No. 16 D.C.C. wire, I can wind only 88 turns in the space of 5¼ in., and of the No. 20 D.C.C. wire, I can wind but 128 turns in the space. The wire is of correct gauge, for I have proved it with a micrometer, so I do not see how it would be possible to wind more turns than stated above. I use a lathe to wind. You will note, with the number of turns mentioned above, the voltage would be about 13½ and 25 as intended (but I would like to know whether it is all right to use but 768 turns on the primary or whether I should add the 7th layer to make the number specified in magazine. I have the coil wound (176 turns on secondary and 768 turns—6 layers—primary), but have not cut the wire, and will await your advices before completing. Ans.—While the smaller number of turns will work the iron at a somewhat higher density than intended, still the core is generous and if you do not intend to use the transformer for greater periods of time than, say, half an hour without turning off the current, the smaller number of turns will give very satisfactory results. We advise, however, that you wind the full number as specified, in order to get the greatest efficiency.

SENDING SECOND-CLASS MAIL BY FREIGHT

A few months since the Post Office Department, inspired by a desire to show an apparent bookkeeping surplus which has never been demanded by the American public or its representatives in Congress, changed the method of transporting second-class matter in an important particular. The larger portion of the monthly mailing of *Electrician and Mechanic* is no longer transported by mail trains, but goes by so-called fast freight to various central distributing points or often to its final destination. The result has been an unprecedented number of complaints in regard to delayed delivery of the magazine. The Post Office Department seems not to know how long it will take magazines to reach a given point, and in some instances from three to six weeks have been required for copies to reach our subscribers in Texas or on the Pacific Coast, whereas the mail trains formerly carried the magazines to the most distant part of the continental United States within a week. The publisher is unable to help out his subscribers in this matter. The present attitude of the Post Office Department to publishers is that they are receiving from the government an unwarranted subsidy, and every effort is being exerted to raise the second-class postal rate. Under these circumstances any complaint as to the quality of service furnished is met with retort that the cheapest that can be given is too good for the price we are paying.

As to the justice of the present postal rate to the government, to the publisher, and to the subscriber, we shall not argue. The subscriber to American magazines receives far more for his money than is the case in any other country in the world, and it may truthfully be said that the whole advantage of the low postal rate has been passed on to the subscriber. If the postal rate is increased many publishers will go out of business and the subscriber will pay a higher price for magazines. It is for the public to decide if they desire this to happen; and the subscribers to this magazine can serve themselves effectively if they will give the publisher definite information of any undue delay in the arrival of magazines. A complaint that a magazine has not arrived is not of so much value as a definite statement of the date of arrival of a magazine which has been delayed. To help the publishers of the magazines, and of this magazine in particular, if you do not receive your magazine within one day of the usual time, will you kindly inform your local postmaster and ask him to start an inquiry and also report the facts to us?

TRADE NOTES

The catalog recently issued by the Ferro Machine and Foundry Co., of Cleveland, Ohio, will especially please anyone who is at all interested in marine engines. This company, which is probably the largest marine engine building concern in the world, have included in their catalog a well written and exceedingly interesting practical treatise on the correct design and construction of marine engines and their equipment. Some of the topics taken up in this treatise are: The Gasoline Engine of the Past and Present; How Marine Gasoline Engines Operate; The Carburetor; Ignition; The Make and Break System of Ignition; Sources of Electrical Current; Installation of Ignition, Cooling System and Lubrication. The Company has attempted to make this treatise as comprehensive as was possible in the space available. Much of the information is entirely new and is the result of extensive search on the part of their Engineering and Experimental departments. Technical expressions, which are likely to be unfamiliar to some of their readers, have been carefully explained in a glossary of terms, which is included in the catalog. Also, believing that photographs would be more easily understood by the laymen than would mechanical drawings, a large number of photographs have been used to illustrate the treatise. Copies of the catalog may be obtained by writing the Ferro Machine and Foundry Company, Cleveland, Ohio, and mentioning the *Electrician and Mechanic*.

The Clapp-Eastham Co., of Cambridge, Mass., are manufacturing a most excellent line of metal mast antennae. These antennae are made from 3 in. galvanized iron pipe coming in 10 ft. sections, having telescoping joints, which may be fitted together to form a pole of any height up to 80 ft., beyond which it is not practical to go. Each pole is fitted with a metal cap for excluding rain and snow, and the pole is insulated from the ground by means of a high-tension insulator supplied with the mast. The pole is extremely light, weighing only about 5 lbs. per 10 ft. section, and is supported by means of guy wires fastened to lugs, which are placed at intervals along the pole. This type of mast is slightly difficult to erect, in that it must be supported by guy wires held at four corners, while being raised, as it will not support its own weight without buckling. However, if a little care is used to keep the mast in a vertical position while going up, it will prove a very substantial antenna, owing to the strength of the supporting guys. The leading-in wire is fastened directly to the base of the pole, and both the pole and guy wires act as a part of the antenna, the latter being insulated by means of strain insulators at their lower end.

This type of antenna may not only be erected on the ground, but also lends itself well to insulation on the roof of a building, in which case the smallest diameter of the building should not be less than the height of the pole which it is proposed to erect.

This company will supply these aerials complete, including mast, antenna wire, insulating base and strain insulators as follows: 30 ft. mast complete, $16.15; 40 ft., $19.30; 50 ft., $25.15; 60 ft., $30.50; 80 ft., $43.15.

Above prices are net cash with order F.O.B., Cambridge, Mass.

It may be of interest to many of our readers to know that the Boston School of Telegraphy is now giving a special course which is devoted exclusively to the training of amateurs on the subject of wireless telegraphy. It is the purpose of this course to especially meet the needs of those who are desirous of entering the Commercial Wireless Telegraphy Service.

Mr. Harold J. Power of Tufts College, a well-known wireless expert, and who was formerly an operator on the *Harvard* and *Yale* boats and also upon Col. Astor's private yacht will give special instructions.

Recent General Electric Bulletins

In Bulletin No. 4924 is described the General Electric Company's Thomson Prepayment Watt-hour Meters for direct and alternating current, Types CP-4 and IP-4.

Bulletin No. 4887 is a rather attractive bulletin —illustrating and describing the General Electric Company's Turbo-Generator Sets in capacities of from 5 to 300 k.w All of these sets are of the horizontal type and can be arranged to operate either condensing or non-condensing, and at any steam pressure above 80 lbs. for the smaller sizes, and 100 lbs. for the larger.

Bulletin No. 4900, just issued by the General Electric Company, is devoted to apparatus used in connection with Series Incandescent Street Lighting, and supersedes in part the Company's previous bulletin on this subject. The bulletin is practically an ordering catalogue and contains no description other than that afforded by the illustrations. It lists lamp brackets of various style (giving their dimensions), tungsten economy diffusers, G-E Edison Mazda street series lamps, constant current transformers, switchboard panels, and lightning arresters.

"Small Plant Direct Current Switchboards" is the title of Bulletin No. 4919, recently issued by the General Electric Company. The bulletin is devoted to a description of panels which are designed for the control of three-wire generators. The panels are arranged for 125 and 250 volts, and in capacities of from 25 to 100 kw.

The General Electric Company's Bulletin No. 4904, illustrates and describes three-phase panels for use in small or isolated plants containing but one generator. These panels are not intended for the parallel operation of generators, but for installation in a switchboard consisting of two or more panels.

Bulletin No. 4905 is devoted to panels designed for general use in small central stations and isolated plants, and are for use with one set of busbars, to which the generators and feeders are connected, and are suited to the parallel operation of generators.

Each of the above bulletins contains connection and dimension diagrams of panels of various sizes.

Bulletin No. 4901, just issued by the General Electric Company, is devoted to Alternating Current Switchboard Panels with Oil Switches attached, and designed for three-phase, three-wire circuits: 240, 480 and 600 volts, 25-60 cycles. This supersedes the Company's previous bulletin on this subject.

Bulletin No. 4907, recently issued by the General Electric Company, contains interesting data relative to the lighting of offices, banks and public buildings by G.E. Edison Mazda lamps. In this connection are shown illustrations of numerous buildings lighted with these lamps and data are included giving the number and sizes of the lamps in each installation. The publication also contains a history of the development of the incandescent lamp, and other information of interest to the consumer of current for lighting purposes.

Without doubt, the lighting of textile mills exerts an important influence on the amount and cost of production, the quality of the product, the amount of spoilage, the safety of the employees and their willingness and ability to furnish the best possible returns in labor.

Bulletin No. 4906, just published by the General Electric Company, is devoted to this subject, and in this connection considers those items briefly. It describes the new drawn wire G.E. Edison Mazda Lamps, which are particularly suited to this class of illumination, contains illustrations of various installations of these lamps, and makes recommendations relative to the illumination of various departments of textile mills. The publication is enclosed in a rather striking cover.

Bulletin No. 4893, recently issued by the General Electric Company, is devoted to a general description of two automatic time switches manufactured by that Company. The bulletin contains diagrams and dimensions.

The General Electric Company has just issued a bulletin, No. 4920, describing its G.E. 203A Railway Motor which is of the box frame, commutating pole type, rated at 50 h.p. on 600 volts, and 40 h.p. on 500 volts. The motor embodies radically new features of construction which have been developed with a view to effecting greater economy in railway motor operation, and it is considerably lighter per horse-power output than the previous design.

The General Electric Company has recently issued Bulletin No. 4918, which illustrates and describes panels designed by that Company for general use in central stations. The list of panels contains both generator and feeder types, and the panels are made for 125, 225 and 600 volts.

Cloth Pinions

The General Electric Company has just issued Bulletin No. 4878, which is devoted to Cloth Pinions. This remarkable and somewhat radical form of machine element is offered for a wide variety of application in mechanical transmission of power where, because of noise or for other reasons, the meshing of metallic pinions, with metallic gears is impracticable or undesirable. The advantages claimed for these pinions are great tooth strength, noiseless operation, freedom from damage by exposure to dampness, dryness or temperature changes, elasticity of teeth, self lubrication and long life. These pinions are made in various styles and sizes which are illustrated in the publication.

BOOK REVIEWS

Telephony. By Samuel G. McMeen and Kempster B. Miller. Chicago, Ill., American School of Correspondence, 1912. Price, $4.00.

An excellent treatise on the theory and practice of all phases of Telephone Engineering, particular attention being paid to the recent developments in automatic systems. The volume contains 960 pages, of a page size 7 x 10 in., and is well illustrated by 700 drawings, diagrams and photos. It is clearly printed on a good quality of paper and is bound in vellum de luxe.

The authors are men who fully understand both the practical and theoretical sides of telephone engineering, and they have put into this book the knowledge gained from their years of experience.

In "Telephony," is covered the installation, maintenance and operation of all types of telephone systems and also an unprejudiced discussion of the relative merits of automatic and manual exchanges.

The Steam Turbine. By Sir C. A. Parsons. New York, G. P. Putnam's Sons, 1911. Price, 50 cents.

There has recently been published a small cloth-bound volume entitled "The Steam Turbine." While but little attempt has been made to explain the theory of the turbine, the subject is treated in a clear and elementary manner. The author describes some of the earlier forms of turbines and also those used at the present time, such as the Curtis, DeLaval and Parsons.

The Kingdom of Dust. By J. G. Ogden, Ph.D., Chicago, Popular Mechanics Co., 1912. Price, 50 cents.

One of a series of handbooks on industrial subjects which are being published by the Popular Mechanics Company. The book is both interesting and entertaining. It is well printed, contains many illustrations and is cloth-bound.

Audel's Answers on Automobiles. By Gideon Harris and Associates. New York, Theo. Audel & Co., 1912. Price, $1.50.

The prospective automobile purchaser, driver or repair-man will probably be unable to find a book which is more suited to his needs than is the book recently published by the firm of Theo. Audel & Company. If one wishes to keep his car in good running order every day in the year, it is necessary that he fully understand each part of the car and in case any breakdown should occur, that he know the proper way to repair it. In this 512-page book the reader is told, in language so simple that even the beginner can understand it, all about the various parts of the automobile and how to keep the machine in perfect adjustment. The book is well illustrated with 380 diagrams and drawings, and the text consists of a collection of Questions and Answers which fully explain the principles of construction and operation of the Motor Car in a clear and helpful way. Some of the topics treated in the various chapters are, Carburetors, Ignition, Gas Engine Operation, How to Run an Automobile, Overhauling the car, Motorcycles and Electric Automobiles and Trucks.

The Most Practical
Electrical Library

The Electrical Engineering Library is part of the International Library of Technology that cost $1,500,000 in its original preparation. It contains the knowledge given from the life experience of some of the best electrical engineering experts in the country, edited in a style that nineteen years of experience in publishing home-study textbooks has proved easiest to learn, to remember, and to apply. There is no other reference work in the world that so completely meets the needs of the electrician as the Electrical Engineering Library. The volumes are recommended by the highest authorities and are used in nearly all the leading universities and colleges. Not only can they be used to great advantage by superintendents, foremen, and engineers as an authoritative guide in their work, but since they can be so clearly understood, even by persons having no knowledge of higher mathematics, they can be used by all classes of electricians that are desirous of advancing to higher positions.

A few of the many subjects contained in the Electrical Engineering Library are as follows: Electricity and Magnetism; Electrodynamics; Electrical Resistance and Capacity; Magnetic Circuit; Electromagnetic Induction; Primary Batteries; Electrical Measurements; Dynamos and Dynamo Design; Direct-Current Motors; Alternating Currents; Alternators; Electric Transmission; Line Construction; Switchboards; Power Transformation and Measurement; Storage Batteries; Incandescent Lighting; Arc Lighting; Interior Wiring; Modern Electric Lighting Devices; Electric Signs; Electric Heating; Elements of Telegraph Operating; Principles of Telephony; Telephone Circuits, Receivers, Transmitters, Apparatus, Bells, Instruments, and Installation; Magneto-Switchboards; Electric-Railway Systems; Line and Track; Line Calculations; Motors and Controllers; Electric-Car Equipment; Multiple-Unit System; Efficiency Tests; Energy Regulation; Central Energy Systems, Main and Branch Exchanges; Common-Battery Signaling Systems; Bell-Energy System; Bell Trunk Circuits; Bell Toll and Testing Circuits; Exchange Wiring; Telephone Cables, etc.

The Electrical Library contains 17 volumes durably and handsomely bound in three-fourths red morocco. Each volume is 6×9 inches in size. They may be purchased in sets of five or more volumes. If you wish to know more about the most practical electrical library in the world,

Send the coupon NOW.

For the Young Draftsman

ELEMENTARY PRINCIPLES OF INDUSTRIAL DRAWING

Presenting the subject of Industrial Drawing in simple and concise form. Especially adapted to the use of the student who has not had time to take an extended course. - **Price, $1.00**

CAMS and the PRINCIPLES of their CONSTRUCTION

A comprehensive treatise on the planning, designing and drafting of this highly important adjunct to modern machinery. - - **Price, $1.50**

These two contributions to the field of literature on drafting by *George Jepson*, Instructor in Mechanical Drawing in the Massachusetts Normal Art School, Master of Columbus Avenue Evening Drawing School, Boston, and Medallist Science and Art Department (Great Britain) should prove a welcome addition to the student's library.

Sampson Publishing Co.
221 Columbus Avenue :: Boston, Mass.

ANY ELECTRICIAN or MECHANIC

Can Make Good Money in Spare Hours

As local representative to demonstrate most successful electrical instrument for the deaf.

Also Installing Ear-Phone Outfits in Churches, Halls and Theatres.

Write or call with references.

GLOBE EAR-PHONE CO.
88 TREMONT ST. BOSTON, MASS.

Telegraphy Taught
in the shortest possible time

The Omnigraph Automatic Transmitter combined with standard key and sounder. Sends your telegraph messages at any speed, just as an expert operator would. Five styles **$2.00** up; circular free.

OMNIGRAPH MFG. CO., 41 Cortlandt St., New York

"BUSINESS POWER" New member of Haddock's Power-Book Library. The Master Builder of Financial Ability and Commanding Business Personality. A practical analysis and instruction book in the underground factors of present-day fortune building, with actual directions for those who seek commercial chieftainship. The greatest book on business power and success ever formulated. Nothing like it in literature. An encyclopedia of rare, scientific rules, methods and result-producing plans for every successful business man, for the leaders in money-making, for those who are big men—or wish to be. This volume is brand new, in a virgin field never before entered; and mark this—it will be the big thing in business literature for years to come. Get the new book at once. Money refunded if book returned registered in five days. Send for circulars. Price $3.25.

POWER-BOOK LIBRARY, Alhambra Sta., Los Angeles, Cal.

PHOTOGRAPHIC BOOKS

Dark Room Dime Series

1. **Retouching for Amateurs.** Elementary instructions on removing defects in negatives, and improving your home portraits.
2. **Exposure Tables and Exposure Record.** Tables for calculating exposure under all conditions, with a note-book to preserve data of exposure conditions.
3. **How to Take Portraits.** Describes the making of backgrounds and apparatus, lighting, posing, exposure and development of home portraits, indoors and out.
4. **How to Make Enlargements.** Simple directions for making enlargements without special apparatus, and instructions for making an enlarging lantern and a fixed focus enlarger.
5. **A Manual of Photography.** A first book for the beginner, but valuable to everybody, because written out of long experience.
6. **Practical Development.** An up-to-date treatise on all the phases of this perplexing subject. Describes the construction of developers and their action under all circumstances.
7. **Popular Printing Processes.** The manipulation of the simpler processes, blue-print, printing-out, and development papers.
8. **Hints on Composition.** Some simple considerations of elementary principles of picture construction.

Photo Beacon Dime Series

1. **Development.** By Alfred Watkins.
2. **Photographic Printing Processes.** By Louis H. Hoyt.
3. **Beginner's Troubles.** By J. Edgar Ross.
4. **Elements of Pictorial Composition.** By F. Dundas Todd.
5. **Isochromatic Photography.** By R. James Wallace.

Any of the above, postpaid, 10 cents each.

Photo Beacon Exposure Card. By F. Dundas Todd. The simplest exposure calculator ever devised. 90th thousand now selling. 25 cents.

First Step in Photography. By F. Dundas Todd. 25 cents.
Second Step in Photography. By F. Dundas Todd. 50 cents.

A Reference Book of Practical Photography. By F. Dundas Todd. 50 cents.

Pictorial Landscape Photography. By John A. Hodges. 75 cents.

AMERICAN PHOTOGRAPHY, 213 Pope Bldg., Boston, Mass.

SALE AND EXCHANGE

Advertisements under this heading, without display, will cost 3 cents per word; 25 words or less, minimum charge of 75 cents. Black-faced type, 4 cents per word; minimum $1.00.

Cash must accompany each order, or advertisement will not be inserted.

AERONAUTICS

5 CENTS brings our new, up-to-the-minute catalog, 40 pages, fully illustrated, includes rules for holding model contests. None free. IDEAL AERO SUPPLY CO., 84-86 West Broadway, New York.

AUTOS AND MOTORCYCLES

AUTOMOBILE ENGINES, boilers and parts. Send for list. J. L. LUCAS & SON, Bridgeport, Conn.

FOR YOU—Ford runabout thoroughly overhauled, has magneto, fair tires. A good little car, price $225. A snap for somebody. HENRY E. HANSEN, Bagley, Minn.

WANTED—A motorcycle engine to attach to bicycle, in good condition. Must be cheap. G. WALLINGFORD, L.B. 92, McLoud, Okla.

BOOKS AND MAGAZINES

I.C.S. AND A.S.C. SETS and odd volumes of books on engineering, wireless, mechanics, shopwork, etc., new and second hand. Few I.C.S scholarships for sale. Bargain. GEORGE F. WILLIAMS, Box 408, New Orleans, La.

MECHANICAL BOOKS—Any subject; catalog for stamp. CRESCENT BOOK STORE, A 329 S. Halsted St., Chicago, Ill.

BRAND NEW—HARPER'S "ELECTRICITY BOOK FOR BOYS," never opened, sent on receipt of one dollar. CLARENCE H. PFEIFER, Ridgewood, N.J.

FOR SALE—Copies "Electrician and Mechanic," September 1907 to January 1909, and others. Also copies "Modern Electrics." HOWARD S. MILLER, 959 Lancaster Avenue, Syracuse, N.Y.

COINS AND STAMPS

STAMPS—All for 10 cents. One set each: 2 Honduras, 2 Ecuador, 2 Nicaragua, 2 Salvador, 500 Quaker Hinges, 50 unused Cuban stamps. QUAKER STAMP CO., Toledo, O.

OLD COINS BOUGHT AND SOLD. Buying catalog, 10 cents. New 55-page 1912 Selling Catalog, to Collectors only, "Free," WILLIAM HESSLEIN, Malley Bldg., New Haven, Conn.

COINS—20 different foreign 25 cent; large U.S. cent 5 cents; 5 different Confederate state bills, 15 cents; 150-page illustrated premium coin book, 25 cents; 1,000 old U.S. stamps 25 cents. F. L. TOUPAL CO., Dept. 57, Chicago Heights, Ill.

ELECTRICAL

ELECTRIC WIRING AND LIGHTING.—Knox-Shad, 192 pages; 150 illustrations. A working guide in all matters relating to electric wiring and lighting. Includes wiring by direct and alternating current by all methods approved by the fire underwriters; and the selection and installation of electric lamps for the lighting of buildings and streets. The utmost care has been used to bring the treatment of each subject within the range of common understanding, making the book especially suitable for the self-taught, practical man. Published by AMERICAN SCHOOL OF CORRESPONDENCE, Chicago, Ill.

ELECTRICAL

FOR SALE—70-watt D.C. dynamo, 15 volts; also set of A.C.S. Automobile Engineering books. Want drawing instruments. TASSO MORGAN, Wilmington, Ohio.

"DON'T WASTE MONEY. Why throw away those old dry batteries when they can be renewed?" Let us tell you how. Formula, 25 cents. STEWART LABORATORY, 631 West End Ave., New York.

WANTED—To exchange an A.C. 1 h.p. 110-volt 2-phase motor for single phase A.C. ½ h.p. motor. SAM NELSON, Dalsville, Ala.

HIGH-GRADE DRY BATTERIES, flashlight and medical. Demonstrate formula. Assembly and completion of ignitors. Endurance and recuperation unexcelled. $25.00. Exclusive territory guaranteed. J. DOUGAN, 254 Smith Street, Winnipeg, Canada.

HELP WANTED

500 MEN, 20 to 40 years old, wanted at once for electric railway motormen and conductors; $60 to $100 a month; no experience necessary; fine opportunity; no strike; write immediately for application blank. Address, S.O.M. Care of *Electrician and Mechanic.*

WANTED—Cabinetmaker's workbench, 6 ft. x 2 ft., two screws, good order, must be cheap. Box 82, Cecilton, Md.

WANTED—Experienced organizers to get members for Order of Owls. Good commissions. Any territory. JOHN TALBOT, South Bend, Ind.

LOCAL REPRESENTATIVE WANTED.—Splendid income assured right man to act as our representative after learning our business thoroughly by mail. Former experience unnecessary. All we require is honesty, ability, ambition and willingness to learn a lucrative business. No soliciting or travelling. This is an exceptional opportunity for a man in your section to get into a big paying business without capital, and become independent for life. Write at once for full particulars. Address E. R. MARDEN, Pres., The National Co-Operative Real Estate Company, L453 Marden Bldg., Washington, D.C.

WE TRAIN DETECTIVES. You can be one. Splendid opportunities. Travel. Earn $100 to $300 monthly. This fascinating profession taught practically and scientifically by mail at a nominal cost. AMERICAN SCHOOL OF CRIMINOLOGY, Dept. 8., Detroit, Mich.

WANTED IN BANKS—Honest, ambitious men. Splendid opportunities. Pleasant work, short hours, good salary. Be a banker. We teach you in a few months by mail. Write for catalog. AMERICAN SCHOOL OF BANKING, 135 McLene Bldg., Columbus, O.

MALE HELP—I will start you earning $4.00 daily at home in spare time silvering mirrors; no capital; free instructive booklet, giving plans of operation. G. F. REDMOND, Dept. A.C., Boston, Mass.

WANTED—A man or woman to act as our information reporter. All or spare time. No experience necessary. $50 to $300 per month. Nothing to sell. Send stamps for particulars. Address SALES ASSOCIATION, 784 Association Building, Indianapolis, Ind. (8)

GATHERING INFORMATION in your locality pays good money. All or spare time. Either sex. Nothing to sell. NATIONAL INFORMATION BUREAU, Sta. A, Columbus, Ohio.

FREE ILLUSTRATED BOOK, tells about over 360,000 protected positions in United States service. More than 40,000 vacancies every year. There is a big chance here for you. Sure and generous pay. Lifetime employment. Easy to get. Just ask for booklet A89. No obligation. EARL HOPKINS, Washington, D.C.

BECOME A DETECTIVE—Earn $150 to $300 per month travelling over the world. Send stamp for particulars. Write FREDERICK WAGNER, 1243 Lexington Ave., N.Y.

ELECTRICIAN & MECHANIC

| VOLUME XXIV | MAY, 1912 | NUMBER 5 |

THE ODDEST ELECTRIC LOCOMOTIVES
GEORGE FREDERIC STRATTON

One of the oddest designs of electric locomotives is designated the Crab; and while the name is perhaps somewhat descriptive of its flat and squatty appearance, it does no justice to its movements. The "Octopus" would be a more denoting term.

◘Gliding into the black entries of coal mines and halting at a crevice in the wall, from which issues the dull ring of distant pick and shovel, and through which may be seen tiny points of light from the miners' lamps, the Crab sends a flexible searching tentacle in for 200 or 300 ft., withdrawing it presently with a car of coal in its grasp. First on one side and then on the other it moves along, feeling into the holes, and never failing to secure its prey. Finally, with a dozen or more cars in its wake, it proceeds triumphantly to the shaft or outlet, and delivers its booty into the insatiable maws of the crusher.

These crabs operate by trolley conductors. They run through the main entries of the mine, but for various economic reasons they are not usually run up into the working faces. Accordingly, each machine is furnished with an electrically-operated drum upon which is carried 300 or 400 ft. of steel cable.

There are several methods of operating this locomotive to advantage, the choice depending upon the system of mining in use. With the double entry system, the locomotive usually hauls a trip of empty cars into one entry and drops them off at the rooms where they are required. Returning on the other entry it stops in front of each room where a loaded car is ready. The trip rider then drags the cable into the room, attaches

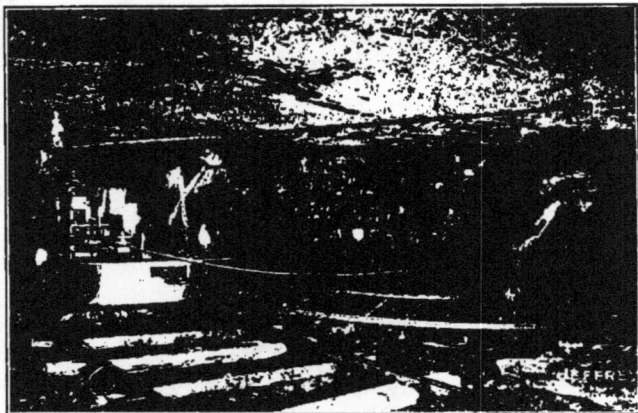

A 5-ton Coal Locomotive in a Coal Mine

its end to the car and signals the motor-man, who starts the crab motor and pulls the car out to the entry track. The locomotive then pulls it to the next room or leaves it standing until as many cars have been drawn from the rooms as are required to make up a trip. Then it pushes them together, and they are coupled up and hauled to the shaft or mine entrance. On the return trip the empty cars are distributed in the entry from which the loads have just been removed, and the locomotive gathers the loaded cars from the entry which was supplied with empties on the preceding trip.

Where the single entry system is used, the locomotive runs in with a train of empties. Stopping successively in front of the rooms in which loaded cars are ready, it hauls out each car to the entry track and pushes it ahead to the next room, dropping off an empty to replace each loaded car taken out. When all empties have been distributed, it then proceeds to the mine entrance with the loaded cars. By either of these methods, a locomotive can gather from 75 to 200 cars per day, depending upon local conditions.

Another type of electric mine locomotive is known as the rack-rail. It is used in mines where the grades are very steep, and, as its name indicates, depends for its tractive power upon a driving sprocket wheel which engages a rack-rail

bolted to the ties. Frequently this rack-rail is used as a current conductor instead of the overhead wiring and the trolleys. The locomotive can also be used in the ordinary way, without the rack-rail, whenever level stretches are reached.

Rack-rail locomotives are made in powers of from 50 to 200 h.p.; and in every case are equipped with unusually powerful brakes for controlling cars on heavy grades. They are also furnished with reels upon which are wound several hundred feet of conducting cable, through which the motors can obtain current while working up into rooms where no permanent conductors are installed.

Mining operations, however, comprise but a small percentage of the great variety of uses for electric locomotives. Like compressed air engines, they are especially valuable in yards and factories and on wharves, where the steam locomotive would be prohibited on account of fire risk. For these purposes electric locomotives are designed to operate by trolley system or by storage battery, or by a combination of both; the batteries being so designed as to run the locomotive for from one to two days on a single charge.

These engines are made in sizes of 2½ tons up to 60 tons. Of course, for regular railroad work very much heavier machines are built; but that is another story. The variety of these electric locomotives is as numerous as the uses to which they are put. Some are built very low to facilitate entrance into low storage warehouses or basements. Some are equipped with hoisting devices to pick up and car y weights but little less than their own. Some are designed solely for hauling cars—others are a combination of car and locomotive, operating as a single unit. They are of course more flexible in their scope of operation than the ordinary steam locomotive. The combination trolley and storage battery engine will dart along under a trolley wire, switch off onto a temporary track, using its own battery power, then, returning to the trolley track, will, while doing its work, absorb juice from the wire to keep its batteries fully charged. Such equipment on great construction operations— dams, bridges and reservoirs—is almost ideal. They are short, compact, set so low and are under such excellent control that stunts may be done with them on curves and grades which make them almost as flexible as the Telpher.

The Telpher is the tiniest locomotive built for work. It runs not upon the surface, but upon a single wire, suspended a few inches below another wire from which it takes its current. It hooks up a barrel of flour or a bale of dry-goods and runs with it, sometimes unattended, to some distant corner of yard or warehouse

Recently, a gasoline-electric tractor has been designed by Massachusetts engineers. This is called the Multiple-

Telpher in use at the Arnold Print Works, North Adams, Mass. It operates among the various buildings, hoisting the load from the road or wagon and carrying it down the alleys and into the various floors of the building.

Unit-Road train, and is comprised of a tractor and two or more cars, each with a capacity of six to eight tons, and carrying its own electric motors, current for which is supplied by generators on the tractor, and driven by a 40 h.p. gas engine. The speed of this train is stated to be, with a 20-ton load, six miles per hour on a macadam road.

7½-ton Locomotive

The Switch Board in the New Hydro-electric Plant of the N.C. Electrical Power Co.

A SIMPLE ELECTRIC LOCK
W. S. LYLE, JR.

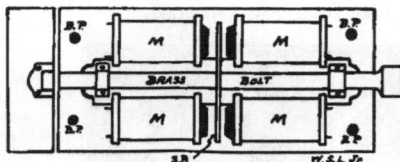

Top View of the Lock

Most electric locks of today are complicated or do not work satisfactorily.

The lock described in this article is simple and very durable.

First purchase a brass push bolt for about 25 cents, then cut the bolt in the center and about half through, with a

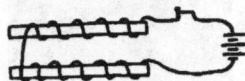

Connections.

hack saw. Cut a piece of iron or steel to fit into this slot and long enough so that the magnets on each side of the bolt will attract the steel bar in the bolt, thus pulling it either backward or forwards.

This lock may be used for secret locks, doors, etc.

THE LOCK COMPLETE.

Emergency Magnifying Glass
HERBERT P. A. HOLDER

Take a piece of small wire or a plant stem and make a loop in the end about the size of a drop of water. In this loop place a drop of clear, pure water. This will serve admirably as a strong magnifying glass.

A NEW HYDRO-ELECTRIC PLANT

N. BUCKNER

The third hydro-electric plant of the North Carolina Electrical Power Company of Asheville, N.C., located on the French Broad River 25 miles northwest of Asheville, which has been in the course of construction for the past two years, has just been completed and put into service. Normal capacity of the plant is 5,000 h.p. The entire plant represents in round figures an expenditure of $500,000, and to construct same it was necessary to raise and rebuild 2½ miles of the track of the Southern Railway which skirts the river at that point. The track at the dam was raised 20 ft. higher than the old roadbed, the total excavation amounting to about 60,000 cu. yds., 80 percent of which was solid granite. The change in roadbed alone cost $75,000, required one year to complete, and all work was done without interference with traffic, there being operated over this line an average of 30 to 40 trains per day.

The dam is 540 ft. long, 30 ft. high, 43 ft. 9 in. thick at the base, and 11 ft. at the top. It is built of cyclopian concrete, the large stones in some instances approximating 5 cu. yds. Approximately 22,000 barrels of cement were used in construction of dam, foundation of power plant and retaining wall for the protection of the roadbed. The downstream face of the dam is curved in such a manner as to insure the water always clinging to the surface and preventing the formation of a vacuum under the falling sheet, since it is generally conceded by engineers that a vacuum on the down stream side is responsible for the trembling often felt in the vicinity of overfall dams.

There are two 7 ft. mud gates in the dam next to the power house, which are operated by hydraulic cylinders and are opened and closed by an electrically-driven pump in the power house. The gates and cylinders are entirely submerged.

Looking North; New Hydro-electric Plant of the N.C. Electrical Power Co. It is located on the French Broad River, near Asheville, N.C.

The Generator Room

The four penstock gates are among the largest cast iron gates made; each gate covers a clear opening of 18 ft. x 7 ft. 3 in. and weighs 13 tons. They are operated in pairs by an electric motor.

The power house, 40 x 76 ft., fireproof throughout, is built of concrete to the floor line, and brick from that point up. Windows are of steel and prismatic glass. From floor to eaves is 31 ft.; from bottom of foundation to comb of roof is 100 ft. A 50-ton electrically-operated traveling crane extends the entire length of the building.

The equipment consists of two 1,875 kw. Westinghouse alternating current generators, 6,600 volts, 3-phase, 60-cycle, 133 revolutions per minute. each directly coupled to two turbines made by the I. P. Morris Co. The units are vertical, with the exciters located on top of the generators. The current, generated at 6,600 volts, is stepped up to 66,000 volts for transmission. The entire control of the plant is from the switchboard, all gates, switches, motors and valves being electrically operated.

Duplicate power lines have been built on private right of way, one on the west side, the other on the east side of the river, to plant No. 2 of this Company, six miles northwest of Asheville, where there is a sub-station for distributing the power to Asheville, Canton and other places.

The other two hydro-electric plants owned by this company develop 4,000 h.p., all of which has been utilized in Asheville. This new plant has been made necessary by the increased demand in Asheville and surrounding territory for additional light and power.

This plant was designed by Charles E. Waddell, member American Society Civil Engineers, Asheville, North Carolina. Capt. W. T. Weaver is President of the Company and Charles Folsom, resident engineer.

METHOD FOR OBTAINING THE EFFICIENCY AND LOSSES OF DYNAMOS AND MOTORS

A. SPRUNG, E.E.
(*Associate Member A.I.E.E.*)

A direct-current dynamo can be operated as a motor, and conversely a motor can be operated as a dynamo. In the dynamo the mechanical energy of the prime mover is turned into electrical energy, and in the motor, the electrical energy is consumed in the production of rotation of the motor armature. The revolving of this motor armature furnishes the necessary mechanical energy available at the shaft. In either of the above cases the energy is produced with losses. These losses are found in the machine itself under any condition of running whatsoever. By machine is meant either the dynamo or the motor.

It is understood from the laws of nature that the amount of work obtained from any mechanical device must be to a certain extent smaller than the amount of energy put into this device in order to obtain the desired work output. This difference of output and input is credited to the energy losses of the device. In other words, the losses cause the output of a machine to be less than the input. Then the amount of work obtained from the machine divided by the amount of work put into the machine is called the efficiency. By means of efficiency the amount of losses is readily determinable. Efficiency = work output ÷ work input.

This is true for both the dynamo and the motor, and is more clearly expressed as follows:

Motor Efficiency = Mechanical Output ÷ Mechanical output + losses

Dynamo Efficiency = Electrical Output ÷ Electrical Output + losses.

It is evident from the above that the output + losses = input. If not for the losses the output would equal the input, which in practice is impossible. Motor efficiency can be obtained also as in the following formula:

Motor Efficiency = Electrical input— losses ÷ Electrical input.

To determine the efficiency in practice, it is necessary to solve the above equations experimentally. This can be done by two methods.

First, by determining the actual input and output of the machine, and second,

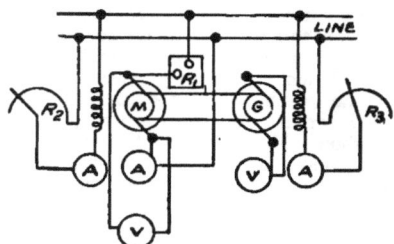

Diagram of Connections

by determining either the input or the output of the machine including the energy lost in the machine.

The accompanying diagram shows the method of connecting for making up the test.

METHOD OF TESTING

In the first case, to determine the input and output directly, an actual load test is necessary. A friction brake load is generally used on motors, and a resistance lamp load is used on dynamos. The above method has its disadvantages in that it is extremely difficult to take accurate mechanical measurements at the brake. The amount of power consumed during the test is wasteful and the total losses only can be ascertained by this method. This does not help the operator in determining just where most of his losses are located.

On the other hand, if the test is made by determining the losses of the machine at no load, the above disadvantages are overcome and we can thereby determine the efficiency at any load, together with a location of various losses.

The losses in an electric machine can be described as follows:

1. Copper losses (C^2R); these losses are found in the armature and field coils.

In order to overcome the resistance a certain amount of energy must be sacrificed or lost in dissipated heat. This energy can be expressed in watts. Hence the watts consumed equals volt drop x current, but volt drop equals CR, then watts or energy loss equals C x C x R = C^2R loss.

2. Iron losses chiefly found in armature core due to hysteresis and eddy currents.

3. Mechanical losses—these are caused by friction in bearings, brush friction and windage or air resistance.

The copper loss can be easily determined just by merely measuring the ohmic resistance of the armature and field winding, by the drop of potential method. Knowing R, the resistance, the copper loss for any current would then be the square of current times R.

The copper losses depend upon the magnetic flux and upon the speed. In shunt machines the losses are approximately constant, since the magnetic flux is constant with a constant field current.

The speed will generally keep constant at all loads.

The mechanical losses depend entirely upon the speed. In direct connected machines the bearing friction is constant, whereas in belt drives the friction is a function of a tensity of the belt.

The iron losses and mechanical losses can be assumed constant at all loads, having the same value at full load as at no load. The losses can then be measured by finding the amount of power necessary to run the machine, as a motor at no load. By noting the amount of current in the armature and field the copper loss (C^2R loss) can be deducted from the total losses, at no load, and the resulting losses will be iron and mechanical losses.

SAVED BY THE MINE TELEPHONE

The public at large has already for some time past appreciated what great reason they have to be grateful for the invention of the telephone; but there are today two miners in Kansas who are more than grateful. They owe their lives to it.

These two miners, or "shot-firers," to be exact, are employed by the Girard Coal Company in a mine at Radley, Kan. The mines of this company have recently all been equipped with Western Electric Company mine telephones, and, according to the rules of the Coal Company, the shot-firers must report to the night engineer, by means of the telephone, the progress of their work as they go through the mine lighting the shots. This enables the engineer to know where his men are, so that if he does not hear from them at certain intervals, a rescue party is sent down.

One evening, after the miners had left, the shot-firers went down as usual to fire the shots which would bring down the coal for removal during the next day. The two men had just entered a refuge hole and one was in the act of ringing the engineer to tell him they had lighted the shots in that particular entry, when an explosion occurred. The force of the explosion was so strong that it blew in the back end of the refuge hole, and the shot-firer did not even get to talk; was immediately overcome by the afterdamp. His partner, who was with him, was likewise overcome. The night engi-

neer, knowing that this was the station from which they should next report, immediately tried to call them, but was unable to get any response and started the distress whistle. In fifteen minutes after the explosion had occurred, a rescue party was in this refuge hole and had the two shot-firers out working upon them, and succeeded in resuscitating them. A little later it would undoubtedly have been impossible to revive them.

There is an employers' liability law in some states, which compels the operator to pay a considerable sum for loss of life or personal injury. The fact that the telephone very often prevents accidents and assists in quick rescue work, saves the operator a great amount of money. In the Girard Company's mines, there were three severe explosions during the winter, but not one of the Western Electric mine telephones was injured, nor was service interrupted.

The business section of the town of Nacozari, Mexico, owes its existence today to the bravery of the Mexican engineer of a burning train in which there were two cars of dynamite. While the train was standing at the depot in the center of the town a blaze was discovered in a box car adjoining one of the cars of dynamite. The engineer, Jesus Garcia, rushed to his engine and pulled the train out of the town. Less than a mile out the dynamite exploded, blowing the engineer to atoms.

THE FUTURE, PRESENT AND PAST OF ELECTROPLATING

CHARLES H. PROCTOR

The future of electroplating, like that of other commercial enterprises, will depend upon the progress made in the art by individuals or by the concerted action or co-operation of men who understand the essential details of its chemical, electrical and mechanical requirements. These will be so interwoven as to create the standard of efficiency that will be necessary for the successful electroplater to attain. Electroplating is so closely identified with commercial enterprises and progress that to keep in line with such advances will require more careful thinking and studying in the electrical, chemical and commercial manipulations than has fallen to the lot of the foreman plater heretofore. For years the art has remained practically dormant chemically. Very little has been accomplished since the days of Elkington, Becquerel, Heeren, Roseleur, Von Leutchenberg, Neidinger and others whose methods and formulas have remained practically standard up to the present time.

But a new era lies in the future. On every hand it is gratifying to note that electrochemists and metallurgists, who have devoted the product of their brains to the mining industries for years past, are now turning their thoughts to the greater possibilities of the electro-deposition of metals. Not only of metals on metals, but also on many non-conducting surfaces, such as wood, plaster, etc., thus obtaining a finish having the appearance of solid metal, combined with extreme lightness of weight and low cost of production. New fields are constantly being opened up for intelligent workmen, so that instead of electroplaters being in less demand in the future, it will be found that there will not be sufficient number of thoroughly experienced men to draw upon to fill positions that will be created by new enterprises. So it behooves the younger generation who expect to take up the art of electroplating in the future to make a thorough study of chemistry and electricity. Adding to these arts such mechanical skill as he possesses, the plater of the future will then be able, with his superior knowledge, to successfully cope with any difficulty that may present itself.

PROGRESS IN RECENT YEARS

The present of electroplating is upon a satisfactory basis, so much has been accomplished in a mechanical way in the past decade. Mechanical plating tanks and barrels of almost endless variety have been put on the market. Mechanical polishing and burnishing methods, that brought forth economical results in manipulations, have saved considerable money in the cost of production, which has heretofore been almost prohibitive in the finishing of small articles. By the application of mechanical electroplating much has been learned that heretofore has not been thoroughly understood. Constant friction by agitation caused greater internal resistance and necessitated denser solutions and greater voltage to produce results as satisfactory as those secured from the still solutions formerly employed. Electro-galvanizing and mechanical plating requiring greater voltages have brought forth the three-wire system. The dynamo developed for this purpose affords a range of from five to ten volts, making the energy created satisfactory for still solutions requiring up to five volts, and for mechanical solutions needing up to ten volts. Such dynamos are replacing the older types because of their particular advantage in developing the high and low voltages required.

The platers of the United States and Canada should feel highly gratified that they have been able to maintain a standard of finishes quite equal, if not superior, to those of any other country. Germany, France and Austria, however, have paid more attention to chemical detail. England has followed the lead of the United States, but pays more attention to the uniformity of deposit. Many finishes are produced in the above countries that are unknown in the United States, and *vice versa*. In the builders' and cabinet hardware industry the United States surpasses the world in the variety of its finishes and designs.

THE TRISALYTE SOLUTION

The introduction of "Trisalyte" for

solutions in the United States by the Roessler & Hasslacher Chemical Company will prove of much value to the plater. The composition of these tri-salts, being uniform in quality and perfectly balanced, will produce more satisfactory results, in the various deposits than has been obtained in the past. Such solutions are prepared for electro-galvanizing, copper, brass, bronze, silver and gold-plating. These salts are used exclusively in Germany, and have been on the market there for several years. In France and Germany bright nickeling has been brought to a successful issue. If a polished surface is immersed in the bright nickel bath the deposit will remain bright even though plated for several hours, and will require no further polishing when removed from the bath.

The Sangamo ampere-hour meter will no doubt prove of great value in determining the actual amount of metal deposited in a given time, so that eventually a system of costs—so much desired—will be installed in the plating department; and, with the introduction of the Rojas method of electrochemical metal coloring the present status of electroplating can be considered as satisfactory.

PRINCIPLE STILL UNEXPLAINED

The past of electroplating must always be interwoven with the present and the future. The secret of precipitation has never been satisfactorily explained. The question is often asked: "What is electricity?" So the plater often wonders what causes those particles of metal to become so evenly distributed over the metal or metallized non-conductive surface, but he cannot explain why. The old myths of unsatisfactory deposits being due to too much sunshine, too much cloudy weather and a hundred and one other imaginative thoughts, have, like the myths of the middle ages, been explained by scientific study of cause and effect. Unfortunately, while many of the old concerns have modernized their mechanical productive departments, they have sadly neglected their plating department. The owners of such plants wonder why they do not produce the same results as their more modernly equipped competitors.

In many instances this neglect of introducing modern methods and equipment revert upon the plater to his dis-advantage; his employer oftentimes thinking he is not as competent as the plater of a competing concern. Recently I paid a visit to one of these modern plants with a plating room linked with the past. The results being produced were unsatisfactory. Several thousand gallons of solutions, consisting of brass, copper, bronze and nickel, were in daily use. The brass deposit varied greatly in color, and was otherwise inferior. In looking over the plant for the probable cause of this variation in color my attention was immediately drawn to the absence of a voltmeter. It is a well-known fact that a uniform deposit from a number of solutions depends not only on the uniformity of the composition of the solutions, but also upon the uniformity of the voltage at the tank terminals, so that unless evenly balanced, the deposit will not be uniform.

Variation in voltage produces variation in color and a variation of internal resistance in the various tanks, and not the incompetency of the plater, is the primary cause of lack of uniformity of deposit.

BECOMING AN EXACT SCIENCE

Give the progressive plater modern methods and he will undoubtedly produce results. This does not only refer to one solution but to the results obtained from any number of solutions in action at the same time. Looking back more than a quarter of a century into the past of plating, one is amused at the ridiculous thoughts and ideas that entered into the mind of the plater as to the probable cause of the troubles he encountered.

The National Electroplaters' Association has been of untold mutual advantage to its members, producing results from exchange of thoughts, ideas and experiences. If Andrew Carnegie or some other great leader in the metal world could understand its requirements, there is no doubt their financial assistance would be forthcoming to maintain the art of electroplating in this country in the foremost ranks of the world. With a scientifically-equipped laboratory and competent men who are experts in their various lines, the solution of many problems could be accomplished and great results accrue, which would be of advantage not only to the individual plater, but to the country at large.—*The Metal Industry*.

MECHANICAL DRAWING

P. LEROY FLANSBURG   L. BONVOULOIR

INTRODUCTION

It is quite apparent to anyone who considers the matter, that thought must always precede each intelligent act; it is therefore necessary that even the most simple of structures must be carefully conceived before it is built. Most persons find it at first extremely difficult to visualize or mentally conceive different physical bodies and their various shapes and relations, either while at rest or in motion. But it is possible to train the mind so that the faculty of visualizing will be developed and that is exactly the service performed by a course in mechanical drawing.

Mechanical drawing is primarily but a means of conveying ideas and the draughtsman should not only be able to make drawings which are both readable and workable, but should also be able to readily read such drawings as are made by other draughtsmen.

It is the purpose of this course to develop these various abilities. While the student may accomplish much by carefully following instructions, still, in mechanical drawing, as in nearly every-thing else, careful and intelligent criticism, especially at the beginning of the student's progress, will save him from forming many careless habits which he might find hard to overcome after they had once been formed. The authors, realizing this, will be glad to criticize any drawings which are sent in and will answer any questions which may bother the student who is taking this course, provided return postage is enclosed, and all the communications are addressed to the Mechanical Drawing Department of the *Electrician and Mechanic*.

In order that the student of mechanical drawing may do his best work, it is very important that he be supplied with a good serviceable set of instruments and certain necessary drawing materials. The man who does but a limited amount of draughting will, however, find little need for the more expensive outfits, and one of the less expensive sets will prove quite sufficient for all his needs. In the following paragraphs will be described the more important of the drawing materials.

THE DRAWING MATERIALS

1. *Drawing-Board.* — The board is generally made of well-seasoned pine, from ¾ to ⅞ in. thick, and for ordinary work, a board 17 x 22 in. in size will be found most convenient. Even though the board be well-seasoned, it has a tendency to warp, and, to prevent this, the maker usually places two cleats, either on the underside of the board or along its shorter edges (*AA*, Fig. 1). One of the shorter edges of the board (*B*, Fig. 1) is used as a working edge; and since the accuracy of the drawing is dependent upon the evenness of this edge, it is made a true plane. It is customary to place the working edge at the left hand, and for most work this is the only edge of

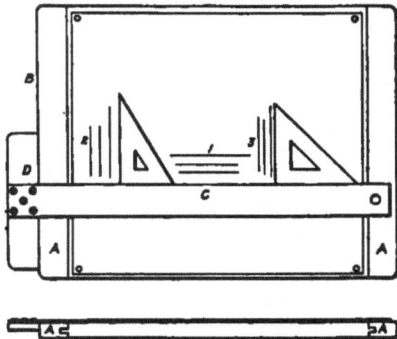

Fig. 1.

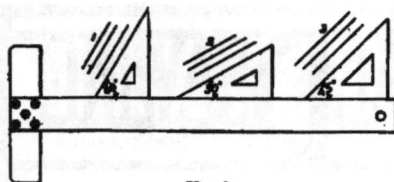

Fig. 2.

the board that need be used. Care should always be taken not to indent the working surface of the board.

2. *T-Square.*—The T-square consists of a blade (*C*, Fig. 1) and head (*D*, Fig. 1); the head being securely fastened to the blade by means of screws. Various woods are used for making the blade and head, pear-wood being the most common. However, on the better grade of T-squares, the blades are edged with either ebony or celluloid. Ordinarily the head is rigidly attached to the blade, so that the inner edge of the head is exactly at right angles with the upper edge of the blade. In some cases the blade is made movable so that it can be adjusted to any angle with reference to the head. In using the T-square, the inner edge of the head slides up and down the working edge of the board and should be held firmly against it with the left hand. Since the working edge of the board is a true plane, it is possible to draw parallel lines (1, Fig. 1), by sliding the T-square along this edge, and all horizontal lines are drawn in this manner. In drawing, only the upper edge of the blade should ever be used, and this edge must not be injured in any manner.

3. *Triangles.*—To draw straight lines other than horizontal lines, triangles are used. These are usually made either of wood or celluloid, are of various lengths and are about ½₂ in. thick. Since the celluloid triangles are transparent and easily cleaned, they are preferable to the wooden ones. Two triangles are used: one a 45 degree and the other a 60 degree. The 45 degree triangle has two 45 degree angles and a 90 degree angle, while the 60 degree triangle has a 60 degree, a 30 degree and a 90 degree angle. When using the triangles, they are placed against the upper edge of the T-square, and since each triangle has a right angle for one of its angles, it is easily

possible to draw vertical lines (2 and 3, Fig. 1), that is, lines which are parallel to the working edge of the drawing-board.

In all mechanical drawing work, the light is allowed to fall upon the board from the upper left-hand corner and the triangles should be so placed that they do not cast shadows upon the lines which are being drawn. When moving the triangle from one position to another, hold the head of the T-square firmly against the working edge of the board with the left hand, and move the triangle with the right hand. When drawing a line, both triangle and T-square should be held with the left hand.

By the use of the T-square and the triangles, it is a simple matter to draw lines which make angles of 30 degrees, 45 degrees and 60 degrees with the horizontal (1, 2 and 3, Fig. 2). Since a line making 30 degrees with the horizontal makes 60 degrees with the vertical, one making 60 degrees with the horizontal makes 30 degrees with the vertical, and one making 45 degrees with the horizontal makes 45 degrees with the vertical, it is a simple matter to draw lines making any one of these angles with either the horizontal or the vertical.

By using both the 60 degree and the 45 degree triangles with the T-square, it is possible to draw lines which make angles of 15 degrees or 75 degrees with either the vertical or the horizontal. The method of drawing such lines is shown in Fig. 3 and 4. In Fig. 3, the edge of the 45 degree triangle makes an angle of 30 degrees plus 45 degrees (or 75 degrees) with the horizontal, and since lines making X degrees with the horizon-

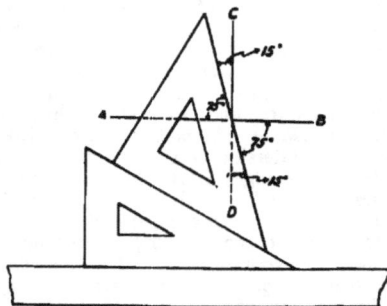

Fig. 3.

tal make (90—X) degrees with the vertical, the edge of the triangle makes an angle of (90—75) degrees or 15 degrees with the vertical. In Fig. 4, the edges of the 45 degree triangle make angles of 15 degrees with the horizontal and 75 degrees with the vertical.

When it is desired to draw parallel lines, other than those that are horizontal or vertical, it may be accomplished by means of two triangles. Place one triangle so that its long edge is quite near and exactly parallel to the given line, and holding the triangle firmly, place an edge of the second triangle against one of the shorter sides of the first. Now hold the second triangle firmly and slide the first either up or down along the edge of the second. It will thus be seen that the long edge of the first triangle will always move so that it is parallel to the original line. An example of this is shown in Fig. 5. DE is the given line

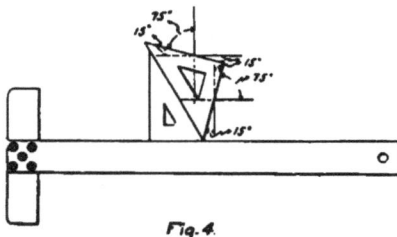

Fig. 4.

and it is desired to draw a line parallel to DE through some given point K. Triangle A is placed along line DE and triangle B is held against one of the edges of A. Then holding B firmly with one hand, A is slid along to position A' and the line FG is drawn through point K.

To draw a line perpendicular to a given line, the following method is used. Place one of the short edges of the 45 degree triangle against the line, and holding it firmly, place the long edge of the 60 degree triangle against the long edge of the 45 degree triangle. Now holding the 60 degree triangle firmly in place, slide the 45 degree triangle along the 60 degree until the other short edge of the 45 degree triangle intersects the original line at the desired point. Then a line drawn along this edge will be perpendicular to the given line at this point. In Fig. 6, the line AB is the given line and it is desired to erect a line perpen-

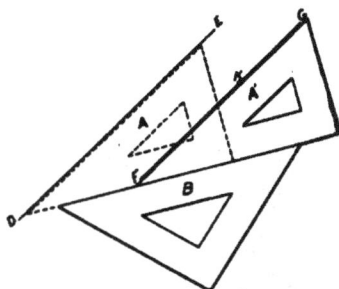

Fig. 5.

dicular to AB at point K. First place one of the short edges of triangle E along the line AB and place one of the long edges of triangle F along the long edge of triangle E. Now hold F firmly in place and slide E to position E'. When in this position, the other short edge of E is perpendicular to AB at point K and the line CD may now be drawn.

4. *Pencils and Penciling.*—It is much easier to learn to draw with a pencil, and finished pencil work will give more effective training in neatness than can be gained by inking-in the drawing. For these reasons, the student is advised to make all of the early exercises carefully finished pencil drawings. After one has gained both speed and accuracy with the pencil, the inking-in of the drawing will prove to be mere mechanical work. The pencils used should be ones having hard leads, and the ordinary writing pencil, being too soft, should never be used. The two grades of lead which are most commonly used by draughtsmen are the HHHH and the HHHHHH. In Fig. 7 is shown the proper manner of sharpening

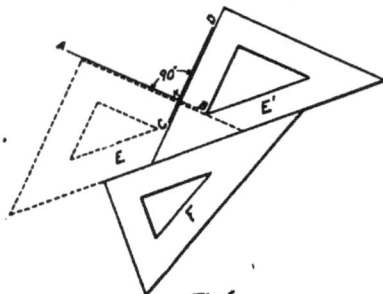

Fig. 6.

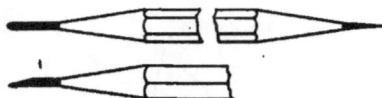

Fig. 7

the pencil leads. For ordinary line work an excellent point is made by cutting the lead clear across at an angle of about 30 degrees. As the outer skin of the pencil lead is much tougher than the center of the lead, such a point wears well. Also, it is a very quick way of sharpening the lead, since all that need be done to obtain such a point is to rub the lead upon a file, a piece of sand-paper or emery cloth. Generally the draughtsman will have a small piece of sand-paper or emery cloth fastened to a block of wood and will sharpen his pencils on this. The second point shown in Fig. 7 is an ordinary needle point, and this type of point is best for use in dimensioning drawings and for laying off distances. It is not only convenient but also good practice to have both ends of the pencil sharpened, one end having the chisel point and the other a needle point.

Be sure to keep the pencils well sharpened at all times, and when drawing hold the flat edge of the pencil against the straight edge. The pencil should be held vertically, and all lines drawn either from left to right or from bottom to top of the board.

All construction lines should be drawn fairly light, since it is therefore much easier to make any necessary corrections in the drawing.

5. *The Ink.*—The ink used for mechanical drawing is called India ink, and can be obtained either in stick form or in prepared form. Either form of ink will give a glistening, jet-black, waterproof line; but since the stick ink must be ground in water before it can be used, the prepared ink is preferable. One good

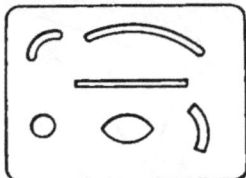

Fig. 8

kind of prepared ink is Higgins' Waterproof Drawing Ink. Red ink is much used for center lines and may be purchased already prepared. It is a much thinner ink than the India ink and will flow more freely, so that more care should be taken when using it, so as not to blur or blot the lines.

6. *Paper and Tracing Cloth.*—For work that is to be traced, either the so-called Duplex paper or a Normal paper is generally preferred. They are very tough and will stand a large amount of erasing without showing any marks. They will also stand inking, but a hot-pressed paper is much better for this purpose and either Whatman's or some other reliable make should be used. Tracing paper is seldom used except in architectural work, tracing cloth being preferred on account of its toughness and lasting qualities. Both the dull side and the glazed side of the cloth are used, but the dull side has the advantage that it causes the ink to flow more freely and also it will not show erasures as much as the other side.

7. *Erasers.*—For erasing pencil lines use a pliable rubber eraser, and for cleaning a drawing or removing light pencil lines which show upon an inked-in drawing, use a kind of eraser known as art-gum. An eraser such as art-gum or sponge rubber will remove the light pencil lines without injuring the inked lines. For removing ink spots or erasing inked lines it is best to first use a scalpel, being careful not to scrape the surface of the paper any more than can be helped. Then finish erasing with a rubber eraser. The scalpel is a small knife, very well sharpened and of a form to facilitate erasing.

8. *Erasing Shields.*—Erasing shields are made of paper, cardboard, celluloid or metal and have slits or holes of various shapes and sizes cut out of them. Such a shield is shown in Fig. 8. When removing an ink spot or a small portion of an inked line, the shield is placed over the spot or line so that only the part to be erased is exposed.

(To be continued)

The real friend is the one who gives you pepper once in a while rather than sugar all the time.—GRIZZLY PETE.

THE INTIMATE THEATRE
C. L. HAGEN

The Intimate Theatre is so termed because of the intimate relation of the audience and the stage. The revolving stage projects into the audience chamber, and is enclosed by the act curtain, just back of which are the footlights; while on each side, arranged vertically, are side lights, and overhead; the border or top lights. All of these lights are concealed, and the light is diffused onto the characters. This arrangement distributes the light equally, and removes the blinding effects of the light filaments upon the eyes of the actors, thus permitting them to see their audience.

An orchestra chamber is provided on one side of the proscenium arch, and an organ loft opposite. The top of the proscenium arch extends over the entire auditorium, returning down to the rear wall, forming an immense sounding board which will reflect sound waves to every part of the audience chamber and permits of a more efficient control of ventilation. The curtain is placed in front of the footlights to permit the stage director to properly light the picture before it is exposed to the audience.

Stage floor coverings, carpets, etc., may be extended to the curtain so that the entire scene is in repose when shown. Movable fire walls separate the stage from the audience chamber, and are arranged as sliding doors suspended from the top and closed in from the sides. This permits the proscenium opening to be closed much quicker than if it were lowered from the top. It also removes the danger of such an enormous weight being suspended over the stage and which might be dropped or lowered upon actors who were trying to pacify an audience in a panic. There is also less danger of obstruction in this movement, as illustrated in the Iroquois Theatre fire. This arrangement also permits of the construction of a light chamber over the proscenium arch and by means of flying or swing bridges, lights and effects can be produced over any portion of the stage.

The Drehbohne, or revolving stage, is surrounded within the sight lines by a horizon wall with a sky dome, preferably of steel construction, rough plastered and of such color as experiments may determine is best adapted to light effects

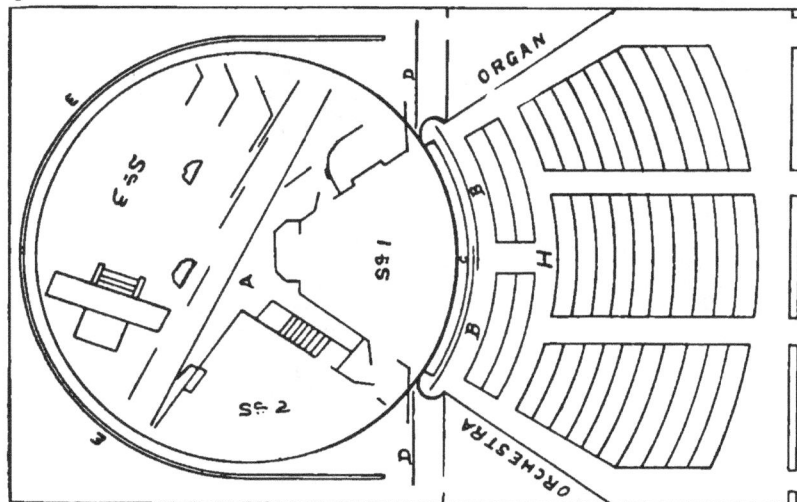

GROUND PLAN

A. Revolving Stage or Drehbohne
B. Act Curtain
C. Footlights, Side Lights, Border Lights

D. Fire Wall Curtain
E. Circular Dome over Stage
H. Auditorium

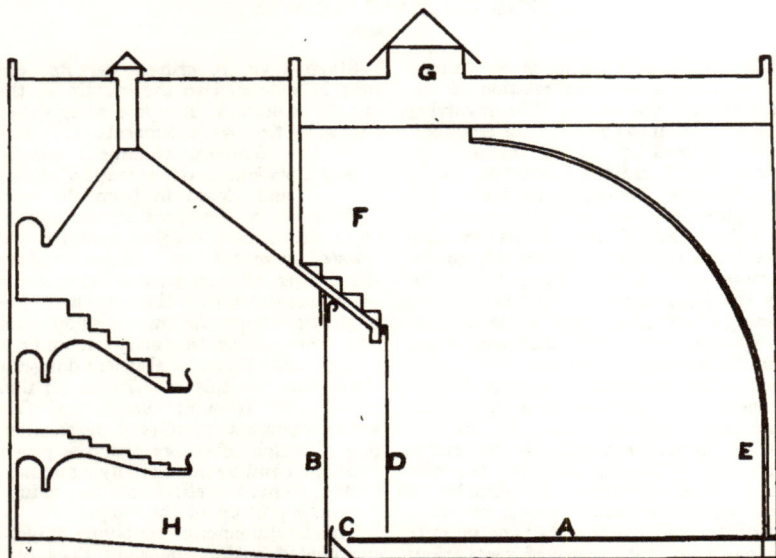

LONGITUDINAL SECTION THROUGH CENTER

A. Revolving Stage or Drehbohne
B. Act Curtain
C. Footlights, Side Lights, Border Lights
D. Fire Wall Curtain
E. Circular Dome Stage
F. Chamber for Reflecting Lights and Light Effects
G. Ventilators
H. Auditorium

projected from the light chamber. In this manner the entire stage may be flooded with diffused light.

The background acts as a sounding board to project sound waves.

Beautiful effects may be obtained similar to those of the artist Mariano Fortuny of Venice, who has invented a new process of stage illumination which closely imitates the conditions of nature, and presents all objects in diffused light. Arc lamps are used exclusively, as their light corresponds in composition more closely with sunlight. The light is reflected by surfaces of cloth and thus is diffused. In order to produce the various tones observed in nature, the reflecting surfaces are composed of a number of strips, some of which serve for the production of colors, and others for the modification of the light by an admixture of black or white (white paper reflects 70 to 80 percent, black velvet $\frac{4}{10}$ of 1 percent). Fortuny has illuminated a stage scene so perfectly that it was photographed without the use of other light

and as clearly as though it had been out in the daylight.

The opportunity provided by the design of the Intimate Theatre permits of a revelation in stage lighting and dissolving stage pictures. The Drehbohne permits of a number of scenes being arranged upon it at one time, with no portion of them extending over—thus permitting scenes to be moved into position both rapidly and silently. Indefinite time may be expended in preparing scene pictures with that care and detail so desired by the director and artist, and with the knowledge that they will appear undisturbed and silently in their proper place in the play. And thus does the mechanical stage play its part in the advancement of the drama.

The forging and tempering of iron and steel can be greatly enhanced, according to recent experiments, by dipping the metal in fused salt. This dipping in salt is also well adapted for annealing steel without the oxidation of the surface.

THE HOME CRAFTSMAN

RALPH F. WINDOES

JARDINIERE STAND

Every home has a place for just such a stand as we here illustrate, so it should behoove every craftsman to make up such a stand. It presents a beautiful appearance when built up of quarter-sawed white oak and finished in one of the modern styles. The stock needed consists in the following pieces, planed and sanded to dimension:

4 pcs. 1¾ x 1¾ x 30 in. quarter-sawed oak
8 pcs. ⅞ x 2 x 13 in. quarter-sawed oak
12 pcs. ⅜ x 4 x 6½ in. quarter-sawed oak
2 pcs. ⅞ x 1 x 11 in. plain oak
1 pc. ⅞ x 12 x 12 in. plain oak
Screws, filler, stain, wax, etc.

First, shape the legs and chamfer their top. They taper at the bottom, starting in the center and working down from 1⅞ in. square to 1 in. square. Next, cut the tenons on the eight side stringers and fit them into mortises cut in the legs. These stringers should have grooves ¼ in. deep and ⅜ in. wide cut into them to receive the side pieces. The craftsman can save himself considerable work if he has these grooves cut out at the mill. The side pieces are given in the bill as being 4 in. wide, but in reality they are 3⅔ in. Of course this is an uncommon dimension, so the builder will be forced to divide an inch into three equal parts and take two of them, in order to get the accurate dimension. The star decoration on these pieces is shown in the detail. It should be laid out on two of the pieces adjoining each other. It represents two squares, the diagonals of which are 2 in. long. All of the lines are drawn at an angle of 45 degrees, therefore it would simplify the work if the T-bevel was set at this angle and the design drawn out with its help. An easy way to set the bevel for such an angle is by means of the steel square. Put the beam of the bevel against one edge of the square and set it so that the blade will pass through the same division on both the body and the tongue of the square, that is, through both 2 in., 3 in., or 4 in. marks. When the design has been cut out of one piece, this may be used as a pattern for the other pieces. After all are cut out and fitted, glue and clamp the parts together, being sure to insert the sides into the stringers before clamping, as it would be impossible to insert them after.

Allow ample time for the glue to set, then screw the two narrow strips of plain oak from the inside, onto two of the bottom stringers, keeping them flush at the bottom and parallel to each other. Onto these fit the bottom piece, which may also be screwed into place.

Scrape off any surplus glue in evidence, sand lightly and apply the finish as before described.

SCONCE

The sconce, illustrated on page 306, may be worked up in a number of ways, but the hammered method is preferable. It is a beautiful piece when finished in dull copper with the reflector highly polished. The materials needed for the hammered method consist in the following, while, if the etching is used it will take heavier and the piercing lighter, metal:

1 pc. copper No. 23 gauge, 8½ x 13 in.
1 pc. copper No. 20 gauge, 2 x 2 in.
1 pc. copper No. 20 gauge, 2 x 4½ in.
1 pc. copper No. 20 gauge 3⅞ x 3⅞ in.
Rivets, steel wool, lacquer.

First, make a complete, full-size drawing of the back on a piece of paper according to the dimensions given. The design shown is suggestive and should not be copied unless the craftsman thinks himself unable to originate a better one. If he does his own designing, he should not forget to plan a reflector directly back of his candle. Second, transfer this design with carbon paper to the large piece of metal, leaving ½ in. all around outside of the cutting edges. Third, drill holes through this margin about ¾ in. apart and insert ¾ in. slim screws through these holes and into a thick softwood board. Fourth, with a 20-penny

-JARDINIERE STAND.-

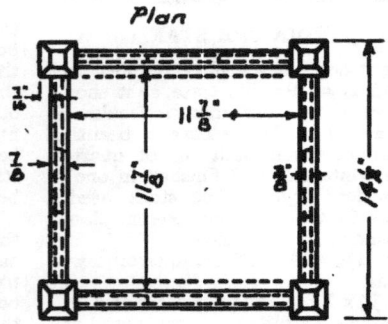

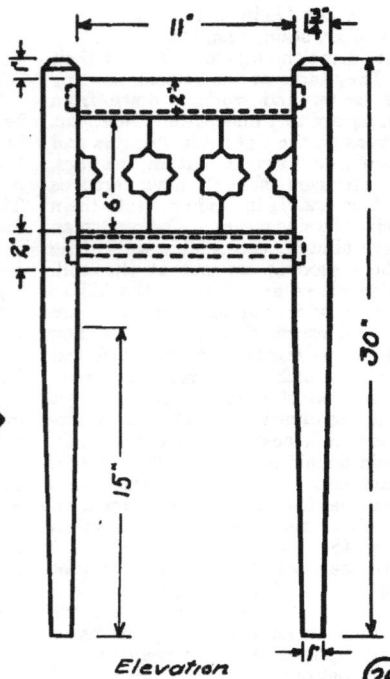

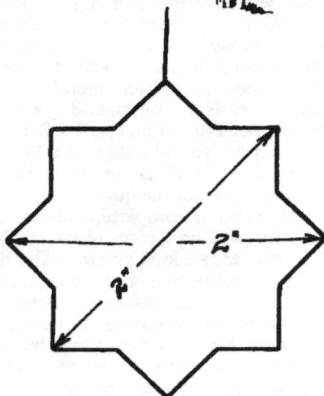

Plan

Elevation

Detail of
decoration
on side strips.

common nail, the end of which has been slightly filed, stamp the background of the design in a manner similar to piercing, but *do not pierce through the metal.* This lowers the background and raises the design. Fifth, sharpen another nail to a blunt chisel edge and stamp along the border of the design and the background. Sixth, remove the metal from the board, cut away the surplus and file the edges. Seventh, polish the reflector with steel wool and lacquer the piece. A high polish can be given it with pumice stone and water. In any event use two or more coats of lacquer. Eighth, shape the bracket as shown and bend on the dotted lines, in each case bending a right-angle. Ninth, draw a circle on the 2 in.

square piece and trim down to this line. With the ball pein of the hammer, pound a hole in the end of a piece of soft wood and form the little bowl in this depression, using the ball pein for all of this work. Tenth, draw the holder out full size on a piece of paper and transfer it to the remaining piece of metal. Cut it out to shape and bend it up on the dotted lines. The upper ends are shaped with the round nose pliers. Finally, rivet the bowl and holder to the bracket shelf and the latter to the back as illustrated, after giving each piece a coat of lacquer.

If the etched or pierced methods are used, the principal directions will still be the same.

(To be continued)

KINKS FOR THE HOME CRAFTSMAN

Every woodworker discovers little short-cuts in his work which materially help him to attain rapidity and perfection. A number of these, from the writer's own experience and the experience of others, will be listed with the hope that they will be of some assistance to the amateur craftsman.

1. In measuring with a rule, always tip it on edge so that the dimension marks are adjacent to the piece being laid out, and in taking a series of dimensions, start from one point only; do not move the rule from one mark to the next.

2. For fine cabinet work, make all lines with a knife and the gauge, but never with a pencil.

3. In setting a gauge, do not rely upon the scale on the beam, but always test with the rule, the end of which can be placed against the head of the gauge, and the dimensions run to the spur.

4. Always tip a plane on its side when laying it on the bench so as not to dull the iron. For the same reason, always raise the plane from the work on the return stroke.

5. In planing end grain, never run the plane entirely across the end, but work from both edges toward the center of the piece. This prevents the splitting of corners.

6. In using an oil stone, there are three things to observe: (*a*) Use plenty of good oil. (*b*) Clean the stone well before putting it away. (*c*) Use the entire face of the stone, not merely the center. If

these precautions are taken a stone should cut perfectly for years.

7. In sharpening plane irons and chisels, always rub on the bevel and never on the back, as this must be perfectly straight at all times to insure perfection in cutting.

8. In boring, never bore entirely through a piece, but reverse the piece and finish the hole from the other side after the worm penetrates.

9. Do not drive a screw into a board with a hammer, as its holding qualities will be greatly lessened.

10. Always drive nails and brads at an angle, as they will then hold more securely.

11. In sand-papering, always use a block if possible, as this will prevent rounding edges where they are not wanted.

12. Sand-paper should be used for cleaning and smoothing purposes only; do not depend upon it for doing the tool work.

13. Sand-papering should not be done across grain.

———

The old custom of leasing land to the highest bidder by the aid of a candle and pin is still being observed at Vadmouth, a village between Reading and Newburg, England. The candle is lighted and a pin stuck into it. Bids are then called for until the pin, owing to the softening of the candle, drops out.

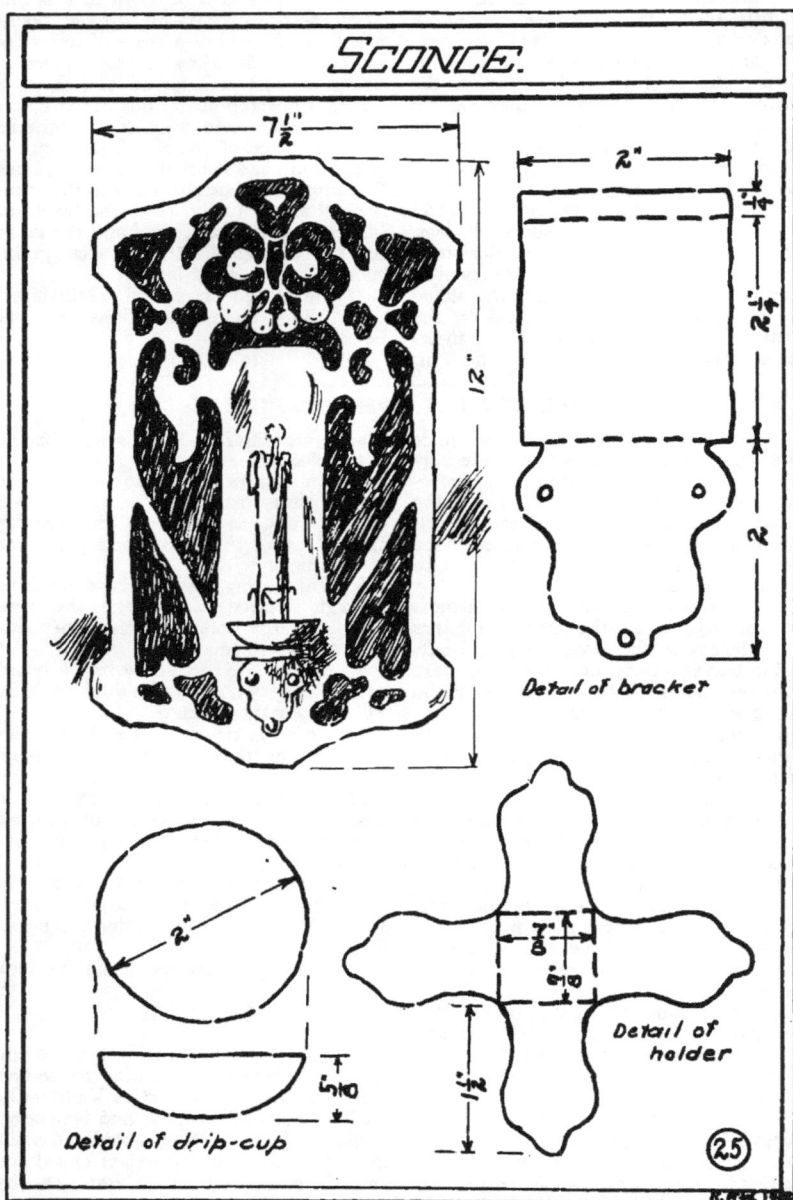

SCONCE.

Detail of bracket

Detail of drip-cup

Detail of holder

25

ENGINEERING LABORATORY PRACTICE—Part VI

The Measurement of Water by Means of Nozzles and Weirs

P. LE ROY FLANSBURG

It often happens that the engineer must know the exact amount of water which is flowing through a given pipe in a given time. When the quantity of water to be measured is small, the simplest and quickest method of procedure is to allow the water to flow from the pipe into a series of tanks known as weighing tanks. These tanks are mounted on platform scales and each tank is filled in succession; then, knowing the weight of the tank when empty, and also its weight when filled, it is a simple matter to determine the cubic feet of water or the gallons of water discharged from the pipe in a given period of time. But such a method is not practical when large quantities of water are to be measured, and the engineer then usually employs either a nozzle or a weir as a means of measuring the water.

The nozzle is but a modified form of the convergent tube with rounded corners, as shown in Fig. 1, and nozzles of this type are inserted in the sides of tanks. Usually, however, the length of the nozzle is three or four times its smallest diameter and it would more closely resemble the one shown in Fig. 2. Such nozzles are attached to the ends of pipes or hose.

By a rather complicated mathematical analysis (and also many experiments) it has been shown that the quantity of water Q, in cubic feet per second, discharged through a nozzle, is directly proportional to the area a of the nozzle, at its smallest cross-section, and also directly proportional to the square root of $2g$ times the head H, where g is the gravitational acceleration, equal to 32.2 ft. per second, per second, and H the pressure head. Expressing this relation mathematically, we have,

$$Q = a \sqrt{2gH}$$

This formula, however, does not take account of the angle of convergence of the nozzle or of the frictional resistance of its walls, and to correct for these two things we must write the formula as

$$Q = Ca \sqrt{2gH}$$

where C is a constant for the nozzle, determined experimentally and called the coefficient of discharge.

Suppose that we have a nozzle inserted in the side of a tank and that the distance from the level of the water in the tank to the center of the nozzle's mouth is 10 ft. Let the cross-sectional area of the nozzle at its mouth be 1 sq. in. Now weigh all of the water discharged through the nozzle during a test of 10 minutes and let it equal 102.6 cu. ft. This gives a value of Q equal to .171 cu. ft. per second.

If we assume a value of C equal to 1, then our formula becomes

$$Q = 1 \times \frac{1}{144} \times \sqrt{2 \times 32.2 \times 10} = .176 \text{ cu. ft. per sec.}$$

However, we found that the actual discharge was .171 cu. ft. per sec. There-

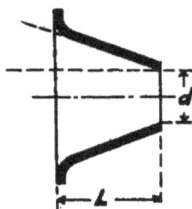

Fig. 1.

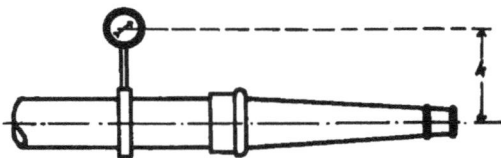

Fig. 2.

Fig.3.

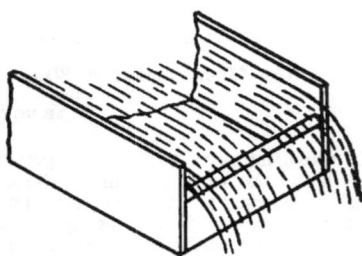

Fig.4.

fore, we see that the actual coefficient of discharge is equal to,

$$\frac{.171}{.176} = 0.97$$

and combining all constants we find that for this nozzle,

$$Q = .0541 \sqrt{H}$$

Therefore, after once calibrating the nozzle, all that need be done to compute the number of cubic feet of water discharged per second is to make a measurement of H and substitute in the formula.

It is important, however, to remember that H must be measured to the level of the smallest cross-section of the nozzle.

When a nozzle is not sufficient to handle all of the water discharged (as when measuring the quantity of water flowing past a given point in a stream) a weir is often made use of.

A weir is simply a long rectangular box having one of its short sides so cut that it resembles the ones shown in Fig. 3 and Fig. 4. This weir box is placed with its beveled edges on the down-stream side, and so arranged that the water from the stream will flow through it. By a process of mathematical deductions (and also by many experiments) it has been found that the quantity of water discharged through a weir, can be expressed by the formula,

$$Q = \tfrac{3}{4} Cb \sqrt{2gh^3} \text{ cu. ft. per sec.}$$

where C is the coefficient of discharge of the weir, b the width of the weir in feet, h the distance (expressed in feet) between the level of the water in the stream and the upper edge or crest of the weir. C has been determined experimentally for different types of weirs and will average about 0.62 for weirs having rectangular openings. To secure accurate results, the

distance h should be measured to thousandths of a foot.

In making measurements of this kind a hook-gauge like the one shown in Fig. 5, is employed. This gauge consists of a graduated rod having a hook on its lower end and fastened by means of a nut and screw to a fixed arm. The rod may be raised or lowered by turning the nut. A vernier is fastened to the rod and passes over a a scale, thus making it possible to make readings of a thousandth of a foot. A pipe is led from the weir box to the hookgauge can and thus the water will stand at the same level in the can as it is in the weir box. To use the gauge the zero reading of the water level must first be taken. Let the water in the weir box be exactly level with the crest of the weir and turn the nut on the hook-gauge rod until the point of the hook is just at the surface of the water in the can. Then read the height of the hook-gauge by means of the scale. After the test has started the water in the hook-gauge can will rise until it is at the level of the water passing over the weir. Take another reading of the hook-gauge (again bringing the point of the hook up to the surface of the water in the can) and the difference between this reading and the zero reading will give the exact value of h.

The following is a test performed by the author to determine the coefficient of discharge for a given nozzle and given weir.

TEST ON NOZZLE AND WEIR

The test was for the purpose of determining the coefficient of discharge for an "Underwriter's Fire Nozzle," and a "Trapezoidal Weir." The method followed in performing the experiment consisted in measuring the volume of water discharged in a given time, the pressure at which the water from the nozzle was discharged and the height of the water above the weir.

In order to know the pressure of the water in the pipe, a piezometer ring was placed around the pipe and to this a gauge was attached. This gauge was read every two minutes for ten minutes and the amount of water discharged was measured in large measuring tanks.

The general sketch of the apparatus is shown in Fig. 6.

DATA
Diam. of nozzle ....1.25 in.
Diam. of piezometer............1.75 in.
Height of center of gauge above
    center of nozzle................5 in.
Length of weir crest...............15 in.
Vol. of water...............391.3 cu. ft.
Nozzle gauge reading...........37.25 lb.
Hook gauge reading..............7.87 in.
Zero reading of hook-gauge=4.30 in.

CALCULATION OF RESULTS
The formula for finding the coefficient of discharge is,

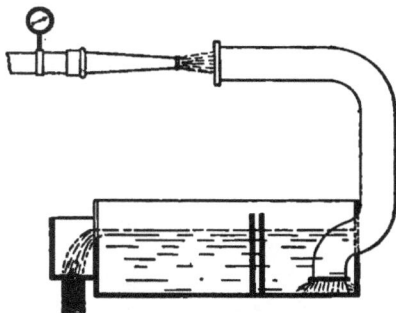

Fig.5.

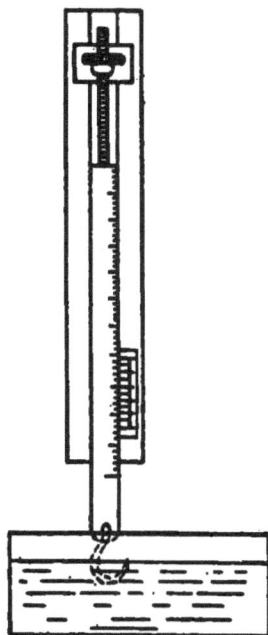

Fig. 6.

$$Q = \frac{0.4374 C d^2}{\sqrt{1 - C^2 \left(\frac{d}{D}\right)4}} \sqrt{h_1}$$

where Q=discharge in cu. ft. per sec.
    C=coef. of discharge
    d=diameter of nozzle
    D=diameter of piezometer
    h_1=total head

When using this formula the size of the piezometer and of the nozzle, together with the height of the center of the gauge above the center of the pipe, must be known.

The formula used for finding the coef. of discharge of the weir was,

$$Q = CLH^{\frac{3}{2}}$$

where Q=discharge in cu. ft. of water
      per sec.
    C=coef. of discharge
    L=length of crest
    H=height of water flowing over
      weir

Weir.

$$7.87'' - 4.30'' = 3.57''$$

$$Q = CLH^{\frac{3}{2}} \quad \begin{array}{l} Q = \text{cu. ft. } H_2O \text{ per sec.} \\ C = \text{coef. of discharge.} \\ L = \text{length of crest in ft.} \\ H = \text{head on weir in ft.} \end{array}$$

$$C = \frac{Q}{LH^{\frac{3}{2}}}$$

$$C = \frac{\frac{391.3}{600}}{\frac{15}{12} \times \left(\frac{3.57}{12}\right)^{\frac{3}{2}}}$$

$$C = \frac{0.653}{1.25 \times .162} = 3.23$$

Nozzle.

$$\frac{5}{12} x \frac{62.3}{144} = .018 \text{ lb.} \quad \begin{cases} \text{Corr. for height of} \\ \text{center of gauge above} \\ \text{center of nozzle.} \end{cases}$$

$$P_1 = 37.25 + 0.18 = 37.43 \text{ lbs.}$$
$$h_1 = \frac{37.43 \times 144}{62.3} = 86.7 \text{ ft.}$$

$$Q = \frac{0.04374 \, C d^2 \sqrt{h_1}}{\sqrt{1 - C^2 \left(\frac{d}{D}\right)^4}}$$

$$\frac{\sqrt{1 - C^2 \left(\frac{d}{D}\right)^4}}{C} = \frac{0.04374 \, d^2 \sqrt{h_1}}{0.653}$$

$$1 - C^2 \left(\tfrac{1}{4}\right)^4 = (9.75)^2 C^2$$

$$C^2 = {}_{1}\tfrac{1}{31} = .828$$

$$1 - 0.26 c^2 = .95 c^2$$

$$C = 0.915$$

$$\text{Total head} \atop \text{at piezom-} \atop \text{eter.}} = 86.7 + \frac{\left\{ \dfrac{\frac{391}{600}}{\frac{\pi 1.75^2}{4.144}} \right\}^2}{2g} = 86.7 + 23.5 = 110.2 \text{ ft.}$$

Total head = pressure head + velocity head.

$$V = \frac{Q}{A}$$

$$\text{head due to velocity} = h_2 = \frac{V^2}{2g}$$

RESULTS

Coef. of weir.......................3.23
Coef. of nozzle....................0.915
Total head at piezometer.......110.2 ft.

CUTTING RUBBER WITH A WET KNIFE
W. E. MOREY

Some years ago the writer was employed in a paper mill in Vermont, and one day saw the machine tender cutting some heavy rubber which covers one of the rolls at the "wet" end of the machine. After every stroke of the knife he would wet the blade in his mouth. Being asked the reason, he explained that it made it cut easier, and then told the following story: A certain firm of rubber manufacturers were experimenting with a new mixture of rubber and were observing the utmost secrecy regarding its composition, with a view to protecting same when a satisfactory mixture had been found.

One day a visitor wormed his way into the works on some plausible excuse or other, and was looking about with apparent indifference and that lack of close observation which characterizes the merely curious.

He was apparently a minister, being of a benevolent mien, with a long black coat of a clerical cut and a white tie. As he came to where some of the finished rubber lay he asked if he could have a piece to use as an eraser and as he looked anything but the technical expert he was told that he might have a piece.

Taking out his knife he took up a piece of the heavy sheet and wet his knife to cut it off. Instantly the man in charge snatched away the sample sheet and with no gentle hand spun the clerical-looking gentleman around with the words: "You get out, and quick, too. You know altogether too much about rubber for us. Now go!" He went.

Although the shark and the octopus share about equally the reputation for being the greatest enemies to mankind in the sea, a much worse creature than either shark or octopus is the devil-fish—a large ray that is common in the warm waters of the Atlantic. The fish grows to a weight of a ton and a half, and besides formidable teeth, is armed with a horrible barbed and poisoned spike in the tail. It has often been known to attack boats. Another nasty customer is the green moray of Bermuda, which resembles a conger eel in form, but is of a very savage nature. The swordfish, sought for its oil and flesh, especially along the Atlantic coast of the United States, is another dangerous creature. Swordfish are harpooned in the same manner in which whales used to be killed. Quiet enough until attacked, the swordfish then seems to go raving mad and fights with unmatched ferocity.

THE USE OF CONCRETE ON THE FARM *

INTRODUCTION

With the rapid decrease of our timber supply and the resulting increase in the price of lumber, there has come a necessary demand for a new building material. Nowhere has this demand been felt more keenly than on the American farm, where lumber has till now been practically the only building material. On account, however, of the farmer's nearness to the timber itself, he has been the last to feel the full effect of the shortage.

A building material has been discovered in concrete that in many instances has proved to be far superior to lumber, brick, or building stones on account of its durability, economy, and safety from fire loss. Moreover, it can very often be used at.the most convenient time by the farmer himself with a little assistance.

SELECTION OF MATERIALS

Frequently cement users have made costly mistakes by not informing themselves properly, before starting their work, concerning the correct methods of making good concrete. For this purpose the following materials are necessary: (1) cement; (2) sand; (3) gravel or crushed stone; and (4) water.

Cement is therefore only one part of a concrete mixture. A far greater proportion of sand and gravel than cement is required. The quantity of cement to be used and the strength of the concrete depend entirely on the quality and size of the sand and gravel, and it is of the utmost importance that these be of the right kind. With, an equal amount of cement a far stronger concrete may be made, if the sand and gravel are of the proper size and correctly proportioned. It is sometimes thought that any kind of soil of a sandy nature, mixed with a small percentage of cement, will make concrete, but this idea is incorrect.

As a guide in the selection of the proper materials, especially sand and gravel, the following suggestions should be observed:

CEMENT

Portland cement is a manufactured product, the principal value of which is its ability to adhere to the various ma-

* U. S. Department of Agriculture. Farmers' Bulletin, 41.

terials used in masonry construction. Most of the American brands of Portland cement are made under the careful. supervision of the manufacturer, w¹ . has his reputation at stake. They have to meet the requirements of a fixed standard which has been, after years of experimenting, adopted by the American Society for Testing Materials and the American Society of Civil Engineers. Guarantees should be obtained from the dealer or manufacturer that the Portland cement furnished meets this standard.

On adding water to the dry cement it becomes a soft, sticky paste, and will remain so for about one-half hour, after which it begins to harden or "set." To disturb the concrete after this initial set has started means a decided loss in strength, while to disturb it after the set is well under way means to destroy the concrete. It should, therefore, be remembered that Portland cement concrete must be placed in position within 20 or 30 minutes from the time it is first wet.

There are also several other minor considerations to be observed. First, the binding power of Portland cement is lessened by exposing the concrete to a hot sun during the first four or five days after it has been placed in position. Then, a green cement mixture, which can be easily frozen at a temperature below 32 degrees Fahrenheit, should be protected from such an exposure. Freezing does not materially affect the binding quality of good Portland cement, provided the concrete does not freeze until after placing, and is not subjected to any load until after it has been thawed out and allowed to "set" in the usual way. It is safest to avoid mixing on days when the temperature is below the freezing point, that is, 32 degrees Fahrenheit. In no case should fresh manure be placed over very green concrete to protect it from freezing, because this will soil the surface of the concrete.

Portland cement is shipped in paper bags, cloth sacks, or wooden barrels. The second means is, for the average user, the best, and, while the manufacturers charge more for this kind of a package, they allow a rebate for the return of the empty bags.

Fig. 1.   Method of Screening Sand from Gravel

Cement is manufactured in twenty-five states in the United States and can be obtained easily in all sections of the country.   Of the various kinds of cements made, Portland cement is without question the best.

Cement must be kept in a dry place. It absorbs moisture from the atmosphere with great readiness and, when kept in a damp place, soon becomes lumpy or even a solid mass.   In this condition it is useless and should be thrown away.   But lumps caused by pressure in the storehouse must not be mistaken for cement that has been wet and has then formed into lumps.   Lumps caused by pressure are easily broken and the cement is perfectly good.

In storing cement wooden blocks should be placed on the floor and covered with boards.   The cement should then be piled on the boards and the pile covered with canvas or a piece of roofing paper.

SAND

In the selection of sand the greatest care should be used, and critical attention should be given to its quality, for sand contributes from one-third to one-half of the amount of the materials used in making concrete.

Sand may be considered as including all grains and small pebbles that will pass through a wire screen with $\frac{1}{4}$ in. meshes, while gravel in general is the pebbles and stones retained upon such a screen.

The sand should be clean, coarse, and if possible, free from loam, clay and vegetable matter.   It was proved in an actual test made by a United States government expert that an exceedingly fine sand required seven times the amount of cement required by coarser sand retained on a 20-mesh screen, without increasing the strength of the concrete.

Sand from the same bank usually varies largely in different places in the bank, and, as sand of such fineness that over 50 percent of the bulk of the sand will pass through a 40-mesh screen is generally unfit for concrete work, it may become necessary to screen out this very fine sand by means of a 40-mesh wire screen. A 40-mesh screen means a screen with 40 holes to the lineal inch of screen surface.   The screen should be placed in an upright position at an angle of about 45 degrees.   The screening should be done at the pit, in order that only the material actually used may be hauled.

Sand with the grains of nearly uniform size does not give as great strength as sand with grains of various sizes.   By testing concrete made from coarse and fine sand of one size of grains, it will be found that the coarse sand gives the stronger concrete. The strength and hardness of the grains of sand are of much importance; and a sand which shows the slightest tendency to dissolve or soften when soaked in water for an hour should be discarded.

The coarseness of the sand can be felt, or can be determined by a screen, and the vegetable matter can be seen, but the amount of clay or loam can not be decided in either of these ways. Four inches of sand should be put in a pint preserving jar, and when the jar has been filled with clear water to within an inch of the top, the lid should be fastened on and the jar shaken vigorously for 10 minutes. The jar should then be rested upright and the contents allowed to settle. The sand will settle in the bottom with the clay and the loam on top, and the water above them. If more than $\frac{1}{2}$ in. of clay or loam shows, the sand should be rejected or washed. The difference in color and fineness shows clearly the line of division between the clay or loam and the sand.

If the sand must be washed, the simplest way is to build a loose board platform from 10 to 15 ft. long, with one end 12 in. higher than the other. On the lower end and on the sides an edge 2 x 6 in. should be nailed to hold the sand. The sand should be spread over the platform in a layer 3 or 4 in. thick and washed with a $\frac{3}{4}$ in. garden hose. The washing should be started at the high end and the water allowed to run through the sand and over the 2 x 6 in. piece at the bottom. A small quantity of clay or loam does not injure the sand, but any amount over 10 percent should be washed out.

GRAVEL

The largest part of concrete is the gravel or crushed stone. This should be clean; that is, free from loam, clay, or vegetable matter. The best results are obtained from a mixture of sizes graded from the smallest, which is retained on a $\frac{1}{4}$ in. screen, to the larger ones that will pass a $1\frac{1}{2}$ in. ring.

For heavy foundation and abutment work, larger-sized pebbles and stones might be used; while for reinforced concrete work pebbles larger than those passing a 1 in. ring should not be used.

If crushed stone and screenings are used, the same care in selecting the sizes must be exercised as in selecting the gravel. In ordering from the crusher plant, the sizes of the stone and screenings should be specified in the order. The crusher dust should always be removed.

Sometimes bank or creek gravel, which will answer the purpose of sand and gravel combined, can be obtained, and it is frequently used on the farm and in small jobs of concrete work just as it comes from the pit or creek. Occasionally this gravel contains nearly the right proportions of sand and gravel, but in the majority of sand pits and gravel banks there is a great variation in the sizes of the grains and pebbles or gravel, and in the quantities of each. This is due to the fact that all the deposits are formed in seams or pockets that make it impossible to secure anything like uniformity. Therefore, to get the best and cheapest concrete, it is advisable to screen the sand and gravel and to remix them in the correct proportions.

Experienced contractors have found that it is not only necessary but economical to pay laborers as much as $2.00 per day to screen the bank material twice, in order to obtain the sand and gravel. First, a $\frac{1}{4}$ in. screen should be used to keep out the gravel, and then the material which has passed through this screen should be screened again over a 40-mesh screen for the sand. All material which passes through this 40-mesh screen should be rejected.

By knowing exactly the proportions of sand and gravel, the exact amount of cement to obtain the desired strength can be determined. Enough cement can be saved to balance the additional pay given to the laborers for screening the sand and gravel.

Dirty gravel can usually be observed without further test, and can be washed in the same way as dirty sand.

WATER

The water used for concrete should be clean and free from strong acids and alkalis.

MANUFACTURE OF CONCRETE

PROPORTIONS

Concrete is a manufactured stone formed by mixing cement, sand, and

Table I.—*Quantities of materials and the resulting amount of concrete for a two-bag batch*

Kinds of concrete mixture	Proportions by parts			Materials				Sizes of measuring boxes (inside measurements)		Water for medium wet mixture (gallons)
	Cement	Sand	Stone or Gravel	Cement (bags)	Sand (cubic feet)	Stone or gravel (cubic feet)	Concrete (cubic feet)	Sand	Stone or gravel	
1:2:4..........	1	2	4	2	3¾	7½	8½	2 x 2 ft. x 11¼ in.	2 x 4 ft. x 11½ in.	10
1:2½:5..........	1	2½	5	2	4¾	9½	10	2 x 2 ft. 6 in. x 11½ in.	2 ft. x 6 in. x 4 ft. x 11½ in.	12½

stone or gravel (*i.e.*, pebbles) together with water. Various amounts of each are used, according to the use to which the concrete is to be put. The mixture in which all the spaces or voids between the stones or gravel are filled with sand and all the spaces between the sand are filled with cement is the ideal mixture. This mixture is rarely obtained, since the voids in each load of gravel and sand vary slightly, and in order to be absolutely safe, a little more sand and cement than will just fill the voids are used. The following illustration (Fig. 2) shows the relative amounts of these various materials for a certain amount of 1:2:4 concrete. It will be noticed that the amount of concrete is only slightly greater than the amount of stone or gravel.

Table I shows the amount of stone, sand and cement used in the various grades of concrete work. The proportions are always measured by volume. A 1:2:4 mixture means one part of cement, twice as much sand, and four times as much stone or gravel, so that the whole mixture consists of seven parts. A 1:2½:5 mixture means one part of cement, two and one-half times as much sand, and five times as much stone or gravel, so that the whole mixture consists of eight and one-half parts.

MEASURING BOXES

It should be noticed that the dimensions given for the measuring boxes for sand and stone or gravel are inside measurements. These boxes are made with straight sides of any kind of rough boards and have no top or bottom.

AMOUNT OF WATER

The amount of water given is only approximated. The amount given in Table I should be used for the first batch; if it is too wet for the use desired, the amount should be reduced, while if it is too dry, the amount should be increased. A bucket should always be used in measuring the amount of water, as this secures uniform results.

VARIATIONS IN MIXTURE

If the sand is very fine, the cement should be increased from 10 to 15 percent.

When the mixture does not have a uniform color, but looks streaky, it has not been fully mixed.

If the mixture does not work well, and the sand and cement do not fill the voids in the stone, the percentage of stone should be reduced slightly, but the concrete should first be properly mixed. Concrete that is poorly mixed may present features that are entirely eliminated by turning it over once or twice more.

METHOD OF MEASURING QUANTITIES

One barrel of Portland cement holds practically 4 cu. ft., or 4 bags of cement. Sand and stone or gravel are measured loose in the boxes and should not be packed. A 2-bag batch of concrete requires 2 bags of cement, while the amount of sand and stone or gravel is measured as shown in Table I. For a 4-bag batch, twice as much sand and stone or gravel and 4 bags of cement should be used.

DETERMINATION OF QUANTITIES

The number of cubic feet of concrete that will be required for the work in question should first be calculated. By multiplying this number by the number under the proper column, as shown in Table II below, the amount of cement, sand, and stone or gravel can be found.

Table II

Quantities of materials in 1 cu. ft. of Concrete

Mixture of Concrete	Cement by bbls.	Sand by cu. yds.	Stone or Gravel by cu. yds.
1:2:4..........	0.058	0.0163	0.0326
1:2½:5..........	.048	.0176	.0352

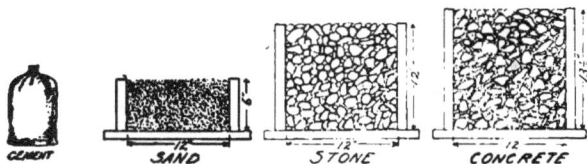

Fig. 2. Required Quantities of Cement, Sand and Stone or Gravel for a 1:2:4 Concrete Mixture and the Resulting Quantity of Concrete

Example.—Let us suppose that the work consists of a concrete silo requiring in all 935 cu. ft. of concrete, of which 750 cu. ft. are to be 1:2:4 concrete, and 185 cu. ft. are to be 1:2½:5 concrete. Enough sand and cement are also needed to paint the silo inside and outside, amounting in all to 400 sq. yds. of surface with a 1:1 mixture of sand and cement. One cubic foot of 1:1 mortar paints about 15 sq. yds. of surface and requires 0.1856 barrel of cement and 0.0263 cu. yd. of sand. The problem thus works out as follows:

Cement:

	Bbls.
For the 750 cu. ft. of 1:2:4 concrete (750 x 0.058)	43.5
For the 185 cu. ft. of 1:2½:5 concrete (185 x 0.048)	8.9
For painting (400÷15 x 0.1856)	4.9
Total amount of cement	57.3

Sand:

	Cu. yds.
For 750 cu. ft. of 1:2:4 concrete (750 x 0.0163)	12.23
For 185 cu. ft. of 1:2½:5 concrete (185 x 0.0176)	3.26
For painting (400÷15 x 0.0263)	.70
Total amount of sand	16.19

Stone or gravel:

For 750 cu. ft. of 1:2:4 concrete (750 x 0.0326)	24.5
For 185 cu. ft. of 1:2½:5 concrete (185 x 0.0352)	6.5
Total amount of stone or gravel	31.0

Thus the necessary quantities of materials are about 57½ bbls. of Portland cement, about 16¼ cu. yds. of sand, and 31 cu. yds. of stone or gravel. It is always wise to order two or three extra barrels of cement if the dealer is at considerable distance, as this avoids any possible trouble that a shortage might cause.

In case a natural mixture of bank sand and gravel is used, the following table should be consulted for the quantities of the mixture:

(*To be continued in the June issue*)

[*Editor's Note.*—This article, upon the "Use of Concrete," will be complete in three instalments, and in the June issue the equipment necessary for the mixing of concrete and the different methods employed will be quite fully and clearly described. It is hoped that this series of articles may prove of much practical benefit to such of our readers who have any concrete work which they wish done.]

Table III

Quantities of materials and the resulting amount of concrete for a two-bag batch, using a natural mixture of bank sand and gravel

NEXT FEW YEARS WILL ECLIPSE ALL AGES IN INVENTION

EDWARD B. MOORE
(Commissioner of Patents)

The age of invention has just begun to dawn. The accomplishments of the last half-century, while marvelous almost beyond conception, will not begin to compare with what will be done in the next half-century.

I base this conclusion upon a definite knowledge of what is being done at present and an appreciation of the great world scope that invention is assuming. There is no evidence of a waning of inventive genius, while greater stores of knowledge, better-trained hands, and both these in vastly greater numbers, are being brought to bear in the field of invention.

The number of patents applied for and the number granted at the patent office last year were greater than at any time in its history. With the increased number there is no decrease in the individual importance, but merely an evidence of increased industrial activity that demands the articles patented.

There are periods of activity and depression along certain lines of industrial art. Some years ago we had ten men at work on bicycles alone, while now one man devotes but half his time to them. Eight men formerly worked on reapers, while they are so nearly perfected that but three are so engaged.

This does not mean, however, that any line of machinery is ever made so perfect that no further inventions will follow, for there are as many patents issued today for the improvement of plows as at any time in the history of the world, and the plow in the form of a forked stick was among the first tools invented. There are, however, certain lines of great activity at present and in the near future.

Electricity offers an unlimited field, and the number of patents bearing upon it is without end, while the flying machine is but beginning to show its possibilities. Wireless telegraph and telephone are just being heard from, while at any time a great basic principle, like that of the Bell telephone, may be discovered that will open up new vistas.

The inventor of today is a different man from the long-haired, erratic genius of a generation ago. He is in nearly all cases an inventor by profession, trained in the best technical schools and devoting his life to the creation of new things. He is, above all, a practical man of affairs.

The people of the United States have gained more than any other nation from their inventions. These have enabled them to enter the markets of the world and force out competition in many grades of machinery. The patent laws of this country have been a greater protection to the inventor than have those of any of the other nations and are being widely adopted.

Treaties for the protection of patents are being universally adopted. Such treaties are now being arranged by the state department with China and all the nations of South America. Japan is but just finding that her people have the same inventive mind that is shown in America.

The awakening of new minds and new nations is going to bring on renewed activity and competition, and matters will go forward at a still greater rate. World's fairs have done much to make this activity world-wide, and the promise is that our children will live in a world that we would not recognize.—*Chicago Journal.*

Concrete in Freezing Weather

In the construction of dams for Huronian Company's power development in Canada, a large part of the concrete work in dams, and also in power-house foundations, was done in winter, with the temperature varying from a few degrees of frost to 15 degrees below zero, and on several occasions much lower. No difficulty was found in securing good concrete work, the only precaution taken being to heat the mixing water by turning a ¾ in. steam pipe into the water barrel supplying the mixer, and, during the process of mixing, to use a jet of live steam in the mixer, keeping the cylinder closed by wooden coverings during the process of mixing. No attempt was made to heat sand or stone. In all the winter work care was taken to use only cement which would attain its initial set in not more than 65 minutes.—*Engineering Contracting.*

THE TESLA STEAM TURBINE

The Rotary Heat Motor Reduced to its Simplest Terms

FELIX J. KOCH

Nikola Tesla, whose reputation must, naturally, stand upon the contributions he made in electrical engineering when the art was yet in its comparative infancy, is by training and choice a mechanical engineer, with a strong leaning to that branch of it which is covered by the term "steam engineering." For several years he has devoted much of his time to improvements in thermo-dynamic conversion, and the result of his theories and practical experiments is to be found in an entirely new form of prime movers shown in operation at some plants in New York.

The basic principle which determined Tesla's investigations was the well-known fact that when a fluid (steam, gas or water) is used as a vehicle of energy, the highest possible economy can be obtained only when the changes in velocity and directions of the movement of the fluid are made as gradual and easy as possible. In the present forms of turbines in which the energy is transmitted by pressure, re-action or impact, as in the De Laval, Parsons, and Curtiss types, more or less sudden changes, both of speed and direction, are involved, with consequent shocks, vibrations and destructive eddies. Furthermore, the introduction of pistons, blades, buskets, and intercepting devices of this general class, into the path of the fluid involves much delicate and difficult mechanical construction which adds greatly to the cost both of production and maintenance.

The theoretically perfect turbine would be one in which the fluid was so controlled from the inlet to the exhaust that its energy was delivered to the driving shaft with the least possible losses due to the mechanical means employed. The mechanically perfect turbine would be one which combined simplicity and cheapness of construction, durability, ease and rapidity of repairs, and a small ratio of weight and space occupied to the power delivered on the shaft. Mr. Tesla maintains that in the turbine which forms the subject of this article, he has carried the steam and gas motor a long step forward toward the maximum attainable effi-

Fig. 1

A 200 h.p. High-Pressure Turbine

This view shows one complete high-pressure unit with the steam throttle above and below it. Note the compactness of the turbine and the many gauges used during the tests.

ciency, both theoretical and mechanical. That these claims are well founded is shown by the fact that in the plant where Mr. Tesla carried out his experiments, he is securing an output of 200 h.p. from a single-stage steam turbine with atmospheric exhaust, weighing less than 2 lbs. per h.p., which is contained within a space measuring 2 x 3 ft., by 2 ft. in height, and which accomplishes these results with a thermal fall of only 130 B.t.u., that is, about one third of the total drop available. Furthermore, considered from the mechanical standpoint, the turbine is astonishingly simple and economical in construction, and by the very nature of its construction should prove to possess such durability and freedom from wear and breakdown as to place it, in these respects, far in advance of any type of steam or gas motor of the present day.

Fig. 2

The Tesla Testing Plant at the Edison Waterside Station

The top half of the casings is removed, showing the two rotors. Each rotor consists of 25 discs, ⅟₃₂ in. thick, by 18 in. in diameter. The steam enters at the periphery and flows in spiral paths to the exhaust at the center of the discs. The driving turbine is to the left, the brake turbine to the right, and between them is a torsion spring. The steam inlets are on opposite sides of the two rotors; the driving rotor moving clockwise. The torsion of the spring is automatically shown by beams of light and mirrors, and the horse-power is read off a scale. At 9,000 revolutions per minute, with 125 lbs. at the throttle and free exhaust, this turbine develops 200 h.p. It weighs 2 lbs. per horse-power.

Briefly stated, Tesla's steam motor consists of a set of flat steel discs mounted on a shaft and rotating within a casing, the steam entering with high velocity at the periphery of the discs, flowing between them in free spiral paths and finally escaping through exhaust ports at their center. Instead of developing the energy of the steam by pressure, re-action, or impact on a series of blades or vanes, Tesla depends upon the fluid properties of adhesion and viscosity—the attraction of the steam to the faces of the discs and the resistance of its particles to molecular separation combining in transmitting the velocity energy of the motive fluid to the plates and the shaft.

By reference to the accompanying photographs, it will be seen that the turbine has a rotor, which in the present case consists of 25 flat steel discs, ⅟₃₂ in. in thickness, of hardened and carefully tempered steel. The rotor as assembled is 3½ in. wide on the face, by 18 in. in diameter, and when the turbine is running at its maximum working velocity, the material is never under a tensile stress exceeding 50,000 lbs. per square inch. The rotor is mounted in a casing which is provided with two inlet nozzles, for use in running direct and for reversing. Openings are cut out at the central portion of the discs, and these communicate directly with exhaust ports formed in the side of the casing.

In operation, the steam or gas, as the case may be, is directed on the periphery of the discs through the nozzle (which may be diverging, straight or converging) where more or less of its expansive energy is converted into velocity energy. When the machine is at rest the radial and tangential forces due to the pressure and velocity of the steam cause it to travel in a rather short curved path toward the central exhaust opening, as indicated by the full black line in the accompanying diagram; but, as the discs commence to rotate and their speed increases, the steam travels in spiral paths, the length of which increases until in the case of the present turbine, the particles of the fluid complete a number of turns around the shaft before reaching the exhaust, covering in the meantime a lineal path some 12 to 16 ft. in length. During its progress from inlet to exhaust, the velocity and pressure of the steam are reduced until it leaves the exhaust at 1 or 2 lbs. gauge pressure.

The resistance to the passage of the steam or gas between adjoining plates is approximately proportionate to the square of the relative speed, which is at maximum toward the center of the discs and is equal to the tangential velocity of the steam. Hence the resistance to radial escape is very great, being furthermore enhanced by the centrifugal force acting outwardly. One of the most desirable elements in a perfected turbine is that of reversibility, and we are all familiar with the many and frequently cumbersome means which have been employed to secure this end. It will be seen that this turbine is admirably adapted for reversing, since its effect can be secured by merely closing the right-hand valve and opening that on the left.

It is evident that the principles of this turbine are equally applicable, by slight modifications of design, for its use as a pump.

In conclusion it should be noted that although the experimental plant develops 200 h.p. with 125 lbs. at the supply pipe, and free exhaust, it could show an output of 300 h.p. with the full pressure of the Edison supply circuit. Furthermore, Mr. Tesla states that if it were com-
~~pounded and the exhaust were led to a~~

tical turbines connected by a carefully calibrated torsion spring, the machine to the left being the driving element, the other the brake. In the brake element the steam is delivered to the blades in a direction opposite to that of the rotation of the discs. Fastened to the shaft of the brake turbine is a hollow pulley provided with two diametrically opposite narrow slots and an incandescent lamp placed inside close to the rim. As the pulley rotates two flashes of light pass out of the same, and by means of reflecting mirrors and lenses they are carried around the plant and fall upon two rotating glass mirrors placed back to back on the shaft of the driving turbine so that the center line of the silver coatings coincides with the axis of the shaft. The mirrors are so set that when there is no torsion on the spring, the light beams produce a luminous spot stationary at the zero of the scale. But as soon as load is put on, the beam is deflected through an angle which indicates directly the torsion. The scale and spring are so proportioned and adjusted that the horse-power can be read directly from the deflections noted. The indications of this device are very accurate and have shown that when the turbine is running at 9,000 revolutions under an inlet pressure of 125 lbs. to the square inch, and with free exhaust, 200 brake h.p. are developed. The consumption under these conditions of maximum of output is 38 lbs. of saturated steam per h.p. per hour—a very high efficiency when we consider that the heat-drop, measured by thermometers, is only 130 B.t.u., and that the energy transformation is effected in one stage. Since three times the number of heat units are available in a modern plant with superheat and high vacuum, the above means a consumption of less than 12 lbs. per h.p. hour in such
~~turbines adapted to take up the full drop~~

THE MAKING OF DINNER GONGS

GEO. F. RHEAD

The art of making ornamental objects in iron-work, so long as it is kept within certain limits, is one that should forcibly appeal to the ambitious home-worker, but in suggesting its addition to the number of home-crafts, we do not at all mean to imply a resetting of our old friend, "The Village Blacksmith," and wield the hammer, and attempt to forge, without a properly constituted workshop.

There are, however, some simple forms of metal work that require no especial outfit—no forge is required, and while the productions may be correctly termed wrought iron, they do not come under the category of the heavy work of the blacksmith.

To turn out useful objects for the home in metal-work, the amateur craftsman requires only a knowledge of soldering, riveting, and simple repoussé. If the worker is familiar with these three processes, and possesses the necessary tools to accomplish them, there are a large number of articles of the home, both useful and artistic, which he will find infinite pleasure in making. He will probably be happier in the construction of objects of utility in the home, than merely be "decorating." Fretwork, poker-work, chip-carving, etc., are all excellent pastimes, but structural wood and metal craft is work of a higher order.

In this enlightened age, we are all of us fond of good ornament, as it adds interest to construction, but the construction of an article for the mere sake of displaying some hand-work upon it is a plan not to be recommended. The utilitarian principle should be the first consideration—it is no use whatever ornamenting anything, and by doing so, lessening its practical value. Very often this is done, but with what a sad result— we have a door handle that one cannot properly grasp, because of its awkward shape, over-projecting surfaces that come constantly in one's way, and a host of devices such as wall pockets, racks, etc., that serve little real purpose than exhibit some handwork upon them. It is therefore always well that the repoussé worker can make simple objects in wrought metal, so that he may execute his work throughout. A fret-worker, or carver, has to have a knowledge of structural woodwork, so that he may employ his talent to the enrichment of objects of use.

Among the articles of the home of considerable utility comes the dinner-gong, which may be made to form a very attractive ornament, while being useful in no small degree. Those shown by our illustrations, Figs. 1 and 2, are examples where the utilitarian principle has been observed, no excrescences having been introduced in the design, the various

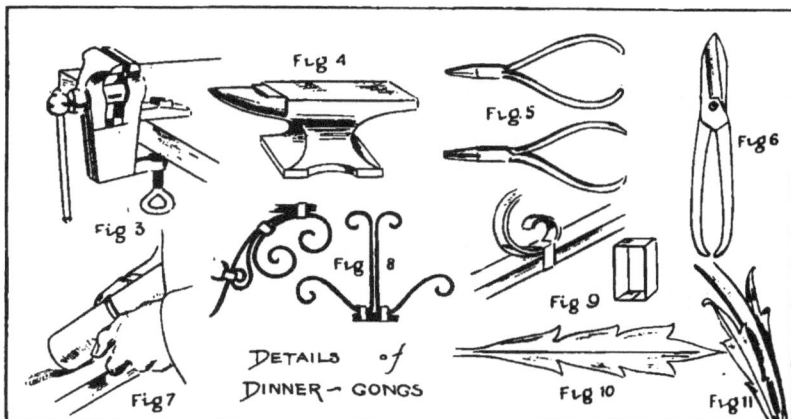

Fig 3
Fig 4
Fig. 5
Fig 6
Fig 8
Fig 9
Fig 7
DETAILS of
DINNER ~ GONGS
Fig 10
Fig 11

parts merely holding the article together, and making a complete and serviceable article. The making is so straightforward and simple that the veriest amateur may fetch out his tool-box without the least fear, for success in the act of forming curves in thin strip-iron comes easily after a little practice.

The designs illustrated are made up of thin strips of iron, copper or brass, $\frac{1}{4}$ to $\frac{1}{2}$ in. in width. In the case of design, Fig. 1, the scroll work is further enriched by a few polished pieces cut out of thin brass or copper, preferably soldered in position. The angle-pieces of both articles should be of thicker metal than the rest—about $\frac{1}{2}$ in. x $\frac{1}{8}$ in., while the back may be a flat piece of wood, polished, or a piece of beaten copper.

Let us now turn to practical considerations, and see what is necessary as stock-in-trade for anyone wishing to carry out such simple designs as these.

First of all, one should work it on a good solid table, with a deal top, such

SETTING OUT THE PATTERN

The design is first of all drawn the full size of execution, either upon a piece of board direct, or upon cartridge paper pasted down upon the working table. In this sketch, all the curves have to be accurately produced, and all portions that touch clearly indicated, or difficulties will crop up later which will not be easy to remedy. A piece of string is then carefully and accurately laid over any given line in this drawing, and indicates the length of iron strip necessary for that portion. It is a convenient method to cut off all the pieces at one time, keeping them labelled with letters showing their place on the design. The strips are then bent up, guided by the preliminary drawing, which is of the thicker metal, followed by the longest curves, these latter being most truly formed by bending over a piece of a wooden roller for a start, Fig. 7. A number of these, of various sizes, will be found useful. All the small curves are bent up by means of the pliers alone,

porarily, by means of wire twisted round the strips, at or near the places where the future ties or clamps are to go. This allows modification and alteration as the work proceeds.

Fig. 9 shows the pattern for a leaf cut from thin sheet brass or copper, ten of these being required, two being inserted opposite one another on each side of the bracket. They are neatly soldered on, and the ends bent to produce a varied effect, Fig. 10. In the case of Fig. 2, the design consists of scroll work alone, but a few foliated ornaments might be included if desired.

Ordinary 12 in. Burnese gongs may be very often picked up very cheaply indeed, from second-hand stores, and bought new they entail very little more expense. They are made of a specially-tempered metal that sound better than those made from copper. The gong-stick, the end of which is of wood, is muffled with leather, and hung from any convenient portion of the bracket.

On the completion of the work, if it is of wrought iron, it requires a finishing coat of black, not too dull or too shiny. Berlin black, which dries with an artistic egg-shell gloss, being perhaps the best substance for the purpose. If any of our readers have acquired the art of gilding, we can also strongly advise this method of finishing off their work. It is effectual as a preservation against rust, and is, in reality, a method largely adopted by many of our best metal workers.

RULES FOR THE CHAUFFEUR

An English motorist has compiled the following rules for his own driver for town driving, says *Automobile Topics:*

If there is a doubt whether you can "get through," don't try. Remember, if any accident occurs it is a discredit, and a bad job for someone.

Don't go too quickly near the pavement, in case a deaf person, or one engaged in other thoughts, steps off into your track. When passing a tram, face on, toot loudly, to provide for a person walking across from behind the tram, who might be bewildered by the confusion. Go too slowly rather than too quickly. If you make an error, make it on the safe side.

Always remember that any useless revolutions of the engine—that is, when the engine is running light, are so many moments less life to its existence. It is an unnecessary cost of petrol, lubricating oil, and wear and tear—a noise, a discomfort, and an irritation to people and mechanism.

A revolution saved is a revolution gained. There is an economic, a durable, and a pleasant speed to an engine, just the same as there is to a living person; a speed at which a person can walk and run without destroying the tissues or over-exerting the muscles of the system, so with the piston of an engine.

And an engine working the car, and running light, is under two distinct differences. Working, the car has the

by a ton in motion with itself, and it is thus held "steady." Running light, it has no staying power; it has no one-ton fly-wheel. Therefore, before declutching, throttle down your engine. Before starting your engine, throttle down. Before starting your car, throttle down to the extent that the engine will easily "take hold."

Never draw up with your brake if you can do so without; it is a penny wasted on tires every time you do so. Withdraw your clutch in anticipation of the place to stop at, and just bring the "stand still" with the brake. It is an act of bad driving to rush up to a stopping place and then apply the brakes—

Because it scares the people about, and the people inside may think perhaps the brakes won't act; because it savors of a wish to draw attention and give an impression of ability, which is not becoming; because it costs as much in tires to stop by brake power as it does to start with the same quickness. In the case of starting or quick acceleration, the engine is the motive power. In the case of slowing down suddenly by fierce brake power, the momentum (of, say, $1\frac{1}{2}$ tons) is the motive power, and the brakes are the retarding power. In both cases the tires in contact with the road surface have to communicate the power, and they depreciate accordingly—

Because the power of retarding is transmitted through the gears and reduces

NEW AND SECOND-HAND MACHINERY

H. W. H. STILLWELL

There seems to be a universal second-hand craze in this country, and it is on the increase There is no particular risk in purcha.ing second-hand machinery tools and shop equipment if ordinary caution is used by the purchaser, but, experience should be hand in hand with the caution, as paint and putty sometimes cover a multitude of mechanical sins. Perhaps there is no class of articles which are doctored up as much or made as pleasing to the eye as second-hand machinery and the unwary, inexperienced purchaser is easily drawn into the trap and repents in sack cloth and ashes for a long time thereafter.

In purchasing material of this sort, if the dealer is a man of known integrity and enough of a mechanic to know the actual value of the material which he is handling, there is little risk in buying from him. It is seldom that the honest dealer and the practical mechanic are combined. Machine sales, which are common in many districts, should be avoided, unless the purchaser, or he who desires to purchase, be a very good judge of machinery. I would not have the readers believe from what I have just stated that all machinery sales are fakes, or that the goods offered in these sales are all junk, as such is far from the case. There are many times when exceptional bargains can be secured where some shop for any good reason is about to be sold out at bankrupt sale, or for other good causes. There is much material disposed of in this manner, which is as good and often better than some new machinery would be, but, as before said, there is a great risk unless you know your ground.

There are many small concerns which have a very limited amount of capital and must count every cent expended; to these, good second-hand equipment is a great boon. There are also many amateurs who own a small workshop and have a second-hand lathe or other tool which may or may not be all they hoped it to be when purchased. It is to the inexperienced class that this warning is dedicated.

The writer knows of several concerns operating plants of considerable size who make a practice of using their machinery until it is only fit for junk and the scrap heap, and then patching, painting and puttying it up until it looks like new to the inexperienced and selling it for a good price. These concerns have an excellent up-to-date shop equipment and are better able to keep it in this condition by imposing their cast-offs upon the guileless public. Experienced mechanics when in the market for machinery never consider anything which is known to have come from these concerns.

One shop known to the writer is second-hand from the foundation to the roof, even the bricks in the walls are mostly second-hand, I have been informed, and the machinery equipment is entirely second-hand. The machine shop equipment is a varied assortment, a sort of patchwork of all makes of machinery, and few, indeed, are those which are worth the space they occupy. The management of this shop consider themselves very keen and economical in their operations, but as none of them are mechanics enough to grind a cold chisel, this is not to be wondered at. A few years ago the business made it necessary to increase the size of a department devoted to drop-forging, and another drop hammer was required. One of the heads of the company took it upon himself, he being a country storekeeper before he invested in the manufacturing line, to visit a well-known eastern firm which had been advertising second-hand machinery. This man was advised to send one of his best men to look the goods over, but he poo-pooed the idea, and said that he knew what he was at. The purchase was made and the drop hammer, in due time, was run in on the railroad siding, and, after much trouble, was taken from the car. The base of the hammer weighed six tons without the columns and head. When the hammer was placed in position in the forge room, it was found that the columns were not parallel, and that they were wider at the bottom than at the top; it was impossible to line up any drop dies for anything like accurate work, so the head was taken off and the columns removed to the machine room, where they were placed on the planer and the job given to an old man who was better at giving the directions than performing the work. Many weeks were consumed in planing the columns and several other parts; then it was found necessary to make a new shoe in the base as the old one was cracked. When all this was completed, new boxes had to be made for the head, and when the columns were set in place, the slots in their base, by which they were secured to the hammer base, were not long enough to close the columns in on the hammer. These slots were therefore made longer and the columns again set in place; all seemed to be fairly good until the hammer was started, when it was found that the workman who had planed the columns had not set them properly, and that they were too wide at the top and too narrow at the bottom; just the reverse of the original trouble. A drop hammer should be a little more free at the top of the stroke, and should fit the ways fairly snug at the bottom; but in this case, the hammer would stick and stop when about 12 in. from the die when the columns were bolted tight. The time taken to get things in this condition was something like six weeks from the time the machine was received. If this time be figured at the average rate of wages for the time of the men employed in connection with these various repairs, and the cost of the materials figured, and the result added to the original cost of the machine, it is soon discovered that the company would have been money in pocket if they had purchased a new hammer, or allowed one of their best mechanics do the purchasing. There is an old saying that "a burned baby keeps away from the fire," but the baby in question has done many things as bad as this since the above took place, and is constantly burning his and the company's fingers and their pocket-books as well.

A short time ago, a small manufacturer wanted an additional boiler for his shop, and, visiting one of these dealers with a questionable reputation, he selected one which had been coal-tarred inside and out, and looked as good as new. He complimented himself upon getting such a good boiler and so "dirt cheap." When it was placed upon its mountings, and the pressure run up to about 60 lbs., it leaked like a sieve. He had it closed down and the seams calked and poured in a lot of rye flour. A soup formed inside, a combination of coal tar and rye flour, and this soon found its way into the cylinders of the engines; and after weeks of this sort of thing, he ordered it all blown out, and hired a good man who was well experienced to look it over. The result was a great amount of patching and repairing; and the cost of those repairs, added to the original cost, would have purchased a good boiler, and he would have been saved the vexatious outlay of money and patience. At small cost this man could have hired a practical man to have accompanied him to make the purchase and thereby saved no end of annoyance.

There is one great advantage in purchasing new machinery or anything else, if the article is good and made by well-known manufacturers, there is a reasonable guarantee accompanying the article, and if for any reason it should not prove satisfactory, due to poor workmanship or defective materials, the makers will replace parts or the whole free of charge.

There is another point too little taken into consideration, and that is the proper handling of new or even good second-hand machinery. There are good and bad mechanics everywhere, and every shop has a good supply of the latter, if not the former, and even if the company starts out with a good equipment and employs many of these unskilled or careless men, it is only a question of time before the machinery, which should, with proper care, have given many years of good and efficient service, will be reduced to a lot of scrap.

An article devoted to the slip-shod methods as employed by some of the self-termed mechanics would be mighty interesting, but a good-sized book would be required, and possibly the good effects would be very slight. It is to be hoped that if any of the readers are about to purchase anything in the line of second-hand shop appliances, that they will profit by this article and save themselves no end of trouble and annoyance, to say nothing of the money.

Don't envy the man who, is riding around in an auto until you know how big his mortgage is.

METAL DRILLING. Part I
Drill-Making, Tempering and Using
M. COLE

There are two principal classes of metal drills: the Cylindrical and the Spear-Headed. In the cylindrical drill the body of the drill is the full size of the hole to be bored, the body acting as a guide to keep the point of the drill in its right place, so that a hole bored by a cylindrical drill is quite straight. In the older forms two flats or grooves were cut in the body of the drill to allow the drillings to escape. The twist drill is another form of this, and has two spiral grooves instead of the straight ones, these bringing the drillings to the surface. The twist drill is certainly the best shape for getting through the work well and quickly, but it has its drawbacks, as it cannot be ground except by special appliances, and any attempt to do so by hand spoils the drill. Another difficulty is that unless run very true it jams, and if small, breaks. To these drawbacks must be added a much higher price, so that unless used in a good lathe or true running drilling machine, the other styles are preferable.

SPEAR-HEADED DRILLS

This is the ordinary shape of metal drills, and until comparatively recent years the only one. The body of shank, of such size that it fits the chuck or drilling machine, is narrowed down to form a neck, then widened and flattened to a head, all that part that goes into the deepest hole to be drilled being smaller than the width of the point. Good well-made drills are so cheap that it is hardly worth while to make them, though so much is to be learned by making a drill that the time is well spent.

Selecting the Material.—Drills, like other cutting tools, should be made from the best tool steel, and it is not dear, but it are perfectly straight, so that if the end is worked into a spear head and a length cut off, a drill can be very quickly made; but they are a very mild steel and should not be used if the proper stuff is to be had. Steel varies in the amount of carbon it contains, some being as low as $\frac{1}{2}$ percent, others up to $1\frac{1}{2}$ or more, and the quality suitable for a razor would not do for a drill.

Shaping the Drill.— The spear-head drill is of three parts: the head, or cutting part; the body, or shank, being the part held in the chuck or socket of the drilling machine; and the tapered neck connecting the two others. The head, the most important part, as it does the work is usually 90 degrees angle, and should be tapered at the point according to the size of drill; $\frac{1}{8}$ in. for 1 in. The neck is the part just above the head and is smaller than the hole the drill will bore. If the neck, as far as it penetrates, were the size of the hole, the drill would be cylindrical. The body or shank is the part that fits the chuck or socket of the drilling machine by which motion is imparted to the drill, and may be either square, cylindrical or taper.

Square body is required only when the drill is used in a joiner's brace, a very convenient method when lathe or drilling machine is not handy; but a large hole should not be attempted, not larger than $\frac{5}{8}$ in. brass, or $\frac{3}{8}$ in. in iron. If required larger, the hole should be opened out with a reamer used in a brace.

CYLINDRICAL

This is the most useful form for the body of a drill when used with a self-centering chuck or in the socket of a drilling machine; but in the latter case

Taper.—The body, or that part of it that is inserted in the socket, is tapered, and the socket itself is bored of an equal taper. This gives the best of all fits, and is used in the best makes of twist drills.

HARDENING AND TEMPERING

Steel may be hammered more than iron, but with light quick blows, gradually with less force as it gets cooler, and not with much force after it has lost its redness. It must not be heated higher than a blood-red heat, and care must be taken that the fire is clean and free from sulphur fumes, coke being preferable to coal. A length of iron pipe may be used in the fire to keep the tool clean while heating. Steel is improved by hammering at the right heat, but spoiled if hammered at too low a heat. Frequent heating also spoils it. Sharp angles must be avoided, as that is where cracks usually start.

Tempering the Drill.—For some reason that has never been explained, when a piece of steel is heated to redness and quickly cooled, it is hardened, so much so that it will break like glass if dropped on a stone; but if the same piece of steel is then re-heated to a lower heat than before, and cooled, it is no longer so brittle, but more or less springy, being then tempered. If it were not for its brittleness, a drill that is dead hard would be the ideal one for cutting, as it would keep its edge. As it is, we are compelled to use a tempered drill, which, though less hard, is tough enough to stand working. Steel changes color with the degree of heat to which it is subjected, but it should be remembered that while the degree of temper denoted by a certain color is always the same for the same quality of steel (provided it has previously been equally hardened) the same color would show various degrees of temper for different qualities of steel. Again, color does not prove steel to be

tempered. If a piece of soft steel be heated to straw color and quenched, it will not have the same temper as if it had previously been hardened. The color is caused by a thin film of oxide. (See Table A.)

Process of Tempering.—Heat the tool to blood-red when looked at in a dark corner, then dip in cold water, holding the drill perfectly straight, point downwards, keep there till cold; it will then be quite hard and brittle; now clean up one of the surfaces of the head with a flat stone, and reheat till it shows the desired color; at once dip in water and hold there till cold, keeping the tool perfectly straight; it is then ready for grinding. It is as well to rub the surface that is to show the color with a piece of stone while hot, to clean it.

Hardening and Tempering at one operation.—Grind a smooth place on the body of the drill, and another on the head, heat to red, and then dip the head and half of neck till the body shows light blue color, lift out of the water and hold till the heat of the body of drill spreads to the head and shows the right color, say light straw, then quench the whole drill. Once in the water a tool must not be lifted out till dead cold. The steel is white when dead hard and cold. A good process of tempering much used is to heat the tool in a clear fire and while hot rub on some common yellow soap, then heat to cherry red, and quench off in some common petroleum. Keep away from the fumes. Everyone thinks his own method of tempering the best. Below are opinions of practical men on the best way to do it. They are all useful, though contradictory:

"Boil the water before using."

"Spray the water on the drill."

"Use water heated to 100 degrees F."

"Put a little oil on the surface."

"Put plenty of salt in the water."

"Use pure, clean water quite cold."

"Keep the drill perfectly still in the water."

"Move the drill gradually lower in the water till cold."

"Dip the head and hold 10 seconds before dipping the remainder."

Good work has been done by all the above methods, and they are all the best way, according to those who use them.

TABLE A

Compiled 100 years ago and still unsurpassed.

430° F.	Faint Yellow	Lancets
450° F.	Pale Yellow	Razors
470° F.	Full Yellow	Penknives
490° F.	Brown	Scissors, cold chisels
510° F.	Brown, with purple spots	Axes and planes
530° F.	Purple	Shears and table-knives
550° F.	Light blue	Swords and watch-springs
560° F.	Full blue	Small saws

ing, use a block of iron, large in proportion to the size of the drill, heat to redness, then lay the drill on it, the head overhanging; the block will heat the drill. A very tough result is got by "blazing off," not so useful for drilling, but very useful where it is necessary to have a drill that will spring rather than break. After heating drill, dip in cold oil to harden it, then put the oily drill in the fire till it blazes up; allow to blaze till the right color is reached, then quench off. When tempering a spring allow to blaze till burned out, then dip in water. When large quantities of articles are to be, tempered they are often done in an oil bath, which can be kept at the required temperature by a thermometer. Linseed oil boils at 600 degrees F., and mercurial thermometers can be had graduated to 600 degrees, which is much higher than is required for tempering.

Grinding is a very important part of drill making. The small cutting edge is the part that does the work, and the most elaborate lathe, chuck, drilling machine, etc., is only a means of moving the unseen and often forgotten cutting edge. The usual angle for a drill is 90 degrees. When required for cutting very hard metal or glass, the angle is smaller, while for thin metal a much larger angle is used, sometimes nearly straight. The object of grinding a drill is to give it a cutting edge, so that if one side is ground away more than the other, only one side will cut, and the drill will only be doing half its work; a little clearance must be given, but only enough to prevent any part except the cutting edge pressing on the work, so grind away only a little more than the depth the drill is intended to cut. A 100th of an inch is a good cut for drill of moderate size. If too much is ground away, the edge is weakened and breaks. When a drill is properly ground the drillings should come away in curls; in one

A grindstone is the safest to grind drills with, but it must be kept true, and not allowed to wear into irregular shapes, as many grindstones do. An emery wheel does the work much quicker and runs truer, but there is always the danger of burning the edge of the drill, unless it is kept flooded with water; merely wet is not enough to ensure safety. It is worth while to finish off with an oilstone, the superior edge being worth the extra trouble.

Testing the Drill.—To see if correctly ground, run the drill very slowly so as to drill into a bit of brass, and watch if both sides cut. If only one side cuts, the work will take twice as long to do and the drill be forced sideways; if small, the drill breaks; if large, the strain injures the chuck.

Drills ground by hand in the ordinary way are never accurate except by accident, so the greatest care should be used to get the error as small as possible.

Making Very Small Drills.—Drills can be obtained ready for use down to the smallest required by watch-makers. To make them it is best to use good steel wire; failing that, they may be made from needles—sewing-machine needles being the best, as they are well-shaped and only require a head making, the body being large enough to be held in a drill chuck. To soften: heat a large piece of iron to redness, lay the needle on it—it will soon reach the heat of the block—allow both to cool together, the larger the block the softer the needle will be. Now break or file off the point, and with one blow of a hammer spread the end to form a head. It will stand one knock, but repeated hammering will spoil it, file and dress to size, then to harden hold the drill with pliers, heat to redness in the flame of a tallow candle and instantly plunge the point in the tallow. It then only requires setting on an oil stone. Mercury may be used to quench the drill to get a

HAND SCRAPING IN THE MACHINE SHOP

STUART K. HARLOW

Hand-scraping is that process of scraping a metallic surface, by the use of a hardened steel chisel, so as to obtain a smoother and more level surface than is otherwise possible in ordinary machine shop practice. It is an example of one of the many instances where hand-work far excels that performed by a machine.

The learner or apprentice first starts with a piece of cast iron that has been molded roughly to the dimensions of 3½ x 2 in. The sand from the mold that still clings to the bottom side in the deep hollows should be scraped out with the sharp tang end of a file. If the block is too thick, it should be tightly fixed in the shaper and planed down to within a 1-32 in. of the desired thickness, ¾ in. to ⅝ in.

The block is next fixed securely in a vise (fitted with copper jaws). Now proceed to bring first one side then the other to a smooth and level surface, as shown by test with a small steel square. When the sides and ends are filed square, turn the block and file the top surface till it is as nearly smooth and level as is possible to attain by hand filing. When the top is finished, the sides and ends only (not the top) should be brought to a high polish by the use of, first a coarse emery cloth, then the finest emery cloth, mounted on a stick for a handle.

The next operation is the most important, as it constitutes the first steps at hand-scraping. The smooth die-plate should be carefully covered with a very thin layer of red lead paint, rubbed on with the finger. The block is next laid face downwards upon the die-plate and is given a circular motion. When it is lifted from the die-plate, the face of the block is found to be marked by the red lead appearing only on the high spots of the metal. Now lock the block securely in the vise and proceed to apply the chisel scraper only on these high spots and in a series of short strokes, equal in length to the width of the scraping chisel. The scraper should be gripped firmly, with both hands, and held at the angle at which it cuts the best, and a great deal of pressure should be applied on the tool to make it take hold and cut.

The first cut should be started at a 45-degree angle with the end. When the high spots have been gone over in this position, at this angle, it should be immediately reversed by stepping to the left-hand side of the vise and taking a cut directly across the others at an angle of 90 degrees. Now remove the block from the vise and mark the high spots again by rubbing it on the die-plate. When the face of the block is as nearly smooth as is practicable, it is shown by rubbing the block on the die-plate, when the whole surface is covered with red lead.

The final operation is to make a series of parallel cuts across the face (leaving a small space between each), parallel to the sides of the block. Then a second cut with the chisel is taken directly across the face from side to side and perpendicular to the other cut. A finishing cut is taken at an angle of 45 degrees in the remaining spaces, when the block is finished.

Encourage the Workers

Men need a word of encouragement now and then just as much as they need food. For as the food is to the body, so is encouragement to the mind and heart. A worker who is discouraged is not half a man. Fear of disfavor often holds back valuable information. Even the most liberal compensation cannot take the place of a word of appreciation and encouragement given in the right spirit at the right time. Men and women crave the assurance that their work is meeting with satisfaction. To withhold that assurance when it is due is not merely a poor business policy—it is an injustice. Part of the compensation of every worker is the satisfaction of knowing that he is accomplishing something, and to withhold that satisfaction is often more harmful than to hold back money duly earned. More and more must those in business recognize the human element in men and women, the part the heart plays in business. It is possible, of course, to say too much to a man, giving him an over-elated sense of his value, but the tendency seems rather in the other direction.

THE PLANIMETER

C. W. WEBBER AND T. C. FISHER

The object of this article is to describe the planimeter in its simplest form so that it may be understood by the general reader and by those who do not use the instrument but would like to know of its workings as a matter of general information.

The planimeter is an instrument by means of which the area of plane figures, no matter what their shape, can be measured easily and accurately. Its early history seems rather vague, it being thought that the first one appeared in Germany about the year 1814. The first one, however, of which we have record appeared in Paris in 1836.

There are four essential parts to the planimeter, namely: (1) polar arm; (2) tracer arm; (3) wheel; and (4) vernier. The relative positions of these parts are quite clearly shown in Fig. 1. The polar arm has a length of from 5 to 8 in., and a square cross section of about ⅛ in. on a side. At one end is a fixed needle-point with a weight directly above it. The material used in the construction of these parts is German silver, as that metal will not rust as iron or steel would under similar conditions of constant handling.

The tracer arm, like the polar arm, has a length of from 5 to 8 in. with the same cross section as before. To one end is fastened the tracer point, while near the other is the wheel. The tracer and polar arms are connected at their ends by means of a pivot. The wheel, another one of the essential parts, has a diameter of about ¾ in. To it is attached a scale of less diameter which is divided into 100 equal parts. The wheel is mounted upon

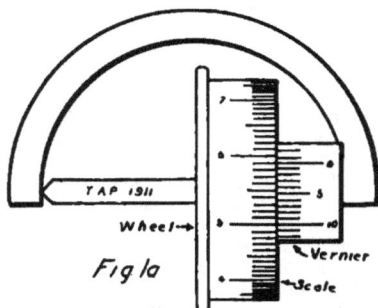

Fig 1a

steel bearings and is plated with nickel to prevent rust and injury due to use.

The vernier, which plays such an important part in the reading of the scale, is the invention of Pierre Vernier, from whom it takes its name. It is a device for reading easily and accurately a fraction of a scale division. The relative positions of the vernier, scale, and wheel are shown in Fig. 1a. Besides this use, the vernier is found on barometers, surveying instruments, and other appliances where accuracy of reading is required. In the case of the planimeter the vernier is mounted flush with the scale on the wheel, and is graduated into ten equal spaces for nine on the scale.

If we let

W = length of one wheel division
V = length of one vernier division

we have

$$10V = 9W$$
$$V = \tfrac{9}{10}W$$

that is, the length of one vernier division is ⅒ less than the length of a wheel division. In Fig. 2 let the zero of the

FIG. I

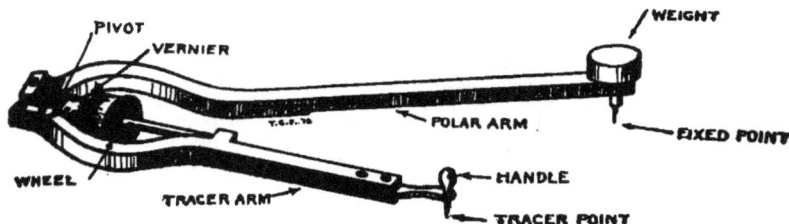

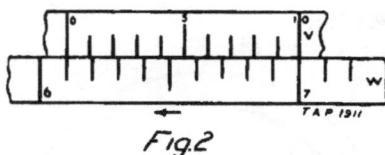

Fig.2

vernier be the index. The reading as shown is 6.10; for the 0 of V coincides with 6.1. Let W slide by V in the direction of the arrow, and, since $V = \%_{10}W$ when the first small division of V coincides with the second small division of W beyond 6, the zero of V will have moved $\frac{1}{10}$ of a division of W. The reading would, therefore, be 6.1 plus $\frac{1}{10}$ of .1, or 6.1 plus 0.01, which is equal to 6.11. To read the vernier, we look first for the index. In Fig. 3 it lies between 6.3 and 6.4. Looking along the vernier, we see that the fifth division coincides with a division on the wheel, meaning that the index has traveled $\%_{10}$ of the way between 6.3 and 6.4. Thus the reading is 6.35.

METHOD OF USE

Let the position be as shown in Fig. 4, it being desired to obtain the area of the figure. When the tracer is at the center of the area, a line from the fixed point a to the rim of the wheel makes a right angle with the tracer arm bc. Place the tracer point on the outline of the figure at point x, the intersection of the arc de with the outline. The radius of the arc is equal to ac. It is not necessary, however, to draw the arc, as x can be estimated nearly enough by eye. A slight indentation is made in the paper and a reading taken of the scale. Call this reading x. Now the outline is traced in a clockwise direction. The point should always be guided with the thumb and forefinger and the rest of the hand be steadied on the paper. When the indentation is again reached another reading is taken. Call this reading y. Then

$$y - x = \text{area of the figure.}$$

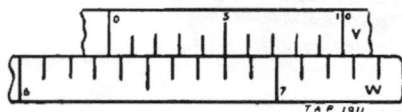

Fig.3

To find the area of the figure given in Fig. 5, start at point b. From there go to point a by the straight line and then back to b by the curve; from there go to c by the straight line, and then back to b by the curve. The difference between the first and last reading will give the area.

The planimeter should be used on a flat surface covered with drawing paper of a uniform medium smooth surface. The paper should have no wrinkles, as the working of the instrument depends upon the constant friction between the surface and the wheel. The wheel should always remain on the surface of the paper and should never run up on the figure being measured.

The planimeter finds many uses, a few of which are: the measurement of hystere-

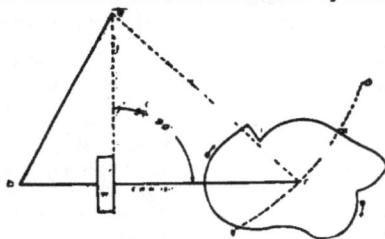

Fig.4

sis curves in electrical engineering; the measuring of the volume of earth works in civil engineering, and the measuring of steam engine indicator diagrams in mechanical engineering.

Hysteresis is the tendency of a magnetic substance to persist in any magnetic state that it may have acquired. When a coil of wire is placed around a magnetic substance and a current is sent through the coil, the substance is magnetized. Some energy has, however, been expended in overcoming this persistence in the substance. This energy appears as heat. If observations are made and plotted, a diagram is obtained, the area of which represents the energy lost in heat. Fig. 7 shows such a hysteresis curve.

The methods of use as applied to civil engineering are too long to be taken up here. The horse-power of an engine can be measured by means of the indicator diagram. A recording mechanism is placed on the cylinder of the engine in such a manner that the steam pressure is recorded for every point of the stroke

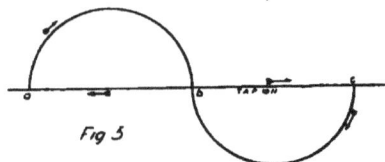

Fig 5

of the piston. This gives a diagram, as shown in Fig. 6. Let AB represent the length of stroke (drawn to some scale; in practice, this length is made about 3 in.), and let distances laid off on AF equal steam pressures. Thus, for a position C of the piston the pressure of steam for the forward stroke is represented by the point E, and the pressure for the return stroke by the point D.

Now, horse-power is given by the formula

$$\text{Horse-power} = \frac{PLAN}{33,000}$$

where

P = mean effective pressure
L = length of stroke (this is the actual length and not the length of the diagram)
A = area of the piston
N = number of revolutions per minute.

It is seen by the diagram that the pressure is different for every position of the piston. To get the mean we find the area of the figure by means of the planimeter and divide by the length of the base AB. The result is the average height of the diagram. The diagrams are automatically drawn to scale. One scale used is where 1 in. measured on AF represents 40 lbs. pressure. Thus, multiplying the average height obtained by 40, we have the mean effective pressure. From this we can substitute in the formula and solve for the horse-power.

THEORY

It may be of interest to some readers to know why the planimeter measures an area by tracing its outline, so for the

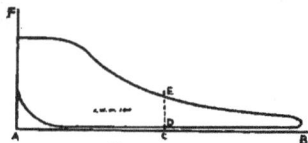

Fig. 6

benefit of those, the following graphical proof is given. This proof while not as satisfactory as a purely mathematical one, can be easily understood by the average reader.

In Fig. 8, let ab be the length of the tracer arm of a planimeter and let c represent the wheel. It is easily seen that if we move the arm along its own axis to b', the wheel will slip, but will not roll, thus not changing the reading. Now if we let ab be moved parallel to itself to the position $a'b'$, Fig. 9, it will be seen that the wheel will roll, but will not slip. Now, if we know the circumference of the wheel, the distance aa' equals this

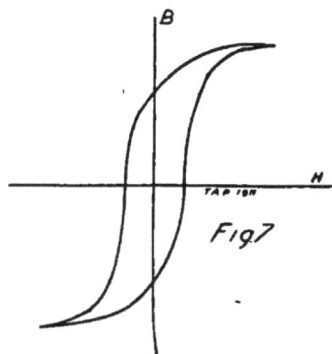

Fig. 7

circumference times the number of revolutions. Then by geometry the area of the figure $abb'a'$ equals ab times the circumference times the number of revolutions; or calling

K = circumference
n = number of revolutions
A = area

we have

$$A = ab \times K \times n$$

Now combining the two diagrams, Fig. 10, we have ab moved to $a'b'$, and then along its axis to $a''b''$. By geometry we know that area $abb''a''$ = area $abb'a'$ = $abKn$; therefore, area $abb''a''$ = $abKn$. If now we go through both motions at the same time, that is, moving directly from ab to $a''b''$ along the lines aa'' and bb'' the effect will be the same as before. Now passing to the next step, Fig. 11, in which aa'' and bb'' are curves, we have

$$A = abKn.$$

In the above we have considered the

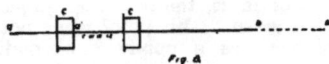

Fig 8

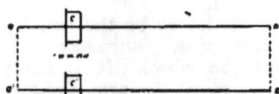

Fig 9

tracer arm moving parallel to itself, which, of course, it does not do in practice, so let us now consider Fig. 12, in which oa is the polar arm and ab is the tracer arm. We now can move ab parallel to itself along the arc aa', which has its center at o; then area $abb'a' = ab\,Kn$. Now let a' be fixed and move b' to b'', causing the wheel to rotate. We will pass by this motion for a moment and come back to it later. Now in the same manner as before, $a'b''$ can be moved parallel to itself to position ab''', the area $a'b''b'''a$ being equal to $ab\,Kn'$: where n' is not equal to n.

To complete the movement, b''' is moved to b, causing the wheel to rotate in the reverse direction from that in covering the arc $b'b''$. By geometry, since $a'b'$ is equal and parallel to ab; and $a'b''$ is equal and parallel to ab''; then angle 1= angle 2 and arc $b'b'' =$ arc bb'''; Thus the distance the wheel moves when the tracer goes from b' to b'' equals the distance the wheel moves when the tracer goes from b''' to b. But the wheel moves in opposite directions, therefore, one motion cancels the effect of the other, and the only motions which have any effect on the final reading of the wheel are those when the arm moves parallel to itself.

As the tracer travels from b to b' the wheel records a certain amount; call this u. Now if we so choose the circumference of the wheel that when a square

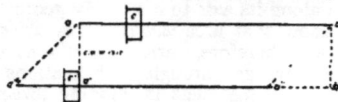

Fig. 10

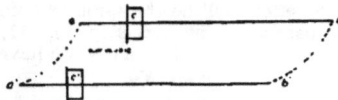

Fig 11

inch has been passed over the reading of the wheel will be 1, it is seen that u will be the area passed over. From b' to b'' another amount, call this v, will be recorded, but the wheel will now revolve in the opposite direction, so that the reading at u will be lessened by the amount v. Therefore, the reading at b' equals $u - v$.

In going from b'' to b''' the wheel revolves in the same direction as from b' to b'', thus reducing u still more; call this amount w. Then reading at b''' will be $u - v - w$.

In going from b''' to b the wheel again changes direction of rotation, thus adding to u; call this amount x. Then the final reading will be $u - v - w + x$. But we have seen that $v = u$, therefore, the final reading will be $u - x - w + x = u - w$; where u, of course, equals $a'b'ba$ and w equals area $a'b''b'''a$

In order to find the difference of these

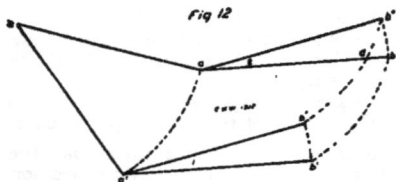

Fig 12

two areas we proceed as follows: Imagine the area $abb'a'$ cut from cardboard. Cut out area $a'b'b''$ and transfer to position abb'''; this does not change the area of the entire figure, for it has been shown that area $a'b'b'' = abb'''$. Thus far we have taken the area u subtracted from it the area v and added to it the area x. Our work thus far can be represented by $u - w + x$.

Now, if we cut away the figure $a'b''b''a$, we have left the area $b'bb'''b''$, but area $a'b''b'''a = w$, so our work can be represented by $u - v - x + w$, but $v = x$, so the result is $u - v$.

In the movements we have subtracted area x from area u leaving area $b'bb'''b''$. Therefore the reading of the wheel equals area $b'bb'''b''$, and we have found its area by tracing the outline with the tracer point.

Now let us carry this still farther and take the irregular area, Fig. 13. Consider it divided into a number of areas similar to area $b'bb'''b''$, Fig. 12. By

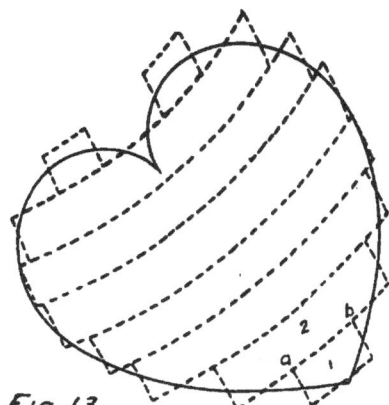

Fig 13

tracing the outline of the several figures we have their areas by the reading of the wheel. By measuring all of the areas and adding them together, it is seen that we have approximately the area of the irregular figure. Again, take the line ab

common to areas 1 and 2. It is seen that in tracing area 1 the motion will be from a to b, but in tracing area 2 the motion will be from b to a, for the motion around a figure must always be in a clockwise direction, as given under "Method of Use" above. The wheel will, therefore, in going from a to b turn in the opposite direction from that in going from b to a, and the motions will cancel each other. The same is true for all other lines common to two areas. Therefore, if all common lines cancel each other the final area can be obtained by tracing around the outline of the figure which is made up of the separate areas.

Now, it is seen that if we make these areas narrower, of course increasing their number, their combined area will come nearer to coincidence with the irregular area. Finally if we make them so narrow that they have no width their number will be infinite and their combined area will be the area of the figure. Thus it is seen that to find the area of any irregular figure we trace its outline and read the area from the wheel.

THE ORIGIN OF THE DIAMOND

The fascinating problem of the genesis of diamonds receives further attention, remarks *Knowledge*, from Dr. O. H. Derby (Journ. Geol., Oct.-Nov., 1911), who puts forward a new speculation as to the origin of the gem. As is well known, diamonds occur, at least in South Africa, in pipes of volcanic origin which are filled with a peculiar ultra-basis rock called "kimberlite." This rock is invariably much fragmented and altered, and contains numerous foreign inclusions (xenoliths), both of igneous and other origin. The weight of evidence is in favor of the diamonds being assigned to the eruptive rock proper, and not to the xenoliths included in it. Dr. Derby believes that a positive, and perhaps genetic, relation exists between the diamond and the fragmental condition of its matrix, basing his opinion on the experiments of Gardner Williams, who crushed 20 tons of the eclogite boulders or segregations from the Kimberley Mine without finding a single diamond.

The association of diamond with fragmentation means that the origin of the diamond is to be assigned to reactions between the rock constituents, made possible by the explosive and disintegrating action of the agency that formed the Kimberley pipes. Under this view the extensive hydration and carbonation of the Kimberley rock is due to deep-seated pneumatolytic action rather than to atmospheric weathering. Kimberlite from the deepest part of the De Beers Mine (2,040 ft.) still contains 6.81 percent of combined water, and it is improbable that this can be due at that depth to atmospheric weathering.

Dr. Derby presents a new hypothesis of the origin of the diamond on the assumption of the deep-seated origin of the alteration of the diamond matrix. He believes that the Kimberlite pipes were saturated with hot (possibly superheated) gases and liquids, and constituted huge crucibles in which carbon would be present at least in the form of carbon dioxide, and probably in other gaseous forms. Thus the material and some of the physical conditions for unusual carbon segregation would be present, and it is possible that, under these conditions, diamonds would be formed.

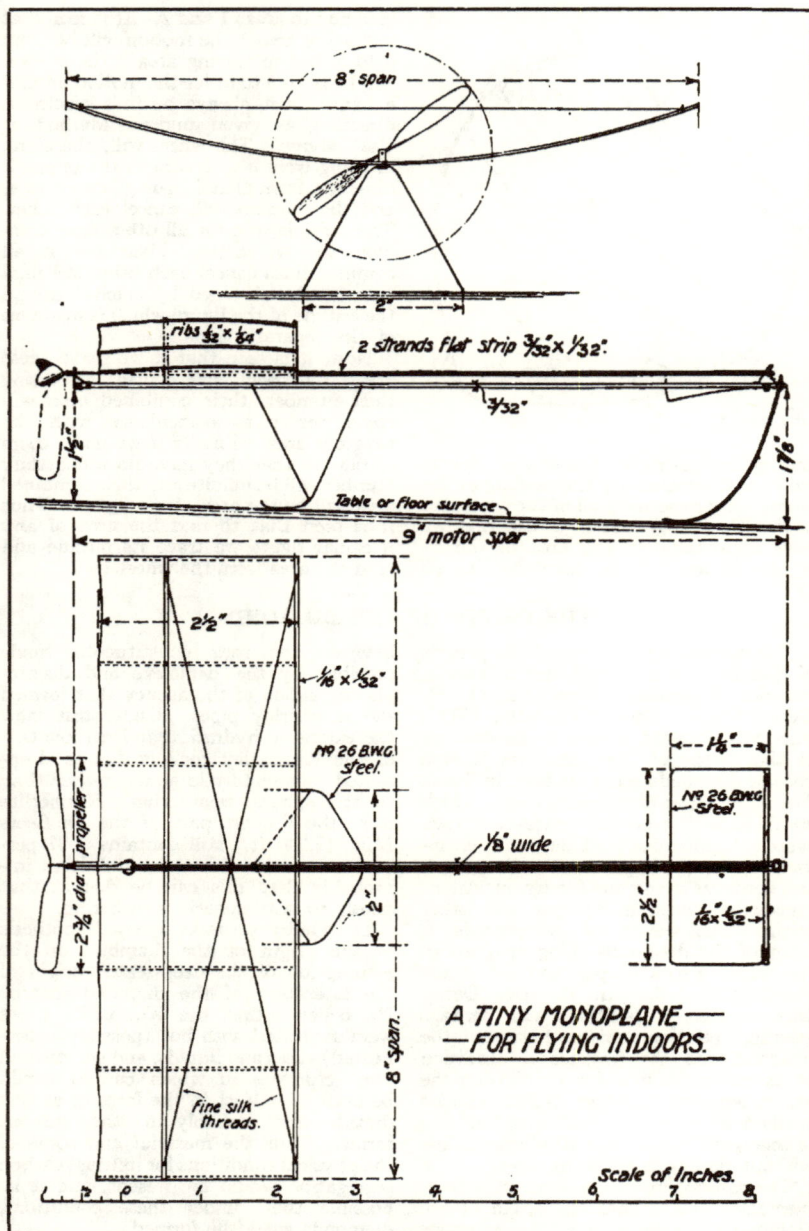

8" span

2"

ribs ⅛" x ¼"

2 strands flat strip ³⁄₃₂" x ½₃₂".

⁷⁄₃₂"

½"

Table or floor surface

1⅞"

9" motor spar

2½"

¹⁄₁₆" x ½₃₂"

Nº 26 B.W.G. steel.

⅛" wide

1½"

Nº 26 B.W.G. Steel

2½"

2⅜" dia. propeller

2"

¹⁄₁₆" x ½₃₂"

A TINY MONOPLANE—
—FOR FLYING INDOORS.

8" span.

Fine silk threads.

Scale of Inches.

0 1 2 3 4 5 6 7 8

THE SULPHUR INDUSTRY IN THE UNITED STATES

CARLETON HAIGIS

For more than one hundred years prior to 1895 the world's production of sulphur came from Sicily, where it is mined from the craters of extinct volcanoes. It, however, occurs in many other countries in small amounts, either as native sulphur or in compounds such as iron pyrites and other sulphides. These compounds are, however, rarely mined, as a source of sulphur, since sulphur itself is found free in such large amounts that any extraction process would be prohibitive on account of the cost of reduction.

In the United States prior to 1868 the element was not known to exist in the free state, and its discovery happened in a rather peculiar way. It was about the time that the petroleum industry was becoming important; and in several places in Louisiana, a peculiar viscous liquid was noticed oozing from the ground. Wells were bored, but no oil was found; instead, at a depth of 1,000 ft., in nearly every case, sulphur was found. Investigation has proved this deposit to be located in the mouth of an extinct geyser, which became silent far back in the geological ages. In boring, the drill passes through 200 ft. of clay, 200 ft. of quicksand and also through a cone of limestone which proves beyond doubt that this immense geyser was covered by the alluvial deposits at the time of the glacial floods. An approximate section through the deposits is shown in the illustration.

At the time of the discovery many methods were attempted in order to raise the sulphur to the surface, but none were successful. The ordinary method of sinking a shaft was out of the question as the sub-surface water of this region, which is nearly at sea level, could not be controlled. After many fruitless attempts the works were abandoned; and not until 1895, when a novel scheme was invented to bring the sulphur to the surface, were the mines reopened. The inventor proposed to send down a series of concentric pipes, 6, 3 and 1 in. respectively. The largest pipe is enlarged at the lower extremity to a diameter of about 10 in. Superheated water at a temperature of 335 degrees Fahrenheit is forced down in the annular space between the 6 and 3 in. pipes from a battery

An Approximate Section through the Sulphur Deposits

of boilers and allowed to flow through the perforations in the enlarged end. The water, because of its great heat and pressure, forces its way through the cracks and crevices of the sulphur-bearing limestone, causing the sulphur to melt out and drain down to the bottom of the well, where it enters another series of perforations communicating with the annular space between the 3 and 1 in. pipes. Since sulphur is twice as heavy as water, columns of the latter and sulphur will stand in equilibrium, the water column being twice the height of the sulphur. Therefore, when the top of the water column stands at the surface, the sulphur is elevated only one-half that distance. It is, of course, impossible to raise the melted liquid the remaining 500 ft. with pumps. This difficulty was overcome by forcing compressed air down in the 1 in. pipe. The sulphur is thus aerated until its weight becomes less than that of water, and it then rises to the surface through the central pipe kept above its melting point by the superheated water surrounding it. The diagram illustrates clearly the action of the sulphur at the bottom of the tube.

It will be seen that the amount obtained from one well depends on the area which can be melted by the water. This

Section through Bottom of Pipe

is usually not more than 50 or 100 ft. in diameter, and when one well is exhausted another is drilled and the process is again repeated. One well has been known to deliver as much as 400 tons per day, and another has actually given 60,000 tons of pure sulphur.

Such large amounts of pure sulphur have been excavated that the ground in the vicinity has sunk to an average depth of 30 ft., and many thousands of carloads of dirt have been used for refilling.

At the surface the sulphur is conducted into large vats constructed out of rough planking. These vats are about 250 ft. long, 350 ft. wide and 40 ft. high. After it solidifies it is broken up with the pick and shovel and loaded into cars ready for shipment.

Prior to 1907, when the plant first began successful operation, the United States produced less than one percent of her total consumption, which is 200,000 long tons. In 1908 the Louisiana mines produced a sufficient amount for the United States market and exported 27,894 tons, having a value of $561,538. In 1909 37,142 tons were exported having a value of $736,000, and in 1910, 350,000 tons were produced, placing the United States in a position to supply the European market, as well as her own, with a product 99 percent pure against Sicily's inferior article containing only 90 percent.

The "Ten Demandments"

A business concern at Steveston, which is away up in western Canada, has the following worldly wisdom conspicuously posted in its shop. While this may be a bit arrogative, it is nevertheless straight from the shoulder:

First—Don't lie. It wastes my time and yours. I am sure to catch you in the end, and that will be the wrong end.

Second—Watch your work, not the clock. A long day's work makes a long day short, and a short day's work makes my face long.

Third—Give me more than I expect, and I will give you more than you expect. I can afford to increase your pay if you increase my profits.

Fourth—You owe so much to yourself you cannot afford to owe anybody else. Keep out of debt, or keep our of my shops.

Fifth—Dishonesty is never an accident. Good men, like good women, never see temptation when they meet it.

Sixth—Mind your business and in time you'll have a business of your own to mind.

Seventh—Don't do anything here which hurts your self-respect. An employee who is willing to steal for me is willing to steal from me.

Eighth—It is none of my business what you do at night. But if dissipation affects what you do the next day, and you do half as much as I demand, you'll last half as long as you hoped.

Ninth—Don't tell me what I'd like to hear, but what I ought to hear. I don't want a valet for my pride, but one for my purse.

Tenth—Don't kick if I kick. If you're worth while correcting you're worth while keeping. I don't waste time cutting specks out of rotten apples.

Among the curios preserved in the Bank of England is a bank note that passed through the Chicago fire. The paper was consumed but the ash held together, and the printing is quite legible. It is kept carefully under a glass. The bank paid the note.

IMPRISONED IN BOILER WITH FIRE UNDERNEATH

Imprisoned in a big boiler, underneath which a fire was gradually heating the flues to a point which would have meant a horrible death if his escape had been delayed but a few minutes longer, is the experience recently undergone by Arthur McDonal, a young boilermaker of Arkansas. He has just left the hospital, a nervous wreck. His hair, which was coal black, now hangs over his forehead, a soft, glistening white.

The experience occurred at a saw mill at Hope, Ark. A new set of boilers had been put in, and negro firemen were relied upon to attend them. Something went wrong, and McDonal was called upon. After fixing the first boiler, he ordered the firemen to fill it with water and build a fire under it, when they had finished the work they were then doing. McDonal then entered the second boiler, and had been working about an hour, when he noticed his candle growing dim, and started to investigate. He found that the manhole cover had been replaced, and, sick with horror, realized that the negroes had misunderstood his orders and were building a fire underneath him. A moment later he heard the rush of water and frantically called for help and struck his hammer against the sides of the boiler,

hoping to attract their attention. Soon the heat began to be felt. He touched a flue and started back with a gasp. It was warm—ever so slightly, but warm, nevertheless.

With hands torn and bleeding, and eyes almost bursting from their sockets, the now thoroughly crazed man crawled back and forth in his prison, panting, praying and moaning. The flues became so hot they burned his feet, and his head swam with the heat.

At last, more dead than alive, he threw himself down on the flues to hasten the end, and then at almost the last moment a way of escape dawned upon him. Grasping his chisel, he placed it against one of the flues under water and dealt it terrific blows, driven by frenzy. The first blow missed, and, striking his little finger, smashed it off. The other blows fell true, and the chisel broke through the flue, letting the water follow. The negroes heard the water when it struck the fire, and, believing that the boiler still leaked, opened the water plug and raked out the fire.

McDonal had a faint recollection of a patch of daylight when the manhole was opened, but knew nothing more for five days.

New Way of Making Cutting Tools

Some remarkable results have recently been obtained by the Bureau of Ordnance of the United States Navy Department with cutting tools produced by a new and interesting process.

Instead of making the tool from high-cost steel tool, containing the carbon and other elements in its entire mass, the new tools are made of soft steel, easily shaped into the proper form, and then treated by the so-called "infusion" process, the carbon and other elements being placed in contact with the metal in the form of a special powder and subjected to a heat treatment which causes the soft steel to become hardened to such a depth as to convert it into cutting material even superior to the far more costly tool steel.

The tests at the Ordnance Bureau showed that milling cutters made by the infusion process cut deeper, faster and

further than tools made of the best carbon tool steel, and fully as well as cutters made from modern high-speed tool steel of far higher cost. The chief of the Bureau says that the infusion process "appears superior to any hardening process now in use at the naval gun factory," so that it appears that we are now in possession of a method of making tools of the highest grade for cutting metal out of ordinary soft steel.—*Cassier's Magazine.*

Perverted Proverbs

The engineer is worthy of a higher hire.

A rolling stone gathers no cement.

A touch from a live wire is breakfast for a coroner.

A survey goeth before construction, and a power plant before a fall.

Too much anchor-ice breeds lament.

The flat wheel makes the greatest sound.—*Power and Transmission.*

In this department will be published original, practical articles pertaining to
Wireless Telegraphy and Wireless Telephony .

THE OSCILLATION TRANSFORMER

RICHARD U. CLARK

At the present stage in the development of the art of wireless telegraphy, one naturally expects to find in modern equipment, only instruments of the highest efficiency and most advanced design.

While it is undoubtedly true that in some cases every instrument is of the most efficient and approved type, yet the majority of medium-powered stations generally are lacking in some one point. Nine times out of ten this deficiency is due to the improper handling of the oscillations produced by the sending coil or transformer, which, in turn, is caused by the absence of an oscillation transformer.

Fig. 1

It seems very strange that while no one would think of being without a receiving transformer, when it comes to the oscillation transformer a few turns of wire are considered a sufficient substitute.

This condition of affairs is undoubtedly due to the fact that this special branch of the art has been greatly slighted as far as comprehensive and practical advice on the subject is concerned. It is, therefore, the purpose of the author to give a short description of the oscillation transformer.

In order to explain clearly the principle involved in its use, it is necessary to

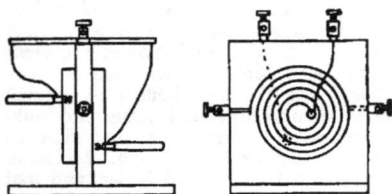

Fig. 2

start with the sending coil or transformer. When an oscillation is produced in the secondary circuit of the coil, it is of a highly dampened nature, that is, it produces a wave which quickly dies out. In order to prevent this dying out of the wave, a condenser is employed, which, in conjunction with the spark gap, gives a wave of greater sustaining power.

Now, when an inductance or a straight sending helix (not a transforming helix) is introduced into the circuit with these instruments, the dampening effect is still further reduced, due to the storing up

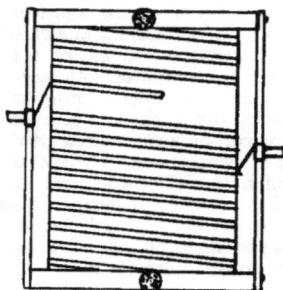

Fig. 3

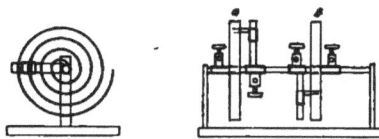

Fig. 4

of energy in the helix itself, and when the proper amount of resistance and self-inductance has been introduced (by varying the amount of wire on the helix) the wave emitted will be properly dampened, and therefore as persistent as possible under the given conditions. However, when a helix of this type is used the full efficiency is not realized, this being due to the following facts.

In order that the energy stored up shall be of the greatest value in producing electro-magnetic waves, it must properly energize the antenna. As an example, consider an aerial capable of using a given amount of energy. Now, any amount in excess of this is wasted in the form of heat. It is therefore necessary that the aerial should be made capable of using all the energy given it, and that it shall be properly handled.

It is not possible to do this with a common helix, which makes a good reservoir of power, but a poor distributer. That is, when employed as a reservoir it can not act as a controller or adapter of energy at the same time. In other words, being directly connected with the antenna, the power stored up is at once dissipated, and in this way the aerial does not receive its energy constantly.

However, upon the addition of a secondary circuit, which can readily be brought into resonance with the circuit already mentioned, the proper results may be obtained. In this case, in the primary circuit (which is really the secondary circuit); or the oscillatory circuit, the energy is stored up by the primary of the oscillation transformer, while it can readily be consumed by the aerial

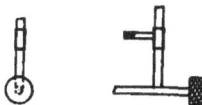

Fig. 5

by bringing the aerial circuit into resonance with the closed circuit (constituting the spark gap, condenser, etc.). This is accomplished by varying the resistance and inductance.

In the design of such an instrument as an oscillation transformer, great care must be exercised in order that short-circuiting may not occur between the consecutive turns of conducting material, as the difference in potential, even in a few turns, is very great, sometimes amounting to several thousand volts. However, if of the proper design and construction no trouble will be experienced in this line.

Among the many prominent and efficient types of oscillation transformers placed on the market today, Fig. 1 illustrates perhaps the most common. It is composed of a primary of copper ribbon, which is about 1 in. wide and is wound as shown in the drawing at a, the convolutions being separated by corrugated

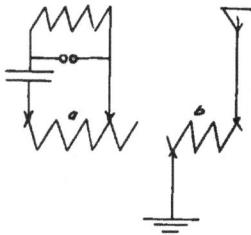

Fig. 6

cardboard strips about ½ in. wide. The variable contact is made by the rotary slider b, in Fig. 1.

The secondary c is composed of copper stripping ½ in. wide and No. 28 gauge. One end is connected to a binding-post as in usual practice, while contact is made by a slider similar to that of a tuning coil, only a little more massive in construction.

In Fig. 2 is seen a much simpler type, which, however, is not quite so efficient, as it cannot be adjusted as simply as the type shown in Fig. 1. This can be wound with wire or ribbon, the latter, when procurable, is always preferable, as it offers more surface for the oscillations to travel upon.

A very simple type is shown in Fig. 3,

being wound on cardboard tubing of a large diameter, say about 1 ft., in the same way as the secondary of the transformer shown in Fig. 1. Connections may be made in most any manner.

Fig. 4 displays another type in which the primary and secondary a and b are similar in form, both being like the primary of the transformer shown in Fig. 1. In this case the primary and secondary are both movable and are very well insulated; however, the conducting material must be at least ⁹⁄₁₆ in. thick in order that it may keep its place unaided. The slider used for this transformer is similar to the

one in Fig. 1, a, and is shown in detail in Fig. 5.

If one has a good straight helix which cannot well be converted to one of these forms, the right effect can be produced by placing another helix parallel to it.

In Fig. 6 is shown the circuit used in conjunction with a loose-coupled sending helix. In this illustration a is the primary and b the secondary of the oscillation transformer.

When this instrument is used a hot wire ammeter should be included in the aerial circuit in order that the maximum efficiency may be obtained.

VARIABLE CONDENSER
H. D. KEMP

Many owners of wireless stations would like very much to add to it a variable condenser, but are unable to make a rotary one on account of the machine work entailed. The condenser described in this article, if well made, will work as well both electrically and mechanically as any rotary one.

The first thing to make is the grooved board which is used to separate the plates. If a circular saw is available, this work should be done on it, if possible, as it is so much more accurately and easily done. If one is not available, the grooves may be carved out with a narrow chisel, or cut with an ordinary saw. There should be two pieces made: one 12 in. long, and the other 6½ in. long. They both have the same cross-section, as is shown in the drawings. They are 4⅛ in. wide, ½ in. thick, and have 15 grooves ⅛ in. wide, ⅛ in. deep and ⅛ in. apart.

The two side pieces may now be made. They are 6½ in. long, 7¼ in. wide and ½ in. thick.

The frame may now be put together

and nicely sand-papered and given some kind of a finish.

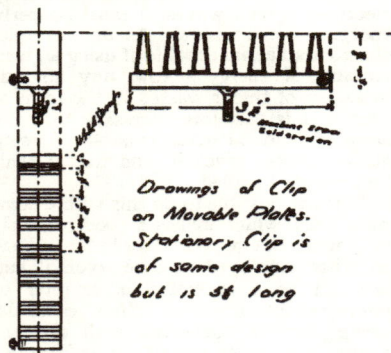

Drawings of Clip on Movable Plates. Stationary Clip is of same design but is 5⅝ long

We now come to the plates. There are 15, of which 8 are movable and 7 are stationary. These plates are made of glass and are coated with tin-foil. This method of construction is used, because if metal plates were used the dust settling between them would form a high resistance conductor, which would affect the system badly. Glass plates of this size are also very easy to obtain, are easy to work, and are much less expensive than brass or aluminum ones of the same size. The maker should go to a photographer and obtain about a dozen and a half old, used 6½ x 8½ photographic plates. This allows two plates for breakage when they are cut. Before cutting they should be cleaned of the gelatine on them. If three or four tablespoonfuls of common

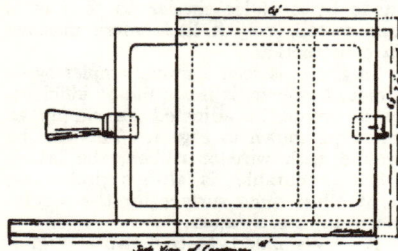

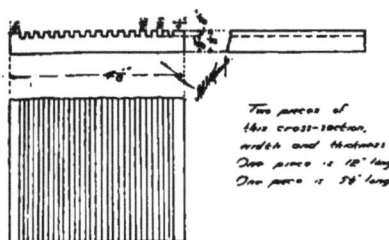

Two pieces of
this cross-section.
width and thickness
One piece is 12" long
One piece is 58" long

washing lye are put in a basin of hot water and the plates allowed to remain in it a few minutes, the film will either be dissolved or removed from the plate. Great care should be taken not to get any of this solution on the hands, as it is rather painful. After the plates are dry, they should be cut off so that they are a little less than $6\frac{1}{2}$ in. long. This is so that they will slide easily.

After the plates are cut to size, they should be covered on each side with tin-foil. A piece of tin-foil $5\frac{1}{2}$ in. square is stuck to the center of each side of each of the movable plates, and to both sides of all the stationary plates, except the two outside ones which are covered on only one side. White shellac is a very good thing to use to cement the tin-foil on with. They should be carefully smoothed out before being cemented on, and should be put on rather carefully.

After the case and plates are finished, the clips to hold the plates and connect the tin-foil sheets should be made. This will probably be one of the hardest things to do. Some rather heavy strip brass $\frac{3}{4}$ in. wide is cut up in $1\frac{3}{4}$ in. lengths and is bent into the shape shown in the drawings. After 15 of these clips are made they should be soldered to two pieces of heavier brass. The heavier pieces are $3\frac{1}{2}$ in. and $5\frac{1}{8}$ in. long, and $\frac{3}{4}$ in. wide and $\frac{1}{4}$ in. thick. It will be seen that the clips are spaced $\frac{1}{2}$ in. apart, and there are 7 of them on the movable plate and 8 on the stationary one. A flat head machine screw should be soldered to the movable clip to screw the rubber handle to. The stationary clip should have two holes bored through it, $\frac{1}{4}$ in. from the ends, so that a small wood screw may be used to fasten the clip and plates to the frame.

When assembling, it should be seen that each side of each clip is touching a sheet of tin-foil, or a sheet is wasted for each non-contact.

This condenser, if carefully made, will prove of great help in tuning, and will work as well as most professionally made ones.

THE NATURE AND PRODUCTION OF HERTZIAN WAVES

ERNEST C. CROCKER

Hertzian waves are the messengers which carry our wireless messages, just as are light waves the messengers which carry heliograph or bonfire-signalling messages. The relation of Hertzian waves to light waves is closer than even the comparison would indicate, for they are really one and the same thing. Both are vibrations of the ether; in the one case very rapid (light), and in the other case relatively slow. Light and Hertzian waves, being ether waves, are rather intangible, but they bear a very close resemblance to sound waves in the air, with which we are more familiar.

Whenever we have any body vibrating, and the rate of vibration is between the limits of 16 and 10,000 vibrations per second, there is produced in the air a vibration which effects our ears, and which we call sound. Although the air is set vibrating by even slower or faster vibrations than those of the nine octaves of audible sound, our ears are incapable of responding. Even in the range of nine octaves (doublings of the rate of vibration) there is a marked difference in the behavior of the sound waves; for instance, sounds of the highest octave which is audible cannot go around corners to any great extent, while low-pitch sounds penetrate and permeate almost everything. It is with surprise that one first notices that even a small obstacle, such as one's hand, interposed between an extremely high-pitched whistle and one's ear, will entirely shut out the sound, so prominent is the shadow effect.

In the case of light and Hertzian waves, we have a special atmosphere, the ether; but in this case we do not have vibrations of solid bodies, but vibrations of electricity, which acts in this case like a special kind of solid body and the only kind which can set the ether into vibration. In this case, as in the case of sound, the vibrations may have almost any frequency, but only certain frequencies are recognizable. If the vibrations are between the limits of 400 trillion and 750 trillion per second, there is produced "light," a vibration of such rapidity as to effect our eyes. If a little slower than that indicated, we can still detect the vibrations by the sense of touch and not with our eyes, but this time as heat. If instead of trillions, we have a frequency of a few hundred thousand or a few million vibrations per second, we have Hertzian waves. If the rate of vibration is slightly higher than that indicated for light, we have "ultra-violet," but if a little slower, "infra-red light." The "heat" waves commence where light waves leave off, and extend well down the scale towards Hertzian waves, but there is still a small gap which has yet to be bridged in order to have a steady progression of frequencies from those of Hertzian waves, through heat and light to ultra-violet light, as we have in sound between the lowest and the highest frequencies.

As with high-pitched sounds, high-frequency ether vibrations travel only in practically straight lines, producing the well-known light-shadow effect; but also, as with low-pitched sounds, the relatively low-frequency Hertzian waves are not so limited as to travel only in straight lines, but can bend around, and so have great penetrative power. It is this flexibility which makes it possible to use Hertzian waves for purposes of wireless telegraphy, for straight-line transmission of messages could never carry even 100 miles. Hertzian waves can bend around any large body and on this account alone is trans-oceanic or circum-terrestrial telegraphy possible.

According to the statements of physicists, light is produced by the vibration of electrons (little pieces of negative electricity) within a molecule (small particle of matter). The space through which they vibrate (less than one millionth of an inch) is so small and the particles themselves so small, that it need not surprise us that they vibrate at the rate of 500 trillion vibrations per second,

a quantity which is here expressed in figures to give an idea of its immensity; 500,000,000,000,000 (or 5×10^{14}).

Since light is caused by the vibration of small particles of electricity in a very limited space, it follows that if we are to produce the similar, but slower-vibrating Hertzian waves, we must cause movement of a large body of electricity in a relatively large space. A vibration of a large mass of electricity to and fro is an alternating electric current, a thing with which we are already familiar.

Since Hertzian waves are always produced whenever we have an electric current alternating at the right frequency, we can now best study the production of Hertzian waves by studying the production of their parent, the alternating current. For wireless telegraphy we need Hertzian waves vibrating at the rate of from 70,000 to 1,200,000 times per second, hence we must use a current vibrating at these high frequencies.

Dynamos can be and have been constructed which give considerable outputs of electrical energy at the proper frequency, but as a general producer of alternating currents for Hertzian waves, the dynamo is practically out of the field, since it is so expensive and troublesome and since we have better methods of producing alternating currents of the required high frequencies.

It was found many years ago that if a Leyden jar was charged to a high potential and then suddenly discharged, that the discharge was alternating, and afterwards, it was found that the frequency of the alternations could be controlled and that the current itself could be used. A great many systems have been produced which utilize condenser discharges for the production of rapidly alternating currents for wireless telegraphy, but they all reduced to two types: the simple antenna circuit (no tuning apparatus) and the various methods of inductive coupling of condenser circuit to the antenna circuit. In both of these types we have sub-divisions as to whether the "spark gap" is long or short or is an arc, and also sub-division as to how the condenser is charged. Although in some cases the difference between two systems may appear to be slight, there may be a very fundamental difference in the character of the Hertzian waves produced.

To understand why a condenser discharge is alternating, we must bear in mind that electricity has a property analogous to "inertia" in matter, and that once electricity is set in motion it must continue in motion until the energy which was imparted to it in setting it in motion is all absorbed. When we lift a billiard ball to a considerable height we store up a considerable amount of energy in the ball. When now we release the ball and allow it to fall, it tries to give back the energy which we put into it by heating the air through which it falls, and in producing heat by impact when it hits anything. If we allow it to fall on a hard surface, but not from a height great enough to cause it to break, there will be very little energy used up in the air, and on account of the almost perfect elasticity of ivory very little energy will be used up by the impact, and the ball will bounce up again to a height nearly as great as that from which it fell, and this bouncing will continue for some time, until the energy of the ball is all used up. Similarly, when we charge the two plates of a condenser to a considerable potential and then allow the whole charge to fall from the plate of high potential to that of low potential. There will be a bouncing back and forth for some time, just as with the billiard ball, for a considerable amount of electrical energy is to be absorbed, and on account of the small absorption of energy by the condenser (due to the imperfect dielectric, the resistance of the metal plates and spark-gap, and to whatever current is radiated, leaks away, or is led off and used) the energy is usually not used up until after many swings. If we could have a perfect condenser and had a discharge circuit of no resistance or radiation, the oscillating or alternating discharge of a condenser when once started would keep up forever.

In what has gone before, we have considered the relation of Hertzian waves to things like light and sound with which we are already familiar. The waves themselves are disturbances in the ether as are light waves, and radiate in all directions as do both light and sound waves, but beside these properties, they possess the additional property, shared only by low-pitched sound waves, of being able, to a considerable degree, to deviate from the straight line in order to penetrate into an undisturbed or "shaded" region. It was shown that Hertzian waves are identical, in the essence, with light and heat waves and it may be further stated that if, instead of the customary few hundred thousand alternations per second employed in a wireless antenna we had 500 trillion, would have emitted instead of Hertzian waves a stream of pure light waves, free from other kinds of radiations.

We considered how Hertzian waves were produced whenever we had rapidly alternating currents. In this case, of course, the electricity must vibrate in a wire, but such is not the case with light or heat waves which may be radiated by either an electrical insulator or conductor, for the electrons producing the disturbance or vibration which we call light can always move within the molecule itself. We also considered how it was possible to produce rapidly alternating currents by condenser discharges, and at another time we shall take up the subject of the different methods of utilizing the alternating currents which the condenser discharge produces.

GUIDED RADIO-TELEGRAPHY

Experiments with Major Squier's system of employing telephone and telegraph wires to guide Hertzian waves were recently made between the Bureau of Standards, in Washington, and the New York office of the Postal Telegraph Cable Company. The object of these experiments was to determine the best frequencies for long distances. It was found that frequencies of from 10,000 to 25,000 cycles per second would serve best for long distance transmission, while frequencies ranging from 25,000 to 100,000 cycles would be used on shorter circuits. Various methods of connecting the transmitters and receivers to the line were tested. Inductive coupling was found to give the best results, and was more flexible, for it permitted the operation of a number of lines from a single generator. It was shown conclusively that guided radio-telegraphy was far more economical in the consumption of power than free radio-telegraphy.

WIRELESS TELEGRAPH RECEIVING SET

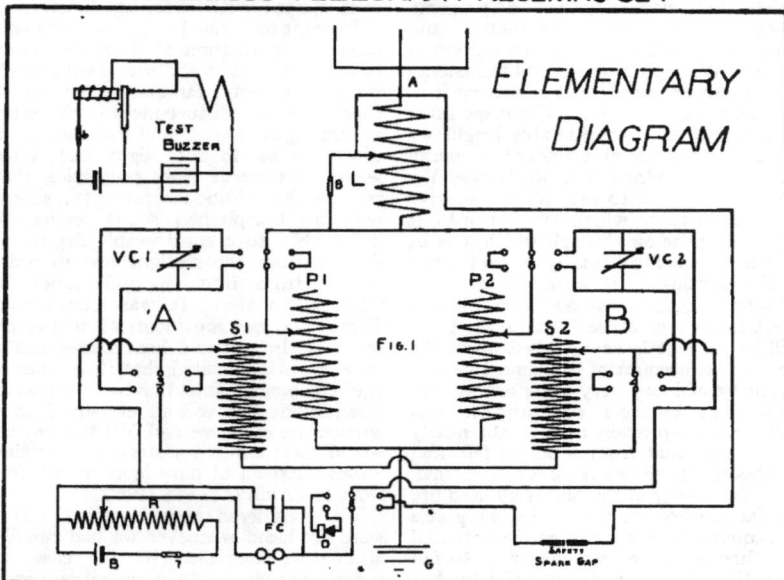

ELEMENTARY DIAGRAM

Fig. 1

LOCATION OF APPARATUS

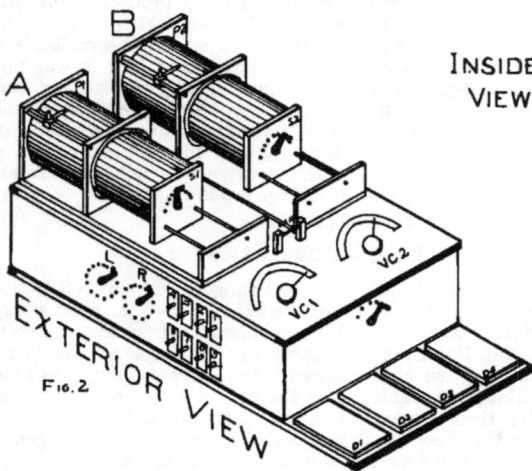

EXTERIOR VIEW

Fig. 2

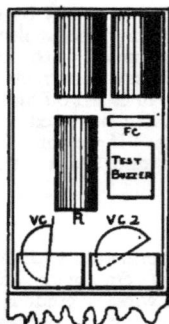

INSIDE VIEW

Fig. 3

A WIRELESS TELEGRAPH RECEIVING SET

H. P. RICHMOND

In the designing of a wireless telegraph receiving set the experimenter occupies a rather peculiar position. On one hand he is beset with the necessity of having his set compact and as serviceable as possible, while on the other hand it must be of such a nature as to permit of considerable experimenting. It must be selective, yet it must not be complicated. To meet these difficulties the following set was designed. The actual dimensions have, for the most part, been omitted, because the experimenter usually desires to try for himself, and to make the set suit his own needs.

The theorist demands a wide range of variable inductances and capacities, while the practical operator turns from such a set to one containing a single tuner and condenser. In the following diagram both demands have been conceded to. Fig. 1 shows the electrical connections. The set consists of two receiving transformers, two variable condensers, four D.P., D.T. telephone key-switches, and the usual apparatus for the detector circuit. By means of switches 1, 2, 3 and 4, it will at once be seen that the circuit may be varied from one containing but a single loose coupler and no variable condensers to the more complicated Fessenden Interference Preventer. A few illustrative examples will best show the merits of this set. For the above-mentioned switches the "down" position has been chosen for the one in which a variable condenser is introduced into the circuit, while the "up" position simply places a loop across the circuit. To obtain the Fessenden Interference Preventer it is necessary to have two loose couplers with their primaries in parallel and their secondaries in series; in addition to this, two variable condensers must also be placed in the primary circuit. A glance at the diagram will show

that in order to accomplish this result switches 1 and 2 should be "down," while 3 and 4 remain "neutral."

Perhaps the most commonly used interference preventer is the IP-76 set, used extensively in the U.S. Navy. Here the oscillation circuit contains no other apparatus than the primary of a loose coupler, while in the secondary circuit the secondary winding is bridged by a variable condenser. In this set we have two chances to accomplish the results of the IP-76 system. Referring again to Fig. 1 it will be seen that in order to use the A set, switch 1 must be "up," while in order to cut out P2 of the B coil, switch 2 must be left at neutral. Since across SI, VCI is needed, switch 3 must be "down," and in order to cut out S2, switch 4 must be "up." Using the B set the reverse is true. The advantage of this combination set is that when listening-in the less selective systems may be used, while when working with stations with which much business is done the more selective systems may be used; the actual changing from one to the other taking on an average three seconds, once the operator is familiar with the set. For convenience a short table, such as has been started below, may be made which will greatly help the operator in using the switches, although once familiar with the set it has been found that the switches can be shifted in less time than it would take to find the positions from the table. Such a table, however, will be of great help to the operator who is using the set for the first time. The following shows a convenient method of arranging the table; the names used for the systems being those by which they are most commonly known.

The various pieces of apparatus are so well known that they do not need any detailed description. It is well to re-

TABLE

System	Switch 1	Switch 2	Switch 3	Switch 4
Fessenden I.P.	Down	Down	Neutral	Neutral
Pickard "A"	Up	Neutral	Down	Up
Pickard "B"	Neutral	Up	Up	Down
Etc.	Etc.	Etc.	Etc.	Etc.

member, however, that with a loose coupler the distance between the windings is an important factor. Since there are two loose couplers in this set it is well to make one with the least possible distance between the windings and the other with ¼ to ⅜ in. clearance. The former will be the more efficient, but less selective than the latter. Their primaries should contain about 900 microhenrys of inductance, while their secondaries should contain from 3 to 5 millihenrys, depending on the wave length desired to be obtained. The variable loading coil L should contain about 3,000 microhenrys of inductance. Although it is possible to get along with ones of smaller capacity, it is best to have the variable condensers of a maximum capacity of .003 microfarads. The remaining apparatus is all of the standard type, so no mention of it need be made.

Fig. 2 shows the exterior view of the set when completed. This arrangement need not be followed, but it is a very convenient one. The switch numbers are the same as were used in Fig. 1; 1, 2, 3 and 4 being for the control of the various systems, while 5, 6, 7 and 8 are the usual ones connected with any set. No form of detector has been suggested because every operator has his own preference. The detectors are switched in by means of the switch located directly above them. Fig. 3 shows a very convenient inside arrangement, leaving plenty of room for the connections from the potentiometer R and the loading coil L to their respective switches.

A few over-all dimensions may be of some assistance to the designer of a similar set, so it may be mentioned that a cabinet with a base 27½ in. x 14 in. forms a very suitable case. An inside depth of 5½ in. is allowed, while a space of 6¼ in., extending across the front is reserved for the detectors. With a little thought on the part of the operator as to the purpose of the different pieces of apparatus, the actual operation of the set will be found to be much simpler than it first appears.

Mr. B. ANDERSON'S WIRELESS STATION

Mr. B. Anderson of Kansas City, Kan., has recently sent us some interesting photos of his wireless station and his description of the station is as follows:

Transmitting Apparatus:—½ k.w. transformer with controlling re-actance coils; helix for tuning oscillatory circuit with aerial circuit; zinc spark gap; special

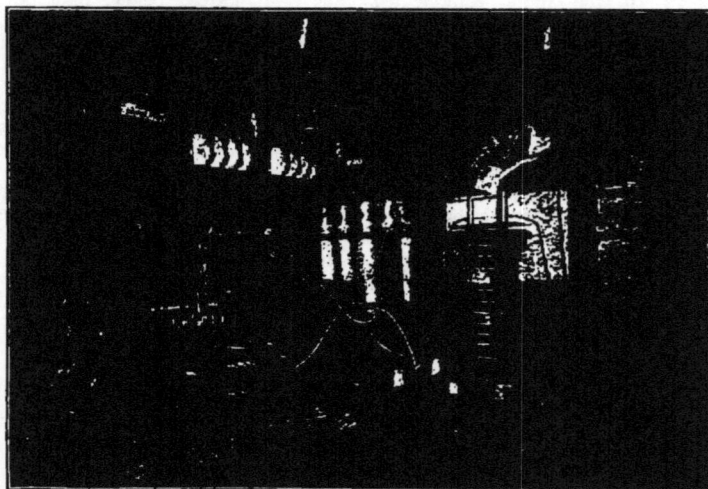

Transmitting Apparatus

Receiving Apparatus

type key of my own design and other accessories, such as antenna switch, switchboard for power circuits, etc. Receiving Apparatus.—Two loose-coupled tuners, silicon detector, two different types of variable condensers, and 2,000 ohm telephones. The antenna, at time photograph was taken, consisted of wires arranged for directive radiation. This has later been changed to a "lateral inverted V type," also arranged for a directive effect. Height of mast 75 ft., built of 4 x 4 in. yellow pine. This mast was put up in a most economical way and no engineering laws were put in practice in the erection of same. In fact an interested spectator may feel justified in making remarks to the effect that it would not endure a "mild wind-storm," such as are prevalent in this part of the country, but it has already breasted several "miniature" tornadoes with wind velocities of 80 miles an hour. Before

I close the description, I may mention the fact that the mast is guyed by No. 12 B.&S. gauge galvanized iron wire, 3 sets of 4 wires each. This mast has been in service about four years; and I am now beginning to get worried as to how to take it down, owing to the fact that the neighborhood has increased its population considerably since the mast was erected. Have had some remarkable results with this equipment, which is nearly all home-made, and Colon, Panama, New Orleans, Chicago, Milwaukee and Leedington are only a few of the long distance stations heard. My transmitting distance with the aerial and transformer as described has been over 75 miles, while at one time a transmitting distance of 160 miles was covered in day-time with a more efficient aerial and mast made by my co-worker, Mr. Arthur G. Kepkinger, to whom I am indebted largely for the success attained.

A NEW USE FOR WIRELESS

The idea of using a wireless telegraph station for the purpose of aging cognac or clarifying champagne is, at first thought, fantastic, but it is being tried with success. Many years ago chemists conceived the idea of utilizing the action of electric currents of high frequency in the per-

fumery industry, producing a kind of electrolysis, which, in a way as yet unexplained, unites and compounds the diverse essences that enter into the composition of a scent. This phenomenon inspired some experiments undertaken in France.—Youngstown, O., *Telegram*.

ANOTHER GALENA DETECTOR
H. L. THOMAS

The detector of the "up-to-date" wireless experimenter is galena.

There are several forms of this detector, such as the copper point, and in combination with zincite. There is, however, a better form than either of these. This is the form using a tungsten wire for the contact point. A piece of this wire (of the size used in 25 watt lamps) must be obtained. It cannot be taken from an old lamp, however, as the wire is brittle after it has been heated. The wire should be curled so as to form an open spiral spring, the end being bent perpendicular to the mineral. It may be soldered to the adjusting apparatus, but an easier way is to drill a small hole and fasten it with the alloy given at the end of this article.

The contact must be very light. This is very easy to obtain with this extremely fine wire.

As a general rule, the detectors using a light contact are easily "knocked out"

of adjustment, but the author has seen the following test applied:

A detector of this type was securely fastened to the operating table without felt or springs. A station was tuned in, and an assistant pounded the table with his fist, but produced no effect on the detector.

Very few of the crystals are sensitive. The best way is to obtain several ounces of the material, break it into suitable sizes (no smaller than $\frac{1}{4}$ x $\frac{1}{4}$ in.) and test each piece. When a good crystal is obtained, this detector will be found considerably more sensitive than any other form known at present.

The following alloy will melt at 176 degrees Fahrenheit, or 36 degrees below the boiling point of water. It may be used to fasten the galena into a brass cup:

Bismuth.................8 parts
Lead.....................5 parts
Tin.....................3 parts
All parts by weight.

STANDARD THREADS
W. W. BRIDGE

While reading articles showing how to build machines, electrical appliances, etc., one will become confused as to what sizes to use, and what thread for a given wire. I take the liberty of stating the standards which have been adopted by the American Screw Co., one of the largest makers of machine screws, and one of the leading makers of cap screws and set screws; also the makers of taps and dies.

MACHINE SCREW SIZES

Wire Threads: 2 x 56; 4 x 36; 6 x 32; 8 x 32; 10 x 24; 12 x 24; 14 x 20; 16 x18; 18 x 16; 20 x 16.

Avoid the use of other threads as much as possible; also the use of sizes such as $\frac{1}{4}$ x 20 and $\frac{1}{2}$ x 13, which are cap and set screw sizes. These sizes are made in machine screws, but one will find that they are very hard to obtain in any hardware stock. The use of fine threads cannot be avoided as adjusting screws where very close results are required, but in other places it would be better to use only standard sizes.

CAP AND SET SCREW SIZES

Wire Threads: $\frac{1}{4}$ x 20; $\frac{5}{16}$ x 18; $\frac{3}{8}$ x 16; $\frac{7}{16}$ x 14; $\frac{1}{2}$ x 13; $\frac{9}{16}$ x 12; $\frac{5}{8}$ x 11; $\frac{3}{4}$ x 10; $\frac{7}{8}$ x 9; 1 x 8.

The use of $\frac{7}{16}$ x 24 should be avoided, as this size is not carried at all hardware stores. At one time $\frac{1}{2}$ x 12 was standard but this has been changed to $\frac{1}{2}$ x 13.

The size drills to be used in connection with taps can be found in most any catalog of drills or taps and dies. When drilling in steel for a tap to follow, use a drill from one to five thousandths larger than given in lists. This will save a great many broken taps.

Correction

On page 262 of our April issue we published an article entitled "A Square Cut Chair." We regret to state that inadvertently the credit for this article was given to "Hobbies" when it should have been given to our esteemed contemporary "Popular Mechanics."

WIRELESS NEWS

The Alexander Bill

March 30, 1912.

To the Honorable House Committee on Merchant Marine and Fisheries:

Washington, D.C.

Gentlemen:—I beg to protest against the adoption of the Alexander Bill for the regulation of wireless telegraphy. I am hardly a schoolboy amateur, for I began telegraphing as a business in 1894, in railroad work, and, though not now so employed, I have not lost interest in the art: I find wireless an entertaining and instructive pastime, and I know this bill in its present form will be a serious detriment to wireless development. There is much to be learned yet before the science is perfected, and improvements in the past have generally come, not from paid or professional operators, but from experimenting amateurs. This bill gives power to suppress every such station to a government department which is confessedly antagonistic, and is a much more drastic measure than is necessary for the conduct of wireless business. That some government regulation might be beneficial I admit; but a general law that would make wilful interference with an operator, or the refusal to keep apparatus quiet for a reasonable time, when requested by any operator actually engaged in working with a distant station an offense punishable as are other misdemeanors (something like the laws relating to the interference with other government employees, such as revenue officers, mail carriers, etc.), is all that is necessary; no restrictions are necessary regarding a station for receiving only, for such a station can in no way whatever interfere with any other. If the divulgence of private messages should prove an annoyance, a clause could be included making the offender liable to the injured party, or subject to imprisonment if civil action would not hold.

No one could or would object to absolute government control of all wireless apparatus in war time or other emergency; but the possibilities and convenience of wireless communication are too great to be hampered or throttled almost at the beginning by unnecessary seizure and absolute control by the government. A law covering the matter of wilful interference with business, applicable to all operators alike, with a clause protecting private messages, would do more to expedite the handling of wireless business of all classes than any regulations regarding wave-lengths or specified hours for each class of business could do, and it would do it without depriving some eighty thousand of our citizens and embryo citizens of a most fascinating and educational recreation, to say nothing of the value of private installations for communication in a business way.

The eighty thousand amateur and experimental stations now in operation in this country represent an investment in the line of over a million dollars.

I notice in the newspapers and the electrical magazines, that in support of this bill the Navy Department lays particular stress on alleged interference with messages of distress from the torpedo boat *Terry*, when it was disabled off the coast of Virginia on January 7th last; my station is but two or three miles distant from New York City Hall, in a district that is probably as busy, wirelessly, as any in the country: in justice to the amateurs of this vicinity, and in reply to whatever statement on that incident may have been submitted to you regarding the matter, I respectfully submit a copy of my record of the occurrence; I keep a careful daily record of what I hear in the time other work allows me to give to wireless, and I would ask that you compare these notes, and in particular the *time* given for each remark, with any that have been submitted to you. When I copied this out of the air, I had no expectation that it might be used in this way, consequently there is no "padding" in the matter at any place, and I am entirely willing to submit it in the form of an affidavit as a "true and correct copy of what passed between the various wireless stations around New York, on the date and at the times mentioned," if you would prefer it that way. I have approximately 300 pages of this kind of matter, taken in the past two and one-half years, and in many places it shows that some of the worst of the interference trouble came from some of those now most in favor of passing this bill.

Please note on the attached, that the first notification of the trouble of the *Terry* was given at 6.23 p.m., January 7th, *by an amateur station in Bayonne, N.I., to a New York commercial office,* who repeated it to the Brooklyn Navy Yard.

Thanking you for your attention, and awaiting your further wishes, if I can be of any assistance to you,

Very truly yours,
CHAS. E. PEARCE,
748 Albert St., Steinway, N.Y.

Copy of Daily Record of Wireless Messages Received at Steinway, Long Island, by Chas. E. Pearce.

(In this copy, conversation is here written out fully, instead of the telegraphic abbreviations used by operators).

January 7th 1912.

"Sat in" (began listening) at 6.15 p.m.

6.20 p.m.—JD (steamship *Northland*, Maine S.S. Co.) calling NY. (NY is the New York office of the United Wireless Co.)

6.22 p.m.—NY answers, and JD reports passing Hell Gate at 6.15 p.m.

6.23 p.m.—WD (an amateur station in Bayonne, N.J.) calling NY; NY answers, and WD says:

"Say, did you get that SOS from RNS?—WD." (SOS is the call for help signal, and RNS is the steamship *Tagus*, Royal Mail Steam Packet line).

NY says: "Yes—can you get the rest of it?—NY."

"Am trying to.—WD" says WD.

6.26 p.m.—NY to QY (QY a New York amateur): "Keep out—there's a boat in trouble and I can't hear a thing through that clatter of yours.—NY."

QY immediately *stopped working*, and I did not hear him again during the evening.

NY calling RNS, and I heard him answer, though my outfit is too small to get what he said, but NY asked:—

6.32 p.m.—"RNS: What's your trouble?—NY."

I could hear RNS working, but too faint to read.

6.36 p.m.—NY calls NAH (NAH is the Brooklyn Navy Yard.)

NAH answers at 6.40 p.m., and NY says:

6.40 p.m.—"Say, the RNS reports NUI in trouble about 500 miles east of the Virginia capes. Says the Revenue Cutter *Onondaga* has word. Latitude 38.21 north; longitude 67 west.—NY."

NAH to NY—"O.K. Thanks.—NAH."

NY to NAH: "Say, who is the NUI?—NY."

Ans.—"The torpedo boat destroyer *Terry*.—NAH."

Nothing more was said of the matter, except NY handled a few commercial messages, working with the *Tagus*, and other ships for some time. NAH called the *Tagus* and *Norfolk* a few times, but evidently was unable to hear them if they replied. Was working with the *Tagus* to the extent of telling him he was too far away, he couldn't get his signals.

7.10 p.m.—RCM (Revenue Cutter *Mohawk*) called NY, and gave this message when NY answered him:

RCM to NY: "Noon. NUI lat. 38.21 north, long. 67 west. Turbines, engines and pumps out of commission. All stores ruined and wireless not working.—RCM."

Now, if the wireless on NUI was not working at noon (as reported by the RCM), how could the Brooklyn Navy Yard expect to work with the disabled vessel in the evening?

7.12 p.m.—JD calling WN (Wilson's Point, Conn.), and sends a message.

All big stations silent until 8.10 p.m. when WD calls NY, and gives the following ship position reports:

8.10 p.m.—WD to NY:

"8.00 p.m.—FH *Saxannah* 324 miles south Sandy Hook."

"8.00 p.m.—YM *Yucatan* 806 miles south Sandy Hook."

"8.00 p.m.—CB *Carolina* 340 miles south Sandy Hook."

Nothing more was said about the disabled boat during the evening, as long as I was listening, probably about 10.30.

January 8th.

6.20 p.m.—DR (a large New York amateur station) calls NAH, and says he just heard the following message:

6.20 p.m.—DR to NAH: "from Norfolk Naval Station to NRZ."—"(NRZ is the U.S.S. *Salem*): *Terry* expected to reach Hampton Roads tonight. Continue search for others. Missing destroyers Department ignorant of whereabouts Drayton, McCall, Mayrant, Patterson, and Burroughs."

Winthrop,
Acting.

CANADIAN CENTRAL WIRELESS CLUB
Winnipeg, Manitoba, Canada

This Club has a number of men of all ages, from the boy in his teens to the middle-aged business man. Meets every two weeks at the home of a member. Most of the members have their own instruments, some of which have been bought, and others made by themselves, but all persons interested in wireless telegraphy or telephony receive a cordial welcome to the meetings, whether they have instruments or not. The president of the Club, Alex. Polson, 94 Cathedral Ave., has an aerial outfit. So also has A. St. Louis, 819 McMilan Ave.; S. Scorer, B. Lazarus (secretary), E. Duma, A. Scott, Reginald Davis, Fred Golmer and Bertram Hill, others have their indoor receiving and sending outfits.

OREGON STATE WIRELESS ASSOCIATION
6th and Taylor Sts., Portland, Ore.

The officers are Chas. L. Austin, president; J. R. Kelley, secretary; Ed. Murray, sergeant-at-arms; C. L. Bischoff, treasurer and corresponding secretary, Box No. 73, Lents, Ore.

The corresponding secretary's call letter is MV, 1 k.w., rotary gap, 200 miles radius. Murdock loose coupler, single slide tuner as loading coil; also have double slide tuner on different circuit. Either straight or loop-aerial as desired. Sixty feet high at one end, 45 at the other, 350 ft. long. Get pretty near everything on the Pacific Coast from Sitka, Alaska to San Francisco, also ships at sea.

Mr. Joyce Kelley and Mr. Charles Austin's call letters are RPC and C2 respectively. These two stations were among the first installed; they are the highest, each having two poles, 150 ft. to 125 ft. in height. The two owners are first-class operators, both having worked on ships, and understand the art perhaps better than any amateur in the United States and better than over half of the licensed operators. Mr. George Swartz also has a high-class station, call letter GE. Each of the above own over a 2 k.w. transformer. Mr. Kelley has a small set which he uses around town and very seldom uses his large set except when handling business or sending a considerable distance. The two first-mentioned stations are better than a good many commercial stations that I have seen.

Yours truly,
CLARENCE BISCHOFF,
Corresponding Secretary.

QUESTIONS AND ANSWERS

Questions on electrical and mechanical subjects of general interest will be answered, as far as possible, in this department, free of charge. The writer must give his name and address, and the answer will be published under his initials and town; but, if he so requests, anything which may identify him will be withheld. Questions must be written only on one side of the sheet, on a sheet of paper separate from all other contents of the letter, and only three questions may be sent at one time. No attention will be given to questions which do not follow these rules.

Owing to the large number of questions received, it is rarely that a reply can be given in the first issue after receipt. Questions for which a speedy reply is desired will be answered by mail if fifty cents is enclosed. This amount is not to be considered as payment for reply, but is simply to cover clerical expenses, postage and cost of letter writing. As the time required to get a question satisfactorily answered varies, we cannot guarantee to answer within a definite time.

If a question entails an inordinate amount of research or calculation, a special charge of one dollar or more will be made, depending on the amount of labor required. Readers will, in every case, be notified if such a charge must be made, and the work will not be done unless desired and paid for.

1757. Electroplating. Spark Coil. O. M. P., Yoakum, Tex., asks: (1) for solution for electroplating with (a) copper, (b) nickel. (2) Will the use of iron in the construction of a detector affect its sensitiveness or interfere with its action in any manner? (3) I am building a spark coil for a 2 in. spark with the dimensions given in June, 1910, issue of *Popular Electricity*: Core ⅞ x 7 in.; primary 184 turns No. 16 d.c.c. wire; secondary, eight sections, using 1 lb. No. 36 enameled wire; condenser 1,400 sq. in. tin-foil; and I want to know if I can fix it so that I can use one-half the secondary, one-half the primary and part of the condenser for a 1 in. spark or both parts at once for two 1 in. sparks, and if so, how to connect it. Ans.—(1)(a) For the preliminary bath use a double cyanide of potassium and copper solution, the method of making this solution being given in answer to question 1745 published in the April issue of this magazine. For the body of the deposit use as an electrolyte a solution of cupric sulphate in very dilute sulphuric acid. (b) A good solution for nickel-plating is as follows: nickel and ammonium sulphate, 10 parts; boracic acid, 4 parts; distilled water, 175 parts. Use a sheet of nickel for your anode. Be very sure that the surface which is to be coated is absolutely clean as even finger marks will render the deposit liable to peel off. If the metal which is to be plated is iron, nickel or zinc, it is difficult to make metallic deposits upon them adhere. To overcome this difficulty, first give them a coating of copper in a solution made as follows: potassium cyanide, 2 parts; copper acetate in crystals, 2 parts; sodium carbonate in crystals, 2 parts; sodium bisulphate, 2 parts; water, 100 parts; moisten the copper acetate with a small quantity of water and add the sodium carbonate dissolved in 20 parts of water. When reaction is complete add sodium bisulphite dissolved in 20 parts of water. Lastly add potassium cyanide dissolved in remainder of water. (2) This cannot be answered by a simple statement. Iron is essential in magnetic detectors, and sometimes useful or necessary in coherers and "carbon-steel" detectors. In the perikon detector, there is iron in the zincite and in the chalcopyrite. In the silicon detector, iron may be used, instead of brass, for the contact point. In an electrolytic detector, iron is as a rule harmful, as the platinum is more readily ruined when iron is present. As a rule it cannot be said that the presence of iron has anything to do with the sensitiveness of a detector, for the currents flowing in a detector are too small to produce perceptible magnetic effects. (3) It is not desirable to make primary or condenser any smaller. Simply use less secondary. Less than one-half the amount will give one-half the spark.

1758. Wireless without an Aerial. W. R., New York, asks: If there is any wireless system which can successfully transmit and receive messages without using an aerial wire. A system which only uses the ground connection for sending and receiving. Ans.—There are several experimenters in Germany who are now working out a practical wireless system which does not make use of the ordinary antenna, either for sending or receiving. It is claimed that the "ground circuit" method (using Hertzian waves) is very successful, although, strange as it may seem, there is considerable trouble from "atmospheric" electricity, especially when very long waves are used. Mr. Marconi is also interested in carrying out experiments on the Sahara Desert in Africa, where the soil is sufficiently insulating so that the two horizontal wires may be laid directly on the ground. It seems probable that in a few years some of the earth current methods will be put into actual use.

1759. Wireless Aerials. J. C. W., Topeka, Kan., asks: (1) Will the aerial shown in diagram give good results with connections shown? (2) Is the insulation of the aerial shown in diagram sufficient for a ¼ k.w. transformer? (3) If a loop aerial is used with the set described in the August number of *Electrician and Mechanic*, may the other aerial wire be run through a single slide tuner to the ground? Ans.—(1) No, it is not effectively connected. It would be much better to connect the lead-in wires at the place where the sloping part joins the horizontal part. The horizontal part, in the place you now have it, adds very little to what the sloping part alone will do. (2) Yes. (3) Yes. There is only a doubtful advantage in using a loop aerial, however.

1760. Teleautograph. H. M. A., Cambridge, Mass., asks: (1) about the teleautograph mentioned in the February number, on page 136, under the heading of "New Wireless Apparatus." (2) The formula for finding wave length for wireless. Ans.—(1) The full details of this teleautograph are not yet available. It may be months before the information is given to the public. (2) The wave-length of a wireless circuit depends upon two quantities: the inductance and the capacity of the circuit. When the capacity C is expressed in microfarads (the customary unit) and the inductance L is expressed (in the C.G.S. units) in "centimeters," wave-length (in meters)=60 $\sqrt{CL}$. If the capacity is in microfarads, and the inductance in millhenrys, wave-length (in meters)=60,000 $\sqrt{CL}$.

1761. Detector Minerals. R. D. M., New Orleans, asks: (1) The address of the firm that sells the best silicon for long distance work. (2) The address of the firm that makes and sells the best minerals for a perikon detector for long distance work. (3) What is the best material for the point to rest on the silicon, in a silicon detector for long-distance and local work? Ans. —(1) It is difficult to state what concerns sell the best silicon for wireless work. The Wireless Specialty Company, which controls the detector patents, presumably has the best silicon obtainable. (2) Wireless Specialty Company, 81 New St., New York, N.Y. (3) Brass or phosphor-bronze is the standard material.

1762. Condenser. T. S. V., Los Angeles, Cal., asks: (1) What will be the necessary capacity of a condenser to be used on an induction coil of this size: core 9½ x 1 in., two layers No. 16 d.c.c. wire and 4½ lbs. of No. 34 enameled wire wound in eleven sections? (2) Is there any difference in thickness of tin-foil used in this condenser? (3) What would be the voltage for the above coil in volts and amperes when using the coil with a Machenzie-Davidson type interrupter described in your April, 1911, magazine? (4) If I use 110 volts A.C. to 110 volts D.C., can I get the necessary amount of electricity for the above coil? Ans.—(1) From 1 to 1½ mf. capacity. The exact amount is not of particular importance. (2) The thickness of the tin-foil used in condensers is of no importance so long as it is great enough to allow a good connection to be made to the plates. (3) The above coil should work at its best on from 8–12 volts, taking, perhaps 3 to 5 amperes. (4) Such a coil is not suitable for use on 110 volt circuits, either A.C. or D.C.

1763. Tesla Transformer. L. R. C., Beverly, Mass., asks: (1) For data on oil-insulated Tesla transformer to give 12 in. sparks or better, when used in connection with a ¼ k.w. transformer. (If copies of the *Electrician and Mechanic* are referred to, please state whether obtainable or not). (2) If glass is a satisfactory insulator between the primary and secondary of a spark coil giving a very heavy spark, and if lead glass is less efficient than regular flint or crown glass. (3) Whether or not a thoroughly efficient station should have sharp points, or corners on the sending end. I refer to such things as a spark gap with a screw adjustment, sharp-edged binding posts, etc. Ans.—(1) We believe that the *Scientific American Supplement* contained in a recent number, if we recall correctly, was published some time in January or February, gives the information you desire. (2) Yes. It is more usual, however, to use ebonite, paraffined paper or "empire cloth." Lead glass contains lead salts which are insulating and which do not prevent good insulation. The only glass to be used with caution is manganese glass—this usually has a pink or violet color. All bluish, greenish or crystal-white glass is usually safe. Very often, flint glass contains lead. (3) Sharp points and edges are only harmful where there is a very high electrical potential. As a rule, there will be no perceptible loss from points

except near the top of the antenna. At the spark-gaps, etc., the loss is entirely negligible.

1764. Wireless. N. A., Kenton, Ohio, says: (1) When constructing an aerial 100 ft. long, 4 aluminum wires, how should the lead in wire be connected to the aerial and at what place should it be located on the aerial? What size of wire is best, and should it be bare or insulated? (2) With the following instruments, 1,000 ohm receiver, Wm. Murdock $3.00 variable condenser, fixed condenser, silicon detector (ferron type), Dawson & Winger double-slide tuner, and aerial 100 ft. long, four aluminum wires, mast 84 ft. high, how far had I ought to receive? (3) How far had I ought to send with ¼ k.w. transformer helix, zinc spark gap, four 2 quart Leyden jars and same aerial? Ans.—(1) As a rule, the station is located at the bottom of the aerial "grid" of wires and connection is made at this point. Use good rubber-crossed wire No. 14 or No. 16 B.&S. gauge, and where the joint is made with the aluminum part, be careful to solder well with "aluminum solder" (obtainable in most auto supply stores) and wrap well with rubber tape to keep moisture away from the joint. (2) It is very difficult to state your receiving distance, so much depends upon the strength of the sending stations. You should be able at night to hear a 2 k.w. station 100 miles away. (3) Probably 25 miles.

1765. Coupling. J. A. L., Maynard, N.D., asks: (1) Is the demand very great for automobile chauffeurs and repairmen? Is a competent one (one who is a graduate of a good auto school) sure of steady employment? (2) Is the demand very great for wireless operators? (3) What is the meaning of "coupling" as used in wireless? Ans.—(1) There is always a good demand for thoroughly-trained chauffeurs. The supply seems to be equal to the demand. (2) Not at present. (3) By "coupling" in wireless, we mean the "closeness" of connection between two circuits, the facility with which one circuit can unload its electrical energy into another circuit.

1766. Fleming Oscillation Valve Detector. X. D., Baltimore, Md., says: (1) I have a Fleming oscillation valve detector and the following directions to use same: (a) Connect 4 to 6 dry cells in series to the lamp filament; (b) Connect one side of a pair of 2,000 or 3,000 ohm phones to the second negative from the negative end of the battery of dry cells; (c) Connect the other side of the phones to one terminal of the secondary of the inductive tuning coil; (d) Connect the other terminal of the secondary of the

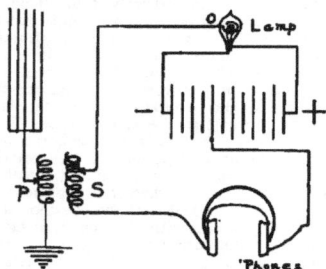

inductive tuner to the cold plate of the valve through the coiled wire on the lamp. I do not grasp the meaning of this, will you explain more fully the operation of this detector? (2) I enclose a hook-up of my receiving set, which is self-explanatory; can you tell me if I can use the valve detector in this circuit, and how, with the best results? (3) Can you tell me of any other circuit in which I can use this detector, using double-slide tuner of 6,000 meters, 2,000 ohm phones, fixed condenser and aerial of about 800 meters and necessary switches, etc.? Ans.—(1) The circuit which is meant by the description is here given; it will be seen that one side of the phones is connected to the "second negative from the negative end of the battery of dry cells," and that the other side of the phones is connected to "one terminal of the secondary of the inductive tuning coil S." The other terminal is connected to the cold plate of the valve through the coiled wire O on the lamp. (2) The most effective way of using the oscillation valve is that shown in the diagram with the addition of a variable condenser across S. It is a very good scheme to connect a potentiometer, of about 1,000 ohms resistance, across the battery and connect the wire from the phones to the sliding contact of the potentiometer. (3) The valve, in series with the battery and phones, may be used with any tuning circuit in the position which is usually occupied by the detector in series with its fixed condenser.

1767. **Wireless Telephone.** R. H., Springvale, Me., asks: (1) If the accompanying circuit is practical for a wireless telephone transmitting outfit. (2) If not, please give diagram which should be used with same instruments. (3) Should the sound of the arc be heard as far as the sound of the voice and how far should sound of voice be heard with said instruments? Ans.— (1) No. The arc for wireless telephony must be supplied with a perfectly steady direct current and not with a rectified alternating current. The telephone transmitter is placed in the aerial circuit and not in series with the arc. Furthermore, the arc cannot deliver energy to the antenna in an efficient manner unless one uses an oscillation transformer and tunes the circuit carefully. (2) Much difficulty will be experienced by a beginner in wireless telephony. The adjustment of inductances, resistances and capacities (condensers) is very trying. (3) No; the arc itself makes no sound in a well-regulated circuit. The greatest commercial distance covered by the arc method is about 20 miles.

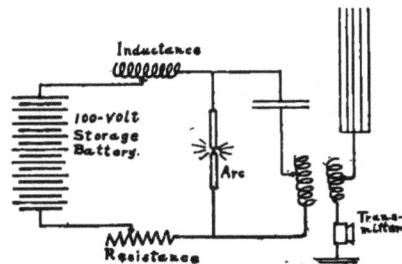

1768. **Coil-making.** W. A. T., Memphis, Tenn., has about 5 lbs. of No. 26 copper wire, and wishes to utilize it in making a transformer for wireless telegraph purposes. He asks for directions. Ans.—This size of wire is quite useless for the purpose. You need considerable coarser for a proper primary, and very much finer for a secondary. Do not go to considerable expense just to use the wire. Unless you really have some motive worth while, do not embark on a course of experimenting that will cost $50.00 just to get rid of $2.00 worth of wire. If the lot is old, the chances are that the insulation is poor and unreliable. If you really wish to make a wireless coil, we could refer you to several good articles already published.

1769. **Induction Coil.** L. W., St. Louis, Mo., proposes to make an induction coil of the following dimensions: core of Norway iron wires, bundle 12 in. long, 1¼ in. diameter; varnished muslin and fiber tube to cover or contain these wires, and then wound with two layers of No. 14 copper wire; double insulation of this sort between primary and secondary, latter consisting of 5 lbs. of No. 34 wire wound in the manner described by Stanleigh in the March, 1911, issue. Condenser of 160 sheets of tin-foil, 6½ x 9 in., separated by paraffined blotting-paper. He asks what voltage the coil should give, and what strength of battery is necessary. Ans.—Your proportions and dimensions appear good. Do not fear that you have provided too much insulation between primary and secondary. A glass tube ⅛ in. thick, or its equivalent, is none too much. Blotting-paper for insulation of condenser will be altogether too thick. The effectiveness of a condenser is proportional to the thinness of the dielectric. Double thicknesses of bond or rice paper, paraffined of course, will be much better. We do not know what voltage the secondary will generate, but sparks about 4 in. in length should be obtained. Use six storage cells on the primary.

1770. **Phase.** W. G. C., Fall River, Mass., asks: (1) What is meant by the "phases" of a generator? (2) Is malleable iron good for making the field magnet of a small motor? (3) Are twist drills for wood the same as those for iron? Ans.—(1) The number of groups of windings that are acted upon by any pair of poles represent the number of successive waves or phases associated with the currents. There are usually as many terminal wires leading from the armature of an alternating current generator as there are phases. The idea is somewhat like a consideration of the number of cranks a steam engine may have. (2) Yes, but annealed cast iron will do just as well and be cheaper. (3) Tools for working in wood are usually left sufficiently soft to be readily sharpened with a file. Metal cutting tools must, of course, be left just as hard as possible without danger of cracking. Drills for wood should be ground at a much sharper angle than those for iron.

1771. **Coil Operation.** H. W. S., Spokane, Wash., has a large induction coil that readily gives 5 in. sparks when energized by direct current through the medium of an electrolytic interrupter. When connected to the 60-cycle alternating circuit, though no resistance is interposed, the coil gives no spark whatever. What is the reason? Ans.—Any "interrupter," and especially˜one of the electrolytic sort, gives an exceedingly sudden break to the primary current. An alternating current is supposed to follow a sine curve for its wave form. Now this purposely gives a gradual and not a sudden change in strength, and the secondary generates only a comparatively feeble electromotive force. The proper and necessary condition for the induction coil to have is that while the primary current is established only after a considerable struggle against the counter electromotive force of self-induction, the break is almost instantaneous. The case is something like getting a heavy fly-wheel started; after some expenditure of effort extended over an appreciable length of time the desired speed may be attained. The motion may be annihilated, however, in an instant, by poking a stick of cord wood in between the spokes. A resounding crack will result, and something is likely to be broken. So when the lines of force are suddenly wiped out, the resulting induction in the fine wire may be enormous, and perhaps the insulation may break down. The steadily rising and falling alternating current would resemble the wheel stopped by the same stages as used at the start.

1772. **Induction Coil.** R. B. P., Cooperstown, Ill., is proposing to make a coil of the following dimensions: core, 1 in. diameter, 9 in. long; winding space 7½ in. long, 4½ in. in diameter. Primary, 2 layers of No. 12; secondary, 5 lbs. of No. 32, emanel insulated. He asks: (1) Is enamel better than s.c.c. paraffined? (2) After secondary has been run through melted paraffin, is it well to boil the whole assembled secondary in paraffin? (3) What length of spark should be secured? Ans.—(1) Yes, in that you can get more turns in a given space. Still you should put paraffined paper between the layers, insulating the end sections much more heavily than those in center, as was described in the March, 1911, magazine. (2) No. (3) Your winding is rather out of proportion, if length of spark is what you wish. For wireless use, the size of secondary is right, but we think 3 lbs. should suffice. About 1 in. sparks will be obtained.

1773. **Dynamo Design.** J. A. E., writes a kindly worded letter calling attention to the wide variations in proportions of wire for apparently the same size of machine as described by different writers. He asks if some of the extreme cases have been tried out in practice by the actual building? Ans.—The designing of a dynamo is not altogether unlike that of an architect's problem in designing a house. One person may devise a much more elegant and economical arrangement than another. Part of the discrepancies to which you refer may be traced to quite extensive differences in specifications. One machine may be made with a cast iron field magnet and a smooth core armature, while another for the same speed and horse-power may be of wrought iron with a toothed armature. Such considerations will greatly affect the quantity of wire. You will find the book on "Electric Motors," by Hobart, instructive for commerical motors of recent design; one on "Dynamo Electric Machines," by Wiener, for earlier types; and one on "How to Build Dynamo Electric Machines," by Trevert and Watson, for explicit directions covering cases of machines actually built and used for a great variety of purposes.

1774. **Induction Motor.** A. S., Fergus Falls, Minn., is proposing to make a motor resembling that given in connection with the vacuum cleaner outfit described in the September, 1911, magazine, only ¼ h.p., however, being desired. He asks for dimensions and for rather clearer instructions for connecting the coils than were given in that article. Ans.—It will be rather inexpedient to make the diameter of the parts less than those mentioned in the article referred to, but the stack of sheet iron can be 2 in. or 2½ in. in thickness rather than the 3 in. specified. Also, in order to assist the motor to start instead of permitting rotor and stator teeth to "lock," it will be better to have an odd number of rotor holes and rods, say 29, rather than the 36 specified. It may be of assistance to you to adopt the "concentric" form of coils, winding them in place instead of trying to put on the "formed" coils described in the text. Also you will find it easy to wind starting coils after the "Heyland" plan. You should make a diagram first, and making it yourself will fix the scheme in your mind. Draw a stator sheet full size and number the slots from 1 to 24 inclusive. Represent a coil that can completely fill slots 1 and 6, folded back onto the surface of the sheet so as not to pass in front of slots 2, 3, 4 and 5. (In actual winding, a wooden form ½ in. or more thick should be fastened over these slots at each end, over the back edge of which the wire can be passed as successive turns are put on. After the coil is wound the blocks can be removed and the coil bent away from the iron sufficiently to allow taping the turns compactly together, and then again pressed back into place.) Represent similar coils extending between slots 7 and 12, 13 and 18, and finally, 19 and 24. (Coils should be wound in exactly similar manner, with start in relatively same position, and when connecting, join the inside ends of two adjacent coils, then the outside ends of one of these coils and the next, then the inside end of this third coil with the inside of fourth, and finally there will be two outside ends to lead to the connection board.) Now represent a coil filling slots 5 and 8, and spreading also into slots 4 and 9; inside end will be at slot 5, outside end at slot 9. Similarly a coil occupying slots 11 and 14, and then filling 10 and 15; 17 and 20, along with 16 and 21; finally, 22 and 2, along with 21 and 3. As before, couple first two ends inside, whereby one pair of these coils will be joined in series with the adjoining pair, then two outside ends, then two inside ends, leaving two outside ends for the leads to connection board. For 110 volts we should advise nothing finer than No. 18 wire, using this size for both windings, getting in all the turns you can. We shall be glad to learn of your success.

1775. **A.C. Generator.** L. O., Minneapolis, Minn., is proposing to make a model of a slow speed multipolar machine, with revolving armature, and asks for some general directions. His sketch shows a design with 16 radial poles, a piece of thick 7 in. steam pipe forming the outside frame and magnet-yoke. Armature is 4 in. in diameter, but axial length is not specified. A speed of 260 revolutions per minute is preferred, and while the winding should consist of the diamond shape of formed coils, no particular voltage or number of phases are important.

Ans.—We would suggest that you have the stack of sheet iron about 1¼ in. thick, and cut with 48 slots, making them as large as possible, yet not requiring the teeth to be less than ⅟₁₆ in. thick at bottom. This number will permit a 3-phase winding to be used, and be much more valuable for experimental purposes than the plain single phase sort, yet permitting single phase currents quite as well as one limited to that kind only. When you have progressed far enough to know the dimensions of field cores, space for their winding, length of air gap, and size of armature slots, we can advise you as to the requisite number of turns to produce any desired voltage. We should advise you, however, not to try anything above 50 or 110 volts, and if you needed higher, obtain it through the use of transformers. If you can get three different colors of insulation on wire, or when winding the coils, can mark their ends in some distinctive manner, you will run less risk of error in connecting them, and we could the easier give the requisite directions. For instance 16 of the 48 coils could have their outside ends dipped in orange shellac, another 16 in black, the rest in white. The distinction between inside and outside ends you can hardly confuse. Whereas all 48 coils would need to be put on together, as if you were winding an ordinary direct current drum armature, you would confine the first run of connections to every third coil. Join two inside ends, then two outside ends, then two inside ends, and so on, having finally two outside ends. Then take coil No. 2 and No. 5; join their inside ends, then advance to two outside ends, and so on, resulting in two more final ends. Then with coils 3 and 6 proceed in similar manner, getting six outside ends to the final grouping. If you wish a Y connection, join the ends of coils 1, 2 and 3; ends of 16, 32 and 48 will lead to collector rings. If you wish Delta connection, join 1 with 48, and lead that to one ring; join 16 with 17, and lead that point to another ring; finally, 32 with 33 for the last ring. The cycles resulting from the number of poles and speed you propose, 35 per second, are rather out of the common standards, 25 being largely used for power transmissions, 60 for general purposes. However, there are installations operating at 40 cycles in the central part of the State of New York, and in some Southern States. Fifty cycles is the standard European frequency.

1776. **Ignition.** S. V. K., Cooleyville, Mass., asks: (1) Can direct current at 110 volts be used for operating the "make and break" (flash) ignition coil? (2) What other metals than platinum are suitable for the contacts? (3) What quantity of No. 24 iron wire should be used in constructing the heating coils for an incubator? Ans.—(1) Not directly, for the flash would be too severe for the life of the contacts. A highly satisfactory makeshift is to open the cord circuit of some incandescent lamp that may be convenient, and insert three cells of some simple storage battery. Let the coil connect to these cells. Let the lamp run most of the time, and the cells will always be fully charged. (2) There is practically no substitute for platinum. (3) We do not know what size of incubator you desire to heat, but instead of making a special resistance coil, why do you

not use incandescent lamps? They form the cheapest and most convenient form of resistance, and of course for a given number of watts liberate the same amount of heat as from any other sort of circuit. One or more lamps may be used, and of different sizes, so you can get as wide a range of heating capacity as desired. By letting the lamps be mostly immersed in a dish of water there will be great regularity in the temperature.

1777. **Electromagnet.** L. J. A., Jacksonville, Ill., is proposing to make an electromagnet 2 in. in diameter and 3 in. long. He asks what should the winding be to produce the greatest lifting power when connected to a 110 volt circuit, and how many pounds will that be? Ans.—We are not sure whether you mean that the iron is 2 in. in diameter or that this is the size of the spool. Neither do you explicitly state that the shape is the familiar U. Is the magnet required to lift a weight some distance, or are you interested merely in the weight that can be sustained? You should make a sketch of just what you have in mind and then we can advise you quite definitely.

1778. **Hardwood Floor.** H. S. M., Syracuse, N.Y., asks: How should a hardwood (maple) kitchen floor be oiled, so the floor will retain its light color and not be darkened as so many oiled floors appear to be? Kindly give constituents of the dressing, quantity required for given surface, manner of application and frequency of application. Ans.—The floor should be scraped and cleaned until all roughness and dirt has disappeared. First, the cracks should be filled. A very good filler is made in the following manner: to one part of white lead in oil, add two or three parts of bolted whiting, and enough coach varnish to form a stiff paste. This should be applied with a putty knife. It will resist moisture, and when dry may be sand-papered and rubbed. Next, give the floor two coats of white shellac, allowing ample time for each to dry; and finally apply a very thin coat of paraffin oil with a brush or a rag, and thoroughly wipe off any surplus remaining on the surface. This oiling should be repeated every week, or whenever the owner thinks the floor needs it. This method of floor treatment has a number of advantages over the wax or varnish finishes. It is ready for use as soon as applied, withstands the hardest wear, is easily put on, and it costs but little after the floor is once put into condition to receive it.

1779. **Telephone Receiver. Dry Battery.** E. C. K., Lawrenceburg, Ind., asks: (1) How to construct a simple telephone receiver. (2) Can a dry battery be restored and how? Ans.—(1) In the June issue of the *Electrician and Mechanic* there will be an article describing the method of constructing a simple telephone receiver and transmitter which will probably meet all of your needs. (2) A dry battery can be restored, but it is seldom a satisfactory operation. For a temporary restoration of the batteries, you might try the following method. Punch nail holes in the zinc containing vessel and stand the battery in a jar (such as a fruit jar); now fill the fruit jar with a sal-ammoniac solution to within 2 in. of the top of the battery. This method is said to work successfully.

TRADE NOTES

The Wm. J. Murdock Co., of Chelsea, Mass., have made several changes in some of their instruments, and have added a new receiving transformer which they believe should prove popular among wireless experimenters. Among the changes are the improvement of their small spark gap which has been a favorite in the past, together with the designing and construction of a new small helix which ought to be of great interest to those using the smaller powers for transmitting. The new loose coupler will be put out at a very low price, in fact at the lowest price that any reliable instrument has ever been put on the market for before. In addition to these instruments, the Murdock Company is now making in their own factory, and of their own material, three sizes of antenna insulators which they think should appeal to the amateur trade. These instruments and the insulators are shown in the latest bulletin of the Murdock Company, which is being distributed this month.

An Improved "Red Devil" Haven's Clamp

Following the policy laid down some months ago by the Smith & Hemenway Co., No. 150-152 Chambers St., New York, N.Y., to make the year 1912 memorable for the productions of entirely new or improved tools, they are now putting on the market an improved Haven's 'Red Devil" Clamp, as shown herewith.

This Red Devil Clamp, the makers say, is the only drop-forged steel clamp of its kind on the market for stretching telephone, telegraph and farmers' wires. Heretofore, these clamps were made with a bent-over loop, but the Red Devil Haven's Clamp is now made of drop-forged steel, doing away entirely with the bent loop.

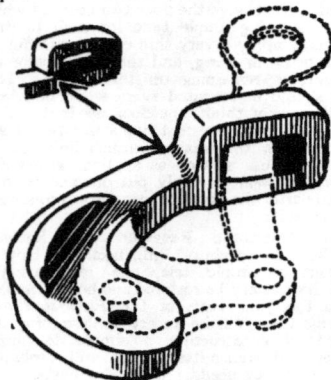

In the illustration the dotted lines show the solid steel forgings, and the makers broach a hole and pass the slip-neck through the forging, thereby making it practically an impossibility to break it. During the life of the patent on the old-style tool, it was never made in this way, and the Smith & Hemenway Co. announce that they have met with wonderful success with this new

Red Devil Clamp, having had it adopted by all the leading telephone and telegraph companies.

The clamps are made of a high-grade drop-forged steel, are gun-metal finished, and are packed one-half dozen in a box. They are all branded with the Red Devil trade-mark which carries with it the Smith & Hemenway Co.'s usual guarantee as to high quality.

The Blaw Steel Centering Co., Westinghouse Building, Pittsburg, Pa., are offering six prizes for the best plans and specifications for small concrete residences. The designs must be submitted by May 15, 1912, and the company will then select the ten which in their opinion possess the greatest merit. The ten designs selected will be submitted to Professor A. D. F. Hamlin of Columbia University, who will make the final selection. The prizes offered are as follows:

$100.00 for the best set of plans.
$75.00 for the next best set of plans.
$50.00 for the next best set of plans.
$25.00 each, for the three next best sets of plans.

The competition will be conducted under the rule of the American Institute of Architects and is open to all. It is the desire of the Company to receive suggestive designs that will give the greatest value for the expenditure, and also new ideas that will tend to stimulate the construction of poured concrete houses. It is immaterial in this contest whether the value is secured by utility, beauty, novelty, fireproof qualities, or by combinations of these.

Following the award of the prizes, the plans of the successful contestants, with names and addresses of the designers, will be published in a booklet, which will be given wide circulation among prospective builders to encourage the construction of concrete houses.

The conditions governing the competition may be obtained by writing the company at their Pittsburg, Pa., address.

We are in receipt of an interesting catalog from the South Bend Machine Tool Co., of South Bend, Indiana. In the catalog are described and illustrated their line of screw-cutting engine lathes. These lathes may be driven by foot-power, engine-power or by an electric motor. All parts of the South Bend lathes are interchangeable and each machine when finished is put under belt, operated and tested.

The General Electric Company is just issuing its 1912 catalog of electric fans. The publication is an attractive one, printed in colors, and contains descriptions and illustrations of the fans manufactured by that Company for use in the home, office and public places. The line of fans comprises those suitable for use on desk or table, and which are manufactured with fan diameters of 8, 12 and 16 in. All of these fans are convertible without the use of tools or additional parts, so that any one may be used on a horizontal surface or attached to a wall. They are made in two styles: fixed and oscillating. The publication contains also illustrations and descriptions of the ceiling and column fans manufactured by the Company, and a line of supply parts for these.

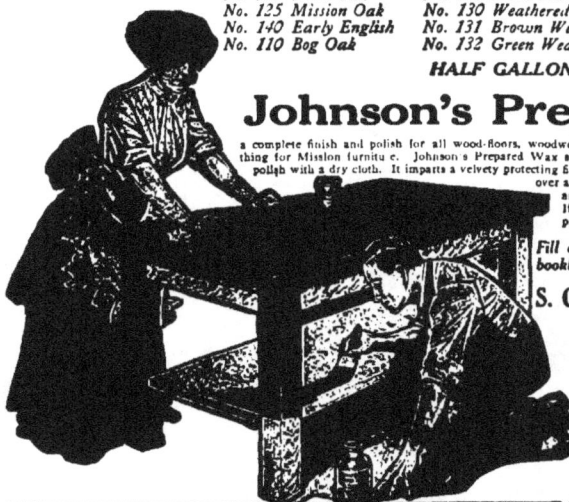

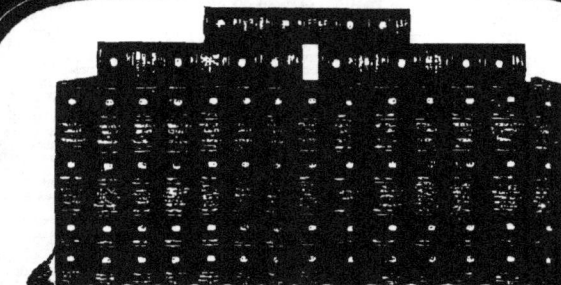

AUTOS AND MOTORCYCLES

AUTOMOBILE ENGINES, boilers and parts. Send for list. J. L. LUCAS & SON, Bridgeport, Conn.

BOOKS AND MAGAZINES

MECHANICAL BOOKS—Any subject; catalog for stamp. CRESCENT BOOK STORE, A 329 S. Halsted St., Chicago, Ill.

"HYDE'S TELEPHONE TROUBLES AND HOW TO FIND THEM"; also phantom toll circuits and how to install. New 15th edition. Price 25 cents. HYDE BOOK CO., Telephone Bldg., 183 5th St., Milwaukee, Wis.

BUSINESS OPPORTUNITIES

"ELECTRIC DRY CELLS," flashlight and medical. Written demonstration how any one can make ignitors; peerless for endurance and recuperation, $95.00. Exclusive territory guaranteed. JOHN DOUGAN, 234 Smith St., Winnipeg, Can.

BIG PROFITS—Open a dyeing and cleaning establishment; little capital needed. We teach by mail. Booklet free. BEN-VONDE SYSTEM, Dept. A-F, Charlotte, N.C.

COINS AND STAMPS

STAMPS—All for 10 cents. One set each: 2 Honduras, 2 Ecuador, 2 Nicaragua, 2 Salvador, 500 Quaker Hinges, 50 unused Cuban stamps. QUAKER STAMP CO., Toledo, O.

COINS—20 different foreign 25 cent; large U.S. cent 5 cents; 5 different Confederate state bills, 15 cents; 150-page illustrated premium coin book, 25 cents; 1,000 old U.S. stamps 25 cents. F. L. TOUPAL CO., Dept. 57, Chicago Heights, Ill.

ELECTRICAL

ELECTRIC WIRING AND LIGHTING.—Knox-Shad, 192 pages; 150 illustrations. A working guide in all matters relating to electric wiring and lighting. Includes wiring for direct and alternating current by all methods approved by the fire underwriters; and the selection and installation of electric lamps for the lighting of buildings and streets. The utmost care has been used to bring the treatment of each subject within the range of common understanding, making the book especially suitable for the self-taught, practical man. Published by AMERICAN SCHOOL OF CORRESPONDENCE, Chicago, Ill.

HELP WANTED

500 MEN, 20 to 40 years old, wanted at once for electric railway motormen and conductors; $60 to $100 a month; no experience necessary; fine opportunity; no strike; write immediately for application blank. Address, S.O.M. Care of *Electrician and Mechanic*.

LOCAL REPRESENTATIVE WANTED.—Splendid income assured right man to act as our representative after learning our business thoroughly by mail. Former experience unnecessary. All we require is honesty, ability, ambition and willingness to learn a lucrative business. No soliciting or traveling. This is an exceptional opportunity for a man in your section to get into a big paying business without capital, and become independent for life. Write at once for full particulars. Address E. R. MARDEN, Pres., The National Co-Operative Real Estate Company, L453 Marden Bldg., Washington, D.C.

HELP WANTED

WE TRAIN DETECTIVES. You can be one. Splendid opportunities. Travel. Earn $100 to $360 monthly. This fascinating profession taught practically and scientifically by mail at a nominal cost. AMERICAN SCHOOL OF CRIMINOLOGY, Dept. 8., Detroit, Mich.

WANTED IN BANKS—Honest, ambitious men. Splendid opportunities. Pleasant work, short hours, good salary. Be a banker. We teach you in a few months by mail. Write for catalog. AMERICAN SCHOOL OF BANKING, 135 McLene Bldg., Columbus, O.

MALE HELP—I will start you earning $4.00 daily at home in spare time silvering mirrors; no capital; free instructive booklet, giving plans of operation. G. F. REDMOND, Dept. A.C., Boston, Mass.

WANTED—A man or woman to act as our information reporter. All or spare time. No experience necessary; $50 to $300 per month. Nothing to sell. Send stamps for particulars. Address SALES ASSOCIATION, 784 Association Building, Indianapolis, Ind. (8)

FREE ILLUSTRATED BOOK, tells about over 360,000 protected positions in United States service. More than 40,000 vacancies every year. There is a big chance here for you. Sure and generous pay. Lifetime employment. Easy to get. Just ask for booklet A59. No obligation. EARL HOPKINS, Washington, D.C.

BECOME A DETECTIVE—Earn $150 to $300 per month travelling over the world. Send stamp for particulars. Write FREDERICK WAGNER, 1243 Lexington Ave., N.Y.

SELL INFORMATION from your district. We explain how free. Good money for bright, alert people. Splendid "side line." Free booklet. NATIONAL INFORMATION BUREAU, Station A3, Columbus, Ohio.

MISCELLANEOUS

FREE TUITION BY MAIL—Civil service, mechanical drawing, stationary engineering, electric wiring, agriculture, poultry, normal, bookkeeping, shorthand and typewriting courses. For free tuition, apply CARNEGIE COLLEGE, Rogers, Ohio. (tf)

YALE pocket adding machine, 30 cents, postpaid. YALE MFG. CO., Dept. F, Newark, N.J.

FREE—"Investing for Profit" magazine. Send me your name, and I will mail you this magazine absolutely free. Before you invest a dollar anywhere, get this magazine. It is worth $10.00 a copy to any man who intends to invest $5.00 or more per month. Tells you how $1,000 can grow to $22,000. How to judge different classes of investments, the real power of your money. This magazine six months free if you write today. H. L. BARBER, Publisher, Room 446, 20 W. Jackson Boulevard, Chicago, Ill. (12)

3 SHIRTS TO MEASURE, $5.00. Express prepaid. Better grades, $2.50 and $3.00 each. Superior quality. High-grade workmanship. Faultless laundry work and perfect fit assured. Send for samples with measuring instructions and booklet, "Shirt Tales." Money returned if not satisfied. FRANK W. HADLEY, Mfr., Dept. E, Norwalk, Conn.

PERFECTION POCKET ADDING MACHINE—Lightning seller. Agents wanted. CINCINNATI SPECIALTY MFG. CO., Dept. O, Cincinnati, Ohio. (12)

$50 per week and up; how far up depends on you; $21,500 in three years by one manager; $6,000 in one year by another; $4,000 by another in six months; must have $500 to invest. WESTERN OXYGENATOR CO., Beatrice, Neb.

MISCELLANEOUS

5 FORMULAS, 24 cents. How to make rubber cement, bluing, artificial gas, and liquid for fish bait; makes fish bite fine, 10 cents each. J. B. WILLIAMS, Hobard, Okla.

HAVE you read "How to Succeed"? It is the greatest satire ever written on the get-rich-quick mania; 15 cents postpaid. A. RUHNAU, Dept. 80, 280 Columbus Ave., New York.

FOR SALE—Roll top desk, 4 ft. 6 in. long, almost new. Will sell for half its value. H. SLOCUM, 145 Powers St., Brooklyn, N.Y.

FOR SALE—Marine gasoline engine, heavy duty, 6 h.p., 4 cycle reversing propeller complete, in perfect condition. H. S. HUSBAND, 13 Priestly St., Wilkesbarre, Pa. (s)

FOR SALE—Hammond typewriter, price $10.00; cost $75.00. R. C. NEWMAN, 2,424 Orchard St., Chicago, Ill. (s)

FOR SALE—Plating Dynamo, 6 volt, 100 amperes, with nickel and silver outfits. Dynamo will work 500 gallons, solution. $35. HARRY STRANG, West Unity, Ohio.

LEARN to make hatpins and brooches of rosebuds and butterflies. Pleasant, profitable employment. Beautiful sample and booklet, 12 cents. A. E. BENSON, Stewartville, Minn.

NO MORE fussy polarity indicators. Send dime for enough "Pol-Test" paper for 30 to 90 cheap, simple, certain tests. Mail order department, 390 N. Euclid Ave., St. Louis, Mo.

EXCHANGE combination billiard and pool table, good condition, size 3 x 6 for induction motor to party living near Seattle. W. M. HINELINE, 512 Maple Leaf Place, Seattle, Wash.

HOW TO RENEW OLD DRY BATTERIES at a cost of 1 cent each. Send 15 cents coin to R. T. KALB, 2631 Sutton Ave., Maplewood, Mo. (5)

AMATEURS LET ME QUOTE YOU on parts. Anything you may be making. Lathework a specialty. C. M. CURTIS, 15 N. 6th St., Fulton, N.Y. (5)

FIFTY VIEWS OF ROCKY MTS. or State Capitols, all different, in colors, glazed finish, prepaid for 25 cents in stamps. Address F. R. BLAIR, post card jobber, Harrickville, Pa. (5)

PATENTS

PATENTS THAT PROTECT: For facts about prizes, rewards, etc., send 8c stamps for our new 128-page book of intense interest to inventors. R. S. and A. B. LACEY, Dept. 94, Washington, D.C., established 1869.

MONEY IN GOOD IDEAS—My patents get it for the inventor. Twenty years' experience. Fees low-payments liberal. "The Truth about Patents," sent free. JAS. R. MANSFIELD, Dept. D, Washington, D.C.

C. L. PARKER, Patent Attorney, ex-examiner, U.S. patent office, 952 G St., Washington, D.C. Inventor's handbook "Protecting, Exploiting and Selling Inventions," sent free upon request. (4)

PATENTS SECURED or fee returned. Send sketch for free expert search and report as to patentability. Books on Inventions and patents, and reference book, sent free. JOHN S. DUFFIE & CO., Dept. 4, Washington, D.C.

THE PATENTOME tells all about patents and how to get them. Free on request. Established 1865. ANDERSON & SON, Patent Solicitors, 731 G Street, Washington, D.C.

PATENTS

PATENTS OF VALUE. Prompt and efficient service. No misleading inducements. Expert in mechanics. Book of advise and Patent Office rules free. CLEMENTS & CLEMENTS, Patent Attorneys, 707 Colorado Bldg., Washington, D.C.

PATENTS BOOKS on How to Obtain and Sell Patents, containing exhaustive information on these subjects with 100 mechanical movements, mailed free on request. F. G. DIETERICH & Co., 604 Ouray Bldg., Washington, D.C.

PRINTING

1,000 IMITATION TYPEWRITTEN LETTERS, heading black, body purple, good bond paper, $3.00; 6 x 9 circulars, $1.50 (1,000 3 x 6 Art Ad. Slip, 150 words, $1.00 postpaid). All other printing low price. Elegant set samples free. GOOD'S PRINT, Harrisonburg, Va.

PHOTOGRAPHY

KODAKS, CAMERAS, LENSES—Everything photographic, we sell and exchange. Get our latest bargain list; save money C. G. WILLOUGHBY, 814 Broadway, New York. (tf)

WE BUY, SELL AND EXCHANGE. Bargains in microscopes, telescopes, field glasses, cameras, etc. Catalog and bargain list free. KAHN & CO., established 1850. 26 John St., New York, N.Y. (s)

A GRAFLEX camera wanted; also jeweler's lathe with slide rest. Will give 6 in. Marconi induction coil, costing $125.00, in exchange. SEITZ, 460 N. Main St., Springfield, Mass.

FIVE EXCELLENT formulas for sensitising silk for photo sofa pillows, and producing enamel pictures on watch-cases, chinaware, etc., only 25 cents. A. E. BENSON, Stewartville, Minn.

WIRELESS

FOR SALE—Complete wireless outfit. Will sell at sacrifice. Write for photograph and full information. Address WIRELESS, 1209 East Capitol St., Washington, D.C.

IF YOU ARE UP-TO-DATE, you will have a pad of our Radiogram blanks before you. Send 25c coin for a pad of 50. WIRELESS MFG. CO., Canton, Ohio.

SLIDERO for tuning coils; will not wear wire. 15 cents per pair, postage paid. Satisfaction guaranteed. M. S. HOWE, Edghill Road, East Milton, Mass.

ANY RECEIVER REWOUND to 1,000 ohms, with No. 50 copper wire for 75 cents. Returned prepaid within ten days. Satisfaction guaranteed. "Arystalpaste" holds your minerals in place without applying heat. Large sample for dime. Amateur's Wireless Supply, 38 Overlea Ave., East Saugus, Mass. (5)

MODEL AEROPLANES AND WIRELESS INSTRUMENTS for sale at reasonable prices. Address E. GREGORY, 261 East 15th Street, Brooklyn, N.Y. Rubber strand. Propellers and parts. (5)

WRITERS

I AGREE TO FIND A PUBLISHER FOR EVERY MANUSCRIPT that I deem worthy of publication. Manuscripts critically read and revised by me, typed and otherwise properly prepared for the publisher by my experts. Technical works given correct form. MODESTE HANNIS JORDAN, 615 W. 13th St., New York City. Send 10 cents for Writers' Leaflet of Instruction. (5)

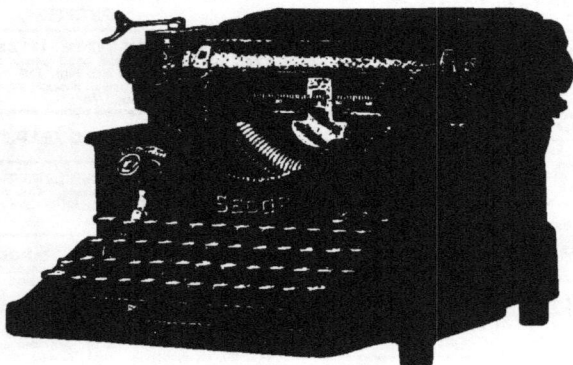

ELECTRICIAN & MECHANIC

| VOLUME XXIV | JUNE, 1912 | NUMBER 6 |

ALTERNATING CURRENT AND ELECTROLYTIC RECTIFIERS
EDGAR BERGHOLTZ

Occasions frequently arise when, for some reason or other, one desires the use of direct current, and yet the only source of supply is alternating current. Instances of this difficulty are noted in all electrolytic work, in supplying current for charging storage batteries of launches and automobiles of the electric type, and of telephone systems, as well as for copper and silver plating.

The first question which one naturally asks is: why cannot alternating current be used instead of direct? The answering of this question requires an explanation of both kinds of current: the direct is had when the flow is in one direction, with a magnitude varying periodically or remaining constant; the alternating occurs when the flow is first in one direction, and then in an opposite direction, with the magnitude varying periodically.

Direct and alternating currents have been aptly illustrated by likening them to a stream flowing into the ocean. Some distance above the mouth, where the tides have no effect, the water flows continuously in one direction, is unidirectional like the direct current. Down at the mouth of the river the tides cause the water to flow first up stream and then down, changing four times in twenty-four hours. This change due to tides is very much like the alternating current, except that the latter varies many times in a single second.

Again alternating currents are very much like a double-action pump (Fig. 1),

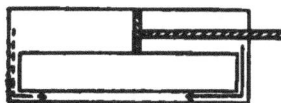

Fig. 1

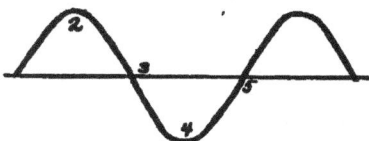

Fig. 2

consisting of a cylinder with a valve at either end; to both valves being connected the same pipe. The water in the cylinder is forced by the pushing piston in the direction of the dotted arrow; when the piston is pulled back it forces the water back in the oppostie direction, represented by the other arrow. This flow back and forth represents the alternating current.

This current can also be graphically illustrated as in the following diagram (Fig. 2). Suppose an alternating current begins at 1: it rises in intensity until it reaches its maximum at 2, and then decreases to zero. This first rise and fall represented as being above the line is called the "positive phase." After passing zero at 3, the current increases in magnitude again, but this time below the line until it reaches its maximum at 4, and then it decreases to zero at 5. This second rise and fall illustrated as being below the line has received the name of "negative phase."

When an alternating current has passed through both phases it is said to have completed a "cycle," and the number of cycles completed in one second is called the "frequency" of the current.

Now the reason why alternating current cannot be used instead of direct current in electrolytic work is because all' that is done in electrolytic solutions

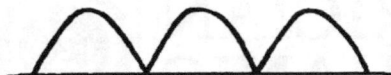

Fig. 3

by the positive phase is interrupted and destroyed by the negative and contrary phase and no more work has been accomplished after the current has passed through than before its introduction.

To overcome this difficulty, to convert alternating current into direct, the modern inventive mind has devised and introduced the "current rectifier." The real difference between the positive and negative phases of an alternating current has been seen to lie in the fact that the negative current flows in a direction opposite and contrary to that of the positive. Now if the negative phase can be changed to conform in direction to that of the positive, a direct, though pulsating, current results. To perform this operation is the purpose of the rectifier.

The converted current can be represented by diagram (Fig. 3), the negative phase being swung above the line to indicate change in direction.

The most usual frequency of an alternating current is 60 cycles, i.e., the current varies through positive and negative phases at the rate of 120 pulsations a second, and in this state can be used for all electrolytic work.

The rectifier which brings about this change is divided into several classes. Excluding the rotary converter, which is, properly speaking, not a rectifier, since it does not change the direction of flow of the current itself, but produces a direct current from a direct current dynamo run by an induction or alternating current motor, the rectifier is of three types: mechanical, vapor (Hg-arc) and electrolytic.

The mechanical rectifier, as is evident from the name, is a machine constructed for current conversion, and is a purely physical contrivance. It is exemplified in the commutator of a direct-current dynamo.

Both vapor and electrolytic rectifiers are based on the principle discovered some fifty years ago that aluminum has the property of "asymmetry," i.e., when a current is introduced into a strip of aluminum immersed in any one of a certain class of electrolytes (bodies which transmit current and at same time undergo chemical decomposition by it), great resistance is offered; when, however, the electric current attempts to pass in the opposite direction, i.e., from the electrolyte to the aluminum, very little opposition is offered.

At first it was thought that the resistance was due to the formation of an oxide upon the aluminum electrode, but it has been lately shown that the more probable explanation is found in Schultz's theory: that the resistance is caused by the formation of a gas lying within the pores of the oxide.

For many years after its discovery, the property of asymmetry was not utilized, and it is only within the last few years that men of electrical genius have been able to build upon it an artificial contrivance for converting the alternating current.

As a consequence of these endeavors, two styles of rectifiers have been evolved, the vapor and the electrolytic.

In the case of the former, the vapor itself is not an electrolyte, since it is not decomposed on the passage of the current, but it has the property of asymmetry. The great efficiency of 98 percent has been attained by this device, and it is interesting to note that it is the most efficient artificial contrivance hitherto devised by human agency. But as this style of rectifier is difficult to make without special materials, somewhat hard to procure, and hence costly, and since special tools and apparatus are required for its production, the electrolytic rectifier is used instead, in many cases, on account of its comparatively lower cost and simpler construction.

Electrolytic rectifiers have been placed on the market, but they can be made

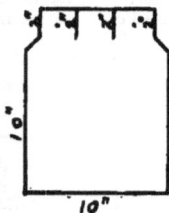

Fig. 4

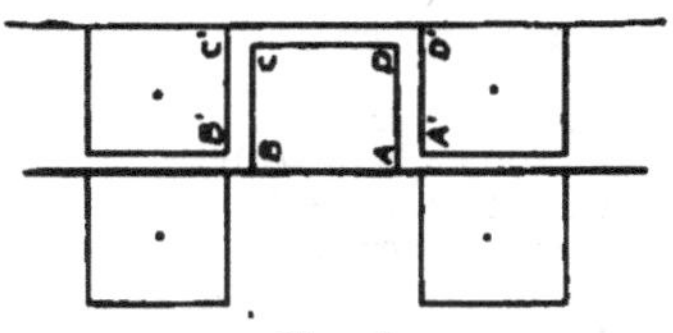

Fig. 5

very cheaply by anyone with a little care. The following is the description of one used at present in a leading American college: In this rectifier the alternating current of 60 cycles with a voltage of 108, produces a pulsating direct current of 25 amperes, with a voltage of 40, or of 1 ampere with a voltage of 100, according as may be desired. Hence the efficiency is something over 50 percent.

To hold the solution, four large earthenware jars were used, the capacity of each being 6 gallons. In the lids were drilled four holes to admit $\frac{5}{16}$ in. bolts, which fasten the plates to the cover. This was done with an ordinary steel drill, with a lubricant in the form of a solution of turpentine, ether and camphor. It may be noted that as this process is somewhat tedious and lengthy, circular lids of wood, preferably white pine, might have been substituted, their diameter being slightly greater than that of the jars. Care should be taken, however, if wood is used, to have the covers soaked well in paraffin to render them thoroughly insulated.

The plates, four of aluminum and four of zinc, measured 10 x 13 in. They were $\frac{5}{32}$ in. in thickness, but $\frac{1}{16}$ in. would have done. At a distance of 1 in. from either side of the plate (Fig. 4), on one end, were made slits 2 in. long; then after measuring off 10 in. on either side from the other end, two slits were cut in such a fashion as just to meet the inmost end of the first slits.

After the corners at the end of the plate had been removed, the end was divided into three equal parts and then cuts 2 in. in length were made toward the center of the plates at the two places marking the division. In the two outer sections were drilled two holes to admit $\frac{5}{16}$ in. bolts; these outer sections were bent back evenly and the middle one forward until they formed an angle of 90 degrees with the plates. The purpose is to make the plates rigid when they are

bolted to the lid, so that they will not sway enough to come in contact with each other and form a short circuit.

The position of the zinc with regard to the aluminum when both are bolted, can be well represented by a cross-section view of the bolted end, illustrated in Fig. 5.

The edges BC, AD, B^1C^1, A^1D^1 were then filed down so that $BCAD$ would fit into the space $B^1C^1A^1D$ without the touching of the plates when they were set in the cover.

The plates were then bolted to the lids, as in Fig. 6, with $\frac{5}{16}$ in. bolts. Next to the head was placed a washer (1), and the bolt was then put through the hole in the plate and tightly clamped with a nut (2) to obtain a good electrical and mechanical contact between the bolt and the plate. The whole was then fastened securely to the lid with a second nut (3) and a third nut (4) was screwed on to act with the second (3), as a binding post. In each lid one zinc and one aluminum plate were fastened in this manner.

When the plates have been duly secured, they should be bent toward each other so that they are not more than $2\frac{1}{2}$ in. apart. The efficiency of the rectifier can be still further heightened by bending them as close as possible, together, provided some insulation is put between them to prevent contact and consequent short-circuit resulting in the production of alternating current instead of the desired rectified current.

The electrolyte which is in use, is a saturated solution of ammonium orthophosphate ($NH_4H_2PO_8$), in amount about 22 liters (5 gallons) for each jar, i.e., as much as would bring the solution level, $9\frac{1}{2}$ in. from the submerged end of the plates. Ammonium orthophosphates is used in preference to other phosphates, i.e., sodium, potassium and magnesium, for although in itself, it does not perform

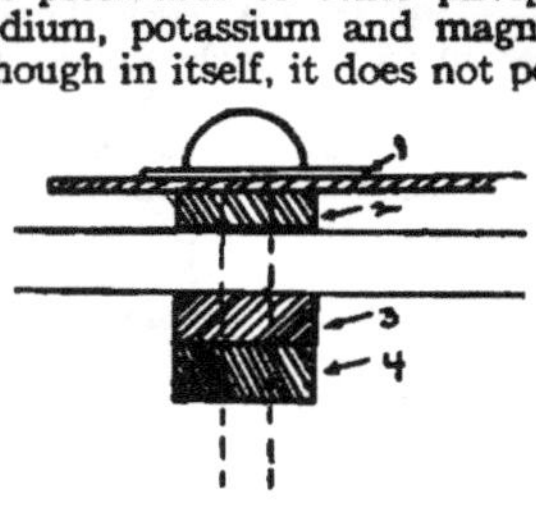

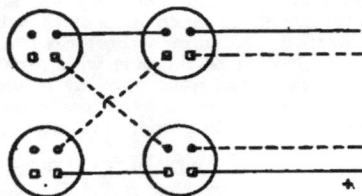

Fig. 7

its function as efficiently as the others, yet it does not undergo chemical decomposition in the electrolytic reaction as quickly and as easily as do the others.

When solution had been made and poured into the jars, the covers were put on, the lids being in solution. The plates were then connected with No. 14 wire, as in Fig. 7. The dotted line represents wire carrying alternating current, and the straight black line, the wire carrying the rectified current. The squares representing the binding posts on the lids are connected to the aluminum plates and the circles represent posts connected to the zinc plates.

The rectifier, being ready for use, was connected in short-circuit through a varying resistance of a few ohms to form the oxide on the aluminum electrode, thereby giving it the property of asymmetry. After the oxide is formed, the rectifier may be connected up to perform its work.

The rectifier as above described will run for over an hour without perceptible heating, but on a ten-hour run, if there be no system of cooling, the temperature will rise almost to boiling point.

The current which comes from the rectifier is of a pulsating character, varying 120 times in one second, when the alternating current is one of 60 cycles.

Sometimes it may happen that a steady current is needed, and the difficulty of changing a pulsating current into one of steady magnitude must be solved. This can be easily overcome by connecting the contrivance which is to receive the current to a set of storage batteries which will give the desired voltage. These, in turn, are connected to the rectifier. In this way a steady current is gained from the batteries and their loss in electrical energy is repaired by the introduction of the pulsating direct current from the rectifier. Of course, even here, some variation still continues, but it is reduced to an almost imperceptible minimum.

This rectifier will not operate indefinitely, as there is a tendency to electrolytic decomposition and to evaporation on the part of the solution and to corrosion of the electrodes. In consequence both solution and electrodes must be renewed now and then.

A STUDY LAMP
C. H. SAMPSON

A convenient arrangement for a study or drawing table lamp is shown in this sketch. The up-right F may be changed where desired on the table; the angle arm A raised to the desired elevation and turned to the angle wished for; the hollow arm E pulled out to any distance over the table, and the shade C adjusted by using the thumb-screw at D. The wire for the lamp may be introduced through a hole bored through the upper part of arm A and passing through E.

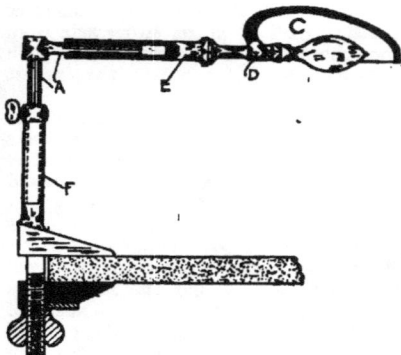

A genius makes his mark because he *hits* the mark; he applies himself to but one thing until he makes that thing the *best*—by a master stroke.

HOW TO MAKE A RHEOSTAT

H. W. H. STILLWELL

One of the most useful forms of apparatus to be found about the experimenter's shop or laboratory is the rheostat. Almost every boy having some knowledge of electricity has one or more about his shop, although they may not be very practical for various reasons. A good practical instrument of this sort is not as easily constructed as at first supposed. It is the purpose of this article to describe the construction of one or more types of these instruments that will, when completed according to these instructions, give excellent service and well repay the maker for his painstaking in the constructing of same. These instruments will be found excellent for controlling the current in connection with a medical or other coil, regulating small battery lamps or the speed of small motors.

Graphite is used extensively as a resistance, especially in medical apparatus, where it is directly in contact with the arm of the rheostat, an uneven wearing of the graphite surface is sometimes the result, which causes the current to be jerky, and more or less serious consequences might result where the apparatus is being used upon some delicate part of the body, as is very often the case with medical galvanic, faradic or other apparatus. It is to overcome this uneven regulating of the current that this instrument was designed. This tendency is entirely eliminated in the rheostat here described, as the arm of the instrument comes in contact with metal studs or buttons, which in turn are in contact with the carbon or graphite which is placed underneath.

Two pieces of well-seasoned hard wood will be required, each about 8 in. square and ½ in. thick; find the center and with a pair of compasses describe two circles, one 6¼ in. in diameter, and the other 5¾ in. Now with a small keen compass saw cut around these lines, following them as closely as possible to insure a neat job. The smallest circle should be completed first, and the round center block, 5¾ in. in diameter, should be removed and laid aside for future use. Then saw along the line of the larger circle 6¼ in.; the wooden collar formed will be of no use and may be destroyed. The

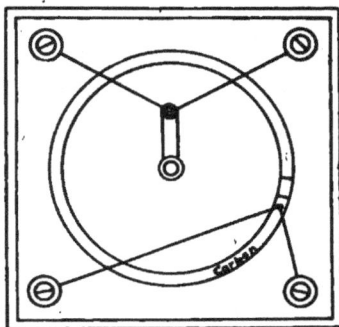

Fig. 1

board will have now a circular opening 6¼ in. in diameter. All rough edges resulting from the sawing should be smoothed off with sand-paper. Place the second piece of wood upon a flat surface and lay the piece just completed upon it and fasten to same by screws or by gluing, which is more desirable but requires more time. Now place the 5¾ in. circular piece exactly in the center of the 6¼ in. circular opening and secure as before; this will form a groove ¼ in. wide and ½ in. deep. Contact buttons must be arranged so as to enter directly in the center of this groove, which will be described later. The contact buttons may be purchased from any large electrical or experimental supply house, and can be had in a variety of sizes and shapes, or can be made by the workman himself, using brass filister headed screws ⅝ in. long under the head. 8 x 32 or 10 x 32 machine screw sizes may be used. File down the head until the slots have been taken out (see sketch), being careful to file the head as evenly as possible to get a uniform thickness. If a lathe be at hand, the process of cutting down the heads may be accomplished much quicker and more neatly than by filing. A lathe is not always at one's disposal, therefore the filing should be done carefully, and if done so, there will be little left to be desired. When the contact buttons have been finished, proceed to lay off the top board of the instrument with a pair of small dividers, which should be set so

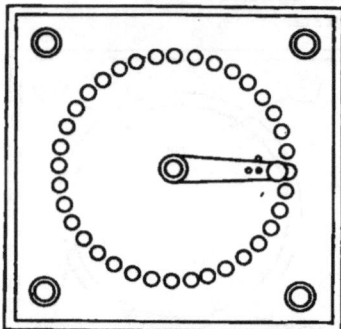

Fig. 2

that there will not be more than ⅛ in. or a little over between the buttons. When the board has been spaced off, proceed to drill the holes to receive the shanks of the buttons, care being taken not to get the holes too large, which should be dipped in a little shellac before being forced into the hole they are to occupy; this will make the buttons stay where they belong and will not make it necessary to use nuts on the bottoms to hold them in place. When all the buttons have been secured in their places, and the coating of shellac dried, care should be taken to remove all dry shellac from the ends of the shanks, which should project into the groove about ⅛ in. or a little less. This shellac may be removed by emery or sand-paper. The slot should now be filled in with graphite or carbon to a depth of not more than ⅛ in. A piece of pasteboard should be cut out to fit in the slot and upon the resistance material; this should fit closely. When this has been fitted some paraffin should be melted in a pan or kettle having a spout; if sealing-wax is at hand, this will give a much better appearance, and should come almost up to the edge of the groove. (If sealing-wax be used, an old receptacle should be used in heating same, as it is liable to be ruined by the wax adhering to it.)

The amount of resistance will vary with the requirements of the builder. It is, therefore, a wise plan not to fill in with the paraffin or wax until the instrument has been about finished, so that it can be tried, after which a greater amount of the resistance material may be added

if necessary, or if the resistance already in place is found to be too small, the pasteboard should be removed and a little paraffin removed, care being taken to keep the amount as even as possible all the way around the groove. When the proper proportion has been determined upon, the wax may be filled in the groove and the instrument will be ready for use. The foregoing paragraph has been inserted as a safeguard, as it is very easy to add too much of the graphite or carbon, or on the other hand, not enough, and if the wax be left until last, the mistake can be easily remedied.

There are four binding posts shown in the drawings; but two may be used if more desirable. By using the four, two different connections may be made with the instrument, with more ease than with the two posts. This part of the construction is largely a matter of taste. If the location of the posts are not desirable as illustrated, they may be placed in a row and near each other at the top or bottom of the board. Connections can be made as shown in the drawings, or may be varied to suit the requirements of the workman.

When the location of the binding posts has been determined, and they are in position, our next consideration will be the arm. It will be noticed from the drawings that there is a spring contact in this style of instrument, a feature which is seldom found on the ordinary rheostat and one well worth the little extra trouble involved in its construction. Medium heavy brass should be used for this arm, about ⅛ in. in thickness; the length will be about 3⅝ in. The laying out of this arm is a simple matter: a small pair of steel dividers, a center punch, a scale, light-weight hammer and one or two good files are the most necessary tools. If the arm is to be taken from a large piece of brass, it will be necessary to lay out the dimensions given on the sketch, and then saw the piece out with

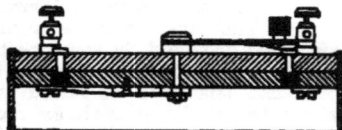

Fig. 3.

a hack-saw. The spring contact or slider will be the next to be considered, the brass used for its construction being a good grade of spring brass. The thickness should be about No. 21 Birmingham gauge .032; the bending may be done in a vise; the small shoe which slides over the buttons may be turned up at the ends, as shown, in the same manner, or a small iron or steel block can be used to hold the brass in the vise, and the shaping can be done with the hammer. It is better to allow a little in the length of this spring for the bending, as this operation will shorten the spring considerably. When the spring has been shaped to suit (the final bending can be best made when the instrument is all assembled), the holes for rivets are now drilled. These can be any convenient size to suit brass rod or rivets on hand. The rubber knob for turning the arm should be located first before the spring is riveted in place, and hole drilled and tapped for same. A very neat hard rubber knob, or one made from composition, can be purchased from any electrical supply house, or a white porcelain knob can be used, although care must be taken not to allow any metal parts to project through the knob. The center bolt, upon which the arm swivels, can be constructed by the workman, or may be obtained from some dealer handling experimental supplies; the same may be said of the heavy washers and the bridge shown in the sectional views.

The small bridge shown in the sectional view of the completed instrument is designed to prevent the wires from wearing or slipping off the center bolt, which is often the case in other types, and causes much trouble and annoyance. Two brass nuts should be used upon the bottom end of the center bolt to keep it from getting loose. If close attention is given to drawings herein, no difficulty should be experienced in making the various parts of the instrument, and when completed and finished properly, will be very serviceable and add to the other equipment, both in appearance and utility.

When completed, the edges of the boards should be carefully trued, and a neat molding should be placed about the four sides of the instrument to add to its appearance, and raise the bottom board higher, so that the bridge and other

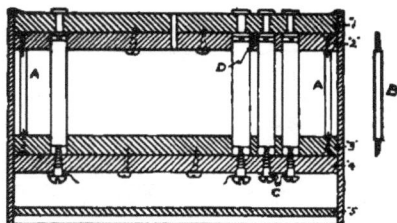

Fig 4

under parts will not come in contact with anything which might injure them. If desirable, a bottom board may be made and secured in place by small screws or nails; this will protect the instrument still more and will hide the under construction from view.

CARBON ROD RHEOSTAT

The rheostat just described is an excellent one, but there are some more or less difficult parts for the amateur to manufacture, especially if his tools are crude or not suited to such work. The carbon rod rheostat, which is here described, is an excellent instrument, and one which will not require as much labor as the one previously described, and the resulting instrument will give excellent and highly efficient service. In building a rheostat for very fine work, or measurements which require any degree of accuracy, the resistance of each step of increase or decrease in the amount of resistance must be definitely known. For such an instrument as this, a great amount of care is required in the making, and a reliable standard form of a resistance would be necessary to test each coil of wire, or whatever form of resistance was employed. The Wheatstone bridge is the standard used for such work, and is extremely sensitive and accurate and costly when purchased. For all ordinary work, however, the instruments here described will answer the purposes for which they were designed.

In the start, this instrument may be constructed much the same as the one first described, and many of the dimensions may be followed. The front board or the one containing the buttons and arm can be made exactly the same; the second lower board, however, is not cut

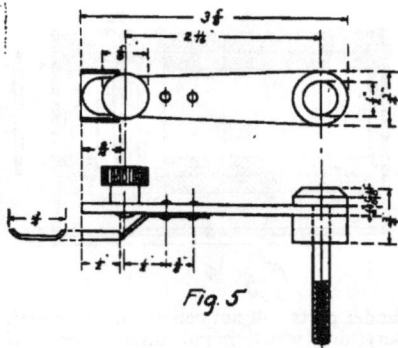

Fig. 5

out with a compass saw. The number
of buttons can be varied to suit the needs
of the builder. In the drawing, there
are 36 points; less may be made if desirable, by spacing off the circle with a
pair of dividers to the desired number.
When the top board has been completed,
a second board will be required. This
should be spaced with exactly the same
number of points as the top board, and
these points must come directly beneath
the ends of the buttons. Now procure
some carbon rods, such as are used in
city arc lamps, or better still, a smaller
carbon pencil, such as may be purchased
from any experimental electrical supply
house. These may be had in several
diameters, the size required depending
upon the amount of resistance required
in the completed instrument. If arc
light carbons are used, it is best that
they be copper-plated on one end, which
is often the case, so that the wires can be
soldered to them when the connecting up
is done. This copper-plating is not necessary, however, as the connections can be
made at the lower ends of the buttons
where necessary; this is clearly shown
in the drawings. In the carbon rod rheostat, the total thickness of the finished
instrument will be considerably more
than the other. This is caused by the
carbon rods being placed upright and
beneath the top board, as shown in Fig. 4.
In drilling the second or under board,
care must be taken to have the holes
directly in line with the holes drilled in
the upper board for the buttons, so that
the lower ends of the buttons will come
in the center of the holes. It is best to

point the ends of the buttons, as shown
in Fig. 4, so that when assembled, there
will be a better contact between the brass
of the button and the end of the carbon.
The carbon rods or pencils should be cut
to nearly the same length, so that the
steps between the points of the instrument
will be as near uniform as possible. The
holes in the board marked "2" should be
drilled just large enough to admit the
carbons and completely through the
board. A second board will be required,
3 in same drawing; this should be made
at the same time as the other, as much
depends upon getting the holes all in
line. A very easy method is to secure
the two boards 2 and 3 together with a
thin nail or two, not driven all the way
in, so that they may be removed when
no longer needed. When the desired
number of divisions are spaced off, the
two boards may be drilled together and
the holes will then be directly in line.
Board 4 should be spaced off the same
as the others; the same care being taken
to get the points in line with the others.
In drilling these holes, the size will be
smaller than those required for the carbons, and should be made small enough
so that a wood screw about size No. 6
or 8 and about ¾ in. under the head will
catch a good thread. These wood screws
are employed to force the carbons up to
the end of the buttons, and to make contact from below, as shown.
    In the assembling of the instrument
little trouble will be experienced if directions are followed. When the top board
is completed with buttons all in place
and secured, and the arm and binding
posts in place, some sort of case should
be constructed to contain the instrument;
and a neat molding may be placed about
the top, projecting slightly over the edges
of the top of board, as shown in Fig. 4.
This outer case may be polished, enameled
or painted to suit the taste of the builder.
The interior of the case should be just a
nice fit for the other boards, 2, 3, 4 and 5,
so that they may be removed at any time
should repairs be necessary. The top
board should now be placed upper side
down, and the second board 2 placed in
the proper position and secured with two
or three small round head screws, which
should not be too long, as they would
then break through the upper board of

the instrument and mar the appearance of the job. It is a little difficult to assemble the various parts in the case, so this should be done before placing it in its final position. Boards 1, 2, 3 should be assembled first, 2 and 3 having a spacer *A* in each corner to hold boards in position. These spacers may be made from wood or a brass rod cut to proper length, and drilled and tapped at each end to receive a machine screw. If brass tubing is at hand, the spacers can be made from several pieces, with a rod through them, and threaded on each end, as shown at *B*, Fig. 4. The arrangement of the spacing is left largely to the workman. When 2 and 3 boards are assembled, the carbons may be placed in position, and when all in place, board 4 may be placed in position and be secured by three or four round head wood screws, as shown. The other screws may now be screwed down until the carbons are in good contact with the end of each button, as shown. A bottom board may be made, as shown at 5, which will hide all the interior of the instrument when completely assembled, and may be placed as shown in sketch, when instrument has been placed in case.

The simple wiring which is required must be done before the instrument is placed in its case. The first button may be left *dead*, if desirable, or *off*, and be marked such; no connection will be necessary with this one. Button 2 should be connected to 3 at the bottom, where the wood screws are placed; a small piece of copper wire may be soldered or wrapped about the head of the screws. From 3 to 4, the connection should be made at the top of the carbons, as shown at *D*, those following being made in the same order, bottom to top, all the way around the circumference of the circle. When completed, each button will allow one rod of carbon resistance to be added or decreased, as the arm is moved forward or backward. The complete instrument will be considerably thicker and heavier than the first one described, but the working of same will be more accurate, and if space is not any objection, we should recommend this in preference to the other.

The top board or plate may be made from hard rubber, slate, or marble, if desirable and the necessary tools be at hand to work such material. If made from such material, the finished appear-

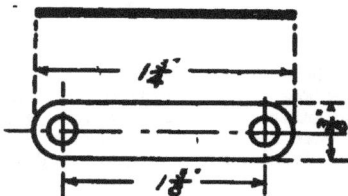

Fig. 6

ance will be very much nicer than the wood, although the wood may be finished with a fine smooth surface if properly filled and rubbed, after which it may be enameled.

There is a satisfaction which comes from putting one's best into any work which may be at hand, which is not to be considered lightly. A good instrument is more or less a difficult proposition to make, but when good service and high efficiency are desirable, this is of secondary importance. The many cheap and makeshift pieces of apparatus which are described from time to time in the magazines are good in their way, and will answer the purpose after a fashion, but if the experimenter wishes to get the most out of his work and desires to have a highly efficient equipment, he must begin with this idea in his mind and construct his apparatus from the best materials that he can procure, and tax his mechanical ability to the utmost, and the result will more than repay him for the extra labor and expense involved.

Paper Bag as a Kettle

"I had no hot water for shaving at the little country hotel, and accordingly heated some in a paper bag."

"Heated hot water in a paper bag?"

"Sure."

"How can that be done?"

"You will take a stout paper bag—or an envelope will do as well—fill it with water and hold it over a gas flame or lamp. The water heats readily. The paper doesn't burn, because it is wet, and wet paper is a singularly tough and non-combustible substance.

"Many and many a time have I heated over the gas jet an envelope or a paper bag of hot water for my shaving, and not once have I had an accident."— *Kansas City Independent.*

A HANDY LIGHTING CIRCUIT

H. P. CLAUSEN

When electric lamps are placed in telephone booths, areaways, basements and other places where a light is only required for a short length of time, the lamp is left burning practically all of the time, and it is for the purpose of providing a simple, cheap and effective arrangement that the following article is presented.

As it will be observed, the electric lamp L is connected to a contact spring K^1, the opposite contact of which connects to the electric supply line L^1, L^2, of the lighting circuit connecting to the lamp. D represents a normally closed door spring arrangement which is installed upon the door through which it is necessary to pass in order to enter the space lighted by the lamp. Push button or switch P is placed on the lamp side of the doorway; that is to say, the door must either be opened in order to permit the push button to be pressed, or the operator must be on the lamp side of the door when the door is closed. R represents a simple relay arrangement, with contact springs K^1 and K^2, these springs being so arranged that when the relay draws up its armature the normally open contact springs are closed. B represents a battery equipment, say, consisting of two or three cells of dry batteries.

The operation of the device is as follows: Assuming that the lamp is suspended in a basement, or, say, within a telephone booth, the operator passes through the doorway and, after closing the door, presses the push button P. This bridges across the break between the K^2 and K^3 contacts so that current now flows through the winding W of the relay R over wire 1 to spring D^1, through the closed contact to spring D^2 to negative side of battery B and back to the push button where the contact PX, of course, is closed while the button is being pressed. The current flowing through the winding W energizes the relay, and operates contact springs K^1 and K^2, resulting in two distinct operations: first, current from the lighting mains flows over the K^1 contact circuit and through the lamp L, lighting it. Obviously, it is impracticable to keep on press-

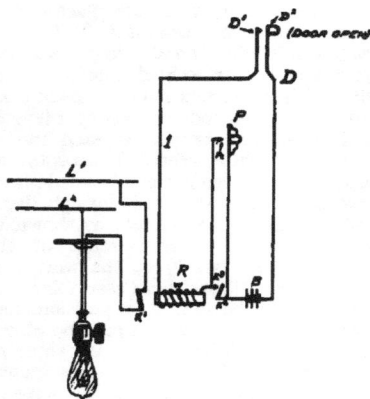

Diagram of Connections

ing the button; therefore, the operator, after giving it one short push, will find it unnecessary, for the reason that when the relay R is energized the current from battery B continues to flow over the circuit, for the relay is now locked up and will keep the circuit of lamp L closed. either until the switch of the lamp socket is operated or the door D is opened.

It is the object of the system, of course, that when the party who caused the lamp to be lighted passes back through the doorway, the door spring D opens the circuit, comprising the relay R, battery B and contacts D^1 and D^2, resulting in the armature of the relay dropping back and unlocking itself, extinguishing the lamp L and permitting the door D to be closed without again lighting the lamp.

The arrangement as illustrated and described may be assembled without calling for any special apparatus other than to have a relay provided with contact springs, as shown, and wound to a resistance of about 50 ohms.

Under practical working conditions it will readily be understood that the successful operation of the device depends upon the assumption that the push button P cannot be pressed from the outside with the door closed, and that the lamp L cannot be lighted unless the push button P is pressed with the door closed.

HOW TO MAKE A USEFUL HOUSE TELEPHONE

It is proposed in this article to describe a more efficient telephone apparatus such as would be of service in a house or shop. The exterior of the instrument itself is shown in Fig. 1, and as this case contains all the working parts of the system, save for the battery, readers will see that the apparatus is extremely neat.

The first step in fitting up a system such as is about to be described is to make the two instruments, one for each end of the line, which in an ordinary house would be perhaps the dining-room or bedroom and kitchen. The cases can be made from any wood to suit the surrounding furniture or woodwork. The wood used should finish to ⅜ in. thick, and the cases should be jointed and fitted together as workmanlike as the reader can possibly manage, and finished, if desired, in an artistic manner. The sizes and shape of these cases are shown in Figs. 1, 2 and 3, which are drawn to scale, and any further dimensions can easily be taken from them. In the front view, Fig. 1, at the bottom will be seen the speaking hole or mouthpiece, and this should be finished exactly as shown.

To give the reader an idea of what the internal construction of the complete instrument is like, a view is shown in Fig. 4, with the back taken off and all the pieces of sheet brass are shown separately in the detail, Fig. 5, the letters there given corresponding with those given in Fig. 4.

The first fitting to make is the diaphragm holder A, which is cut out of sheet brass and stamped to the size and shape shown, and when finished should be screwed into position exactly concentric with the speaking hole, as shown in Figs. 1 and 4. We next require the brass piece B, 1 in. long and ⅜ in. broad, screwed on in the position indicated by the center screw.

Over the piece B is a spring piece C, 2 in. long by ⅜ in. broad, and on referring to the detail, it will be seen that the end is bent up, so it can make perfect contact with the piece B. This piece must be screwed to the case, as shown in Fig. 4, so that the bent or turned up end, above referred to, lies over the lower end of C for at least ¼ in., as shown in detail, Fig. 5, to be operated by the center stud P, Fig. 1. The angle piece D must next be cut from sheet brass, ⅜ in. wide, and of a length in the long arm X so as to allow the spring piece C to make perfect contact when it is pushed from the outside by a small push through P, Fig. 1. Normally, the arm X of this angle piece should rest about ¼ in. away from the spring piece C. The next fitting is the receiver hook E, two views being given

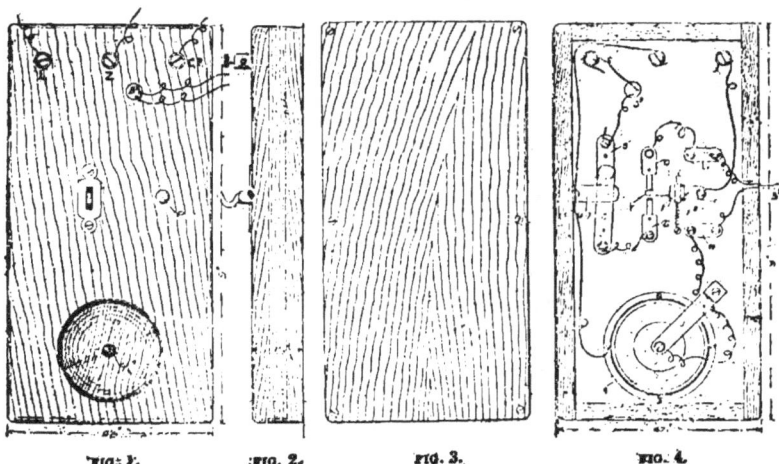

FIG. 1.   FIG. 2.   FIG. 3.   FIG. 4.

in the detail, in order to show the long pin Y. This hook is for the receiver to rest on, and is kept in position by passing through the center slit in Fig. 2 and 4, and then screwed to the angle brass piece F, by a screw F^1, the $\frac{1}{8}$ in. hole Z in the piece F being tapped to take the screw F^1.

In connecting these last two fittings a small spiral spring, shown at F^2 in Fig. 4, is also employed. The angle brass piece F is secured by its screw to the wooden case, the switch hook E, passed through and then the screw F inserted to make all secure. Two brass pieces G are then cut to the size and shape shown, and fixed as indicated in Fig. 4, so that the projecting pin of E can touch one or other of the curled ends of G. The small spiral spring F^2 is now fitted, as shown in Fig. 4, so as to keep the pin always in touch with the upper piece G, and only the weight of the receiver on the other end of the hook will cause the pin to make contact with the lower piece G.

Two other plates must next be screwed on at H, and then the transmitter is ready to be finished.

Fig. 6 shows a section of the "Hunnings" transmitter, which is made by fixing a thin carbon diaphragm U in position under the brass ring A, with just a thin ring of blotting-paper V inserted, as shown enlarged in Fig. 7. At the back of the carbon diaphragm is a block of carbon L, with a serrated inner edge and kept away from the diaphragm by means of rings of cotton wool or soft felt, the felt being glued to both.

The bent arm J is then fitted in place,

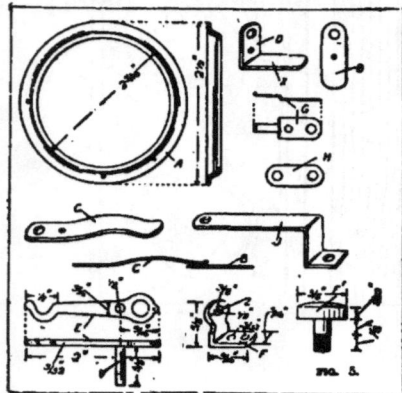

FIG. 6.   FIG. 7.

as shown in Fig. 4, and the bell fitted onto the front in a convenient place. This bell should preferably be one in which the works are under the gong itself, and these can oftentimes be picked up in the market places. However, any bell mechanism will answer, simply fixing the working parts to the front instead of to the backboard of the bell, the cover being affixed as at present adopted, or the bell can be put away in a convenient place.

The connections of the instrument are as shown, and in fitting it is of course essential that as the wire is naked in the body of the instrument, it is best to fix carefully and not to allow short-circuiting. The wires leading to outside should, of course; be covered, except where going under binding screws.

One wire starts from underneath the screw on the rim A, passed along the side of instrument and fixed once round under the head of the screw, fixing the angle piece D to the side of the case. From thence the wire goes to the center screw at the top.

Another naked wire passes from the screw of the angle piece F, holding switch hook E to the terminal at the top, as shown. Two short lengths of wire also connect G and H, and G and C, as shown, as well as a length of wire from H to the center carbon block at the bottom. The other wires lead outside, as shown.

We now come to the construction of the receiver. This is of the watch pattern, and is constructed briefly as follows: The outer case is cup-shape in form, as shown in Fig. 8, about 2 in. diameter, $\frac{5}{8}$ in. deep and made from hard

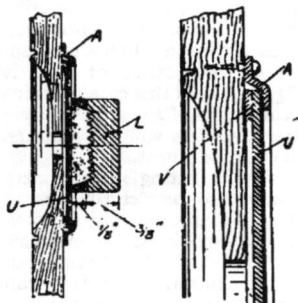

FIG. 5.

wood or iron, about ⅛ in. thick, holes being bored in the bottom about ⅞ in. apart for the connecting cords. The front, Fig. 9, is screwed on Fig. 8, to which is afterwards fixed the mouthpiece, Fig. 10. The circular magnets, Figs. 11 and 12, are of sheet steel; these should then be inserted, fixed to the outer case by a small screw through the top, Fig. 13 is a cross section, which should make all clear.

The pole pieces, Fig. 14, are made from soft iron, and the bobbins, Fig. 15, covered with No. 38 B.W.G., each coil being wired in opposite directions, commencing with one coil and finally finishing onto the other coil.

The terminals are placed onto a piece of vulcanized fiber, Fig. 16, with holes for the fixing screws, the flexible cords having their ends bared where connected underneath the heads of such screws, but covered where passing through the case (viz., through the holes bored at the bottom of case, Fig. 8).

The diaphragm is fitted as shown in the side view, a ring of paper being put round the periphery at each side. To take the strain from the connecting cords, a ring is fixed at the bottom of the case for another cord which can be plaited into the other two.

The action of the instrument is as follows: Immediately the push in front is depressed, the pieces B and C make contact and the circuit to the distant bell is completed thus: Starting from carbon of battery, the current passes along up to the middle terminal at the top, from thence down the wire inside to D, through C (because contact is made) and to G and F¹ (because receiver is on hook) to the line wire L (see Fig. 1). This line wire, of course, passes to neighboring or second instrument. The current, on reaching the second instrument, passes to C through exactly the same channels, but instead of passing to D (as this push is of course not depressed) passes to B through bell, which it rings and thence to earth and back to original battery.

The bell has rung and the second receiver is taken from hook, as also is the first one, so that conversation can be started, and by so doing spring F² pulls end of E down and thus the pin Y makes contact with the upper plate G. The current is now conveyed through both batteries, and conversation can be heard quite plain at the receivers.

Thus only one line is necessary connecting the L terminals (see Fig. 1) of instruments. The middle terminal (Fig. 1) is connected to the zinc of a two-cell battery, while the other terminal CE (Fig. 1) is connected to the carbon of the battery and to earth. The two terminals of the bell are connected through B¹ (see Fig. 4) and the receiver cords through B². A suitable back board should be fitted to protect the working parts, and a nice brass plate for the receiver hook E to work in.

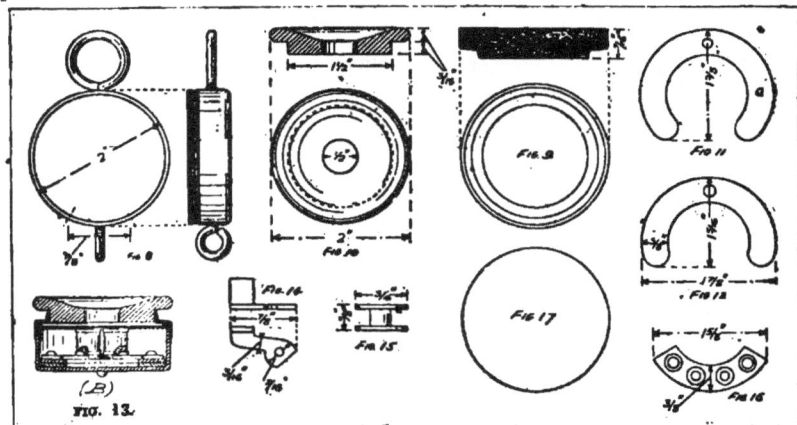
FIG. 13.

A SWITCHBOARD FOR EXPERIMENTERS

M. C. MORRIS

This switchboard is designed for electrical experimenters, and can be used for testing wireless aparatus, motors and coils. It can be used on 110 volts A.C. or D.C., or batteries of any kind, but when used on house-lighting circuit the main-line fuses should be increased to 25 amperes and the circuit feeding the board to 15 amperes.

The material needed for constructing this board is as follows: One board, 16 x 22 in. of fiber, pressed asbestos, oak, slate or hard rubber (this should be about ¼ or ½ in. thick); 10 porcelain Edison base lamp receptacles; one 15 ampere double-pole double-throw knife switch; one small single pole knife switch; one piece of brass (spring) about the thickness of cardboard and cut to the shape and size shown in diagram for rotary switch; 10 binding posts which can be taken from worn-out batteries. Binding posts with large heads or round head machine screws (brass) may be used as

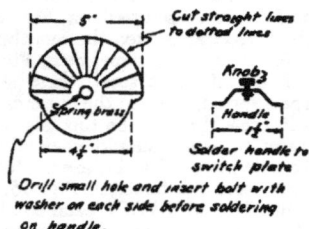

Drill small hole and insert bolt with washer on each side before soldering on handle.

Rotary Switch

contacts for the rotary switch, but be sure that they are all equally high above the surface of the board.

Use No. 14 rubber-covered wire when wiring the board. Connect as shown in the !sketch. Mark the double-pole, double-throw switch D.C. on one side and A.C. on the other. Mark single pole switch 110 volts, and be sure that this switch is kept open except when you wish to test a high resistance, otherwise you will get a short-circuit, and blow a fuse. If 16 c.p. or 60 watt lamps are used in the receptacles, the switch on all points will give 5 amperes at the lower terminals, and if 32 c.p. lamps are used it will be 10 amperes. The rotary switch controls one light at a time so that any amperage up to the full capacity may be had at the terminals, and if 110 volts is desired straight, without the lamp bank, throw in the single-pole switch.

If batteries are connected to the switchboard, throw the double-pole double-throw switch over to the side to which the batteries are connected, and then throw in the single-pole switch. Three sets of terminal binding posts are shown on the switchboard, which will be found very 'convenient, although only one set is enough if not convenient to put on more.

It would be a great convenience to mount a switch and fuse-block on the board or near the board and let the feed wires come through them, so that in case of a short-circuit the house will not be in darkness until you fix a new fuse in the main line or branch circuit. Be sure to have the switchboard fuses lighter than either the main line or branch circuit fuses.

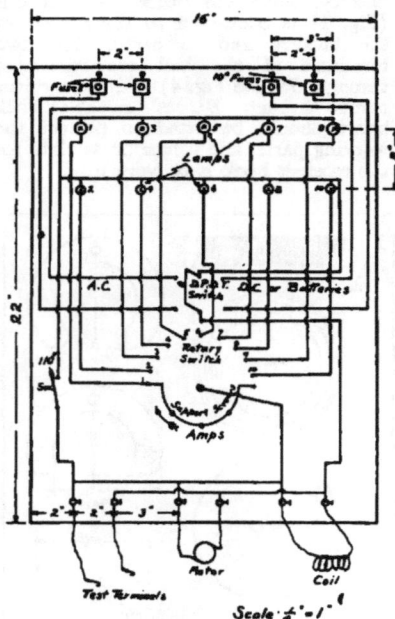

Diagram of Connections

THE CONDUCTION OF ELECTRICITY IN METALS

FRANCIS HYNDMAN B.SC.

In recent years our ideas with regard to the mechanism of metallic conduction have undergone considerable change—or, rather, there are several very various methods of explaining the same fundamental facts. These methods start from entirely different conceptions, but really arrive at the same stage of exactness in explaining and foretelling the observed phenomena.

Until recent years all observations on absolute or relative conductivity were made at ordinary or higher temperatures. These show that metals differ very largely in their conductivity, both for electricity and for heat, and that, taken generally, the order of conductivity would be the same in the two cases.

In every known case the conductivity decreased with rising temperature, and in such a proportion that it might reasonably be presumed that the conductivity would gradually increase with lower temperatures, until at the absolute zero it would be infinitely good, or, at any rate, very large.

This property has been of late years used for the purposes of thermometry. The resistance (the inverse of conductivity) being measured at certain definite temperatures well known by other means, such, for instance, as the melting-point of ice, the boiling-point of water, or of other substances, such as oxygen, sulphur, etc.

The results of many accurate measurements have shown that the resistance of even the purest obtainable metal (mercury) is not simply proportional to the temperature, but that it increases more rapidly, so that if R_t is the resistance at any temperature, $t°$ centigrade and a and b are quantities which are constant for the particular wire, nearly constant for different wires of the same metal in the same state of purity, and not very different for most pure metals, then

$$R_t = R_0(i + a\,t + b\,t^2)$$

will very nearly give the value at any temperature.

There are three metals which are worth special attention, and attention may be confined to them. First, mercury, because it can be obtained almost absolutely pure, by distilling it in vacuo at temperatures of about 60 degrees to 70

degrees C., and condensing it at the temperature of liquid air – 290 degrees C. Secondly, gold, because, of the metals which can be drawn out into wire, it can be obtained purer than any other. Finally, platinum, because of its durability, and it is, in consequence, almost universally used for thermometry.

The following small table gives the relative resistances of each metal against its value at zero. The temperatures are given in what are known as "absolute temperatures," that is, temperatures as if measured from the absolute zero, which is here taken to be -273.1 degrees C. In the next column the absolute temperatures are divided by 273.1, as it will be seen that the relative resistances are proportional to these numbers.

From observations with gold with various degrees of purity, it appears that the resistance is less the purer it is, and that the differences are proportional to the amounts of foreign metals present. This would lead to the conclusion that perfectly pure gold would have a zero resistance about 4.3 degrees Ab. It is not possible to obtain platinum at any great degree of purity; but here, again, observations on various wires of known composition would lead to the conclusion that in any case the resistance would nearly disappear at 4.3. These results are very remarkable, but their full significance is yet hardly realizable, until observations have been made on the other properties of the same metals at these extremely low temperatures. Although decreasing nearly to zero and then remaining constant, gold and platinum do not present any abnormal behavior; the case of mercury is, however, quite different. It behaves as the others until

Temp. Absolute	How Obtained	T-273.1	Platinum	Gold	Mercury
373.86	Boiling water..	1.3	1.405	1.397	—
273.1	Melting ice....	165	1	1	1
169.29	Methyl chloride	0.617	0.581	0.593	—
97.93	Oxygen.........	0.285	0.199	0.219	0.279
20.18	Hydrogen......	0.074	0.014	0.008	0.056
13.88	Hydrogen......	0.054	0.010	0.003	0.033
4.30	Helium	0.016	0.009	0.002	0.002
1.30	Helium	—	0.009	0.002	0.000

a temperature of about 4.30, being a worse conductor than either gold or platinum, from its freezing-point downwards. At this point the resistance decreases so suddenly that at 4.2 degrees Ab. its value is 0.115×10^{-5} ohms, while at 4.2 degrees Ab. it is practically zero, and it continues to have this value to the lowest measured temperatures.

These results were quite unexpected. Under various theories the resistance would be infinite at Absolute zero and under others would vanish. Neither appears to be true, and the minimum point is at practically the boiling point of pure helium under atmospheric pressure. Why this particular temperature is not known modern theories of matter and electricity suppose that the material atom consists of a negative electron enclosed in an envelope of positive electricity. The electron is free to move in the envelope, and the nature and extent of its motion are of fundamental importance. In conductors such as metals there are also free electrons, which have power to move, their passage giving rise to the phenomenon known as an electric current.

Further investigation under these conditions will certainly lead to the most fruitful results, and cannot fail to have a most important bearing on the conception of the passage of electricity in metals. It is not too much to say that they could not be carried out at all, except in one or two places in the world, owing to the impossibility of obtaining these conditions.

Not only is this the case, but they represent part of the result of twenty-five years' continuous work with one object in view, and require in themselves extreme care, patience and knowledge.

The greatest credit is due to Professor Onnes, of Leiden University, who not only first liquefied helium, but has reduced the labor of accurate measurements in it to one of ordinary laboratory routine. Such work, condensed into a few pages a year, is worth volumes of the so-called research work turned out in some other places, as it is framed on a definite scheme, and gives results which are of the greatest scientific and technical value.—*English Mechanic and World of Science.*

THE TELEPHONE AND THE FARMER

The telephone has become as great a factor of farm life as the harvester and reaper and other labor-saving devices that contribute to the farmer's prosperity. The farming implements help prepare the crop for market, but without the telephone the farmer cannot sell the crop to the best advantage. His handicap used to be that he was too far away from all agencies of business, now the agencies of business are no farther away than the telephone box on the dining-room wall.

An apple grower in an eastern state had an experience last fall that illustrated how large may be the return on the money invested for a telephone. One day when his crop was just in the right condition to market, a traveling agent offered $1.00 a barrel for the 1,000 barrels he had to dispose of. The buyer insisted that the price was as high as anybody was paying, and the deal was almost closed, when it occurred to the apple grower that it would do no harm to see how much somebody else would offer.

So he called up the city commission house that had bought his fruit in previous years, and they not only offered him $1.50 a barrel, but agreed to send a man to the farm to do the packing. The additional $500 made on that transaction will pay telephone bills for a good many years to come and leave a snug little sum over.

The coming summer will no doubt see great development along electrical lines in St. John, N.B. The Intercolonial Railway has given the St. John Railway Company the right to cross their tracks at Haymarket Square. The street railway will be extended to Kane's Corner and to Fernhill. It is understood that about nine miles of track will be laid. The Railway Company, which supplies the light and power to the city, have also made arrangements for the extension of their power line to Rothesay, where many of the houses have been wired, and it is understood that several factories will operate there.

NATURAL MAGNETS

WM. A. MURRAY

Magnets may be grouped into two classes: natural magnets and artificial magnets. The particular class to which this discussion applies is the first-mentioned, *viz.*, natural magnets.

The term magnet usually is applied to any substance which possesses the property of attracting pieces of iron or steel when such pieces are placed in their immediate vicinity, and which substance will, when freely suspended, come to rest pointing nearly north and south.

The term natural magnet can be correctly applied only to a mineral substance known commercially as oxide of iron, and then only when the specimen possesses the properties mentioned above, as it has been found that some good specimens of oxide of iron do not possess the properties mentioned, and should not, therefore, be termed natural magnets.

It is generally supposed that this ore and the properties which are peculiar to it were first discovered and experimented with by the ancients who first found it in or near the town of Magnesia in Asia Minor, and derives its name from the place of its supposed origin, Magnesia.

During recent years excellent specimens have been found in various parts of the world, some having been found in the state of Arkansas.

The term lodestone, meaning leading stone, is sometimes used in referring to this ore (oxide of iron). This term derives its name from the fact that it was sometimes used by the early navigators as a guide for steering ships, its uses for this purpose being adapted to practical navigation due to the fact that it would, when freely suspended, come to rest pointing in a certain definite direction, and always with the same end pointing in the same direction. It will be readily understood from this that a navigator, by its use, could get a somewhat definite idea of the direction in which his ship or craft was sailing.

The origin and practical application of its use for the purposes just described is generally accredited to the ancient Chinese.

The commercial use of this ore is to make iron, the same in kind as is made from any other kind of iron ore, and it is said to make a very good quality of iron.

As heat alone, of sufficiently high temperature, is capable of producing the destruction of the magnetic properties contained in either the natural or artificial magnets, there is no doubt but that any magnetic properties existing in the oxide of iron are destroyed during the process of making iron from this ore.

Another property which was found to be possessed by this ore was that it could be used to impart its magnetic properties to other pieces of iron and steel, thus creating artificial magnets.

"When a bar or needle of hardened steel is rubbed with a piece of lodestone, it acquires magnetic properties similar to those of the lodestone, without the latter losing any of its own magnetism, such bars are called artificial magnets."

Lodestone is not generally used for the purpose of making artificial magnets, the principal reason for this being that its own magnetic force is not very strong, and it has been found that a more convenient means and better results may be obtained by the use and proper manipulation of the electric current.

The following described tests may be made for the purpose of determining accurately and positively whether or not a given substance is a magnet either natural or artificial. Plunge the substance which it is desired to test into a quantity of iron filings, small iron or steel tacks, or similar materials, when it will be found that a small quantity of the filings or tacks will have clung to the end of the substance under. test if it is a magnet. Suspend the piece to be tested freely by means of a string or thread, and it will be noticed that if the substance is a magnet it will come to rest always with the same end pointing in the same direction; if not a magnet no certainty of its position when it comes to rest can be foretold.

Approach the piece to be tested to the points of a horizontal magnetic needle, when it will be found to attract or repel the needle according to which pole of the needle is approached; as according to the laws of repulsion and attraction: like poles repel each other; unlike poles attract each other.

FLIGHT

NEW AERONAUTICAL SPEED RECORDS
Arranged by Austin C. Lescarboura, Member of the Aeronautical Society.

The French aviator, Vedrines, on the 22d of February at Pau, France, established remarkable speed records eclipsing all speed flights made thus far in aeronautics. On January 13th, he had already established speed records for the 50 and 100 km. distances, but the aviator Bathiat had broken these records ten days later. After breaking both Bathiat's records for the 50 and 100 km. flights, Vedrines continued his flight and broke all speed flights for distances up to 200 km. The following table illustrates the comparison of the flights by the rival aviators.

Distance	Bathiat's Record (Jan. 27th)	Vedrines' Record (Feb. 22d)
50 km.	20 min. 43 sec.	19 min. 3 sec.
100 km.	41 min. 29 sec.	37 min. 58 sec.
150 km.	....	56 min. 41 sec.
200 km.	....	1 hr. 15 min. 20 sec.

The former record for the 200 km. distance was held by Tabuteau, who had made this distance in 1 hr. 54 min. and 21 sec. The speed figures made this year thus far indicate that enormous speeds will be witnessed before 1913. Already, the records in kilometers per hour around a pylon course have been broken as follows:

Vedrines (Jan. 13th) 145 km.	Bathiat (Jan. 27th) 147 km.	Vedrines (Feb. 22d) 169 km.

The records made in aviation are kept with the greatest exactness, so that they may be referred to at any time. At the end of each year, the best record in each feature of aviation competition is recorded, so that the following table is obtained from these records. This table illustrates the remarkable progress made in aviation more fully than numberless words of praise.

These figures are the official ones of the International Federation of Aeronautics, and are given in the metric system measurements. The reader may convert the kilometer figures to miles by multiplying them by .621, and the meters by 39.37 to obtain the American inches.

		DISTANCE		
Year	Aviator	Date	Country	Record
1906	Santos-Dumont	Nov. 12	France	0 kilometer, 220
1907	H. Farman	Oct. 26	France	0 kilometer, 770
1908	Wilbur Wright	Dec. 31	France	124 kilometer, 700
1909	H. Farman	Nov. 3	France	234 kilometer, 212
1910	Tabuteau	Dec. 30	France	584 kilometer, 745
1911	Gobe	Dec. 25	France	740 kilometer, 299
		DURATION		
1906	Santos-Dumont	Nov. 12	France	21⅕ sec.
1907	H. Farman	Oct. 26	France	52⅗ sec.
1908	Wilbur Wright	Dec. 31	France	2 hrs. 20 min. 23⅕ sec.
1909	H. Farman	Nov. 3	France	4 hrs. 17 min. 53⅗ sec.
1910	H. Farman	Dec. 18	France	8 hrs. 12 min. 47⅘ sec.
1911	Fourny	Sept. 1	France	11 hrs. 1 min. 29⅕ sec.
		ALTITUDE		
1906				None recorded
1907				None recorded
1908				None recorded
1909	H. Latham	Dec. 1	France	453 meters
1910	G. Legagneux	Dec. 9	France	3100 meters
1911	R. Garros	Sept. 4	France	3910 meters
		SPEED PER HOUR		
1906	Santos-Dumont	Nov. 12	France	41 kilometer, 292
1907	H. Farman	Oct. 26	France	52 kilometer, 700
1908				Same record
1909	L. Bleriot	Aug. 28	France	76 kilometer, 955
1910	A. Leblanc	July 10	U.S.	109 kilometer, 736
1911	E. Nieuport	June 21	France	133 kilometer, 136

AERONAUTICS IN THE FRENCH ARMY
The New War Establishment

For the first time in history, air craft will be definitely incorporated in the war establishment of one of the great armies. It is true that the French Army has already, and for some years now, been in possession of a splendid aerial organization, but, with the exception of the new military dirigibles completed during the last six months, it has never hitherto possessed any air craft intended for actual warfare. Steps have now, however, been taken to proceed at once with the creation of a war fleet composed of craft no longer intended for instructional purposes, but for the operations of war itself, and maintained in a state of instant preparation for mobilization.

The newly-formed French Ministry, through the War Minister, M. Millerand, have decided to ask the Chamber for an appropriation of 22,000,000 francs, or $4,400,000, for the purpose of military aviation during the present year.

It may be recalled that by the famous aeronautical program of February 25, 1910, a sum of 20,000,000 francs ($4,000,000) was voted for the construction of a fleet of war dirigibles and the accompanying sheds. This program of construction is to be completed by the end of 1913. It has already equipped the French Army with a fleet of 10 dirigibles of the most recent type—all of them launched during the last eight months—in addition to the three old training vessels, and with 11 modern dirigible sheds scattered along the eastern frontier. Further, the Army possesses three transportable dirigible sheds, while ten further sheds are privately owned and would be available in case of necessity.

Within the last few weeks the French General Staff has strongly reported in favor of the continuation of the dirigible program, an opinion largely based, no understood in the light of previous events.

The French Army purchased its first aeroplanes in 1909. Since then the aeroplane fleet has grown according to the following figures:

1909 .5 aeroplanes
1910 (Feb.) . . . . .20 aeroplanes
1910 (Oct.) . . . . .32 aeroplanes
1911 (Dec.) . . . . .254 aeroplanes

The growth in numbers is remarkable, but—and this is the important point— the machines hitherto acquired are not military machines in any special sense, they are aeroplanes of the ordinary commercial types, and military only in so far as they belong to the Army. Now, the whole of this imposing and unquestionably efficient fleet has by one stroke of the pen been struck off the active list. Henceforward these vessels will simply be used for instructional and training purposes, and will be relegated to the schools.

Simultaneously the authorities have set to work to build up a fleet of war machines constructed solely for military purposes. Towards the attainment of this end the vote of $4,400,000 for the present year, which will undoubtedly be readily granted by the Chamber, forms the first step. This sum will suffice to equip the Army with 322 aeroplanes with the necessary shed accommodation and repair trains. These machines will be distributed among the various forts and attached to the different army corps.

A military pilot will be placed in charge of each aeroplane, which he will navigate himself on mobilization. None of these machines will be used for instructional purpose during their year on the active list. They will only be maintained in a state of efficiency and readiness by their respective pilots. At the end of the year they will be transferred to the training

THE DETERMINATION OF OCEAN CURRENTS

The officer of a vessel, who, while at sea, steps to the rail and hurls a well-corked bottle overboard, in which there is a slip of paper, is not bent on a romantic venture. Instead of a request for a letter of acknowledgment and possibly an acquaintance from the seaside maiden who might find it, perhaps a year later, that slip of paper contains the name of the vessel and its master, the date and the latitude and longitude at which the bottle began its voyage. Below this there are blank spaces for the name of the finder, the date, locality and the post office address.

At the bottom of the slip, printed in eight different languages, are these instructions: "The finder of this will please send it to any United States Consul or foward it direct to the hydrographic office, Washington, D.C."

The whole operation represents a little step in a great task upon which work has uninterruptedly gone forward for the past thirty years. It is one means of determining the direction and movement of the great as well as the smaller currents of the ocean, and it will perhaps be due to the continual varying of the latter currents that this task may never be ended.

There has always been a knowledge of the existence of ocean currents but until the early 70's it was limited. Mariners knew that on the west coast of Europe there was a force which tended to carry them southward, and that on the east coast of North America a similar force which carried them northward, but they knew neither the true set or direction, nor the drift or rate of flow.

It was the object to determine the direction and rate of flow in originating this system of bottle voyages, and while each bottle is but a little step toward a great end, those little steps have so far succeeded in establishing beyond question the true direction and speed of the great currents of the five oceans.

The advantage of such knowledge is not limited to the mere use of the direction and velocity of the ocean currents. The maritime world does, in fact, depend mostly on this knowledge in laying the courses of its vessels from and to the great ports of the world. For instance, vessels leaving New York will follow the Gulf Stream on its great curve north-eastward across the Atlantic to Europe, and thus gain the advantage of its six-knot-an-hour flow, while those ships returning from the other side will lay a course much farther to the south to escape the retarding effect of the same stream.

It has been found also, that these great ocean streams have a powerful effect on the course of storms after they leave the mainland of North America. The knowledge of the direction of these currents gives an insight into the general track of these storms. The mariner, therefore, has gained much from the tale told by the little voyagers, because all of the great ocean-sailing routes which, for the past score of years have been followed by both steamers and sailing vessels alike, were mapped out with the aid of the date secured by this means.

There are twenty-seven permanent currents in the oceans of the world, and there are nearly as many more of the semi-permanent variety existing at one time. Several causes tend to originate and maintain these drifts. Uniformly directed winds have the greatest influence, and differences of temperatures, storms, polar ice and eddies each have some effect, creating usually the currents of semi-permanent variety.—*Washington Star.*

Letter Delivered after Fifty-four Years

After it had been more than fifty-four years in the mails, a letter was delivered recently to a woman in Newark, N.J. One morning the Newark newspapers reported that a letter had been received at the post-office addressed to Miss Elizabeth Garthwaite, the postmark on which showed that it had been mailed in New Orleans on December 30, 1854. The letter proved to have been written by a cousin who attended the same school more than half a century ago. The ink inside the letter was so badly faded that the letter was hardly decipherable, but the address was plain. No one was able, or willing, to offer any explanation of the whereabouts of the letter during the years that elapsed since it was mailed.

KEROSENE AS AN ENGINE FUEL

JAMES A. KING

The power obtained from any fuel depends on two things. These are the fuel itself and the engine. No more power can be obtained from any fuel than that which is stored in it. The percentage of the power stored in a fuel which is available at the crank-shaft or the draw-bar of an engine depends largely upon two factors. These are the ability of the engine to burn the fuel and the amount of power required to operate its working parts.

Kerosene contains more power to the gallon than does gasoline. But most of the internal combustion engines on the market today are not capable of burning kerosene economically or successfully. This is because they are built as gasoline engines and not as kerosene engines.

The manufacturer is up against two problems when designing a kerosene engine. He must so build it that it will properly spray the kerosene and mix it with air. He must so build it that it will prevent the deposit of free carbon when the charge of mixed kerosene and air is exploded. These problems arise from the nature of the fuel itself.

Kerosene is heavier than gasoline. It vaporizes at a considerably higher temperature. It consists of a mixture of much more complex compounds which burn at a wider range of temperature. In the presence of high heat some of these more complex bodies break up into simpler ones which contain smaller proportions of carbon, therefore free carbon is deposited.

In order to burn kerosene successfully, an engine must work with its cylinders hot enough to vaporize the fuel properly, yet the cylinders must not be hot enough to bind the pistons or give pre-ignition. This means that the manufacturer must carefully design his cooling system with reference to the capacity of his engine so as to produce these conditions. To obtain the required temperature it is necessary to start an engine with gasoline and run it until cylinders and cooling system are sufficiently hot to properly vaporize the kerosene. Then the gasoline is shut off and the kerosene is turned on.

But this is not the end. If it were,

then any engine that would burn gasoline could also successfully burn the heavier fuels. This is where the composition of kerosene begins to show its effect. Many engines will run successfully on this fuel for a short time, but usually a carbon deposit soon forms on the piston head and the inner walls of the cylinder. It takes but a short time for this deposit to become disastrous. All engine users know how annoying carbon deposits in their engines can be.

The reason why carbon is deposited may help one to understand how it may be prevented. Much is known about the results which are obtained when certain things are done with carbon compounds, but there is some difference of opinion as to the exact changes gone through by these compounds in reaching the final results. Consequently, I offer the following explanations as indicating what probably takes place during combustion and not as indisputably authoritative statements of what actually does take place.

It is well known that when certain complex bodies found in kerosene are subjected to temperatures as high as those found inside the cylinder of a working engine they break up into simpler compounds. Thus

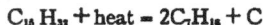

$$C_{15}H_{31} + \text{heat} = 2C_7H_{16} + C$$

It is quite probable that most of the more complex bodies contained in kerosene go through some such process of breaking down as this before combustion takes place. So that when kerosene is used in an engine under the same conditions as gasoline it is probable that free carbon is deposited as above explained.

But it has been found that when a mixture of air and vaporized kerosene is exploded in the presence of a small quantity of water vapor the combustion of the kerosene is much more rapid and complete than when the water is absent.

Recognizing this fact, some manufacturers of kerosene burning engines prevent the deposit of free carbon by spraying a small quantity of water into the cylinder with the charge of fuel, using a separate water cup in their carburetor, built the same as the fuel cup.

Just how the water acts in this case is not fully known, though plenty of evidence is at hand regarding the beneficial results obtained by the use of water vapor. The following chemical equations look good to me as an explanation of the chemical actions that take place from the time the charge, consisting of a mixture of air, kerosene and water spray, is drawn into the cylinder until combustion has ceased.

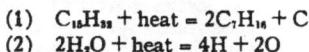

(1) $C_{15}H_{32} + heat = 2C_7H_{16} + C$

(2) $2H_2O + heat = 4H + 2O$

Reactions (1) and (2) take place as sort of a preliminary to final combustion, since the temperature of a working cylinder reaches round about 3,000 degrees, sufficient to cause the splitting up of the carbon compounds and the dissociation of water into the free elements—hydrogen and oxygen.

Nascent atoms which have just been released from combination are more highly active than those which are not nascent. As a young man who has just been thrown over by one girl is caught up by another more readily than he would be had it not been for the throw-down, so the nascent oxygen from the dissociated water combines more readily with the nascent carbon from the broken down carbon compounds than would the normal oxygen of the air. Thus it seems to me that the further reactions may properly be written as follows:

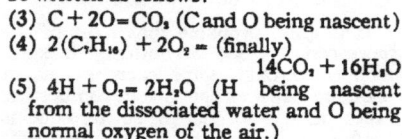

(3) $C + 2O = CO_2$ (C and O being nascent)

(4) $2(C_7H_{16}) + 2O_2 =$ (finally)
$$14CO_2 + 16H_2O$$

(5) $4H + O_2 = 2H_2O$ (H being nascent from the dissociated water and O being normal oxygen of the air.)

Another series of equations that have been advanced in explanation of what takes place in the presence of water vapor is as follows:

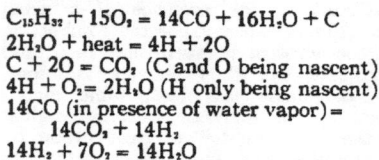

$C_{15}H_{32} + 15O_2 = 14CO + 16H_2O + C$

$2H_2O + heat = 4H + 2O$

$C + 2O = CO_2$ (C and O being nascent)

$4H + O_2 = 2H_2O$ (H only being nascent)

$14CO$ (in presence of water vapor)$=$
$$14CO_2 + 14H_2$$

$14H_2 + 7O_2 = 14H_2O$

In championing this series of equations it is claimed that the water acts in some way as a sort of catalyzer, much as in the germination of grains diastase aids in changing starch into sugar.

At any rate, when water vapor is present the combustion of kerosene fuel is much more complete than without it. There is another beneficial action resulting from the presence of the water and the high heat of the cylinder.

As said before, kerosene consists of a mixture of complex compounds which burn at a wide range of temperatures, but in the presence of the high temperature found in a working cylinder these more complex compounds break down into simpler ones before combustion takes place. This action results in a larger amount of the charge burning more nearly at the same time and so giving a more powerful explosion than if the combustion were spread out more uniformly through the length of the piston stroke. Also it has been found that a combination of water vapor and high temperature increases the rapidity of the combustion. Thus a combined increase in the quantity of material "combusticated" and in the rapidity of the combustion increases the net power obtained from the explosion.

There is still another beneficial effect of the water. Without the presence of water when burning kerosene in an engine designed to burn either kerosene or gasoline, the combustion end of the cylinder becomes so hot that the piston binds and so reduces the net power delivered to the crankshaft. In fact, I have seen engines become so hot that they "killed" themselves. Now the introduction of water into the cylinder prevents this.

Prior to explosion a great deal of heat has been consumed to dissociate the water. Just as a great deal of heat is used up in evaporating or vaporizing the water so a great deal more is used up in dissociating it or breaking it up into the free elements—hydrogen and oxygen.

This reduces the amount of heat that must be conducted out through the cylinder walls at the combustion end and thus enables them to properly perform this part of their duties. But at the outer end of the stroke where the temperature has dropped lower than it was at the combustion end, the hydrogen, freed by the dissociation of water, combines with the normal oxygen of the air, giving out some of the heat that was used up, or stored up, by the dissociation of the water.

This heat is readily conducted away by the cylinder walls at this end of the stroke which, otherwise, would not be required to conduct so much heat. The cylinder walls in some four cycle engines are aided in this task by placing a relief exhaust port near the outer end of the stroke. This port is uncovered just before the piston reaches the end of its outstroke and some of the gases are exhausted through it instead of through the exhaust valve in the combustion chamber. Thus they take their heat out with them and it does not need to be conducted away by the cylinder walls and the cooling system.

I do not believe the combining of the hydrogen and oxygen at the outer end of the stroke, with the consequent giving up of heat, increases materially, if at all, the sum total of the pressure exerted upon the piston during the full outward stroke. The principal benefit obtained from this is the distribution of the heat of combustion over a larger area of cylinder wall, de-

creasing the temperature at the combustion end and increasing it at the other, thus increasing the capacity and efficiency of the cooling system by distributing the heat more uniformly over the entire conducting surface. As the physicist and the thermo-chemist would say, the endothermic action is equal to the exo-thermic. The heat absorbed by the dissociation of water at the combustion end equals that given out by the formation of water at the other end of the cylinder.

So, to sum it all up, it seems to me that the water accomplishes the following things: It increases the total gross power by increasing the completeness of the combustion of the fuel. It increases the total net power by increasing the amount of the charge of fuel that is burned at a given temperature; by increasing the rapidity of that combustion; and by increasing the efficiency of the cooling system through a more uniform distribution of the heat over its entire surface.—*Gas Review.*

A NEW RAIL BOND

A company is being organized in Pittsburgh, Pa., to manufacture a new type of rail bond for bonding track rails in cases where electric current is used. John J. Jamison, of Wilkinsburg, is the inventor of this method. He has carried out his experimental work over a period of six years, and installations have been made on several coal roads. The method, it is said, will prevent the theft of copper and eliminate practically all maintenance costs.

The bond is a bushing made of soft copper, the outside of which is slightly larger in diameter than the hole into which it is to be placed. One or more of the bonds are installed on either side of the rail joint, depending on the size of the rail and the strength of the current in amperes which is to be transmitted. The reaming insures a perfectly round hole with clean, smooth sides. The bond is forced through the rail and fish plates, and is expanded with a mandrel until the copper comes into contact with every part of the surrounding steel. The final step is to insert and tighten up the bolts.

It is claimed for the new bond that the fish plates and the copper bushing carry current around rail joints more satisfactorily than copper wires or strips. A small amount of copper only is used. It requires no soldering, extra holes or riveting, and is said to reduce the cost of installation almost 80 percent.

Unusual demands will be made upon the electrical supply houses of Pittsburgh during the next few weeks in anticipation of the electrical displays contemplated for the State Conclave, Grand Commandery, Knights Templar. This event is scheduled for the week beginning May 27th, and the electrical equipment necessary to meet the demands of business houses and the municipal authorities will make a very important total. In 1906, when the event was last held here the electrical decorations required upwards of 450,000 lamps, and at the eleventh hour it was necessary to order two carloads of equipment from distant points by express. Plans for the coming event are expected to require nearly double the equipment employed in 1906. The celebration will bring 50,000 visitors to Pittsburgh, and it is proposed to illuminate the entire down-town section of the city in their honor.

A 75 FT. POLE AND ELECTRIC LIGHT MARKS THE CENTER OF POPULATION OF THE UNITED STATES

FRANK C. PERKINS

The center of population of the United States, according to the last census, is located at Bloomington, Indiana, as seen in the accompanying illustration. At this point there has been erected a substantial and attractive platform, with a metal flag staff 75 ft. high, upon which is mounted a 120 c.p. electric light. This points out to visitors the site, and several thousand dollars have been expended in honor of the population center.

The "center" is on factory grounds, just south of the wall of a main building and fronting 8th Street. Now, no one goes to Bloomington without a visit to the famous spot, the place having been made attractive as well as historic.

About the center, a platform 10 x 12 has been constructed of cement 2 ft. high. On the center of this, is a large clock of oolitic stone, made circular; on the front and top of which is cut in letters laid in gold leaf the words "Center of Population, U.S.A. 1910 Census."

The flag staff for the light runs up from the center of this stone and from the top of the staff floats a large American flag to denote patriotism, and beneath it a pennant 8 x 14, which reads "Center of Population, U.S.A., 1910." Located on the front of the platform is a very fine block of oolitic stone and the platform is built so that people can stand exactly on the population center.

On the pretty green lawn just east of the furniture plant on West 8th Street, the exact spot indicating the center of population was mathematically located by Prof. W. A. Cogshall, chief astronomer of Indiana University, who made his observations assisted by Prof. C. A. Drew.

It is proper and in keeping that in the population center monument an oolitic stone is also displayed, and in fact the entire arrangement is in keeping with such an event, and is by far the most pretentious reception and honor that has ever been accorded the center of American population.

"Well begun is half done"; and many people stop right there.

To do work in haste well, you must have first done a large lot of work slowly.

MECHANICAL DRAWING

P. LEROY FLANSBURG L. BONVOULOIR

THE DRAWING MATERIALS—Continued

9. *Scales.*—The scales used in mechanical drawing are usually made of boxwood, and in many cases have white celluloid edges. They are made in two general styles, the one being flat, while the other is triangular. In most cases, however, the triangular form of scale is the one preferred, since it not only has the greater number of edges, but also because when laid upon the paper, the scale divisions will lie close to the paper and the light will fall directly upon the ruled surface. The edges of the scale should be perfectly straight and free from all nicks, while the graduations should be narrow, clean-cut lines. At times, steel scales or paper scales are used. While the steel scale is not subject to the slight variations, such as warping, of the wooden scale, yet one seldom uses the steel scale except in cases where very fine measurements must be made. The chief object in having a paper rule or scale is that the paper rule or scale will expand or contract, under the varying degrees of moisture in the atmosphere, in the same manner as does the drawing-paper.

A scale should never be used as a straight-edge or ruler, since such use would soon batter its edges and thus render it useless for accurate work.

When measuring a line, lay the scale along the line in such a manner that the desired edge of the scale is away from the body and in a good light. Always measure directly with the scale, for if measurements are taken off the scale by means of either compasses or dividers, the surface of the scale will very quickly become marred and scratched, thus rendering it unfit for use. In laying off distances, use the "needle-pointed" pencil or a pricker; a pricker being simply a needle or other sharp point, inserted in a wooden handle. In order to avoid any accumulation of errors, it is best to lay off as many dimensions as possible from the same point, each time adding successive dimensions instead of laying off individual dimensions separately. Of course this applies only when the dimensions are to be measured along the same line.

Fig. 1

For most mechanical drawing work a 12 in. architect's scale is used, and Fig. 1 is an end view of such a scale.

As is seen from the cut the scale is triangular in shape and the grooves on its faces are for convenience in handling as well as for lightness. One of its edges. divided into inches, halves, quarters, eighths and sixteenths of an inch. The other five edges each have two scales, the one being one-half the other. Small numbers at each end of each scale indicate the denomination or size of the scale, as for instance on the scale marked 16, 12 in. = 1 ft., and each inch of scale is divided into 16 equal parts. This scale is called the full size scale. On the scale marked 1/2, 1/2 in. = 1 ft., and this scale is called the one-twenty-fourth size scale. The following list gives the various marks and scales.

Mark	Size	Scale of Drawing
16	Full size	12 in. = 1 ft.
3	1/4	3 in. = 1 ft.
1 1/2	1/8	1 1/2 in. = 1 ft.
1	1/12	1 in. = 1 ft.
3/4	1/16	3/4 in. = 1 ft.
1/2	1/24	1/2 in. = 1 ft.
3/8	1/32	3/8 in. = 1 ft.
1/4	1/48	1/4 in. = 1 ft.
3/16	1/64	3/16 in. = 1 ft.
1/8	1/96	1/8 in. = 1 ft.
3/32	1/128	3/32 in. = 1 ft.

Fig. 2

If, for instance, it is desired to make a drawing one-eighth size, use the scale marked 1½, since on this scale 1½ in. = 1 ft. At one end of the scale, a distance of 1½ in. is divided into 12 parts, each part of which is sub-divided into 4 parts. Each of the 12 divisions represents an inch and the sub-divisions represent portions of an inch. Since the scales are so divided, it requires very little calculation to lay off any required distance. For example, suppose that it is desired to lay off a distance of 3 ft.—7½ in., making it one-eighth size. Use the scale marked 1½ and find the mark corresponding to 7½ in. on the end of the scale. Place this mark over one end of the desired length, as at A, in Fig. 2. Then the point B, which is opposite the mark 3 on the scale, will be the other end of the measured distance. In other words the number of feet is measured or read from the zero mark on the scale toward the right, while the number of inches are read from the zero mark toward the left.

Another type of scale which is used to some extent is the engineer's scale. This scale, like the architect's, is usually 12 in. long, and is graduated into 600 equal parts. Each inch is therefore divided into 50 equal parts, each of which is usually used to represent 1 ft. This scale is chiefly used for civil engineering work.

10. *Scale Guard.*—When the same scale is to be used for some little time, a metallic clip or guard is slipped on the scale and serves to indicate which face of the scale is being used.

11. *Curves, Splines and Lead Wire.*— When it is desired to draw lines which are neither straight lines nor arcs of circles, it is customary to use templates called curves. These curves are made of wood, hard rubber or celluloid, and their use is so obvious that but little explanation is required. One of the curves is laid along the line of points, which determine the position of the desired curved line. Starting with some one of the points, first, draw a smooth, free-hand pencil curve through the remaining points. Second, apply a portion of one of the

irregular curves which seems to fit most closely the free-hand curve. Third, draw the final curve by following the irregular curve. Do not try to save time by going beyond the point where the free-hand curve and the irregular curve coincide, for, if this is done the final curved line will not be a smooth one. The curves must be adjusted carefully for good results, and the drawing of irregular curves gives excellent practice in accuracy. Be sure to always keep the pen or pencil tangent to the curve and perpendicular to the paper.

Splines are long, flexible strips of wood or rubber, from 2 to 5 ft. long. They are held in place by means of weights, and are used for drawing long, irregular curves. Spline weights have projecting metal fingers which fit into a groove on top of the spline. The spline is thus held firmly in place, while one of its edges is left free. This edge can be adjusted to any curve and a line of any length drawn. Lead wire is often used for drawing irregular curves. A special form of this wire is an adjustable curve ruler or "snake." This consists of a lead center having thin, flexible steel strips along the two working edges and the whole covered with a rubber sleeve. Such a rule can be instantly adjusted to a curve of any shape. The working edges of the ruler are made rounded, and thus parallel curves may be drawn by merely inclining the pencil.

12. *Protractors.*—When it is necessary to lay off angles which are not given by the triangles, a protractor, such as is shown in Fig. 3 is used. Protractors are made of metal, wood, rubber, horn, celluloid or paper, but for mechanical drawing work the metal or celluloid ones are to be preferred. They are usually made semicircular in form, having a reference line near the bottom and lines on its rim, which are drawn radially from the middle point of the reference line. The various degrees are marked on the rim. When using the protractor, place the center A at the vertex of the desired angle, as shown in Fig. 3, and lay the base of the protractor so that the reference line will coincide with the base line of the desired angle. Now make a small mark at point C, which is opposite the desired angle; in this case the angle is 35 degrees. Now remove the protractor and draw the line

CA. Then the angle CAB is the required angle. For very accurate work it is best to use a trigonometric method of determining the angle, and such a method will be described later in the course.

13. *Thumb-Tacks.*—Thumb-tacks are tacks having large, flat heads and long, thin points. They are used for fastening the drawing-paper to the drawing-board and since the head of the tack is but little raised above the surface of the paper, the T-square will ride easily over them. Sometimes 1 oz. copper or iron tacks are used for attaching the paper to the board, but such tacks are more difficult to remove from the board, and, therefore, they are not quite as convenient as are thumb-tacks.

14. *Horn-Centers.*—When many circles are to be drawn about a common center, a horn-center is sometimes used. Such a center consists of a small, circular piece of horn, around the edge of which three small points are imbedded. The center is placed as nearly over the required point as is possible, and is held in place by the three small points. The compass point is then pressed into the horn instead of being pressed into the paper, and thus the circles can be drawn without injuring the paper and with no loss in accuracy.

15. *Lettering Pens.*—An ordinary steel pen, having a fine and yet well rounded point, is used for lettering or dimensioning the drawing.

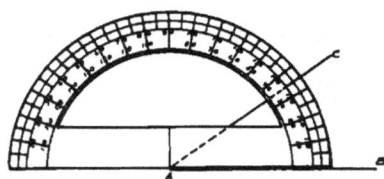

Fig. 3

16. *Two-Foot Rule.*—When making sketches or at any other time when measurements must be taken of a piece of machinery, a 2-foot rule, having each inch divided into 32 parts, will be found most convenient. It is best to use a rule having brass edges. Where, however, it is necessary to secure greater accuracy a steel rule is used.

17. *Calipers.*—Two kinds of calipers are used in mechanical drawing work, one kind being known as inside calipers, the other as outside calipers. Calipers are used for obtaining such measurements as the diameter of a shaft or the diameter of an engine cylinder. After they have been set to the diameter which is being measured, they are placed against the edge of the rule and the distance between the legs of the calipers is read directly.

Author's Note—The preceding paragraphs have described all of the more important drawing materials, and in the next installment the drawing instruments will be considered.

THE CLOCKS OF TURKEY

Fifty years ago, says a writer in *Armenia*, a watch or a clock was almost as rare in Turkey as an aeroplane is in America now. Even today, in the smaller cities and villages, house clocks are a luxury of the rich.

One of their methods of telling time by the sun is to make a kind of sun-dial of their hands. They hold their thumbs so that they touch each other horizontally and extend the forefingers up perpendicularly. Then they divide the thumb and forefinger of each hand into six parts, nominal hour points, one hand representing the morning and the other the afternoon. According to this division, where the thumbs join is 12 o'clock, the tip of one forefinger represents 6 o'clock in the morning and the top of the other 6 o'clock in the afternoon. The hours between 12 and 6 fall at different points between the junction of the thumbs and the tips of the forefingers.

Telling the time by a cat's eye sounds absurd, but it can be done. Everyone, perhaps, is not aware that the shape of the cat's eye undergoes a progressive change during the day. In the morning the pupil is round, but as the day advances it gradually narrows, until at noon it becomes merely a narrow streak. From noon to night it reverses its action, becomes oval at about 3 o'clock, and is again round at about 6 o'clock.

CROSSING THE EQUATOR

When the "Battle" fleet crossed the equator on its way to the Pacific Ocean three-fourths of the crew of 14,000 men experienced their first shave at the hands of Neptune, and although they did not overly relish the treatment received from the monarch of the seas, they took the fun with the best of grace and were thankful to receive the certificate of initiation which in the future will place them among the tormenters, not the tormented. It is a time-honored tradition in the navy that a sailor crossing the line for the first time must be shaved by "Neptune Rex." The origin of the custom is wrapped in mystery, and it is difficult to find any sea-faring man who knows it by a more comprehensive name than "initiation." Ships generally try to reach the equator at noon, and Neptune pays each ship a preparatory visit the night before reaching the line, and this is the way it is done:

A hail is heard, apparently a hundred yards from the bow: "Ship ahoy!" The officer of the deck on the forward bridge replies: "Ahoy! and who may you be?" The reply comes: "Neptune Rex, monarch of all the seas." "Come aboard, sir," says the officer courteously, at which invitation Neptune climbs on board and is met on the quarterdeck by the captain or the admiral, to whom he presents a bag of fish and a warrant of summons to be served on the unfortunate members of the crew whose names he has not yet placed on his rolls. Then with many bows and interchange of florid compliments Neptune goes over the bow again, after informing the captain of his intention to return the following day with all of his court and having received the officer's assurance that everything would be ready for his reception.

The following day the ship crosses the line and shortly after the engines slow down and Neptune comes over the bow, accompanied by his wife, clerk, barber, doctor, bears, and policemen with stuffed clubs with which to belabor the victim who may show fight. Neptune is attired as grotesquely as possible and to an elaborate degree according to the ingeniousness of the crew, and of late years the ceremony has been attended by a showing of splendid costumes and innovations never dreamed of by the older man-of-warsmen. Neptune and Amphitrite wear crowns. The barber carries a bucket and a whitewash brush and a razor usually made of wood or a barrel hoop. Preparations on deck had already been made, there being a large canvas basin filled with water, spilling line, etc. The clerk opens his books, using a coil of rope as a desk, and Neptune calls out the names of the victims, and one by one they go through the ceremony to the delight of the audience which crowds the rails, turrets, bridges and tops. If a man foolishly hides himself away the policemen go after him and whack him so that he wished he had remained to take his medicine. A victim is seated on the edge of a platform, and when he opens his mouth to answer a question a large unsavory pill, made of soap, pepper, etc., is inserted in his mouth. Then the barber gets to work and lathers him, using the whitewash brush, and covers his face and clothes with a composition of sand, molasses, flour, salt and anything that will make a disagreeable mess. With a final shove the almost initiated goes backwards into the tank where the seabears are waiting to welcome him, with brooms and swabs and a hose of running sea water. No matter how roughly some of the men may be used, none ever thinks that the trouble is not worth while the pleasure experienced when the certificate of initiation is handed him by Neptune after the ceremonies. The officers are not exempt from the initiation, and sometimes one will be found who prefers the real experience to paying a forfeit of wine, beer and cigars. Every dog has his day, and so has the enlisted man, for from the time Neptune's flag is hoisted to the truck until pipe down after the ceremonies, there is no restriction upon any reasonable amount of fun, and rank is not excused from the good-natured though respectful pranks of the petty officers and men.—*The Blue Jacket.*

Some New York boys, according to the *Sun*, have a yell which goes like this: Pooh, Pooh, Harvard! Pooh, Pooh, Yale! We learn our lessons through the mail! We're no dummies; we're no fools! Rah! Rah! Rah! Correspondence Schools

METAL DRILLING. Part II
Drill-Making, Tempering and Using
M. COLE

Case Hardening.—In an emergency it is sometimes necessary to make a tool of wrought iron instead of steel; the surface may be converted to steel by case hardening. Take equal parts of prussiate of potash, sal ammoniac and salt; powder and mix them. Heat the iron to a bare red heat, dip in the powder and put in the fire so as to melt the mixture; remove and repeat process several times; then raise to a full red and temper in the usual way. Another process is with leather cuttings; but, although better, it is too slow for emergency work.

CHUCKS, DRILL STOCKS, ETC.

In their earliest forms, metal drills were only pressed against the work and turned to and fro by hand (though a form of bow drill was used by the ancient Egyptians), but it is obvious that this method is too slow for modern times. There are two methods by which a drill may be moved, viz.: continuous and reciprocating. The former is of course the preferable one, as in that case the drill cuts all the time it is moving; while by the latter it cuts only when moving in one direction, the time used in the return stroke being wasted. The fiddle, or bow drill, watchmakers' turns, and Archimedian drills belong to this class. The bow drill in its various sizes is the easiest to make. In its simplest form, the drill itself has a pulley or bobbin on which the cord of the bow works to give it motion, while the upper end of the drill is pointed and rests in a countersink in some suitably placed piece of metal. In small turns this is usually the end of the bench vise, the work being supported on a suitable block to bring it to the right height; or the work may be placed against another point in the same plane as the drill. In larger sizes it rests in the countersink of a plate of metal placed against the breast of the worker, who can thus apply pressure to the drill; this is, however, a very dangerous method, and should be avoided. The Archimedian drill stock is much safer. In all drills of this class, there is some difficulty in keeping them perfectly steady so as to drill a straight hole to any depth; they are, however, useful for thin work. The Swiss drill is used upright, and can easily be fitted with a bracket to keep it in position so as to do accurate work. It is only suitable for small holes, as there

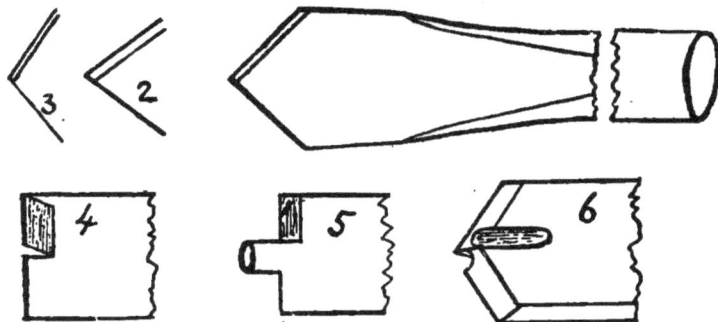

is only the weight of the tool to keep the drill up to its work. The continuous motion drill is in every way to be preferred. All modern drilling machines and drill stocks belong to this class.

Chucks.—Except in the simplest form of bow drills, etc., the drill is separate from the tool that gives it motion, and is held either in a socket that takes one size drill body, or in a chuck that will take various sizes. The Bell chuck is one of the earliest styles of adjustable chuck, and is one of the most dangerous tools ever made. The self-centering chuck is the best and most practicable method of holding a drill, and will soon be the only one.

These chucks are now very much lower in price than they were a few years ago, and can be had in all sizes, from the smallest used by watchmakers to those required for drilling the heaviest work. They can also be used for drilling work in the lathe between the centers, the shank end of the chuck resting against the running center, and carried round by a dog or carrier, the work being held between the point of the drill and the back center. For light castings where heavy work is to be drilled, it must of course be supported in the usual way.

Boring and Drilling in the Lathe.—When the chuck (of whatever form) is fixed to the head of the lathe, the result is really a horizontal drilling machine, but when very accurate work is to be done the drill should rest against the running center, and be carried round by a dog or cramp fixed to it, and driven by a pin on the face-plate. When the work revolves instead of the drill the process is boring, whether the cutter is the full size of the hole as a drill or smaller and attached to a rod as in a boring bar.

Joiner's Brace and Drills.—For odd jobs a joiner's brace of modern pattern fitted with a jaw chuck will do a large

shank and fit well in the chuck. Twist drills should never be used in such a brace, as they are liable to be strained. Holes up to $\frac{1}{4}$ in. in brass, or $\frac{1}{8}$ in. in iron, are easily drilled with a brace. For larger-sized holes, a leading hole should be drilled first, say, $\frac{1}{8}$ in., then follow with a larger drill. Those braces fitted with ratchet have the extra advantage that the hole can be drilled in places where there is not room enough for the sweep of the brace. As a good deal of pressure is required to prevent the drill slipping when drilling the larger holes, it is necessary to rig up some arrangement to get the pressure by a lever. This is easily made from a few lengths of wood.

Drilling Hard Steel.—It is sometimes necessary to drill a hole in hard steel. In tempered steel this can be done with a drill left dead hard, but the utmost care must be used not to put on much pressure or the drill will break. If the steel is too hard for this, a splinter of either diamond or corundum, mounted as a drill, must be used, or it may be done with a copper rod fed with emery powder. In this case a guide is made from a bit of wood with a hole in it the size of the drill, cramped on the work, so as to keep the drill in its place. Feed with emery of moderate fineness, and use oil to keep it in place. The grains of emery will bed themselves in the copper, but the pressure is not great enough to force them into the iron.

Very small or delicate work should be held in the fingers, the hand resting on the work bench or other support. A bow drill is best for this, the pointed end resting in a hollow in the side of the vise or other fixed object. There is so much flexibility in this arrangement that breakages seldom occur. The hole is afterwards rectified with broaches, which can be had in all sizes from the size of a fine needle upwards.

Drilling Glass.—Holes can be drilled in glass with ordinary steel drills as used for metal, kept cool with either turpentine or paraffin oil, but the glass is broken off in fragments rather than cut. To start the hole, rub the glass at the place to be bored with a bit of emery cloth, then tap gently with a sharp pointed bit of hard steel, until a hollow is formed deep enough for the drill to rest in and get a grip of the glass; then proceed as in metal drilling, but using slow speed. When half way through, reverse the work and drill from the other side, to complete the hole, which can be opened to square or other shape with a file kept wet to prevent heating.

Diamond Drills are made from a splinter of the outer crust of the diamond, which is much harder than the inner parts used for jewelry. To mount it, drill a hole in the end of a piece of brass wire, insert the splinter of diamond and fill up with soft solder; then remove as much as is necessary to expose a cutting surface.

TABLE C

Speed of Twist Drills compiled by the Morse Twist Drill Co.

Diameter	Speed for Steel	Speed for Iron	Speed for Brass
1-16 in.	1712	2383	3544
1-8 in.	855	1191	1772
3-16 in.	571	794	1181
1-4 in.	397	565	855
5-16 in.	318	452	684
3-8 in.	265	377	570
7-16 in.	227	323	489
1-2 in.	183	267	412
9-16 in.	163	238	367
5-8 in.	147	214	330
11-16 in.	133	194	300
3-4 in.	112	168	265
13-16 in.	103	155	244
7-8 in.	96	144	227
15-16 in.	89	134	212
1 in.	76	115	191

It must be used carefully, as, although it is the hardest known material, it is very brittle and easily broken. A fragment of corundum is almost as good for the purpose. The diamond drill is usually kept wet with turpentine to prevent the heat produced softening the soft solder.

Drilling Fluids.—Use soapy water for wrought iron; oil for steel.

The Start at Sunset

This interesting photograph was taken by Mr. Ward E. Bryan last July, while he was camping on Keuka Lake. The lake is located in New York, and extends from Hammondsport to Penn Yan, a distance of some 20 miles. The Curtiss aeroplane factory is at Hammondsport, at one end of the lake, and a pupil of Mr. Curtiss was trying out the Hydro-aeroplane, having a naval officer from Washington as a passenger. Mr. Curtiss met them at Keuka Landing, which is half-way up the lake. The trip up the lake was on the surface of the water, but on their return from Penn Yan they flew several feet above the surface.

EXTINGUISHING A BURNING GAS WELL

Extinguishing a burning gas well which was delivering over a million cubic feet of gas per hour at a pressure of about 600 lbs. per square inch, was accomplished last winter at Vanderpool Well No. 1 of the New York Oil & Gas Company, near Independence, Kan. The well had been drilled to a depth of about 1,500 ft., and there was in the hole about 300 ft. of 8¼ in. casing when the gas was struck. The gas was, in a measure, unexpected, and certain fittings not being on hand it was decided to tube the well with 6 in. pipe and set a packer. When 1,100 ft. of 6 in. pipe were in the well and its closure seemed certain, a severe electrical storm occurred and a flash of lightning ignited the escaping gas. Forty feet of 6 in. pipe had been left extending above the ground and from the top of this there rose with a great roar a jet of burning gas 150 ft. high, which destroyed the derrick over the hole.

The problem included not only extinguishing the flame, but also preserving the well so that the gas could be finally controlled. Nine boilers, such as are used in connection with oil well drilling outfits, were collected; and eighteen 2 in. jets of live steam at a pressure of about 120 lbs. per square inch were simultaneously turned on the flame in an attempt to smother it. No impression was made and the scheme was abandoned. The two joints of pipe projecting above the ground were removed by throwing a line round the top, bending the pipe to one side about 45 degrees, and then unscrewing it close to the ground with the same line. The 1,100 ft. of pipe hanging in the well was supported by "elevators" or clamps fastened around its top and resting on the top of the 8¼ in. pipe.

It was determined to cover the well with a conical hood, through the top of which the flames could pass until the bottom should be made tight when the top could be closed and the flame extinguished. This plan was finally carried out successfully, but not until numerous unsuccessful attempts had been made and considerable special apparatus destroyed. The cast-iron hood finally used was conical in form, 3 ft. in diameter at the base and about 6 ft. high. In the top was fixed a 12 in. gate valve upon the stem of which was fastened a reel wound with flexible wire so that two men running out with the end of the wire could quickly close the valve. A crane built of 6 in. steel pipe with a mast about 50 ft. high and a boom about the same length was placed so that when the boom was swung over the well the hood would come directly above the opening. Means were provided for controlling the motion of the hood in all directions. The clamps holding the 6 in. pipe were then pried off and the pipe slipped down, causing the flame to issue solidly from the 8¼ in. casing. This made it possible to approach the well with screens and dig a saucer-shaped cavity around it, which was made muddy so that the bottom of the hood might be submerged. The latter was then lowered to place and the flame shot through the gate valve. The bottom of the hood was made tight with successive layers of earth and canvas kept thoroughly wet and wire cables were thrown over the hood and fastened to dead men buried deep in the ground. When all was ready the men who were to shut the valve were given the signal to run. The attempt was successful, after five weeks of effort, and the flame went out. Less than ten seconds later the gas broke through under the hood, but the fire being extinguished, closing the well was then an easy matter.—*Engineering Record.*

A Cure for Restlessness

(*By William Wallace Whitlock*)

Why they called Bill Meyer "shiftless"
　　Was a question for the wise,
For he "shifted" without ceasing
　　In each business enterprise.
He was first a traveling salesman,
　　Then a patent lawyer's clerk,
When he tired of patent cases,
　　For a bank he went to work.
Teacher, preacher, writer, speaker—
　　He was each and all of these,
But for some mysterious reason
　　Every calling ceased to please.
Till at last the Weather Bureau
　　Made a place for him, and then,
As the weather did the changing,
　　Why, he never changed again.
　　　　　　　　　　　—Judge.

THE USE OF CONCRETE ON THE FARM *—Concluded

Equipment for Mixing Concrete

When the proper materials have been selected, the next step is to mix them properly and with dispatch. On large jobs it is more economical to mix concrete by machine, but for small jobs, using even as much as several hundred cubic yards of concrete, it is much cheaper and more expedient to mix by hand. This is, of course, especially true when only two or three men are available and the work is often interrupted. There are many ways of mixing by hand, all of which have the same good results. The way herein described is believed to be the one best calculated to obtain good results with a minimum of labor. In this description and the accompanying illustrations, a 2-bag batch of 1:2:4 concrete is taken as the basis of the calculation.

The Concrete Board.—A concrete board for two men should be 9 x 10 ft. It should be made of 1 in. boards, 10 ft. long, surfaced on one side, and should be held together by five 2 in. x 4 in. x 9 ft. cleats. If tongue and groove roofers 1 x 6 in. can

*U.S. Department of Agriculture. Farmers' Bulletin 41.

be obtained, fairly free from knots, they serve very well. The boards are surfaced in order to make the shoveling easy. They are so laid as to permit the shoveling to be done in the direction in which the cracks run, so that the shovel points will not catch in the cracks. The boards must be nailed close together so that no cement grout may run through them during the mixing. Knot holes may be closed by nailing a strip across them on the underside of the board. It is a good precaution against losing cement grout to nail a piece of wood 2 x 2 or 2 x 4 in. around the outer edge of the board. Often 2 in. planks are used in making concrete boards, but these are unnecessarily heavy and very difficult to move.

The concrete board is the manufacturing plant, and the advantages of its location should be carefully considered. Generally it is best placed as close as possible to the forms in which the concrete is to be deposited, but local conditions must govern this point. A place should be selected which affords plenty of room

Fig. 3. Concrete Board and Various Tools used in making Concrete

Fig. 4.   Measuring Sand

and is near the storage piles of sand and stone or pebbles. The concrete board should be raised on blocks so as to be level, in order that the cement grout may not run off on one side and that the board may not sag in the middle under the weight of the concrete.

Runs.—The boards for the wheelbarrow runs should be carefully selected. The run should be well built, smooth, and at least 20 in. wide, if much above the ground. It is surprising how this one feature will lighten and quicken the work.

Miscellaneous Tools.—The following is a list of the tools and plant to be used in mixing, giving sizes, quantities, etc.

The lumber for the concrete board for a 2-bag batch, 9 x 10 ft. in size, is as follows:

9 pieces ⅞ in. x 12 in. x 10 ft., surfaced on one side and two edges*
5 pieces 2 in. x 4 in. x 9 ft., rough
2 pieces 2 in. x 2 in. x 10 ft., rough
2 pieces 2 in. x 2 in. x 9 ft., rough.

The lumber for the concrete board for a 4-bag batch, 12 x 10 ft. in size, is as follows:

12 pieces ⅞ in. x 12 in. x 10 ft., surfaced on one side and edges*

* Here any width of 'plank may be used, but 12 in. is specified as the most convenient.

5 pieces 2 in. x 4 in. x 12 ft., rough
2 pieces 2 in. x 2 in. x 10 ft., rough
2 pieces 2 in. x 2 in. x 12 ft., rough

For the runs, planks 2, 2½ or 3 in. thick and 10 or 12 in. wide are needed.

The measuring boxes for the sand and stone or gravel should have the following dimensions:

For a 2-bag batch with the 1:2:4 mixture:
4 pieces 1 in. x 11½ in. x 2 ft., rough (for the ends of the sand and stone boxes)
2 pieces 1 in. x 11½ in. x 4 ft., rough (for the sides of the sand box)
2 pieces 1 in. x 11½ in. x 6 ft., rough (for the sides of the stone box)
(It should be noted that the 2 pieces 4 ft. long and the 2 pieces 6 ft. long have an extra foot in length at each end for the purpose of serving as a handle.)
For a 2-bag batch with the 1:2½:5 mixture:
2 pieces 1 in. x 11½ in. x 2 ft. (for the ends of the sand box)
2 pieces 1 in. x 11½ in. x 2½ ft. (for the ends of the stone box)
2 pieces 1 in. x 11½ in. x 4½ ft. (for the sides of the sand box)
2 pieces 1 in. x 11½ in. x 6 ft. (for the sides of the stone box)
(As in the preceding case, the 2 pieces 4½ ft. long and the 2 pieces 6 ft. long have an extra foot in length at each end to serve as handles.)
For a 4-bag batch (these figures can be obtained by doubling the cubic contents of the boxes as shown above)
Shovels: No. 3, square point.

Wheelbarrows: At least two are necessary for quick work, and those with a sheet-iron body are to be preferred.
Garden rake
Water barrel
Water buckets, 2 gal. size
Tamper: 4 in. x 4 in. x 2 ft. 6 in., with handles nailed to it
Garden spade or spading tool
Sand screen, which can be made by nailing a piece of ¼ in. mesh wire screen, 2½ x 5 ft. in size, to a frame made of boards 2 x 4 in.

METHOD OF MIXING

When the mixing board has been arranged and the "runs" are made, the concrete plant is ready. The sand should first be loaded in wheelbarrows and wheeled on the board. It should then be emptied into the sand-measuring box, which is placed about 2 ft. from one of the 10-ft. sides of the board, as shown in Fig. 4. When the sand box is filled it should be lifted off and the sand should be spread over the board in a layer 3 or 4 in. thick. Two bags of cement should then be spread as evenly as possible over the sand. Two men, stationed, as shown in Fig. 6, should then start mixing the sand and cement in such a way that each man may turn over the half on his side of a line dividing the board in half. Starting at his feet and shoveling away from himself,

each man should take a full shovel load, and in turning the shovel, he should not simply dump off the sand and cement, but should shake the materials off the end and sides of the shovel, so that they may be mixed as they fall. This is a means of great assistance in mixing these materials, and in this way the material is shoveled from one side of the board to the other.

The sand and cement should now be well mixed and ready for the stone and water. After the last turning, the sand and cement should be spread out carefully, and the gravel or stone measuring box should then be placed beside them and filled from the gravel pile. The box should now be lifted off, the gravel shoveled on top of the sand and cement, and spread as evenly as possible. With some experience, equally good results may be obtained by placing the gravel-measuring box on top of the carefully leveled sand and cement mixture, and filling it, so that the gravel is placed on top without an extra shoveling. About three-fourths of the required amount of water should be added with a bucket, and the water should be dashed over the gravel on top of the pile as evenly as possible. Care should

Fig. 5. Spreading the Cement on Top of the Sand

ELECTRICIAN AND MECHANIC

Fig. 6. Mixing the Sand and Cement

be taken not to let too much water get near the edges of the pile, because it may wash away some of the cement as it flows off. This caution, however, does not apply to a properly constructed mixing board, where water can not flow away. Starting in the same way as with the sand and cement, the materials should be turned over in much the same manner, except that, instead of shaking them off the end of the shovel, the whole shovel load should be dumped and dragged back toward the mixer with the square point of the shovel. The wet gravel picks up the sand and cement, as it rolls over when dragged back by the shovel, and the materials are thus thoroughly mixed. Water should be added to the dry spots as the mixing goes on until all that is required has been used. The mass should be turned back again, as was done with the sand and cement. With experienced laborers, the concrete would be well mixed after three such turnings; but if it shows streaky or dry spots, it must be turned again. After the final turning, it should be shoveled into a compact pile. The concrete is now ready for placing.

When a natural mixture of sand and gravel is to be used, the materials should be measured by means of the right measuring box for the proper proportion, as shown in Table No. III. The mixture of sand and gravel should be spread out and enough water added to wet it thoroughly. The cement should be distributed evenly in a thin layer over the sand and gravel and turned over, as described previously, at least three times, while the rest of the water necessary to get the required consistency should be added as the materials are being turned. It requires good judgment to prepare a natural mixture of bank sand and gravel, and, if one is at all doubtful about the concrete made from it, the sand should first be screened from the gravel as described above, and then mixed in the regular way.

For the operation of mixing, only two men are required, although more can be used to advantage. If three men are available, two of them should mix as described above, and the third man should supply the water and help in mixing the concrete by raking over the dry or un-mixed spots, as the two mixers turn the concrete, and in loading the wheelbarrows with sand and stone or gravel.

If four men are available, it is best to increase the size of the batch mixed to a 4-bag batch by doubling the quantities of all materials used. The cement board should also be increased to 10 x 12 ft., In this case the mixing should be started

Method.—After the concrete is properly mixed, it should be placed at once. Concrete may be handled and placed in any way best suited to the nature of the work, provided that the materials do not separate in placing. Concrete may be placed properly by shoveling off the concrete board directly into the work; by shoveling into wheelbarrows, wheeling to the proper place, and dumping; by shoveling down an inclined chute; or by shoveling into buckets and hoisting into place. Concrete should be deposited in layers about 6 in. thick, unless otherwise specified.

Consistency.—The following three kinds of mixtures are used in general concrete work:

(1) Very wet mixture.—Concrete wet enough to be mushy and run off a shovel when being handled is used for reinforced work, thin walls, or other thin sections, etc.; with it no ramming is necessary.

in the middle of the board, and each pair of men should mix exactly as for a 2-bag batch, except that the concrete is shoveled into one big mass each time that it is turned back on the center of the board.

Placing the Concrete

(2) Medium mixture.—Concrete just wet enough to make it jelly-like is used for some reinforced work and also for foundations, floors, etc. It requires ramming with a tamper to remove air bubbles and to fill voids. This concrete is of a medium consistency and would sink under the weight of a man.

(3) Dry mixture.—In this case the concrete is like damp earth. It is used for foundations, etc., where it is important to have the concrete put in position as quickly as possible. This must be spread out in a layer from 4 to 6 in. thick and thoroughly tamped until the water flushes to the surface.

The difference between the mixtures is that the drier mixture causes the concrete to "set up" more quickly. In the end, however, when carefully mixed and placed, the results from any of the above mixtures will be the same. A dry mixture, to be sure, can not be used readily

Fig. 7. Emptying the Gravel in the Measuring Box to the Side of the Sand and Cement

Fig. 8. Measuring the Gravel by Dumping it on Top of the Sand and Cement Mixture

with reinforcing steel, and is both more expensive and harder to handle. It must be protected with greater care from the sun or from drying too quickly, and unless spaded by very experienced hands it may show voids or stone or gravel pockets in the face of the work when the forms are removed.

Spading.—Concrete of any of the three degrees of consistency mentioned above should be carefully spaded next to the form where the finished concrete will be exposed to view. Spading consists of running a spade or flattened shovel down against the face of the form and working up and down. This action causes the stone or gravel to be pushed back slightly from the form and allows the cement grout to flow against the face of the form and fill any voids that may be there, so that the face of the work will present an even, homogeneous appearance. Where the narrowness of the concrete section, such as in a 6 in. side wall, prevents the use of a spade, a board 1 x 4 in., sharpened to a chisel edge on the end, serves as well. This board should be sharpened only on one side and the flat side should be placed against the form. In the case of a dry mixture spading must be done with the greatest care by experienced hands to get uniform results, but with a medium or wet mixture it is very easy to obtain first-class work; indeed, with a wet mixture spading is required only as an added precaution against the possibility of voids in the face of the work and is really necessary in few cases.

Protection of Concrete after Placing.—New concrete should not be exposed to the sun until after it has been allowed to harden for five or six days. Each day during that period the concrete should be wet down by sprinkling water on it both in the morning and afternoon. This is done so that the concrete on the outside will not dry out much faster than the concrete in the center of the mass, and it should be carried out carefully, especially during the hot summer months. Old canvas, sheeting, burlap, etc., placed so as to hang an inch or so away from the face of the concrete, serve very well as a protection when wet. Often the concrete forms can be left in place a week or ten days, thus protecting the concrete during the "setting up" period, and the above precautions are then unnecessary.

FORMS FOR CONCRETE

Concrete is a plastic material, and, before hardening, takes the form of anything against which or in which it is placed. Naturally, the building of the form is a most important item in the success of the work. These forms hold the concrete in place, support it until it has hardened, and give it its shape as well as its original surface finish.

Kinds.—Almost any material which will hold the concrete in place will serve as a form. Concrete foundations for farm buildings require shallow trenches, and, up to the ground line, usually the earth walls are firm enough to act as a form.

Molds of wet sand are used for ornamental work. Frequently colored sand is used for this purpose and provides both the finished surface and color to the concrete ornament.

Cast, wrought, or galvanized iron is used where an extremely smooth finish is desired without further treatment after the removal of the forms. Forms made of iron are more easily cleaned and can be used a greater number of times than those of wood. Rusty iron, however, should not be used. By far the greatest number of forms are made of wood, because of the fact that lumber in small quantities can always be obtained.

Requirements of a Good Form.—Forms should be well planned, so that there may be no difficult measurements to understand. As few pieces of lumber should be used for the work as possible, and these should be fastened together by as few nails as will serve the purpose; otherwise it will be difficult to take the forms apart without splitting them.

Forms must be strong enough to hold the weight of the concrete without bulging out of shape. When they bulge, cracks may open between the planks, and the water in the concrete, with some cement and sand, may leak out. This weakens the concrete and causes hollows in the surface, which have a bad appearance after the forms are removed.

Forms which lose their shape after being used once can hardly be used a second time. A part of the erection cost of forms is saved if the forms are built in as large sections as it is convenient to handle. This saving applies to their removal as well as their erection. Consequently the lightest forms possible with the largest surface area are the most economical.

Plans for Forms.—The first consideration in planning forms is the use to which they are to be put. Neglect of this point means waste of money and time. If they are for work afterwards to be covered

Fig. 9. Pouring Water Over the Mixture of Sand, Cement and Gravel

Fig. 10.   Final Mixing of the Concrete

with a veneer coat, the finish of the surface is of small consideration, while the alignment of the form is all important. On the other hand, if a tank or retaining wall is to be built, the fact that the forms are not in exact alignment will hardly be noticed, but money can be saved, if the forms are rigid in alignment and surfaced.

In planning forms for large structures it should be borne in mind that if the forms are to be used several times the more nearly perfect they are the more often they can be used and the cheaper they become. If forms are to be used only once, as is generally the case on the farm, they should not be nailed so securely as to prevent them from being readily taken apart and the lumber from being used for something else. It is often better to put them together with screws, but if nails are used, they should not be driven in all the way.

Selection of Lumber for Forms.—The selection of the lumber is of importance. If the forms are to be used many times, the use of surfaced lumber, matched, tongued and grooved, and free from loose knots, is economical. If, however, they are to be used only once, almost any plank will do. Forms with bad cracks or knot-holes may be made tight by

filling them with stiff clay mud and then tacking a strip over the crack and on the outside of the form.

Green lumber is preferable to kiln-dried or seasoned material. Seasoned lumber, when wet, either by throwing water on the form before placing the concrete, or by absorbing the water from the concrete, warps, and the shape and tightness of the form are damaged.

Originally, only surfaced lumber was used for forms, and this was depended on to give a finish to the work, while today, since rough surfaces for concrete are the fashion, unfinished lumber may be used. Nevertheless, surfaced lumber has some advantages for use in forms. The forms fit together better and are easier to erect. They are more easily removed and cleaned. All of these items reduce the cost of the work, but the saving effect will, of course, depend on the difference in local price between finished and rough lumber.

Method of Cleaning Tools and Forms.—At the beginning of the noon period and again at the end of the day's work all the tools, and especially the concrete board, should be carefully cleaned. Particles of concrete are also very apt to adhere

(Continued on page 411)

THE HOME CRAFTSMAN
RALPH F. WINDOES

THE MODERN SUBSTITUTE FOR THE APPRENTICESHIP SYSTEM

Perhaps no other tradesman has suffered more by the advent of modern machinery than the metal-worker. The all-around machinist of our fathers' time, the man who could construct an engine from the designing to the painting processes, is now almost a thing of the past. His place has been taken by the "specialist," the mere "hand," who can do but one thing and do it well; the man whose mentality stopped growing when he first pulled a lever on a certain machine, and whose mind will remain dormant as long as he does pull that lever and thinks of no good reason—outside of a remunerative one—why he is pulling that lever. By this I mean that the "hand" of today is given a machine to work at, he is shown how to run that machine, how to make a certain article, and told to keep at it. But he is not told what that article is to be used for, along with others, and very often he does not care. If such is the case he loses his interest in the work, and worst of all, he stops growing, culturally at least.

Let us consider how the conditions developed which placed the worker in such an undesirable position. Years ago we had the apprenticeship system of training. A boy was pledged to work under a master workman until he had learned all of the processes involved in the particular trade he was studying. It was a hard task, but an interesting one; as he could watch the machine grow from step to step, he could figure out the reasons for each step, and the principles upon which the machine worked when completed. He was interested in his work, and interested in his life, hence he was leading a happy existence.

Soon the capitalist stepped in and declared this system a waste of time and material. He opened a large shop, equipped with the best machinery obtainable, and offered the apprentices wages to come and work for him. This looked tempting, so they came. He put each at a machine, made him an expert time-saver, hired a few "old hands" to assemble the separate parts and offered his machine for sale at a much lower price than the "old timer" could. The latter admitted his defeat and retired. Hence was lost to the world the all-around machinist, the deftly trained man who combined his work and his recreation and lived a happy life. In his place came the degraded, evil-voiced "hand" of today, who gets no joy from his work, his only pleasure being found in his "pay-night revels" with others of his class.

We have reached the age of specialization, and now we are clamoring for the "good old days of yore." The original capitalist is the one that is talking the loudest and working the hardest to bring about a change which will land conditions very near the point at which he began revolutionizing them. And best of all he is accomplishing this very thing—through the vocational school.

The man he wants now, the modern machinist, is the man who can run any machine in the shop; read any blue-print in the shop; talk and write intelligently upon current events; who uses no slang; neither smokes, chews, nor drinks; knows something of the raw materials which go into the article upon which he is working, and is an all-around dependable fellow. He realized the impossibility of training such a man in his factory, hence he called upon the school to do it for him. The school responded and told him that it would take the young man part time and give him the scholastic training he needed and leave him to the practice of it in the factory the rest of his time. The capitalist agreed, and we have the continuation school of today as the result.

While in school the embryo machinist

. -BOOK RACK.-

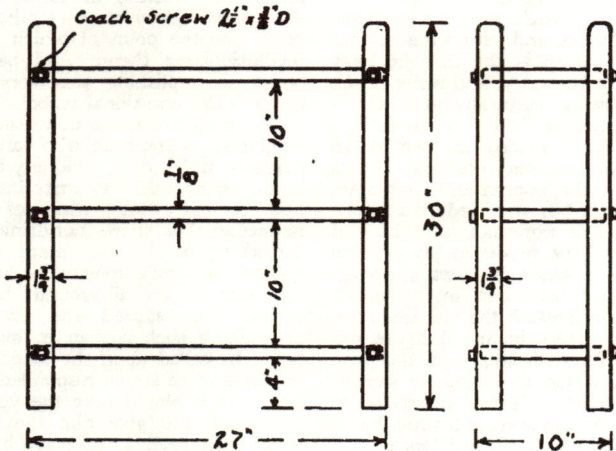

Coach Screw 2½" x ⅜" D

10"

10"

30"

4"

1¾"

1¾"

27"

10"

26

studies English, mathematics, mechanics, physics, chemistry, commercial geography, first aid to the injured, mechanical drawing, civics, American history and business methods. Every subject is given with the end in view of training the student for the practical duties of life, and not for the cultural duties as given in our classical high schools of today. He is clearly shown how the study of English, for instance, will help him when he writes a letter of application for a position, or if he is sent out on a job which will require a written report on results obtained. He realizes the practical value to him, hence he works harder to attain perfection than he otherwise would. So with every other subject on the list—practical, every one of them.

In Fitchburg, Mass., is operated a continuation school for boys, started by the Simonds' Saw Manufacturing Company, one of the largest concerns of this kind in the world. The school is run on the above plan, the boys working in the factory one week and attending school the next. They are given full pay while at work, which makes them independent and able to complete the four years required before graduation, without being forced to quit school and help to support the family, as a great many of our high school students have had to do.

Now let us see what such an education really accomplishes for the boy. In the first place it has made him open-minded; he thinks of his work as he carries it along, "uses his head," as it is commonly expressed. He knows what part his

particular labor is of the whole organization and realizes the necessity of doing his work well. He also is aware of the fact that more pleasure and profit will come to him by being virtuous than by loafing around the saloons after working hours.

And what does it do for the manufacturer? To quote from Mr. W. B. Hunter,* director of the Fitchburg School: "It gives him a better class of apprentices, boys who will make thinking mechanics, not mere machines who require all the foreman's time and attention explaining every little detail of a drawing. They will be able to read a blue-print and go ahead. Foremen on every hand speak in the highest terms of the work, and they fit with the men. They are three years ahead of the ordinary high school graduate; they are working in the plant where they would have to apply for a position if they wanted work; the men know what they can do, and when they become journeymen there is no kick about paying them good wages.

"And lastly, what does it do for society? By this plan the worker is given an opportunity to continue his education, to be a better citizen as the result of his acquaintance with the civic operations of his community and their relation to the worker, to be a contented and happy being, because he can see beyond his daily task into the great storehouse of literature and history of his trade that has made possible the rise of our nation and continue his supremacy as the artisan par excellence."

* "The Fitchburg Plan of Industrial Education."

BOOK RACK

Simplicity in construction is the greatest advantage offered in the book rack here presented, as no joints are used. The three shelves are fastened to the posts with coach screws, whose square heads add much to the appearance desired. As for the utility side of the project, the rack will undoubtedly find good use if it is once constructed. The material list consists in the following:
4 pcs. 1¾ x 1¾ x 30 in. oak
3 pcs. ⅞ x 9½ x 26½ in. oak
12 coach screws ⅜ in. D. x 2½ in. long.
Stain, filler, wax, etc.

First true up the posts and round off their tops. Be sure to plane and scrape

off all of the chatter marks in evidence. Next locate on the posts where the center of the shelves will come and bore holes for the screws, using a 7⁄16 in. bit. The shelves should now be smoothed up and the corners cut out so as to receive the posts. Do not have the edges of the shelves flush with the posts, but back about ¼ in. Place the posts and shelves into position and mark the points where the screws will enter the latter. Bore holes in them and fasten all of the parts together. Fill, stain, shellac and wax the piece as before explained, putting this material over the screw heads so as to give them a slight polish.

- *COSTUMER.* -

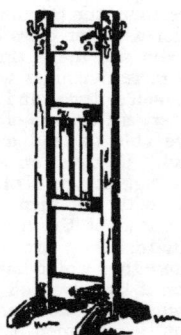

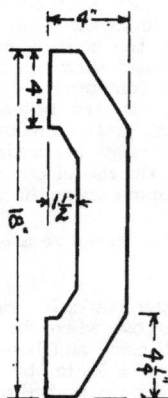

Detail of
Standard.

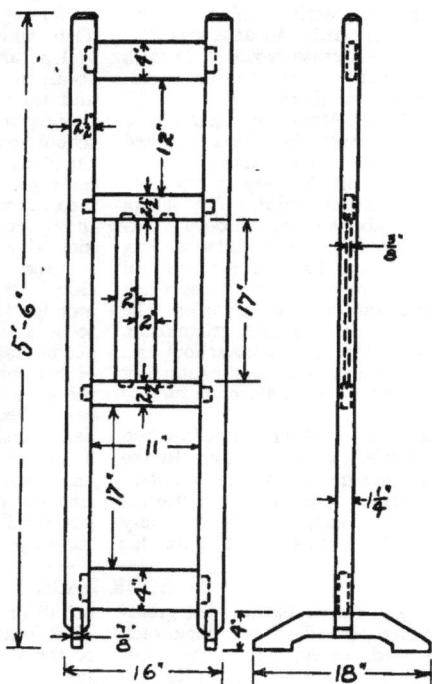

Front Elevation. Side Elevation

COSTUMER

Another article of distinct usefulness is the costumer, or hall tree, as it is sometimes called. It presents no new problems of construction to the craftsman, so little need be said concerning the making of it. The material needed is the following:

2 pcs. 1¼ x 2½ x 5 ft. 6 in. oak
2 pcs. ⅞ x 4 x 13 in. oak
2 pcs. ⅞ x 2½ x 13 in. oak
2 pcs. ⅞ x 4 x 18 in. oak
2 pcs. ⅜ x 2 x 18 in. oak
4 large antique copper hooks
4 smaller antique copper hooks
Stain, filler, wax, etc.

The tenons should be cut on the pieces bearing them and the corresponding mortises where they belong according to the drawing. The ends of the posts are cut out so as to receive the standards, which are shaped according to the detail. In fastening the parts together, the pieces should be glued and clamped, being sure to glue the slats in the center section before fastening the other pieces. The standards may be secured by boring holes through them and the posts and inserting oak dowel pins, whose heads are planed off flush with the sides of the posts. Use two ½ in. pins on each post and glue securely.

Scrape off all of the glue in evidence, and finish the piece. Place the hooks as illustrated, the large ones on the posts and the smaller ones on the top stringer, four on each side.

(To be continued)

THE WANING HARDWOOD SUPPLY

Although the demand for hardwood lumber is greater than ever before, the annual cut today is a billion feet less than it was seven years ago. In this time the wholesale price of the different classes of hardwood lumber advanced from 25 to 65 percent. The cut of oak, which in 1899 was more than half the total cut of hardwoods, has fallen off 36 percent. Yellow poplar, which was formerly second in point of output, has fallen off 38 percent, and elm has fallen off one-half.

The cut of softwoods is over four times that of hardwoods, yet it is doubtful if a shortage in the former would cause dismay in so many industries. The cooperage, furniture and vehicle industries depend upon hardwood timber, and the railroads, telephone and telegraph companies, agricultural implement manufacturers, and builders use it extensively.

This leads to the question, Where is the future supply of hardwoods to be found? The cut in Ohio and Indiana, which, seven years ago, led all other states, has fallen off one-half. Illinois, Iowa, Kentucky, Michigan, Minnesota, Missouri, New Jersey, Tennessee, Texas, West Virginia and Wisconsin have also declined in hardwood production. The chief centers of production now lie in the Lake States, the lower Mississippi Valley, and the Appalachian Mountains. Yet in the Lake States the presence of hardwoods is an almost certain indication of rich agricultural land, and when the hardwoods are cut the land is turned permanently to agricultural use. In Arkansas, Louisiana and Mississippi the production of hardwoods is clearly at its extreme height, and in Missouri and Texas it has already begun to decline.

The answer to the question, therefore, would seem to lie in the Appalachian Mountains. They contain the largest body of hardwood timber left in the United States. On them grow the greatest variety of tree species anywhere to be found. Protected from fire and reckless cutting, they produce the best kinds of timber, since their soil and climate combine to make heavy stands and rapid growth. Yet much of the Appalachian forest has been so damaged in the past that it will be years before it will again reach a high state of productiveness. Twenty billion feet of hardwoods would be a conservative estimate of the annual productive capacity of the 75,000,000 acres of forest lands in the Appalachians if they were rightly managed. Until they are we can expect a shortage in hardwood timber.

Circular 116, of the Forest Service, entitled "The Waning Hardwood Supply," discusses the situation. It may be had upon application to the Forester, Forest Service, Washington, D.C.

ENGINEERING LABORATORY PRACTICE—Part VII

The Pelton Wheel

P. LEROY FLANSBURG

At the present time there is a very rapid increase in the number of hydro-electric plants which are being erected, and more and more of the available water power of our rivers is being utilized.

Whenever water falls from one level to another, it is possible by means of a properly designed machine, to transform the potential energy of the water into kinetic energy. For example: if there are a number of buckets fastened to an endless chain (the chain passing over two sprocket wheels) and water is allowed to flow into one of the upper buckets, the weight of the water will cause the bucket to move downward. Since the bucket is attached to the chain, the bucket in moving downward will cause the sprocket wheels to revolve, and when the bucket passes around the lower sprocket wheel, the water which has caused all of this motion will be discharged from the bucket. By so designing the apparatus, that water will flow into each bucket in turn, as it passes over the upper sprocket wheel, and will be discharged from the bucket as it passes under the lower sprocket wheel, it is a comparatively simple matter to secure continuous power from a stream of water. A water motor which is built upon this principle is known as a gravity motor.

The water motors of today may be broadly divided into two classes, the one the impulse wheel, the other the water turbine.

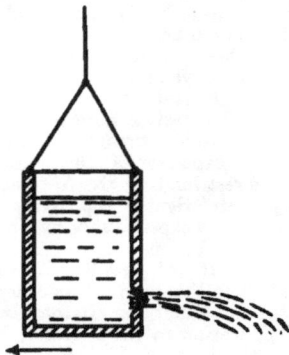

Fig. 1

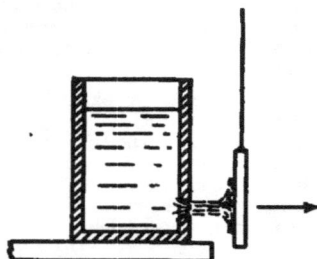

Fig. 2

The impulse type of water wheel is simply a wheel upon whose periphery is mounted a number of small curved vanes or buckets. This type of wheel is driven by a jet of rapidly moving water, directed tangentially to the rim of the wheel, and which impinges upon the vanes or buckets.

In order to more clearly understand the principle of the impulse type of water wheel, let us refer to Figs. 1, 2 and 3.

Fig. 1 illustrates the case of simple reaction. The tank containing the water is freely suspended and the water is allowed to escape through an orifice in the side of the tank. The hydrostatic pressure at the orifice accelerates the water, causing it to flow through the orifice in the form of a jet. Since action and reaction are equal, an unbalanced force equal to the force which accelerates the jet, reacts upon the tank and causes it to be pushed back in the direction indicated by the arrow.

Fig. 2 illustrates the case of a simple impulse. The tank containing the water is fixed in its position and the jet of escaping water impinges upon the flat plate which is freely suspended. The pressure of the jet of water against the plate, causes the plate to move in the direction shown by the arrow. There will be also a reactive force equal in magnitude and acting in the opposite direction to the impulse, but since the tank is immovable the reactive force does no work.

Fig. 3 illustrates the case of combined impulse and reaction forces and is an example of the forces exerted by the water upon the buckets of a so-called impulse water-wheel. The tank containing the water is, as in the last-mentioned case, rigidly fixed in position, but instead of water impinging upon a flat plate, a plate which is U-shaped is used. The jet of water striking against this curved surface is turned back upon itself through an angle of 180 degrees and leaves the bucket while moving in a direction opposite to that by which it entered. From the drawing it is at once seen that the jet acts by impulse when entering the bucket or curved surface and by reaction upon leaving it. Now, since action and reaction are equal, the combined forces which tend to move the plate in the direction indicated by the arrow are twice as great in magnitude as would be the impulse force illustrated by Fig. 2.

In practice it is not possible to turn the jet through an angle as great as 180 degrees, since if this were done, the water discharged from one bucket would be cast back against the one behind and offer a resistance to it. Generally, therefore, an angle somewhat less than 180 degrees is used.

The Pelton wheel is one of the impulse type of water wheels and it is particularly adapted for use where a high head of water is available. Owing to the simplicity and ease of operation of the Pelton wheel, it is a most satisfactory type of wheel to use in power plants. The buckets are similar to the one shown in Fig. 4, having two lobes which are nearly rectangular in form. The sharp ridge of the bucket provides for the gradual deviation of the water from its original path, and prevents eddying and consequent internal fluid friction. The dividing ridge separates the bucket into two lobes.

Fig. 4

To secure the best efficiency from the wheel:

1st.—The bucket should be so shaped or curved as to avoid any sharp angular deflection of the jet of water.

2d.—The bucket surface at entrance should be approximately parallel to the path of the jet.

3d.—The number of buckets should be small and the path of the jet in the bucket short, as this does away with much of the friction loss.

4th.—The speed of the center-line of the buckets should be approximately one-half the linear velocity of the water in the jet, as it leaves the nozzle or orifice.

The Pelton wheel generally employs one or more conical nozzles. To reduce the forces acting upon the wheel you can reduce the size of the jet, reduce its velocity or so divert the direction of the jet that only a portion of it acts upon the buckets.

To reduce the size of the jet an internal conical stopper or needle made of brass is commonly employed. The stopper can be moved parallel to the axis of the nozzle, and by moving it backward or forward, the size of the jet may be either increased or diminished.

To reduce the velocity of the jet it is only necessary to partially close a valve placed in the piping just upstream from the nozzle.

A diversion of the jet is sometimes resorted to, but is very wasteful of water.

The following is a test made by the author upon a Pelton wheel to determine the best speed at which to run the wheel and to find the efficiency of the wheel both at this speed and also at speeds above and below it.

Fig. 3

TEST ON A PELTON WHEEL

The water used was supplied by a double-acting, twin cylinder Blake pump. A piezometer ring and gauge were placed just back of the nozzle to measure the pressure head. The velocity of the jet of water was practically constant throughout the test. The velocity of the wheel was varied by changing the load applied, the load being applied by means of a prony brake. The entire wheel was enclosed in a glass case, and it was thus possible to observe the wheel under operation and at the same time the case prevented the water from splashing out. The water escaped from the case through a cylindrical draft tube, which was connected directly to a weir box. The water discharged from the weir was then meas-

ured by means of a hook-gauge. To measure the revolutions per minute made by the shaft, a revolution counter was directly connected. Five runs each of six minutes duration were made and under varying loads. Readings were taken every two minutes of the revolutions per minute of the wheel, the piezometer pressure and the hook-gauge. In working up the report the average piezometer pressure was used. The mean of the hook-gauge readings was also used. The speeds of the wheel were such that for the first two runs, the rim velocity was less than one-half of the jet velocity, while in the last two runs the rim velocity was greater than one-half the jet velocity.

A general, diagrammatic sketch of the apparatus as used is shown in Fig. 5.

DATA

Diameter through center line of buckets....................3.70 ft.
Diameter of nozzle...0.1147 ft.
Diameter of piezometer.....................................0.50 ft.
Width of weir..2.00 ft.
Circumference of brake arm.................................16.50 ft
Tare load on brake...31⅛ lbs.
Height of gauge above center-line of piezometer............1.00 ft.
Zero reading of hook gauge:................................0.853 ft.
One cubic foot of water weighs.............................62.30 lbs.

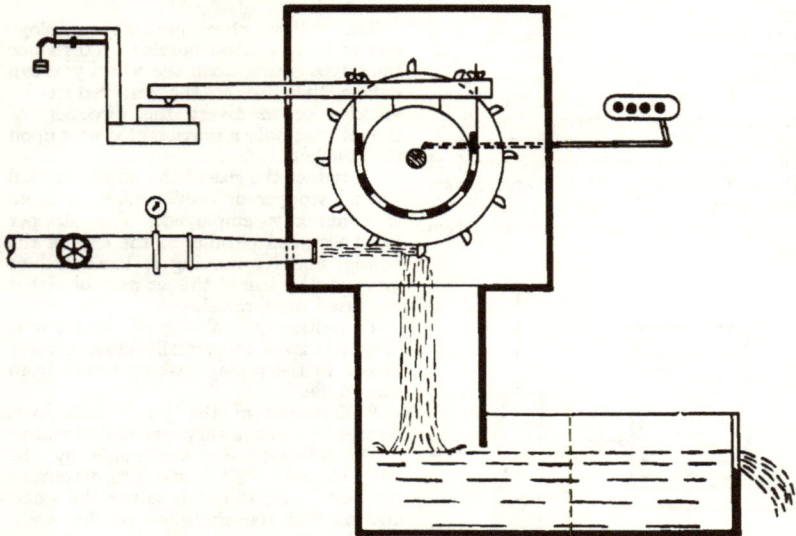

Fig. 5

OBSERVATIONS

Test No.	I	II	III	IV	V
Piezometer pressure (lb.)............,	39.2	39.5	39.1	37.1	36.0
Hook-gauge reading (ft.)............	1.075	1.076	1.075	1.072	1.069
Revolutions per minute............	126.5	147.3	146.5	196.8	237.0
Load on brake (lb.)..............	89	64	59	49	39
Piezometer pressure (lb.) average...........38.2					
Hook-gauge reading (ft.) average.........1.073					

CALCULATION OF RESULTS

$$q(\text{by first approximation}) = C(\tfrac{2}{3})(b)(\sqrt{64.4})(H^{3/2})$$
$$= C(\tfrac{2}{3})(8.04)(H)^{3/2}$$
$$= 10.7\,C\,H^{3/2}$$

Test No.	I	II	III	IV	V
H..............................	0.222	0.223	0.222	0.219	0.216
C..............................	0.624	0.623	0.624	0.628	0.632
q..............................	0.698	0.703	0.698	0.688	0.672

Adding .43 lb. to piezometer reading to correct for height of gauge above the center of piezometer.

Piezometer reading (corr.).........	39.63	39.93	39.53	37.53	36.43

Average piezometer reading............$(38.20 + .43) = 38.63$

MECHANICAL EFFICIENCY

Input. $\qquad\qquad\qquad\qquad\qquad\qquad\qquad\qquad\qquad C = .974$

$$(\textit{Velocity of jet at end of nozzle}) = v = \sqrt{\dfrac{64.4(h_1)}{\dfrac{1}{C^2} - \left(\dfrac{d}{D}\right)^4}}$$

$$v = \sqrt{\dfrac{64.4\,(h_1)}{1.057 - .00275}} = \sqrt{\dfrac{64.4(h_1)}{1.054}} = \sqrt{61.0(h_1)} \ \therefore\ v^2 = 61(h_1)$$

$$h_1 = (38.63)\dfrac{144}{62.3} = 89.3 \text{ ft.}$$

$$v^2 = 61(89.3) = 5450 \text{ and } \dfrac{v^2}{2g} = \dfrac{5450}{64.4} = 84.7 \text{ ft.}$$

$$\text{H.P. Input} = \dfrac{q \times 62.3}{550}(84.7) = 9.6\,q$$

Test No.	I	II	III	IV	V
H.P. Input	6.70	6.75	6.70	6.60	6.45

Output.

$$\text{H.P.} = \dfrac{2\pi RNP}{33,000} = \dfrac{(\text{cir. of brake arm})(N)(P)}{33,000} = (.0005)(N)(P)$$

Test No.	I	II	III	IV	V
H.P. Output.	5.63	4.72	4.32	4.82	4.63

$$Efficiency = Output \div Input$$

Test No.	I	II	III	IV	V
Efficiency	84%	69.9%	64.5%	73.0%	71.8%

VELOCITY OF CENTER-LINE OF BUCKETS

$$Vel. = (r.p.m.)(\pi d)$$
$$= (r.p.m.)(11.62)$$

Test No.	I	II	III	IV	V
Vel. in ft./min.	1470	1714	1705	2290	2760

RESULTS

Test No.	I	II	III	IV	V
Mechanical efficiency	84.0%	69.9%	64.5%	73.0%	71.8%
Velocity of center-line of buckets	1470	1714	1705	2290	2760
Velocity of jet at end of nozzle	Average			73.8 ft./sec.	

PROTECTION OF CHEMICAL GLASSWARE

Undoubtedly many of the readers of this magazine do more or less experimenting with chemical apparatus, and have had much trouble and inconvenience when using glass-ware which had to be heated to obtain a desired result, often causing breakage and considerable expense. It is usually customary to put a coating of clay upon glass-ware that is to be exposed to a temperature that would soften or melt the glass, or where they are liable to be broken or otherwise damaged by draughts of air. It is customary to add cow's hair or asbestos to the clay which strengthens it, but this practice is not always satisfactory, inasmuch as checks and cracks are liable to form and the mass is very liable to scale off and breakage to result. A mixture of infusorial earth and water glass when properly applied will last for weeks, and hence is not expensive, while the strength and protection it affords surmounts all considerations of expense. In mixing this, it is very important to follow these simple directions to obtain the most satisfactory results. A mixture of one part infusorial earth with 4 or 4½ parts of water glass will answer very well. This should form a soft and somewhat elastic, but not liquid, mass.

The vessel to be treated is covered with a coating of from one-fifth to two-fifths of an inch, and dried at not too high a temperature; a drying closet will answer admirably, or the vessel may be supported over a stove. If the temperature is too high at first, air bubbles will form in the mass and the results will not be as good. If bubbles should appear, they should be pressed out with the finger, care being taken not to burn one's self, and if a crack should appear, it can be filled with the mixture and allowed to dry again. It is often desirable to have some parts of the vessel transparent. This may be had and at the same time protection may be obtained by applying several thin coats of water glass alone and allowing each coat to dry before adding the next.

This mixture, although a little expensive, has been used to good advantage upon gas retorts, furnaces, stoves, and for a variety of other purposes. The results have been just as satisfactory as for glass and porcelain vessels.

In using the first mentioned mixture of clay, water, sand and hair, the disadvantages before mentioned may be overcome by adding a little glycerine. This preparation is cheap and easily prepared and applied where desired. The danger from cracking or checking is overcome, and always retains its softness.

WIRELESS TELEGRAPHY

In this department will be published original, practical articles pertaining to
Wireless Telegraphy and Wireless Telephony

A COMMERCIAL AND ACTUAL INTERFERENCE PREVENTER

A. L. PATSTONE

As the number of wireless telegraph stations have enormously increased during the past year, because of the recent Act of Congress, the demand daily increases for a very efficient tuning device capable of eliminating that "bane" of all wireless operators—interference. Although much attention has been devoted to the loose coupler or transforming tuner in past issues, all fail in the excess advantages described herewith, which have become a complete and concluded success with the omission of a variable condenser, and should strongly appeal to the advanced commercial operator.

We know it is almost impossible to find two stations having exactly the same wave length, but on account of the close coupling of many stations, which is always detrimental to sharp tuning, as it produces a widening of the resonance curve, as observed at the receiving station, the tuning out of such stations becomes difficult, and in some instances almost impossible, therefore it is not practicable to completely tune out every undesired station.

Using the interference preventer the signals to be received can, in nearly every instance, be found on from three to five complete different positions, a feat unknown to the modern transforming tuner, consequently during moderate interference we will find our desired signals to be audible and clearly readable on at least one of these positions. This is made possible by very fine and accurate adjustment, by the unmatched alterations of the primary coil and entire hook-up, and by the two possible paths for the signals to pass to the earth, one through inductance A, and the other through

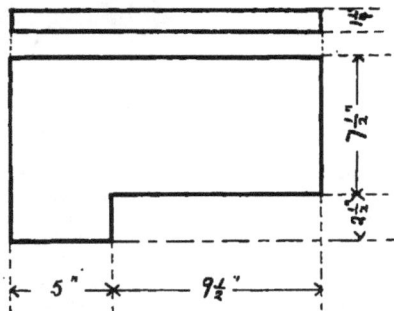

Fig. 1

inductance C. It is supposed that the untuned discharges will divide themselves equally between the two paths in the antenna and will produce no effect on the detector circuit.

One must not expect results unless he undertakes to expel the undesired by manipulating the different sliders, and should not be satisfied until successful, otherwise interference preventing is impossible. To sit there with the sliders in one position and wait for a station to come in on the set tune or wait for some intruder to become cleared before completing business, is to condemn one's self as being an inexpert operator.

After reading an article many hesitate on constructing an instrument, which, for some reason they rightly believe, will prove no more efficient than their present apparatus. By allowing plenty of time and being careful to make every part perfect and by following directions of this article, an instrument which will instantly

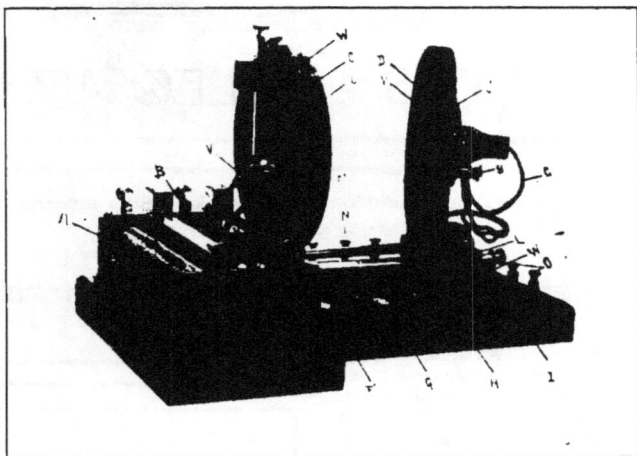

An Interference Preventer

demonstrate its efficiency above the ordinary type, and which will yield surprising results as an interference preventer is certain, and also with which many peculiarities will be noted after being pronounced as learned. The described instrument was designed with that end in view of obtaining the maximum efficiency in a concentrated form, to eliminate interference other than at the expense of decreasing signals, and has gained complete confidence among its operators in these respects.

A general constructional idea can be attained from the photograph, so it will be unnecessary to go into details on some of the parts on account of alterations which may be due to inconvenience.

We must employ the looped aerial or our instrument will be useless.

It should be remembered that perfect insulation is the success of all wireless instruments. The best of wood gives leakage to the weakest high-frequency currents; fiber collects dampness, rapidly rendering it worthless during the slightest damp weather; these should be avoided. To construct the entire base of hard rubber is unnecessary as well as expensive. An excellent base can be built from dry, well-seasoned wood; teak gives an admirable appearance and is easily worked, inlaid with round, hard rubber plugs at least 1 in. in diameter under all binding posts, and a strip sufficiently large enough for the front switches, both plugs and strip having a depth of ½ in. After careful fitting, these should be firmly set without the aid of glue or wedges, and when finished the base should be free from flaws which will be a safeguard against possible efficiency losses. A plane will not give satisfactory results on hard rubber when in this position, but by using a heavy file and finishing with fine sandpaper, the inlaid parts can be brought to be perfectly flush with the surface of the base. Before assembling the parts, the base should receive a smooth polished surface. A substantial finish can be easily accomplished by the inexperienced with the use of refined material from a reliable wood-finishing firm, whose advertisements may be found in the columns of this magazine. If the builder should not have any of the necessities on hand, it will be far better to wait, even a week or more, until they can be obtained, and have a perfect instrument in the end, than to use what might do for the sake of rushing its construction and when complete be dissatisfied.

All connecting wires necessary for the entire receiving apparatus should be of No. 22 flexible rubber-covered cord; those in the base to be enclosed in ⅛ in. rubber piping and embedded in sufficiently deepened grooves. All grooves and holes are to be filled with wax, which will add to its insulation, and wax covering the underneath parts of the switches will protect any of their parts from becoming loosened. Heavy cardboard secured on the base will protect all wires and sustain from injury to the table, etc.

BASE

A and B are cylinder coils made of non-shrinkable tubing 8 x 1⅜ in., each containing approximately 85 ft. of No. 22 S.S. copper wire wound to within ½ in. of their ends. The end pieces are of hard rubber, measuring 4 x 2 x ½ in., thereby insulating the slider rods. Spiral spring sliders have given perfect satisfaction, although any form of slider may be used; it being important that contact is made upon only one turn of wire, since contact on two turns decreases signals. When contact to slider is depended on from slider rod, more or less friction is necessary, and especially during dampness the rods become sticky, causing much annoyance. Most sliders have four small screws on their underneath side; tap one of these to ⅛ in., to which connect a piece of telephone cord under a washer. The tension on slider rod may now be lessened and perfect contact is always certain. Avoid the use of iron or steel screws throughout the entire receiving apparatus; brass is most convenient; copper is still better. Use machine screws to secure parts to hard rubber.

U, the support for the primary, shown in Fig. 2, is of ⅞ in. teak wood, with strip cut out on an angle, leaving opening 1 in. wide on back and ½ in. wide on face. Round hard rubber plugs, having a depth of ¾ in. are inlaid under slider rod sup-

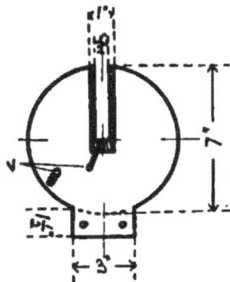

Fig. 3

ports W, preventing any possible leakage; screws holding W should be but ¼ in. long. All open spaces should be temporarily filled to make winding surface smooth. Drill a ¼ in. hole to center to take one end of primary, shellac three or four sheets of good paper on winding surface, then start winding with No. 20 S.S. wire, about 95 ft., each turn being wound on previous turn, using shellac sparingly. Wind until ⅛ in. from edge, giving entire primary a thickness equal to the diameter of the wire. Patience counts here, care being taken to have wire perfectly flat and closely wound. Too much shellac will cause uneven and open winding. Both ends of the primary extend out the back and into the base to their proper connections, the exposed wire being enclosed in ⅛ in. rubber piping, as shown at V. After being wound a heavy coat of shellac should be applied and the coil laid away for two or three days until thoroughly dry. If this is done it will be impossible for the slider to loosen winding. A suitable color paper secured on face of winding will add to its neatness and puzzle the inquisitive.

X, the support for the secondary, shown in Fig. 3, is of ⅞ in. teak wood, and is wound same as primary with approximately 425 ft. of No. 28 S.S. wire. Here the greatest caution should be exercised, taking ample time to make winding perfect. When placed in position the winding is to be in opposite direction to that of the primary. Ten small holes are drilled for taps, which are enclosed in ¹⁄₁₆ in. rubber piping, at such distances as will give all sections nearly equal lengths of wire, as shown by P, Fig. 3, so as each respective tap comes directly to its proper connection under secondary switch J, which will

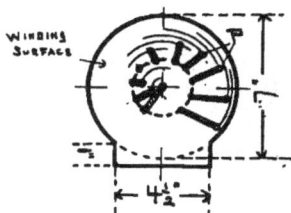

Fig. 2

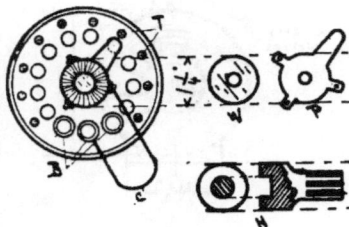

Fig. 4

leave no marks on back of support. When finished the primary winding should be directly opposite the secondary winding; should secondary winding be a trifle high the sufficient amount may be taken off the base of support X.

Secondary switch J, Fig. 4, is an ordinary 10-point switch, having hard rubber base. Three binding posts B are added to the base: W being washer, P contact arm of spring brass, which is fastened to hard rubber handle H on outside. A hole is drilled in the base of the handle to allow free movement over set-screw, which holds contact arm firmly. Small holes T are drilled in switch base just in front of each contact point for the taps to extend through. It will be found possible to tighten the contact points from outside by first placing the taps under contact heads and then drawing them upward until the nut catches while turning. A short piece of telephone cord C is connected to center binding post and moving contact arm thereby insuring perfect contact always, and making it unnecessary to remove the switch base when looking for faults that often arise from constant usage.

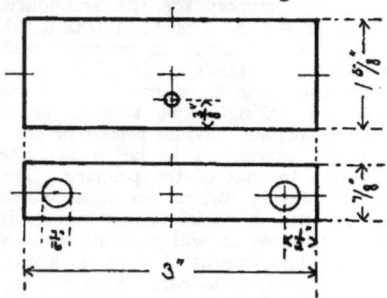

Fig. 5

L, the secondary truck, Fig. 5, is of wood into which two holes 7/16 in. diameter are drilled perfectly true.

The photograph plainly shows the manner in which the secondary support X is secured to truck L. 1/4 in. square brass rod was used and so constructed as to allow the secondary to be placed in any desired position for experimental purposes.

The secondary round brass guide rods are each 3/8 x 7½ in. Two holes, 2 in. apart and 11/16 in. from base to their centers, thereby allowing sufficient clearance from base for truck L, are drilled in primary support U, 7/16 in. diameter into which the guide rods are inserted ½ in., thus giving 1/16 in. play, which will allow free movement should the drilling

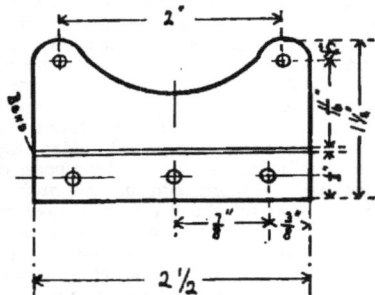

Fig. 6

in truck L be slightly off-centered. At the opposite end the guide rods are drilled and tapped slightly off-center to afford proper adjustment, and are secured by thumb nuts on a brass support W, which is 7/16 in. in thickness. Notice that 1/16 in. is allowed for bending.

W, brass guide rod support.

Y, wood block secured to base and against which is fastened primary support U.

F, 3-point switch controlling coil A, looped, straight and open leg antenna connections.

G, 2-point switch controlling coil B for long wave lengths.

H, 2-point switch allowing discharges to enter at center or outside of primary coil.

I, 2-point switch to alter direction for induced currents in secondary coil.

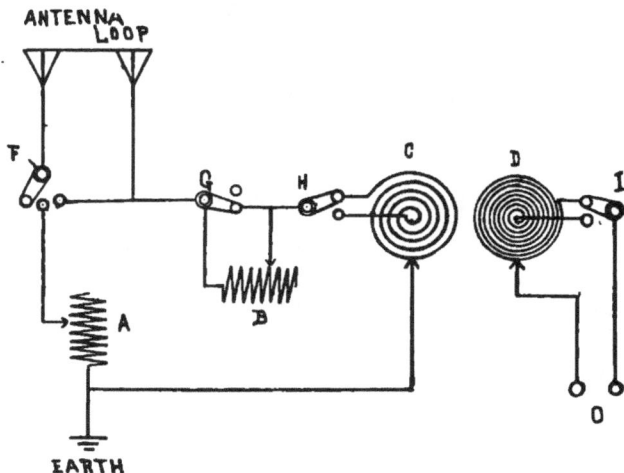

Fig. 7. Wiring Diagram

M, aerial connections.
N, ground connection.
O, secondary terminals for connection to detector circuit.

It will naturally require some little time to master the instrument, and on account of so many possible changes every effort should be put forth to find stations at their loudest, when note should be taken or positions memorized. Placing sliders on positions shown with coil *B* cut out, discharges entering outside winding of primary, and using coil *A* will probably give best results for trial; adjusting only coil *A*, secondary inductance and the degree in coupling. Long wave lengths are best obtained when one leg of antenna is open and adjusting coil *B*. Should any station's signals increase when connection is made from one leg of the antenna to one terminal of the secondary circuit, the right terminals to be determined by experiment, it will indicate that a better tune can be obtained by adjusting the sliders. In some instances interference can be eliminated by this connection when impossible otherwise.

Wireless instruments, like all others, depend for their efficiency on their condition and amply repay good care. All sliding contacts should be clean and bright and free from foreign matter.

At all stations the best results are invariably obtained and the most satisfactory service given by alert and careful operators who take pride in the condition of their instruments.

The Use of Concrete on the Farm

(*Concluded from page* 396)

to the forms. In order to prevent this the surface next to the concrete should be given a coat of oil or soft soap. Linseed, black or cylinder oil may be used, but never kerosene.

Before erecting, the forms should be painted with an oil or soap and then carefully protected from dust or dirt until erected. If chips or blocks of wood fall inside the forms while erecting they must be carefully removed. Upon removal, the forms should be immediately cleaned of all particles of concrete adhering to the surface. A short-handled hoe will remove the worst, while a wire brush is most effective for finishing. In cleaning, care should be taken not to gouge the wood, because this spoils the surface of the next section of concrete.

[*Editor's Note—*Owing to the fact that volume xxiv of this magazine closes with this issue, this article on "Concrete" has been made complete in two installments, rather than in three, as was first intended.]

A REVIEW OF THE CRYSTAL DETECTORS
CHARLES BEALS

In my experiences in operating both commercial and amateur wireless stations, I have found that the best results have been obtained with the following types of solid rectifying detectors, in sensitivity, reliability and in ease and permanence of adjustment. Below is given a description and notes on operating each of these:

Perikon—consists of a crystal of either bornite, chalcopyrite or copper pyrites in contact with a piece of native zincite. A small applied e.m.f. increases the efficiency of this detector nearly 20 percent, the current flowing from the bornite to the zincite. A detector stand that allows the bornite to come in contact with any part of the surface of the zincite with a moderate spring pressure gives best results. This detector is extremely sensitive and easy of adjustment, but is usually "knocked out" in sending.

Pyron—consists of a piece of "ferron" or "pyron" in contact with a sharp metallic point. No local battery is used. This crystal varies greatly in sensitivity, and it is difficult to secure a good piece, but one is amply repaid for his trouble. It is very sensitive and keeps its adjustment for a long time, being but little affected while sending.

Galena—consists of a piece of galena or lead sulphide in contact with a light coil spring of about No. 30 copper wire. No battery is used. It is easily constructed, and nearly as sensitive as the perikon and pyron, but is so delicate that it must be readjusted after each period of sending.

Silicon—consists of a piece of metallic silicon in contact with a sharp metallic point. No local battery is used. I have not found this as sensitive as any of the previous types, but crystals above the 90 percent grade show more uniform sensitivity, and it keeps its adjustment fairly well. It is a suitable detector for amateur use where great range is not required.

Carborundum—consists of a crystal of carborundum or carbide of silicon in heavy contact with a blunt brass point. There are at least three distinct varieties of this material, one vari-colored, blue predominating; one a solid light green, and one jet black. A large piece should be secured and a small piece should be broken off the end that is coated with graphite. This end is packed in tin-foil and the rectifying terminal is the surface opposite this, and at right angles to the parallel of the grain. In practice I have found the green and black crystals to give the best results. Local battery is imperative for long-distance work, and voltage must be very carefully regulated. The polarity of the crystals vary even if broken from the same piece. More skill is required in using this detector than in any of the other types, but I have found it the best crystal for general use. It is not easily disturbed by either mechanical or electrical disturbance, and it compares favorably with the others in sensitivity.

From these results it would seem that an ideal combination for general use would consist of one of the more sensitive types, as perikon, pyron or galena, in connection with one of the more stable types, silicon or carborundum, the former to be used for long-distance receiving and the latter for copying local messages.

A low melting point alloy for mounting these crystals may be made by melting commercial solder and adding about 30 percent of mercury, previously heated in a separate vessel.

With a well-developed piece of any of these minerals I have been able to copy 1,200 miles with a 50 ft. antennae elevation, using a modified Marconi multiple tuner and Sullivan receivers.

Transatlantic Wireless

A notable advance in wireless telegraphy has been achieved in the last three weeks in the adoption by the New York *Times* of a daily transatlantic wireless service for European news.

The *Times*, which was the first newspaper to foresee the possibilities of development in the wireless telegraph, has itself been one of the most active and important factors in that development. For three weeks the *Times* has received no special cables from London, all its news dispatches, approximating 20,000 words a week, being transmitted across the Atlantic by wireless.

THE NEW WIRELESS LAW AND THE TITANIC DISASTER

FRANK ROY FRAPRIE, S.M., F.R.P.S.

We published last month a letter of protest in regard to the so-called Alexander Bill for the regulation of wireless telegraphy. In the meantime the terrible disaster to the *Titanic* has again focused public attention on the utility of wireless telegraphy as a means for the saving of lives at sea and also on the necessity for more efficient wireless service than is now given. Naturally, following the disaster, the first news of which was given to the world by means of wireless telegraphy, there was an enormous amount of activity in the line of wireless communication. The press was filled with accusations and recriminations of all kinds, charges of interference on the part of amateurs, of transmission of false messages, of failure to pay attention to distress signals and especially of failure, on the part of the operators on the *Carpathia*, to transmit press intelligence or to answer questions. In the maze of conflicting accounts, a few facts seem to have been established. In the first place, the *Carpathia* had but one operator, and it was by the merest chance that he received the distress signal which resulted in saving what passengers were rescued; ten minutes later he would have retired for the night, and the *Carpathia* would not have been the rescue ship. While the *Titanic* was fitted with modern high-power apparatus, fitted for long-distance communication, the *Carpathia* had one of the older sets, with a sending distance of from eighty to one hundred and twenty miles, and consequently was out of range of land most of the time after the rescue until she docked in New York. During all this time, however, the operators on the *Carpathia*, one belonging to the ship and the other rescued from the *Titanic*, utilized their instruments almost constantly in sending off lists of the rescued and personal messages from the rescued to their friends ashore. Press matter was not sent, and the operators seem to have paid little attention to messages directed to them asking for details. It is probable that they were justified in thus acting, both by the rules of their company and by the orders of the commander of the *Carpathia*, but considerable ill-feeling arose because of their apparent disregard of questions sent from United States naval vessels at the instance of the President of the United States. The operators later claimed that they answered these questions, but stated that the work of the operators on board the United States ships was so inefficient that they found it impossible to keep up satisfactory communication with them. The Navy Department claims that its operators are of the highest efficiency, but the testimony of the commercial companies and also of government officials, as given in the hearings before the Committee on Merchant Marine and Fisheries of the House of Representatives, seems to indicate that the operators on the naval ships cannot always satisfactorily communicate with commercial operators and are not of a sufficiently high standard to secure employment by the commercial companies after discharge from the navy.

A very unfortunate feature was the bringing out of the fact that the operators on the *Carpathia* had been advised by the chief engineer of the Marconi Company to hold their news for sale to the newspapers at a personal profit; and at the hearings before the Senate Committee this fact was brought out and Mr. Marconi himself admitted that the practice was injudicious and would probably be forbidden in the future. The precedent was set at the time of the disaster to the *Republic*, when Operator Binns was allowed to sell his story to the newspapers; but the general opinion of the press and of the public is that operators should not be allowed to hold back news of such a disaster for the sake of personal profit.

As a result of the facts brought out, public opinion has been aroused to such an extent that new regulations for the handling of wireless on passenger ships will doubtless be made effective, and it seems certain that all passenger ships will be required to carry two wireless operators, to keep apparatus working constantly, and to be equipped with apparatus capable of communicating over a distance of one hundred miles under the most unfavorable conditions, that is, in the day time. Such sets will probably be capable, under favorable conditions, of transmitting messages at night from two hundred to five hundred miles and should be far more effective than the present sets on many of the older steamers.

Coincident with the hearings on the *Titanic* disaster, the report of the Committee on Merchant Marine and Fisheries of the House of Representatives on the Alexander Bill was issued, and on April 20th the amended bill was reported to the House, committed to the Committee of the Whole House of the State of the Union and ordered to be printed. While this by no means insures the passage of the bill, it is probable that owing to the state of feeling on the subject, it will be passed through both House of Congresses and become a law at this session, and it is not likely that serious amendments will be made.

An analysis of the amended bill shows that it is a far more effective measure than that introduced by Mr. Alexander in December, and that it is drawn up with scientific care and with reasonable regard for the rights of all persons concerned. We have not space to reprint the bill or to analyze it in detail, but every reader who is interested may obtain from his Congressman the bill and the reports on it if he will write for a copy of HR 15357 and Report No. 582, to which he may add a request for the hearings before the Committee if he sees fit.

The bill requires that every station, for either sending or receiving, shall be licensed. The license is to be issued by the Secretary of Commerce and Labor, but he has no discretionary power and is required to issue a license to every applicant who complies with the provisions of the law. Every station in operation at the time of the passage of the bill, whether amateur or commercial, will be entitled to a license, provided it is operated by an American citizen or corporation, and the restriction of American citizenship also applies to operators.

The bill goes much further than the original bill, in prescribing regulations for the use of wireless instead of leaving these to the discretion of the President or a government department, although a certain latitude is allowed to the Secretary of Commerce and Labor for cases of emergency. Amateurs will be required to use wave lengths of two hundred meters or less, and cannot use a transformer input of more than 1 k.w. or of more than

½ k.w. within five nautical miles of a naval or military station. They are also, and this is a very important requirement, confined to the use of pure and sharp waves. This regulation will almost completely eliminate the danger of interference among stations and it applies to all stations, commercial, naval and military, as well as amateur. From the reports it appears probable that the Bureau of Standards will issue pamphlets for the instruction of amateurs and experimenters, giving advice on the construction and arrangement of their apparatus to meet with the requirements of the law, and it is certain that amateurs who do adjust their stations and apparatus to the requirements of this law will learn far more and become far more efficient operators than when working with the present inefficient apparatus which may be found in common use.

Another very important clause of the law states that every operator shall be obligated in his license to preserve and shall preserve faithfully the secrecy of radiograms which he may receive or transmit, and for failure to preserve such secrecy his license may be cancelled. This will prevent such cases as occurred on the Pacific Coast recently, where amateur operators intercepted an important message and sold it to a newspaper, and will also do away with other abuses which need not here be described. Willful or malicious interference, such as has too frequently occurred between operators of rival commercial companies, will become a serious offense, punishable by both fine and imprisonment. The sending out of a false message, signal or call, whether a distress signal or not, will also become a most serious offense, punishable by a fine up to $1,000, or imprisonment up to two years, or both. The operation of wireless instruments, whether for sending or receiving, without a license will be punishable by a fine of not more than $500, and the forfeiture of the apparatus.

There are numerous other provisions of the act which are of less importance to amateurs, but which provide for the needs of commercial companies and the government departments. The army and navy have received far less than they asked for, but all that they need to enable

(*Continued on page* 416)

A NEW WIRELESS STATION WHICH OPERATES AT 100,000 VOLTS

FELIX J. KOCH

Wireless telegraphy has become a most efficient aid for commercial as well as for military purposes. The plant described below is one of the strongest stations in the world and was installed lately by the French Army at a cost of several hundred thousands of dollars, at the summit of the Eiffel Tower, about 1,000 ft. above the ground. Its strength

Fig. 3
Room where the messages are received

Fig. 1
Room where the high tension current is produced

allows it to communicate with the Marconi station at Glace Bay in Canada, about 7,500 miles away.

The cuts show the most important parts of this plant, by means of which the Old and New World can be connected in a few seconds.

four doors of this yard being made of metal, and which establish the connection with the different parts of the wireless

Fig. 4
Apparatus, switchboard and other accessories for high tension work

station, have to be closed, when the current of high tension is turned on, in order to prevent any accident. In Fig. 3 is

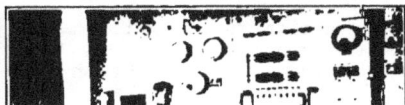

shown the receiving room. A wireless message can be received by four different operators, who thus check each other. The receiving is carried out by an electric apparatus, which was invented by the commander of this military post.

In Fig. 4, the apparatus and accessories of the high-tension chamber are demonstrated, while Fig. 5 shows the moment when the current is turned on and telegraphing is started. In the middle of this room we see a device which is one of the most improved and by means of which the radiotelegrams are sent out. The other connections are used to turn on or shut off the current.

LIST OF INTERNATIONAL RADIOTELEGRAPHIC STATIONS

The official list of radiotelegraphic stations open for international traffic has just been published by the International Telegraph Bureau in Berne. The countries included are the United Kingdom, Germany, Russia, Austria, Italy, Spain, Denmark, Sweden, Norway, Belgium, Holland, Japan, Uruguay and Chili. The catalogue does not include stations in the United States, as the government of this country did not ratify the adhesion of its delegates to the convention. Canada and France are also missing from the list. The particulars of each station whether on the coast or on board ship, are entered in eleven parallel columns, and give its geographical position, call signal, normal range, system employed, wave-length, nature of service, *i.e.*, whether public or restricted hours of working, and charge per word. The number of stations in the list reaches the total of 690, although the war vessels of several of the countries are omitted. Of these 690 installations, 124 are coast stations, the majority being open to the general public, and the remainder to messages from ships in distress only. In the latter class are the naval and military coast stations and those on lightships. The list does not include any inland stations, since these do not come within the purview of the convention. The distribution of the stations on the coasts of the various countries is as follows: Great Britain and Ireland, 35; British West Indies, 4; Gibraltar, 1; Malta, 1; Italy, 23; Germany, 15; Tsing-tau, 1; Russia, 13; Denmark, 7; Japan, 5; Norway, 4; Austria, 3; Holland, 3; Chili, 3; Spain, 2; Uruguay, 2; Belgium, 1; Brazil, 1. Roumania, 1. Even without counting the British stations in Canada and elsewhere, which do not appear in the list, it is clear that Great Britain has realized that wireless telegraphy is of far greater importance to her than to any other country. In the matter of ship stations she is also well ahead, the totals for Great Britain and Germany being: Great Britain, warships, 176; merchant vessels, 86; Germany, warships, 95, merchant vessels, 53. It is interesting to note that among the coast stations in Great Britain there are four which conduct the ordinary telegraphic business of the Post-office, taking the place of a wire or cable. These stations work in pairs, one pair communicating across the Wash and the other between Mull and the Outer Hebrides.

New Wireless Law and Titanic Disaster

(*Continued from page* 414)

them to have uninterrupted communication, without committing them to the indefinite preservation of antiquated apparatus. Their apparatus will have to be of modern types to comply with the requirements of the act, and only by such compliance can there be any possibility of proper functioning of naval and military wireless service in case of war or of deliberate interference by an enemy. The commercial companies are given full opportunity for carrying on their work, and every amateur is entitled to a license, which cannot be denied him, provided he has apparatus of a character which will give others the right to use the ether properly. A set with 1 k.w. input and two hundred meters wave length will carry over the larger part of any ordinary city and give ample scope either for play or for serious experimenting.

ELECTRICIAN AND MECHANIC feels gratified that the rights and needs of the amateur have been so fully met in the conflict of selfish demands brought before Congress, and trusts that this bill will be passed without serious amendment, believing that it will result in a great stimulation to wireless experimenting.

WIRELESS NEWS

CHICAGO WIRELESS ASSOCIATION
Athenaeum Building
18–26 E. Van Buren St., Chicago, Ill.

The present organization is a re-organization of the old Chicago Wireless Club, formerly affiliated with the Electrician & Mechanic Wireless Club of Boston, Mass. The club was re-organized about two years ago, and the name changed to the Chicago Wireless Association. The officers are: J. Walters, president; E. I. Stein, vice-president; C. Stone, treasurer; F. Northland, recording secretary; R. P. Bradley, corresponding secretary; W. J. McGuffage, chief operator; P. Pfiffer, and Geo. Blackburn assisting chief operators.

The regular membership is about seventy members, with a large number of irregular members who attend the meetings frequently.

The members of our association control and operate stations having power ranging from a ¼ in. spark coil to a 2 k.w. transformer.

The rules which are printed in the Call List were adopted after much discussion, and we are enforcing them with very good success. In the evening, there are always two chief operators upon the job, the chief operator being on duty all evenings, while the assistant chief operators alternate. The city is divided into two divisions, one of which is under the supervision of the chief operator, and the other division is under control of the acting assistant chief operator.

The members also transmit messages for each other and for outsiders, free of charge, using a regular message blank and envelope.

The Association is also attempting to get a relay route established between Chicago and New York, in this way being able to transmit messages from Chicago, and points between, to New York and vice versa.

As soon as this route is completed and in good working order, we will, no doubt, co-operate with the Tri-State Wireless Association of Memphis, Tenn., in regard to establishing another relay route to New Orleans, thus giving us a complete route from New Orleans to New York City. And once we have reached New York, it ought to be an easy matter to make up a route from New York to Boston.

At the present time we have no stations on representing the amateur stations in the city, almost a month before a similar one was started by a club in Boston. The map is brought out on meeting nights, so that a member may see where another member lives, in relation to his own location, and so as to tell just how many miles the two stations are apart. The map is 48 in. wide by 72 in. long, on a scale of 2½ in. to the mile.

ROBT. P. BRADLEY,
Cor. Sec'y.

4418 Wabash Ave., Chicago, Ill.

WIRELESS ASSOCIATION OF BRITISH COLUMBIA, VANCOUVER, B.C.

We have received from the above-mentioned Association, of which C. C. Watson is president, a list of about thirty members with the call letters of their individual stations, also the rating power. To anyone interested we will be pleased to give the names and addresses with call numbers, or they may be obtained by communicating with Harold J. Bothel, corresponding secretary, 300 14th Ave. E., Vancouver, B.C. In the Province of British Columbia wireless stations are limited to a power not exceeding 1 k.w. and an aerial length of about 30 ft.

THE MANCHESTER RADIO CLUB
Manchester, N.H.

We are now corresponding with a number of other clubs throughout the United States and are glad to add new ones to our list, and we will appreciate any information you can give us.

Our club has decided to put up a new sending and receiving station and the work is now in progress. The sending outfit will consist of a closed core 1 k.w. transformer, a large, air-cooled, glass condenser, a specially-constructed heavy, key, brass strip oscillating sending helix, and three spark gaps, quenched spark ·gap, rotary spark gap, and a series spark gap.

The receiving station will consist of double-slide oscillating transformer (2,000 meters wavelength capacity) one fixed condenser, one variable condenser, three detectors, a silicon, a carborundum, and an electrolytic detector, potentiometer and battery, and a pair of 2,000

AN EXPERT'S VIEWS

April 18, 1912.

Dear Sir:

The danger of ill-advised legislation by the Federal government, tending to regulate and limit the use of wireless telegraphy in this country, is a constant menace to the development of this new and important art; an art to which is already to be credited the saving of very many lives and numerous vessels at sea.

Since the sinking of the *Titanic* on Sunday night last, those interested in having restrictive legislation enacted have redoubled their efforts and are making use of the fact that there was more or less interference with the operation of certain United States government stations by amateur operators as an argument for wholesale restrictions. The danger of hasty legislation in this regard is now, I believe, imminent.

1st.—There seems to be a probability that our government will join with other powers in subscribing to the so-called "Berlin Treaty," and in imposing certain restrictive regulations upon the use of wireless telegraphy in this country conforming to practice in European nations.

2d.—That there seems to be a likelihood that one or more bills for imposing restrictive regulations upon the use of wireless telegraphy, which have been presented in the House of Representatives, may be enacted into law.

Under ordinary circumstances, the scientific and technical staffs of the departments of our national government might be expected to be fully competent to advise the House and Senate sufficiently upon scientific and technical matters; but, in the present instance, the Department of the Navy appears to have, or appears to feel that it has a strong interest in restricting, almost to the point of suppressing commercial wireless telegraphy, or, in fact, all wireless communication other than governmental messages between its stations on land and afloat.

I am informed that so preoccupied are the wireless companies of the United States in litigation and in recovering from the financial crises through which most of them have passed, that they have, for the first time, failed to take active part in the discussion of the matter at the hearings which have been held on the subject before the Committee on Merchant Marines and Fisheries of the House of Representatives.

The excuse for restricting wireless telegraphy arose some years ago, as you are probably aware, from the much advertised "Interference" between messages. It is in devising and perfecting means for overcoming and preventing this interference that the future utility of the art to mankind chiefly depends. Most of the "interference" at present experienced is entirely due to the use of apparatus which is out of date, in that it makes inadequate provision (a) for excluding extraneous "interfering waves"; (b) for minimizing the power of interference of the waves it sends out.

The provisions of the so-called Berlin Treaty are such as to stifle the progress of the art by making the natural development of non-interfering apparatus useless and unprofitable. It tends to take away from the inventor, the engineer and the manufacturer all incentive to perfect the art to that point at which mankind may be able to profit to the maximum from this wonderful, flexible and economical mode of instantaneous intercommunication. Where many progressive but separate and distinct nations are crowded together on a continent, as are the European nations, such a convention as the Berlin Treaty may, perhaps, to a certain extent, be excused in spite of its unnecessarily restrictive character; but, for the United States, having for itself the greater part of a continent—far remote from these nations—to join with them in such a convention would be, I believe, a grave and inexcusable mistake.

The experts in wireless telegraphy of the United States Navy would like to have such laws enacted as would give the United States Government unrestricted use of all wave lengths between 600 meters and 1,600 meters. This would leave only the least desirable wave lengths for the use of the wireless business of the country, and bring about an altogether intolerable condition in the transaction of business by wireless telegraphy. Moreover, it would free the United States Navy from the necessity of using any but the cheapest and least efficient wireless apparatus and would tend to indefinitely postpone progress in the art.

If the United States Government's naval stations and ships were equipped with as efficient wireless apparatus, *i.e.*, as "selective" wireless apparatus, as that used in the German Navy, they would not be so easily interfered with by every little amateur's station in their neighborhood. In this connection it is to be remembered that the amateur of today will be the inventor and engineer of tomorrow, and if he be prevented from freely experimenting now his attention and interest will probably be permanently diverted from the subject of wireless telegraphy with loss to himself and the community at large.

I ask you to recall the "interference" that used to be experienced between the old grounded circuit telephone lines, in the first ten or twelve years of the history of that art. Circuits a quarter of a mile or more apart used to interfere so much with each other that it was often impossible to tell whether the voice you heard in your telephone was that of the man at the other end of your line or whether it was the voice of someone carrying on a conversation on a totally different line a quarter of a mile or more away from yours. Suppose the government at that time had enacted restrictive laws, prohibiting the establishment of telephone lines within a half mile of each other; would we today have cable lines with 100 telephone circuits all within a circle of 2 in. diameter, so cleverly arranged in metallic circuits and twisted relatively to each other that not a sound passes from one of these circuits to any of the other 99 circuits? I am of the impression that we would not, and that the great benefits which we today enjoy from the commercial development of the art of telephony would have been indefinitely postponed by any such legislation.

I feel that the *Boston Herald* will be doing service to mankind and certainly to our nation if it is instrumental in checking this tendency to stifle progress in this new art which is daily growing in flexibility, in range and importance to us all.

Yours truly,
JOHN STONE STONE.

QUESTIONS AND ANSWERS

Questions on electrical and mechanical subjects of general interest will be answered, as far as possible, in this department, free of charge. The writer must give his name and address, and the answer will be published under his initials and town; but, if he so requests, anything which may identify him will be withheld. Questions must be written only on one side of the sheet, on a sheet of paper separate from all other contents of the letter, and only three questions may be sent at one time. No attention will be given to questions which do not follow these rules.

Owing to the large number of questions received, it is rarely that a reply can be given in the first issue after receipt. Questions for which a speedy reply is desired will be answered by mail if fifty cents is enclosed. This amount is not to be considered as payment for reply, but is simply to cover clerical expenses, postage and cost of letter writing. As the time required to get a question satisfactorily answered varies, we cannot guarantee to answer within a definite time.

If a question entails an inordinate amount of research or calculation, a special charge of one dollar or more will be made, depending on the amount of labor required. Readers will, in every case, be notified if such a charge must be made, and the work will not be done unless desired and paid for.

1780. Engine Drive. J. L. B., Thiep River Falls, Minn., has charge of running a supposedly 200 k.w. 250 volt direct current generator, the drive being by belting to a double engine with 18 x 26 in. high pressure cylinders. The engine has ample power, but evidently the generator is badly designed, for at even ⅛ load heating and sparking are severe. Engine runs at 135 revolutions, and the question is asked if at this speed the operation would be sufficiently steady to permit the removal of one connecting rod, and using of the other half of engine only. This would result in considerable economy of steam. Ans.—Your proposal is not an unknown one, and your correspondent has seen the procedure, years ago, in two different electric light stations. During daylight hours, when the load was light, one-half the double engine was used; then as afternoon came on, the load would be shifted to a somewhat larger engine. The connecting rod of the first would then be put on, and for the peak of the evening load, as well as for caring for the street lights, this engine was ready for its full capacity. An objection is, that the crank-pin bearing is a rather troublesome one to adjust, and once in good condition, an engineer dislikes to disturb it. A little taking-up is always in order, but the fresh set, every day, invites the chance of either a hot box or a loose one, and once the engine is started for its night's run, no stop is to be permitted. Perhaps in your case, with water for the principal power, a short shutdown of the steam engine for readjustment would be perfectly allowable. As a permanent arrangement, the proposition is entirely feasible, for, at the stated speed, there should not be noticeable flickering of the lights, with even a single cylinder acting. If you really wish the full ouput of the generator, the case would properly belong to the dynamo builders to remedy, and any concern that valued its reputation would be anxious to set the machine a-right.

1781. Alternating Current Motor. L. Bernadin, Lawrence, Mass., has a General Electric fan motor, form D, that has been rewound. On 110 volts direct current, the speed is 3,500, but on alternating current, only 400 is reached. He asks if he can wind the motor so it will act synchronously on the alternating current circuit. Field magnet is bipolar and laminated, and armature has 11 slots. Will an induction motor be as good as one of the synchronous sort for operating a rotary spark gap for wireless telegraphy? Ans.—You can try the motor in the manner you propose without rewinding, but if it works the rest of the apparatus satisfactorily, it will be more economical for regular use to substitute such special windings as we will describe when you are ready. Fit two brass or copper rings over the commutator, being insulated from the segments and from each other by fiber or paper. One small headless set-screw is to be provided for each ring by which it is electrically connected to some particular segment, and the other ring to the segment most nearly opposite. Provide a copper brush for each ring, and to these the alternating current is to be led, but with provision for including one or more incandescent lamps for safety. Field is to be separately excited from the direct current supply, lamps being also included in this circuit. Have a small double-pole switch, one blade of which can connect with the d.c., other with the a.c., other two wires being without switch. To start motor, it may be speeded to as near 3,600 revolutions as can be judged, by hand or foot, say through the mechanism of some old sewing machine, and then the switch being closed, the driving belt being at the same time automatically slackened. An induction motor will not be suitable for operating the spark gap if latter is connected to the alternating current supply. If direct interrupted currents are used for operating the primary of induction coil, an induction motor will be as suitable as any other. With the alternating currents it is important to have the impulses come at critical instants, and this result will be attained when both coil and rotary spark device are operated in synchronism.

1782. Fan Motor. W. A., Cambridge, Mass., has a 6-pole, 110-volt direct-current fan motor, five of the field coils of which are wound with copper wire of the size of sample sent, while the sixth is of the same size, but of German silver wire. Everything else appears to be right. What is the reason for this construction? Ans.—This method of winding is not new, but merely makes a field coil take the place of a starting rheostat. If it were not for the undesirable heating produced in the machine, the scheme might be used on other than fan motors. In this case the breeze has not only to keep the user cool, but to keep the fan cool. If you put a pulley in place of the fan and tried to use the motor to drive some light machine, the coils would get intolerably hot. Evidently you have a larger motor than we at first supposed, for it is unusual to have desk fan motors for direct currents with more than two poles. The size of wire is No. 21, and this indicates a motor larger than the common fan size.

1783. Induction Coil. B. H., Wilkes-Barre, Pa., wishes to make a coil for wireless messages, using about 1 lb. of No. 34 d.c.c. wire for the secondary. He asks for directions and dimensions. Ans.—If you follow directions given in various articles that have appeared in this magazine you will be likely to get a reliable coil. Two articles of value are in the February and March, 1910, issues. For your case, the iron core should be about 8 in. long, 1 in. in diameter. Primary coil should consist of two layers of No. 16 wire, secondary of the 1 lb. of No. 34. Condenser may have 100 sheets of tin-foil, each 5 x 7 in. Using three or four large bichromate or small storage cells, you ought to get 1 in. sparks, but we do not know for what range of messages the coil will work.

1784. Gauge. H. J. T., Loudonville, O., asks where he can find directions for making a gauge or instrument that will indicate the reading on another instrument half a mile or a mile away. Ans.—This is not the first time a request of this sort has been made, yet we regret to state we do not know of anything on the market. The Weather Bureau at Washington would probably send on request a description of their distant recording instruments, but these would undoubtedly prove too expensive. Of course two wires at most should do the work. A single wire and the ground would prove unreliable for some sort of signals or currents in the only practical device we could suggest. This might consist, at the operating end, of a float that would raise or lower a slider on a row of electrical contacts, such as are used on a rheostat. Resistance coils would be connected between these contacts, and at the receiving end considerable electrical energy could be introduced into the circuit, say from 110-volt direct current lighting mains. Alternating currents would not work as well. A low scale ammeter could be specially graduated to read in any particular units you pleased, and the varying position of the pointer, dependent upon the current flowing, and that in turn by the particular amount of resistance in circuit, would give the indications desired. There would be the undesired errors introduced into the circuit by the varying resistance of the wire in different seasons of the year, and at different times of the day, due to changes of temperature, by the varying leakages at the insulators in dry and wet weather, and by variations in the operating voltage. In the absence of other sources of direct current, you might use gravity batteries, provided they were always kept in circuit and in good order, with a milli-ampere-meter for indicator, but the readings might be more in error.

1785. Battery Charging. R. H., Newport, Vt., asks: (1) Is it a better plan to recharge storage cells, say those of the "Exide" sort, in an electric launch, after every short run, thus keeping them nearly charged all the time, or use from them until practically exhausted? (2) Can it be shown that a piece of hardened steel that has been run for some time in a bearing without showing signs of wear has actually changed in size? (3) In which sort of vessel will water boil more quickly, and why,—one with dull, rough sides, or one brightly polished? Ans.—(1) Just what you mean by a short run is rather indefinite, but if such use approximates

half a discharge, you will greatly improve the working of the battery by charging it without further delay. The complete discharges are what tell on the life of the cells, 300 such being often as the legitimate limit. Department stores that use electric delivery wagons, with supposedly battery power for a whole day's run, find it highly desirable to put in even a partial charge during the noon hour. (2) Yes, with a micrometer caliper. (3) The rough dish, for it presents a greater heating surface, and also allows points from which the steam can be directed. The case is not so marked with water as with some liquids that have a habit of "bumping." That is, even when the temperature at which the liquid is supposed to boil has been passed, yet no evolution takes place until all of a sudden such a mass is converted into vapor as to drive all the liquid out of the dish, and the experimenter may feel fortunate if he escapes a scalding. Beauty marks on the ceilings of chemical laboratories are silent witnesses to escapades of this sort. To avoid accidents, it is customary to put glass beads or broken bits of glass in the flasks used for boiling such misbehaving liquids, whereby the greater area and numerous sharp points will encourage the escape of vapor when the boiling temperature has been reached.

1786. Molding. L. O., Vernon, Tex., asks for directions for making plaster of Paris or ordinary sand molds for casting brass. Ans.—To give adequate directions in these columns would be quite impossible, and we shall have to refer you to some book, say the "Pattern Maker's Handbook," price, 50 cents. The November and December, 1909, issues of this magazine contained good chapters on this subject.

1787. Wireless Transmitting Station with Induction Coil. W. R. T., Fair View Ranch, Wash., says: A friend and I living two miles apart have wireless outfits. Our aerials are 50 and 125 ft. long. We have double slide tuning coils, 1,000 ohm receivers, silicon detectors, fixed condensers and 1 in. spark coils, but we cannot get each other. (1) How can we tune our transmitting sets? (2) Explain how sending helix is adjusted to change wave-length. (3) Will the difference of our aerials (which are 50 ft. high) in length affect the wave-length? Ans.—(1) With spark coils you cannot use the commercial "hookups" to advantage. The station having the 50 ft. aerial will need a helix consisting of a great number of turns put in series with the aerial and one side of the spark gap, and the ground connected to the other side of the gap. The spark gap should be connected right across the secondary of the spark coil and no sending condenser should be used. With these connections you will have a set that is always in tune. The station with the 125 ft. aerial need use no helix, simply put aerial on one side of the gap and ground on the other. (2) To lengthen the wave put more turns of the helix in the aerial circuit, and to shorten wave take out some of the turns of the helix in the aerial circuit. (3) Yes, decidedly so, because the wave-length is determined by the length of the aerial and the number of turns of the helix in series with it.

1788. Wireless. L. R. J., Lynn, Mass., (1) Can you tell me the nature of calorite? Its temperature coefficient and electrical properties? Who manufactures it? (2) Give call, wave-length, height of aerial, power and make of wire-less station on Board of Trade building, Broad St., Boston. Ans.—(1) Calorite is used by the General Electric Co. in heating devices. It is an alloy that is "practically indestructive" when used for heating purposes. For further informa-tion write to the General Electric Co. Heating Dept., Lynn, Mass. (2) Call "FBN"; 2,200 meters wave-length at present. A 5 k.w. Fessen-den set owned by the National Electric Signalling Co. Towers are 75 ft. high above the roof of the building.

1789. Wireless and Batteries. M. V. P., Chesaning, Mich., asks: (1) In the electrical catalog of J. J. Duck, Toledo, O., page 248, a 1 in. spark coil for wireless is advertised for $4.25. On page 135 is advertised an induction coil (1 in.) for $17.00. What makes the differ-ence? (2) Could the wireless 1 in. spark coil be used with the Tesla coil described in this magazine recently, if protected by the device used in ⅛ k.w. transformer? (3) In Carhart & Chutes high school physics book, on page 338, it says "The copper plate is the positive electrode or cathode and the zinc the negative electrode or anode," and in Newel's Descriptive Chemistry, on page 121, it says, "It is customary to speak of the current as entering the electrolyte by the anode or positive electrode and leaving by the negative electrode or cathode." In a book I have by Alinger Small, it shows a diagram as follows: In this same plate is both positive and negative which would explain the seeming con-tradiction above. Which is right? Please state your authority. Ans.—(1) The induction coil is finished more elaborately, but probably would not be any better for wireless than the $4.25 one. (2) No, it gives too limited a supply of energy to be used for this purpose. (3) That plate or piece of metal by which the current leaves the cell is called the cathode, while that piece of metal or plate in a cell by which the current enters the liquid is called the anode. The plate by which the current leaves the cell is not dis-solved and in some cases receives a deposit on its surface. Authority—Silvanus Thompson. It is customary to speak of the cathode or plate by which the current leaves the cell as the posi-tive plate meaning that the current leaves from the cathode, flows through the external circuit and then returns to the anode. In order to complete the circuit, the current must now pass from the anode through the liquid (or electrolyte) and then to the cathode.

1790. Wireless. J. A. S., Allegheny, Pa., asks: (1) Kindly explain how the telegraph key (as advertised in *Electrician and Mechanic*) called the "Mecograph" works. Are the dots and dashes made by the hand the same as with the ordinary telegraph key, or does the "Meco-graph" make the dots and dashes automatically? I have one of their booklets, but it does not give any information as to how it works. (2) Is there any way to connect a telephone receiver to a telegraph circuit that the telephone receiver will make the dots and dashes like a telegraph sounder? The back click in a telephone receiver is several times louder when the circuit is broken than when the circuit is closed. Would a con-denser be of any use, and if so, how could it be used? (3) There is a wireless station about two

miles from my home. This station has a tower about 80 ft. high. My home is about 300 ft. higher than the top of this tower (wireless station is in a valley.) What instruments would I have to have in order to receive a message from this wireless station? Ans.—(1) If you will write directly to the manufacturer, Mecograph Co., 321 Frankfort Ave., Cleveland, O., they will be very glad to give you literature and information regarding same. (2) You can connect the tele-phone receiver in place of the sounder and get a click every time the circuit is made or broken, but it never can sound much like the clear click of a sounder. A condenser shunted directly across the telephone will help out. (3) A two-slide tuner, silicon detector, fixed receiving con-denser and a pair of 75 ohm telephone receivers will do the work fine.

1791. Wireless. C. C. B., West Dennis, Mass., asks: (1) I am building a wireless receiv-ing apparatus as follows, and would like any suggestion that would improve my receiving station. I have a tuning coil 12 in. long, 4½ in. in diameter with two slides, one condenser, glass type, 10 pieces of glass 4 x 5 with tin-foil alter-nated between each piece of glass, detector elec-trolytic type, and a pair of electro government phones 1,500 ohms each. I want information in regard to a 50-mile sending set. (2) I have a 2 h.p. gasoline engine that I can bring into play if I know what kind of a generator to get to make the amount of current I would need to operate my station with. What other machine would I need to complete my 50-mile out set? Ans.— (1) A much better receiving set would consist of the following: An induction tuner made by winding a layer of No. 22 wire on a tube 4½ x 8 in. and a layer of No. 28 wire on a tube that will just slide easily inside of this one. A sliding contact is placed on the 4½ in. coil and about eight taps brought out to an 8 point switch from the inside winding, which is called the secondary; a variable condenser connected right across the secondary of the inductive tuner, and a silicon or perikon detector. Use your glass plate con-denser across the telephone receivers. (2) You will want a 1 k.w. alternating current generator, a 1 k.w. transformer, spark gap, sending con-denser, helix and key.

1792. Wireless. E. T. D., Cambridge, Mass., asks: (1) Can copies of the "Manual of Wireless Telegraphy for Naval Electricians" be obtained, if so, where and at what price? I understand it is issued by the government. (2) Where can I obtain information as to the instruments and methods of connection employed by the im-portant commercial wireless companies? (3) Where can I obtain specific information concern-ing the Telefunken and Von Lepel systems of wireless telegraphy? Ans.—(1) We can supply this to you for $1.50. (2) The "Manual of Wireless Telegraphy for Naval Electricians," gives diagrams of all the commercial companies' connections. (3) I have seen no "specific in-formation" outside of a few paragraphs in the "Manual of Wireless Telegraphy for Naval Electricians," regarding these systems.

1793. Condenser. J. M. L., Ashland, Wis., asks: I am building a 12 in. coil on the plan of Mr. Stanleigh's 6 in., given in the spring numbers. Would you please answer the following about the condenser? (1) Is this piece of paraffin paper too thick or too thin (for the dielectric)?

(2) Is the tin-foil found on tobacco good for this purpose; if not, would it be all right if several thicknesses were used as one sheet? How many? (3) What advantage has the heavy foil over the thin? Since the charge is only on the surface, it seems as though either would be good. Ans.— (1) While the sample you send us is rather thick for use in making a condenser, still it will undoubtedly be satisfactory providing you increase the number of sheets in order to get sufficient capacity. It is preferable to use two or more thicknesses of very thin paper between condenser plates in order to lessen the danger of breakdown through flaws in any one piece of paper. (2) The tin-foil you mention is quite satisfactory, and it is not necessary to use more than one thickness. (3) For a primary condenser the thinnest grade of foil is electrically all right. It is difficult to handle, however, and the medium grade will be found to be better for this reason.

1794. **Transformer Construction.** H. W. S., Columbus, O., asks: I am having some difficulty in making a ½ k.w. stepdown transformer to step 220 volts from the three-wire lighting circuit to 110 volts to use on my laboratory switchboard. In a back issue of your book *Electrician and Mechanic*, dated September, 1911, page 203, under an article by Stanley Curtis: "For the Laboratory. Experimental High-Frequency Apparatus—Part IV," you give the figures and formulas for calculating a transformer. I tried to design my transformer like this one, but I do not get the full amount of current through it. These are the dimensions of it: core, No. 24 gauge, soft enameled sheet iron, 2 in. square; length of winding space 4 in., outside dimensions are 8 x 6 x 2, window is 2 x 4 x 2, primary 660 turns of No. 16 wire or 10¾ layers. Secondary is 330 turns of No. 13. This is the way I figured it from the formula you gave in the book I referred to. I have been testing the transformer very close. So far I find 40 watts loss. I get the desired voltage on my secondary—110 volts; but when I go to put a load of lamp bank on, I have not enough current to light four 120 watt lamps and on a short circuit I can get 4 and 5 amperes. I can light two 120 watt lights bright. I have a ¼ k.w. stepdown transformer that I bought, and I have found that a short circuit over it will easily blow a 6 ampere fuse, 110 volts or pull about 8 amperes. Of course this is not all good for the transformer, but I have had it to happen once in a while and that was the result. I beg for any suggestion you might give on the subject from the data I here enclose. Also I would like to have some tables telling about the core density per square inch and different cycles and voltages. I am very much interested in this data in your magazine. Ans.—We assume that you have built your transformer with the primary on one leg of core and the secondary on the other leg. This is good practice in the case of the transformer to be used for charging condensers or for an arc compensator where very loose coupling is, technically speaking, poor regulation is highly desirable. In the case of the power transformer, however, the regulation between light load and full load must be as near perfect as possible, the proper method is to place primary and secondary on the same leg or in the case of a core type converter such as you have constructed, to place half of each on the respective legs. The core should also be made as com-

pact as possible and the magnetic circuit as short as is practicable. Better still is the shell type construction shown in the drawing. In this type, the primary is usually divided in two sections which are placed on either side of the secondary. The windings are completely surrounded, except at top and bottom, by the iron core, and such a transformer gives splendid regulation if well constructed. The core is more difficult to cut and assemble, however, and this probably accounts for the fact that amateur builders prefer the simpler core type. The construction shown in Mr. Stanleigh's article in the January, 1911, issue combines both the simplicity of the core-type and the regulation of the shell type transformer, and we suggest that you look up the article. This construction is known as the "Ferranti," and for small transformers it is very useful. For a table giving the core loss at various frequencies, we refer you to Twining's "Wireless Telegraphy," which we can supply at $1.50.

1795. **Tesla Coil.** W. H. L., Napa, Cal., asks: (1) Is it true that distilled water is a good dielectric to use around high-frequency coils of the immersed type? (2) Will a Tesla coil of this description give a 64 in. spark from an alternating current of 20,000 volts: Primary, 15 turns No. 6 wire wound double to form a coil 22 in. in diameter and 28 in. high; secondary wound on open fiber frame placed inside. Primary: 580 turns No. 26 wire, core 17 in. diameter, 28 in. long. (3) What would be the watt consumption of above coil? Ans.—(1) This is a new one on us. It is usually conceded that water or even moisture is best kept away from apparatus of this kind. (2) It most assuredly would not. (3) With a 2 k.w. transformer you might obtain a 30 in. spark, providing the Tesla coil was immersed in transformer oil and properly tuned to the transformer and condenser.

1796. **Wireless Telegraphy.** R. J. B., Eau Claire, Wis., asks: (1) Do you think I would be able to use antenna 60 ft. high at both ends, 6 wire, 75 ft. long, and 2,000 ohm receivers, series connected with aerial and ground to hear the static or splashing of an electrical storm 500 miles radius? Am situated about 70 ft. below the hilltops of this section and the hills are about 2 miles distant in any direction from my station. (2) Are wireless signals weakened by having five or more connections or contacts between antenna and detector? Ans.—(1) You will have to use a detector, preferably either carborundum or silicon, in series with the aerial, telephone receivers, and ground. The hills will decidedly cut down the radius of receiving. (2) Not if the contacts are as near electrically perfect as possible.

1797. **Wireless "Swinging."** (1) G. H. J. Seattle, Wash., asks: Why do wireless stations north of here die out when listening to them? For instance, Tatoosh Island, N.P.D. or Point Esteven U.S.D. come in so one can easily read them, but gradually die out so you can hardly hear them. It is not my aerial or set. I lay it to the location and conditions of the weather. Ans.—This peculiar phenomenon is technically called "swinging." As yet it has had no satisfactory explanation, but is probably due to the fact that the air varies in density at different instants and therefore varies in its reflective ability toward the ether waves, thus making the signals "swing" in and out. "Swinging" is more marked from stations north and south than from those east and west.

1798. Armature Connections. R. J. B., Eau Claire, Wis., asks: (1) How should I connect the armature coils to the commutator in the dynamo-description below: Armature laminated, 12 slot, 2¼ in. wide, 3¼ in. diameter, ½ in. slots, 24 segment commutator. Ans.— (1) If your armature is a single parallel wound armature, that is, has only one coil of wire in each slot, connect the two leads of each coil to adjacent commutator segments.

1799. Transformer Design. H. S. M., Syracuse, N.Y., asks: In the article in the September, 1911, issue of *Electrician and Mechanic*, where the author assumes an efficiency of 94 percent, does he not give the reader the idea that the design as worked out in the article will come up to that efficiency? The author divides the losses into core and copper loss, and then figures his core so that its loss comes within the predetermined figure. If the author desires to base his design on efficiency, should he not design the copper so that its loss comes within the required figure, just as in the case of the core? Using the size and length of wire designated by the author, the primary RI loss at 20 C figures 17.5 watts and the secondary loss 27 watts, making the total copper loss 2.6 times that assumed at the start, and reducing the efficiency of the instrument to below 90 percent. In order to be logical, should not the author assumed an efficiency of 90 percent and taken the copper loss as 75 percent of the total loss, to justify the design as given? The analysis of this problem brings out a very interesting fact. For instance, the transformer as designed actually has an efficiency of about 90 percent, which is very good for a ½ k.w. instrument. Now to bring the efficiency up to the figure where it should have been if the design as assumed were carried out, the copper losses, as we have shown, would have to be reduced to ₁⅟₅ of their present value, which would mean that the weight of wire used would be 2.6 times as great. This would require the use of 37¾ lbs. of wire instead of 14¼. Figure the difference in cost for yourself. Does it not show that when we have an efficiency approaching unity, that to raise this efficiency only a few percent requires the expenditure of a large amount of money in proportion to the results attained? With small apparatus for amateur use it appears from this example that, in general, it is better to sacrifice a few percent in efficiency for the sake of first cost. Ans.—Your points are well taken and the author agrees with you that the average reader might be given a somewhat incorrect impression. On actual test the design as given produces a transformer having an efficiency of about 90 percent when the best grade of silicon steel is used in the core and when the workmanship in assembling the core is up to a certain standard. The principal object of the article in question was, however, not to provide any very complete data on power transformer design, but rather, to give the man with little or no technical education a simple and reasonably accurate method of designing small transformers suitable for wireless telegraphy or high-frequency work. The multitude of questions on the order of "How many turns shall I use?" received by the author of the article suggested that a few, simple instructions on this subject would be very acceptable. That the instructions were acceptable and adequate is evinced by the large number of amateurs who have built transformers from the design given and by the reports of highly satisfactory operation of the instruments.

1800. Armature Construction. C. M. C., Fulton, N.Y.: In reply to query No. 1735 would say that "Dynamo Electric Machinery," Sylvanus P. Thompson; "Dynamo Electric Machines," A. E. Weiner; and Vol. 13 C of the International Library of Technology, issued by the International Correspondence School, will furnish information. The International Correspondence School will not sell a single volume, but W. H. B. may purchase same from Geo. F. Williams whose ad. appears in Sale and Exchange Department.

1801. Motor Design. W. H. H., Fort Wayne, Ind., asks: (1) What changes would be required in the design of the motor as described by Louis Potter in September, 1911, issue of *Electrician and Mechanic* to make a 1 h.p. motor? Could a 1 h.p. motor be made over same diagrams by increasing the length and not the diameter? I would like to have a 1 h.p. 60 cycle 110 volt or 220 volt. What would be the comparative difference in 25 and 60 cycle machines, same horse-power? Ans.—(1) Can be increased in length only, but better by about 30 percent in length and diameter both. Use wire three sizes larger, with about 60 percent less turns. For 25 cycles, the design should have but two poles, and in other respects the design would be considerably changed.

1802. Motor Design. E. B. K., Missoula, Mont., asks: I wish to build a ¼ h.p. motor to run on single phase, 60 cycle, 110 volt A.C.; laminated field having two poles, and armature being wound and having a commutator and brushes. (1) Is there any book published which would tell me how to make this motor? (2) What would be the proper dimensions and sizes of wire for this machine? (3) How should the armature be wound and how connected to the field? Ans.—Your proposed motor would be essentially of the direct-current sort of construction. We wonder if you really mean this or do you wish one of the "induction" type? A commutator motor operated on alternating currents sparks horribly. A machine of the induction type was described in these columns recently under the head of a "Vacuum Cleaning Outfit." We are expecting soon to have a publication in much greater detail.

1803. Dynamo Design. W. H. B., Chicago, Ill., asks: Please advise if you have a book in regard to armature and field calculations. I want a book that will tell me just how to calculate the size and amount of wire necessary for armature and fields of motors and dynamos; also how to calculate how much current a dynamo is generating. Ans.—This question involves the whole field of dynamo design, and no one book or short course of reading will give the information desired. Underwood's book on the "Electromagnet," Hobart's "Armature Construction" and "The Electric Motor" are first class, but perhaps beyond the scope of the average enquirer. I would recommend a course in direct current dynamo design in one of the Correspondence Schools.

BOOK REVIEWS

The Application of Hyperbolic Functions to Electrical Engineering Problems. By A. E. Kennelly, M.A., D.Sc. London, University of London Press, 1912. Geo. H. Doran Company, N.Y., American Agents. Price, $2.25.

To all who are familiar with electrical engineering research, Professor Kennelly's name is well known; with the publication of these lectures, his work should rapidly get from the lecture room and study into the laboratory and field. The use of hyperbolic functions should not deter readers from the book, for these funtions are similar to the trigonometric and in reality are nothing but convenient combinations of the exponential or logarithmic; rather may it be said that the neat formulas which the author derives are ample justification and reward for any time spent in the study of the functions. The fundamental idea underlying the lectures is that the theories of D.C. and A.C. phenomena may be combined into a single theory by the use of vector diagrams and complex numbers, so that the formulas for alternating currents may be derived from those for direct currents by substituting imaginary numbers for real. From the point of view of the computer, whether he works graphically or arithmetically, the fundamental and neatest thing in the book is the manner of replacing an actual line by what Professor Kennelly calls equivalent T's or Π's, which are lines of particularly simple type.

The theory is developed in intimate connection with the applications which include D.C. lines of uniform resistance and leakance, A.C. power transmission lines, wire telephony, and wire telegraphy. To aid the reader of the work there is at the end a complete list of symbols with their definitions, a long and complete bibliography, and an excellent index. This book and Professor Kennelly's theories have a brilliant future before them.

Knots, Splices and Rope Work. By A. Hyatt Verrill. New York, Norman W. Henley Pub. Co., 1912. Price, 60 cents.

A most practical treatise giving complete and simple directions for making all the most useful and ornamental knots in common use, with chapters on splicing, pointing, seizing, serving, etc. The book is well illustrated with 150 original cuts, showing how each knot, tie or splice is formed and its appearance when completed, and it is well adapted for the use of travelers, campers, yachtsmen, boy scouts, and all others having to use or handle ropes for any purpose.

House Wiring. By Thomas W. Poppe. New York, Norman W. Henley Pub. Co., 1912. Price, 50 cents.

This book treats in a clear and non-technical manner the subject of house wiring. It describes and illustrates the most up-to-date methods of installing the wiring, and the methods described do in no way conflict with the rulings of the National Board of Fire Underwriters. It will prove of special value to electricians, apprentices and helpers, while even the advanced electrical worker will find many things in it which will interest him, as many labor and time-saving operations and diagrams are illustrated and described.

Brazing and Soldering. By James F. Hobart. New York, Norman W. Henley Pub. Co., 1912. Price, 25 cents.

This little, well-illustrated booklet of about 50 pages is one of the few reasonably priced books upon the subject of brazing and soldering. It explains how to handle any job of brazing or soldering that may occur; tells what mixture to use, and also how to make a furnace if one is needed. The booklet has recently been entirely revised and enlarged.

Practical Lessons in Electricity. By F. B. Crocker, L. K. Sager, H. C. Cushing, Jr., and Harris C. Trow. Chicago, American School of Correspondence Pub. Co. Price, $1.50.

An excellent working guide to the fundamental principles of electrical science and one especially adapted for purposes of self-instruction and home study. The book serves the purposes of a working guide for the beginner as well as a manual of information abounding in suggestions of much practical value to the more experienced electrical worker.

The authors are men well qualified to write such a book, since they thoroughly understand both the practical and theoretical sides of electrical engineering. They have explained the theory by means of simple language and a copious use of illustrations, adequate explanation being given of the fundamental principles and mechanical instrumentalities involved in the production and transmission of electricity.

The latter half of the book is devoted to the more common applications of electric energy and there are chapters devoted to the construction and use of Storage Batteries, and to the installation of conductors for power, lighting and other purposes.

The Slide Rule Simplified. By Geo. W. Richardson. Chicago, Geo. W. Richardson, 1912. Price, $1.00.

This book, as its title signifies, explains the how and why of the slide rule. In it, is given an excellent description of the Richardson Direct-Reading Slide Rule, and by means of the numerous illustrations and text matter, even the novice can readily grasp the method he should use in solving the various technical or practical problems which may confront him. As our readers are always desirous of learning of any time or labor saving device, a good clear description of the slide rule should prove most interesting to them. It is also well to note that the book is supplied free to all purchasers of either the 5-inch or 10-inch Richardson Direct-Reading Slide Rule, advertisement of which appears elsewhere in these pages.

"Train Lighting Lamps," is the title of a bulletin recently issued by the Engineering Department of the National Electric Lamp Association, which is sustained by certain works of the General Electric Company, covering the description, performance and economy of Mazda and Gem lamps in train lighting service. Copies of this bulletin may be secured free from the Engineering Department above mentioned.

www.ingramcontent.com/pod-product-compliance
Lightning Source LLC
Chambersburg PA
CBHW020905210326
41598CB00018B/1780